All Possible Worlds

A History of Geographical Ideas

FOURTH EDITION

Geoffrey J. Martin

New York   Oxford
OXFORD UNIVERSITY PRESS
2005

Oxford University Press

Oxford   New York
Auckland   Bangkok   Buenos Aires   Cape Town   Chennai
Dar es Salaam   Delhi   Hong Kong   Istanbul   Karachi   Kolkata
Kuala Lumpur   Madrid   Melbourne   Mexico City   Mumbai   Nairobi
São Paulo   Shanghai   Taipei   Tokyo   Toronto

Published by Oxford University Press, Inc.
198 Madison Avenue, New York, New York 10016
www.oup.com

Oxford is a registered trademark of Oxford University Press

Library of Congress Cataloging-in-Publication Data
Martin, Geoffrey J.
    All possible worlds : a history of geographical ideas / Geoffrey J. Martin ;
with the assistance of T.S. Martin.—4th ed.
        p.   cm.
    Includes bibliographical references (p.) and indexes.
    ISBN-13: 978-0-19-516870-9
    ISBN 0-19-516870-4
    1. Geography—History.   I. Martin, T.S.   II. Title.
G80.M38 2004
910'9—dc22
                                                                    2004057559

Printing number: 9 8 7 6 5 4 3 2

Printed in the United States of America
on acid-free paper

The optimist proclaims that we live in the best of all possible worlds; and the pessimist fears this is true.

—*James Branch Cabell,* The Silver Stallion

CONTENTS

FIGURES

PREFACE

The history of geography constitutes an investigation into the way in which geographic subject matter was recognized, perceived, thought about, and evaluated over the course of centuries. Ideas evolved; some were correct, some were erroneous, but all were part of a long-developing body of ideas that helped humankind comprehend its surroundings. Out of this, a discipline and profession evolved. This pattern emerged in several countries, each independent of the other, until recent times, when ease of transport and circulation of the written and spoken word has facilitated the exchange of ideas.

Each national geography has its own necessities, and each country produces its own sages, each with his or her own inimitable point of view that naturally influences students. Nevertheless, since approximately the middle of the nineteenth century an ever-increasing exchange of thought has taken place that permits international recognition of a geographical point of view. The history of this point of view is offered as an account of the successive images scholars have developed concerning arrangements and patterns on the face of the earth. Long before the dawn of written history, men who explored even just a short distance from home were aware of differences that distinguished one place from another. The concept of areal differentiation had been born. Some people were drawn to the challenge of forming mental images of what it was like beyond the horizon and communicating these images to others. The differentiation of the face of the earth is what John K. Wright called "geodiversity." This is what geography is all about.

How can a mental image of geodiversity be formed with sufficient clarity so that it can be communicated to others? What things are combined in different places on the earth to produce the complex characteristics of the world's landscapes? In the first place, the world is much larger than humans are, and on its curved surface our range of vision is narrowly restricted. Most of the fields of study in modern science seek to form mental images of things and events that are much too small to observe directly; but to form a mental image of a world that is too big to see, it is necessary to generalize, to select certain features to build into the image, and to reject other features as irrelevant. Furthermore, no coherent image of our world could be put together unless the observed features are located in relation to a known point. One of the distinctive characteristics of a geographer is a consistent concern about the relative location of things.

Along with the formation of a mental image of geodiversity, however, has existed our need to explain it. Scholars formulated many different explanations to make the mental images seem plausible or acceptable. Their explanations, in turn, often determined the features they chose to observe. Scholars developed ideas concerning sequences of events, or processes of change, that satisfied the need to understand; they also sought and found mathematical regularities separate from the processes of change that nevertheless satisfied the urge to explain the image of geodiversity.

The mental images and explanations of one generation seldom were satisfactory to succeeding generations. There has been a continued search for new and more perfect images and for explanations that accord with contemporary beliefs. These changing images and the theoretical structures established to give them plausibility provide us with the content of the history of geographical ideas.

There are certain built-in biases that should be recognized at the start. The text embraces the whole sweep of the history of ideas from the viewpoint of this time and looks at different nations of the world from the United States of America; inevitably the view is foreshortened. Much attention is given to the twentieth century and to a consideration of geographical ideas in America. However, the existence of scholars in other times has not been ignored, nor have the geographical ideas and developments of the non-Western world been overlooked.

THE CLASSICAL AND THE MODERN PERIODS

Two major periods are defined. The first period extends for thousands of years from the shadowy beginnings of geographical thought to the year 1859. This is the classical period, during which relatively little attention was paid to the definition of separate fields of study. This was the period when the world's knowledge was not so great that a scholar could not become master of it. So it is that almost every Greek philosopher, often listed as a historian in the history books, may with equal justification be called a geographer. Even in the eighteenth century, when the separation of fields of study had begun, there existed scholars such as Benjamin Franklin, M. V. Lomonosov, and Montesquieu who are usually not known as geographers but nevertheless are important links in the history of geographical ideas. The last person who could claim universal scholarship, however, was Alexander von Humboldt; after his death, no one could attain the preeminence in the world of scholarship Humboldt had been able to reach. The year 1859 functions as a significant divide: Humboldt's death was accompanied by that of Carl Ritter. Alfred Hettner was born, as was Darwin's *Origin of Species*. What we call the classical period is discussed in Chapters 2 through 6.

The modern period began in the latter part of the nineteenth century. It is distinguished by the appearance of the professional field called geography—that is, a field of study in which trained students could earn a living by being geographers. To create a professional field, three conditions had to be satisfied. First, there had to be a body of concepts or images that was accepted by the members of a profession, as well as an accepted way of asking questions and seeking answers. In other words, there had to be a model of professional behavior. Second, no such model could be formulated or passed on to later generations of scholars until there were societies, associations, periodicals, and departments of geography in universities offering advanced training in concepts and methods. Third, graduate departments of geography could not develop until paying jobs existed for students who had earned advanced degrees from these departments. But when all three conditions were met, a professional field was born. This constituted a study in progression from a subject matter, to discipline, to profession, to its dispersal. This is what is called the new geography. Chapter 7 discusses what was new about it.

THE NEW GEOGRAPHY

The new geography began in Germany in 1874, when departments of geography headed by scholars with the rank of professor were established in the German universities. Before that time, students could attend lectures and perhaps go on to give lectures themselves, but never before were there clusters of scholars qualified to offer graduate training in geography. When such positions became available in Germany in 1874, there were no trained geographers to take them; those who were appointed to the newly created professorships had to find their own answers to the question: What is geography? This German innovation spread rapidly to other countries—notably France, Great Britain, and Russia. It was also transmitted by various routes to the United States. In each of these five countries, distinctive national viewpoints developed, depending on the answer given to the question concerning the nature of geography. From these five sources, the new professional geography spread all around the world. Chapter 8 reveals something of the development of the new geography in Germany; France is discussed in Chapter 9, and Great Britain in Chapter 10; Chapter 11 discusses Russia and the former Soviet Union. The emergence of a very keen and capable Canadian geography is revealed in Chapter 12; the rich content of Swedish geography is summarized in Chapter 13. Japanese geography is presented briefly in Chapter 14. Chapters 15 through 18 reveal the development of the new geography in the United States. Chapter 19 concerns the new methods of observation and analysis. Chapter 20 embraces the interwoven strands of innovation and tradition. Of course, there are many other countries with both vibrant geographical histories and traditions and abundant accomplishment that are not written of here. Further coverage is not practical or purposeful given the spirit and purpose of this book. Today's innovation is tomorrow's tradition, which is the way in which geography makes its advance.

THE PURPOSE OF THE BOOK

Clearly, this is a study in breadth, not in depth. This is a book about the history of the field. It cannot be concerned with the details of geographies produced in recent times. While the strength of national geography might be assessed *en passant*, there is more concern to develop the history of that geography. What intellectual paraphernalia are borrowed, loaned, or shared, and the point of genesis and dispersal, are matters of interest to the historian of science. And comparisons between national geographies are revealing. At some point someone must put the whole sweep of things into perspective—if only for the purpose of locating desirable places for future detailed studies. And so an attempt is made to identify the major traditions as they appeared in the early classical period and as they reappear repeatedly in somewhat different form throughout the course of history. Ever since Pythagoras there has been a mathematical tradition in geography; ever since Anaximander there has been a cartographic tradition; and ever since Hecataeus there has been a literary tradition. These traditions reinforce each other and serve to guard against a point of view that might be too narrowly specialized. For the seriously minded geographer a knowledge of the evolution of the field of study is prerequisite to a larger comprehension. It is part of the intellectual paraphernalia of the geographer to comprehend the path our forbears trod . . . from the derring-do of ancient Greek travelers, the mental effort pitted against the mysteries of the universe by early thinkers to comprehend a spherically shaped earth (only to have the posture repealed by a flat earth [four corner earth] church), to the Middle Ages replete with learned geographical lore from the Crusades, eventually to the age of discovery, scientific advance, and comprehension of the meaning of evolution, etc. Intellectual inquiry brought us from subject matter to discipline and eventually profession. Understanding the path trod enlarges our comprehension and appreciation of the contemporary phase of the geographical inquiry. Of course, we shall proceed beyond this state just as previous millennia of thought have brought us to the present. We need to understand how growth takes place and the circumstances thereof. And we shall learn that many of the ideas of recent times are not new but are ideas of earlier times refurbished. This is what this book is all about.

This book has been designed as a text. It may function as text for "Classical Geography," or for "Modern Geography." It may also serve as text for a course on the history of American geography (by combining Chapters 7 and 15 to 20). And, of course, the full text may be (and usually is) adopted for a course intended to introduce the student to the sweep of ideas that have brought us to the geography of the present day. Selected references are found at the end of each chapter. These references are in no sense complete, nor do they do justice to the important geographical writings in languages other than English, French, and German. These references are heavily weighted toward books and articles published in the English language that are readily available to American students. In the Index of Names, each entry is accompanied by dates and a brief identification. Some names in the Index of Names do not appear in the text.

Throughout this book there has been an honest attempt to maintain the language free of gender bias. Occasionally, traditional terms such as "man–land" and "man–environment" have been retained as representing the jargon of a discipline as it used to be. Hopefully this arrangement is satisfactory to all readers.

ACKNOWLEDGMENTS

Geographers too numerous to mention have helped with previous editions of this book. With special reference to this, the fourth edition, grateful mention must be made of those who have read parts of the manuscript or otherwise provided assistance. They include Christopher Baruth (American Geographical Society Library Curator), Ronald R. Boyce (Seattle Pacific University, Emeritus), Anne Buttimer (University College Dublin, Ireland), Brian J. L. Berry (University of Texas at Dallas), Paul Claval (University of Paris, Emeritus), Jeremy Crampton (Georgia State University), George J. Demko (Dartmouth College, Emeritus), Maynard Weston Dow (Plymouth State College, Emeritus), Gary Dunbar (University of California, Los Angeles, Emeritus), Eckhart Ehlers (University of Bonn), Peter Haggett (University of Bristol), Donald Janelle (University of California, Santa Barbara), Tom Mels (University of Kalmar, Sweden), E. Joan Miller (Illinois State University, Emeritus), Alexey Postnikov (Institute of Geography, Moscow University), Marie Claire Robic (University of Paris), David R. Stoddart (University of California, Berkeley, Emeritus), and James O. Wheeler (University of Georgia).

Included in this recognition must be the diligent though unnamed librarians of both Southern Connecticut State University and Yale University. Peter Boppert, Director of the Learning Resources Center, Southern Connecticut State University, has assisted in the matter of photographic requirements. My thanks to T. S. Martin, my son, who has assisted notably in the completion of this manuscript.

Institutions including the Association of American Geographers, the American Geographical Society, the library of the U.S. Geological Survey, and the National Geographic Society have been most kind in providing some of the illustrations. The author also wishes to thank production editor Leslie Anglin of Oxford University Press for consideration throughout all stages of production and continued attention to detail. And to William A. Koelsch, fellow traveler in the groves of the history of geographic thought, special thanks for reading the third edition of this work and making suggestions for its improvement. Any error or unhappy presentation or interpretation of the facts is, of course, solely the responsibility of the author.

We may compare the mind of man to a mirror which has the ability not only to reflect but also to retain, record, and interpret more or less imperfectly the images that it reflects. It is not a clean, bright mirror which gives exact images, but too often is warped, clouded, spotted, cracked, and broken. The appearance of the image, no matter of what the reflection may be, is determined very largely by the nature of the mirror itself and by the spots, dust, and other foreign matter that may have accumulated upon it.

—*John K. Wright, "The History of Geography: A Point of View"*

1

A Field of Study Called Geography

. . . to understand the earth as the world of man . . .
—*J. O. M. Broek,* Geography, Its Scope and Spirit

In the beginning there was curiosity. We may assume that some of the earliest questions formulated by our primitive ancestors had to do with the character of the natural surroundings. Humans, as well as many other animals, identify a particular extent of territory on the earth's surface as their living space; and, like many other animals, they are irked by the possibility that the grass may be greener in someone else's living space. Curiosity compels them to find out what it is like beyond the range of hills that form their horizon. But the world they discover is filtered through their own minds, and in the long course of history people have discovered and described many different worlds. There are challenging limitations on the human capacity to observe and to generalize what has been observed. As this capacity to observe and generalize is improved, the result is a new image of the world—yet all possible worlds are far from having been described.

Our world includes what can be perceived on or from the surface of the earth. The earth is a medium-sized planet circulating around a medium-sized nuclear explosion that we call the sun. If the sun were the size of an orange, the earth, at the same scale, would be the size of a pinhead about one foot away. Yet this pinhead is just large enough to provide the gravitational pull to hold the thin skin of gases called the atmosphere close to its surface. And the earth is just the right distance from the sun, so that at the bottom of the atmosphere temperatures occur at which water can remain a liquid. Organic life originated in the seas, and when organisms came out of the water to live on the land they brought their water environment with them inside "space suits" made of skin. Similarly, when the astronauts traveled to the moon, they took their gaseous environment with them inside artificial space suits. All the forms of life that we know about are dependent on water in the form of liquid and on energy received from the sun.

The face of the earth is a zone extending as far down below the surface as humans have been able to penetrate and as far up above the surface as humans normally go. All science and all art are derived from observations made in this zone, which, until 1969, was the human's whole universe. But it is a complex universe: some things (phenomena) on it are produced by physical and chemical processes; some plants and animals are produced by biotic processes. Humans themselves are influenced by their natural surroundings and, acting through economic, social, and political events are also agents of change in their surroundings. All these things and

1

the events of which they are the momentary signs exist in complex association and interconnection, forming what is called the human–environment system.

The Greek scholar Eratosthenes, who lived in the third century B.C., was the first to use the word *geography* (from *ge*, meaning the earth, and *graphe*, meaning description). But humans were investigating geographic questions long before this. The Greek geographers credit the beginning of geographical writings to Homer, but the earliest known map was made by the Sumerians in about 2700 B.C. The history of geographical ideas is the record of human efforts to gain more and more logical and useful knowledge of the human habitat and of human spread over the earth. They are logical in that explanations of the things observed could be so tested and verified that scholars could have confidence in them. They are useful in that the knowledge so gained could be used to facilitate the individual's adjustment to the varied natural conditions of the earth, to make possible modifications of adverse conditions, or even to gain a measure of control over them.

WHERE IS IT?

Even a clear description of what our world is like would not be of great value unless it were possible to answer the question: "Where is it?" Starting from known living space, in what direction do we travel and how far? How do we measure distance and direction? When we reach the place on the other side of the horizon, where are we?

This is no simple question to answer. To illustrate the difficulties involved in establishing location on the surface of a sphere, take a ping-pong ball with a black dot on it and then attempt to describe the location of the dot. Location must always be relative to something else. The ancient Greeks first formulated the theory of locating places with reference to a grid of imaginary lines of latitude and longitude based on the poles and the equator. But the Greeks lacked the instruments necessary to make exact measurements of latitude and longitude. Location still had to be described relative to some observable feature, such as a range of hills, a coastline, a river, or a town. As knowledge of the features of the earth's surface was increased, a greater variety of relative locations could be defined. Furthermore, in any specific question that is raised concerning the arrangement of things on the face of the earth, location relative to latitude and longitude may not be relevant. For example, to say that New York City is located at 40°42′N and 74°W (at City Hall) is meaningless if one is asking questions about the significance of location in the growth of the city. Asking questions about locations is one of the distinguishing characteristics of the field of study we call geography.

WHAT IS IT LIKE?

The phrase "What is it like?" stands for a fundamental thought process. How does one go about observing and reporting on things and events that occupy segments of earth space? Of all the infinite variety of phenomena on the face of the earth, how does one decide what phenomena to observe? There is no such thing as a complete description of the earth or any part of it, for every microscopic point on the earth's surface differs from every other such point. Experience shows that the things observed are already familiar, because they are like phenomena that occur at home or because they resemble the abstract images and models developed in the human mind.

How are abstract images formed? Humans alone among the animals possess language; their words symbolize not only specific things but also mental images of classes of things. People can remember what they have seen or experienced because they attach a word symbol to them. Humans can look at a hill and, if they are familiar with hilly country, they can recognize the peculiar and unique character of each specific example, but they can also develop the mental image of "hill" in general.

During the long record of our efforts to gain more and more knowledge about the face of the earth as the human habitat, there has been a continuing interplay between things and events observed through the senses and the mental images of things and events. The direct observation through the senses is described as a *percept*; the mental image is described as a *concept*. Percepts are what some people describe as reality, in contrast to mental images, which are theoretical, implying that they are not real.

The relation of percept to concept is not as simple as this definition implies. It is now quite clear that people of different cultures or even individuals in the same culture develop different mental images of reality, and what they perceive is a reflection of these preconceptions. In some African cultures, for example, color perception is limited to red and blue because their language contains words for only these ends of the spectrum. As a result, these people do not perceive intermediate colors, such as orange, yellow, or green (Krauss, 1968).[1] An artist trained in the perception of color and with a long list of word symbols can identify a great many colors. The direct observation of things and events on the face of the earth is so clearly a function of the mental images in the mind of the observer that the whole idea of reality must be reconsidered. The idea that one could approach a problem with a blank mind—that is, without preconceptions—is nonsense, for such an assertion means only that the person is not conscious of his or her preconceptions and is, therefore, a slave to them. Nor is it possible to dream a set of images not based on previous perception. Such a situation could apply only to very young infants.

We are faced then with an apparent paradox. Concepts determine what the observer perceives, yet concepts are derived from the generalization of previous percepts. What happens is that the educated observer is taught to accept a set of concepts and then sharpens or changes these concepts during a professional career. In any one field of scholarship, professional opinion at any one time determines what concepts and procedures are acceptable, and these form a kind of model of scholarly behavior. Such a body of doctrine determines what problems are considered worth investigating and what kinds of answers are professionally acceptable. But at all times the accepted concepts and the procedures based on them are subject to challenge. Progress is achieved when one working hypothesis is replaced by another.

---

[1]References are listed at the end of each chapter.

WHAT DOES IT MEAN?

One concept that runs through the history of ideas since the earliest records is that of an ordered, coherent, and harmonious universe. Human beings have instinctively rejected the notion that they and their natural surroundings are the result of random events. Clarence J. Glacken writes as follows:

> What is most striking in concepts of nature, even mythological ones, is the yearning for purpose and order; perhaps these notions of order are, basically, analogies derived from the orderliness and purposiveness in many outward manifestations of human activity: order and purpose in the roads, in the grid of village streets and even winding lanes, in a garden or a pasture, in the plan of a dwelling and its relation to another (Glacken, 1967:3).

There are many examples of human effort to identify meaning as a reflection of the images in the human mind. With the concept of an orderly world firmly established, the perception of evidence to give it support was a natural consequence. The Greek philosophers distinguished between chaos (*kenos*, meaning void) and the cosmos, which refers to the universe conceived as a system of harmoniously related parts. Having accepted, almost without challenge, the concept of an ordered universe, the next step was to find some plausible way to account for it.

During the thousands of years since the earliest records of the history of ideas, learned people have accounted for the order they perceived in the universe in different ways. The accounts range along a continuum from arbitrary rule by humanlike deities, through rule by a deity subject to law, through various kinds of cause and effect relations, to abstract mathematical law. These do not represent successive stages of increasing sophistication, for all of them can be found in the thinking of ancient Greek philosophers as well as in the contemporary world.

Rule by a deity or deities is a very ancient concept. In Sumeria the religious leaders saw a world ruled by living beings like humans but endowed with super-human powers and with immortality (Glacken, 1967:4–5). Each of these beings was responsible for the control and maintenance of some feature of the world, such as the flow of rivers, the rise and fall of the tides, the shift of the winds, the productivity of the harvest, and the abundance of game animals. The deities competed with one another and reacted arbitrarily and often vindictively to human acts. Other cultures explained matters in terms of a single deity whose acts were frequently subject to the bestowal of human favor.

A very different way of accounting for an ordered universe is the recognition of cause and effect sequences that take place in accordance with general law. In some cases the notion of a single deity is retained, but the acts of this deity are not arbitrary. Some would say that this god is the law. The idea of law itself is an anthropomorphism—that is, a reflection of human experience. Those who break divine laws are subject to punishment, but those who act in harmony with the law are rewarded. Of course, there is a great difference between human law and scientific law: human law governs the behavior of things, and events are subject to law, but scientific law is a general description of events.

There are two different ways of interpreting cause and effect sequences, each represented by one of the two great philosophers of ancient Greece: Plato and Aristotle.

Plato conceived of the world as having been created in perfection but now in the process of decline from perfection. Aristotle, on the other hand, conceived of the world as striving toward perfection. Aristotle was the father of the teleological concept, which sees the universe planned by its creator in the manner of a carpenter drawing up plans in advance for the house he is building. This doctrine of final cause differs from the idea that cause must be antecedent to effect. David Hume in the eighteenth century argued that cause and effect sequences could never be perceived or tested except in terms of human experience, yet the use of the concept continues (Brown, 1963; Ducasse, 1969).

At the other end of the spectrum is the use of mathematical regularities to explain the order in the universe. Order was conceived as mathematical even by Pythagoras, some six centuries B.C. Mathematical order can be illustrated by the regularity of the octave in sound, the principle that the circumference of any circle is the same as the length of the diameter times 3.1416, or that the force that attracts one body to another is proportional to the total masses of the two bodies divided by the square of the distance between them. Mathematical formulas describe the motions of planets, the flow of rivers, and the predictable flow of telephone calls between two cities. So all-pervasive are the principles of mathematics that one astronomer was led to remark that the universe was like a dream in the mind of a mathematician. It was Plato who insisted that to explain anything it was necessary only to identify the laws it obeys.

THE STUDY OF GEOGRAPHY

The study of geography for the purpose of gaining more logical and useful knowledge regarding the human habitat and human interrelations with it goes back to the beginnings of human scholarship. Certain repetitive sequences appear in the interplay between concepts and percepts, hypotheses and observations. There are occasional periods marked by flashes of brilliant intuition when new and challenging concepts are set forth. New concepts presented as hypotheses result in a burst of new empirical observation, for the new concepts always broaden the range of human perceptions. The new observations may demonstrate the inadequacy of the hypothesis, which is then withdrawn in favor of a new one or is substantially modified. These are periods of great progress. Then, when a conceptual structure becomes widely accepted and a paradigm of scholarly behavior is established, there is a period when observations increase so fast that they must be stored away for future use. These periods of intellectual stability are not periods of notable progress. Eventually, a new and sometimes radically different concept of the meaning of the observed data is set forth, and the sequence is repeated.

The first amazing period of intellectual ferment that is part of the written tradition of the Western world took place in ancient Greece, culminating in the fourth and third centuries B.C. The scholars of Babylonia had already collected a large number of data on the motions of the stars and planets, and they developed the concept that the position of the celestial bodies had a fundamental effect on human activities. In astrology, if observed facts disagree with general principles, the facts are explained as exceptional and the principle remains unchanged. The Greeks developed the procedures we describe as the scientific method. If observed facts differ from general principles, the principles must be revised. The Greeks developed the science of astronomy, a tremendous step forward in the whole history of ideas.

Many of the basic procedures were first elaborated in the writings of Plato and Aristotle. Quotations from their works are as relevant today as they were thousands of years ago. Plato, who developed the deductive procedures, is the most often quoted by those who prefer to give theory the position of chief importance. Aristotle, who developed the inductive procedures, preferred to formulate his concepts as generalizations of empirically observed "facts." Aristotle insisted on the importance of direct observation. Instead of making logical deductions from theory, he taught his students (including Alexander the Great) to "go and see."

Among the ancient Greek philosophers, almost all of whom made contributions to the study of geography, two basic traditions of geographic study are to be found. One is the mathematical tradition, starting with Thales, including Hipparchus (who formulated the theory of locating things by latitude and longitude), and summarized by Ptolemy. The second is the literary tradition, starting with Homer, including Hecataeus (the first writer of geographical prose), and summarized by Strabo.

A long period of decline set in during the Middle Ages, when geographic horizons contracted. The wide-ranging Greek and Phoenician explorers and the Greek geographical concepts were largely forgotten, except among Arabic scholars. Observations piled up in Christian monasteries, but the intellectual climate was not favorable for the formulation of new interpretations. Then the Age of Exploration began in the late fifteenth century, and the geographic horizons were again pushed back. The arrival of all these new observations in Europe was enormously stimulating and started a sequence of events that continues to the present.

Two fundamental innovations began to shake the world of scholarship in the sixteenth century, and this ferment reached its full proportions in the second half of the nineteenth century. In the first place, the concepts derived from a literal reading of the Scriptures were challenged, and the battle to establish the principles of what we call academic freedom began. This is the right of professionally qualified scholars to seek answers to questions, to publish their findings, and to teach what they believe to be the truth, free from any controls except the standards of scholarly procedure established within their own professions. The battle started with Leonardo da Vinci and Copernicus in the sixteenth century and reached full fury as a result of the tradition-shattering concepts developed by Charles Darwin in the latter part of the nineteenth century. In 1809 the world's first university was founded at which neither faculty nor students were required by law to accept any particular religious creed or political doctrine. This was the University of Berlin. The idea of the university as a community of free scholars continued to gain acceptance throughout the world until the whole concept was challenged in the contemporary period.

The second fundamental innovation began in the seventeenth century and reached widespread development in the second half of the nineteenth century. This was the separation of the academic world into distinct fields, or disciplines, each devoted to the study of a specific group of related processes and each regulated by its own paradigm based on its own theoretical structure. From the time of ancient

Greece, scholars were not concerned about restricting themselves too narrowly. Herodotus made major contributions to both history and geography, and he is often called the father of ethnography. But the last person who could claim to be a universal scholar and who could speak with respected authority on the whole wide range of the world's scientific knowledge was Alexander von Humboldt. Humboldt and Carl Ritter, both of whom died in 1859, represent the culmination and the conclusion of ancient scholarship.

The rise of experimental science was a major cause of this separation of the disciplines. Such fields as physics and chemistry developed around the study of particular processes. Processes were isolated in laboratories, and as a result the knowledge of these processes was greatly expanded. This fragmentation of knowledge had a major impact on geography, which became known as "the mother of sciences." The movement started with the physical sciences, then with the biological sciences, and most recently with the various social sciences. Where processes could not actually be studied in laboratories, as in economics, they were isolated statistically; the laboratory was represented symbolically by the phrase "other things being equal."

Was there any place left for geography after Humboldt and Ritter? By 1870 it had become clear that some aspects of knowledge were not covered by the new sciences. Especially in the social sciences, processes never actually took place in isolation: they were modified in particular places by all the other aspects of the total environment that were not equal. Geography moved into the field of particular places, where a variety of things and events were examined in their natural but unsystematic groupings. Geography also undertook to bridge the widening gap between the physical and biological sciences on the one hand and the social sciences on the other. But an academic discipline cannot exist until there is a body of trained professional scholars formed around university faculties, where new generations of scholars can receive advanced instruction. Institutes of geography headed by professors came into existence for the first time in Germany in 1874, but the professors selected to head these institutes were people with training in geology, history, biology, or other fields. During the period from the 1870s until World War II, geographers all around the world were seeking to establish the status of geography as an independent discipline, distinct in its concepts and procedures from other disciplines (Dunbar, 2001).

World War II had a greater impact on the world of scholarship than is generally recognized. Scholars were called upon to study the issues involved in the very complex problems of policy; and when they went to work on these pressing questions, they discovered that workers in any one discipline were usually ill-prepared to wrestle with essentially interdisciplinary matters. The workers in the separate fields were too narrowly trained and specialized—in some cases so specialized that they could not even appreciate the importance of work done in other disciplines. Geographers made important contributions in the form of maps and analyses of the significance of location that other people were accustomed to overlook.

One outgrowth of the experience of the war was the development of General System Theory (Bertalanffy, 1968) in which the existence of complex associations of interconnected and interdependent elements was recognized, and methods were developed to solve problems of multiple causation, where predictions had to depend on probability theory. Fortunately, just at this time the electronic computer was developed. Without the computer the complex computations involving a great variety of factors could never have been carried out with speed and precision. Furthermore, new electronic devices for scanning the face of the earth from orbiting satellites revolutionized the methods of data gathering. These innovations, which came largely after midcentury, have introduced another major period in the history of ideas.

What then does geography do? It is important to understand that since World War II this question does not call for a definition of geography that would establish its boundaries. The trend now is for all fields of study to come together around specific problems. The process of separation has been replaced by a process of integration, in which each professional field brings its own special skills and concepts to bear on such major difficulties as poverty, overpopulation, race relations, and environmental destruction. The special skills of geography are those related to the significance of location and spatial relations of things and events. Geography has always had a holistic tradition, so that it comes as no intellectual shock to study systems of interconnected and interdependent parts of diverse origin. Geography is closely involved with cartography in the development and use of maps, which are ideally suited to the study of complex location factors. A geographer is a person who asks questions about the significance of place, location, distance, direction, spread, and spatial succession. The geographer deals with problems of accessibility, innovation diffusion, density, and other derivatives of relative location.

MANY NEW WORLDS

There are many new worlds yet to be discovered, and the number of people engaged in discovering them has increased at an unprecedented rate since the end of World War II. As with many other fields of learning, it can be said of professional geography today that of all the scholars who ever lived and contributed to geographic knowledge, a large percentage lived on into the second half of the twentieth century. The volume of published material is vastly greater than the materials with which Humboldt and Ritter had to work. The electronic data bank has arrived just in time. But with all these new scholars entering the field and with the variety of pressing problems to which geographic concepts and methods can be applied, there is no reason to narrow the field or to establish just one paradigm of scholarly behavior.

As a general assessment of the field of geography in the early 2000s, we may say that there is still freedom to be curious, to inquire, to offer novel conceptual structures that reveal the outlines of yet another of all possible worlds. The mathematical and literary traditions are still being vigorously developed. The use of mathematical theory has been so clearly useful that graduate training in geography often involves preparation in mathematics through calculus and linear and matrix algebra. The literary tradition, threatened by the neglect of language training in the schools, continues to attract those who can command the skills of rhetoric and the manipulation of language. It holds appeal for those interested in traditional geographies, the philosophy of geography, and the history of geographical thought.

Cartographic skills are so interconnected with the analysis of location that even when maps are made by computers, concepts regarding the use of maps as analytic devices have lost none of their importance.

At least five different kinds of questions of geographic character can be investigated: (1) *Generic* questions that have to do with the content of earth space but that cannot be effectively answered without a framework of concepts to guide the separation of the relevant from the vast complexity of the irrelevant. (2) *Genetic* questions that have to do with the sequences of events leading from past situations through geographic changes to present conditions; these are studied by the methods of historical geography. (3) *Theoretical* questions that deal with the formulation of empirical generalizations or of general laws, perhaps even with basic theory, and with the methods of drawing logical deductions. (4) *Remedial* questions that have to do with the application of geographic concepts and skills to the study of practical economic, social, or political problems. (5) *Methodological* questions that have to do with experiments in new methods of study, new techniques of observation and analysis, or new cartographic methods.

Although many in each generation of geographers are inclined to establish their own methods of work as the only acceptable way, these efforts have been resisted. The freedom to inquire wherever curiosity leads has been preserved. There are still new worlds to be discovered by the enthusiastic scholar.

REFERENCES: CHAPTER 1

Bertalanffy, L. von. 1968. *General System Theory: Foundations, Development, Applications.* New York: George Braziller.

Bowman, I. 1934. *Geography in Relation to the Social Sciences.* New York: Charles Scribner's Sons.

Brown, L. A. 1949. *The Story of Maps.* Boston: Little, Brown. Republication unabridged by Dover Publications, New York, 1977.

Brown, R. 1963. *Explanation in Social Science.* Chicago: Aldine.

Claval, P. 1972. *La pensée géographique, introduction à son histoire.* (Publications de la Sorbonne. Series N.-S. Recherches, 2.) Paris: Société d'Édition d'Enseignement Supérieur.

Clozier, R. 1972. *Histoire de la géographie.* 5th ed. ("Que sais-je?," 65.) Paris: Presses Universitaires de France.

Corley, N. T. 1973. "Geographical Literature." *Encyclopedia of Library and Information Science.* Vol. 9. New York: Marcel Dekker, Inc.

Crone, G. R. 1964. *Background to Geography.* London: Museum Press, Ltd.

Dickinson, R. E., and Howarth, O. J. R. 1976. *The Making of Geography.* Oxford: Clarendon Press, 1933. Reprinted: Westport, Conn.: Greenwood Press.

Ducasse, C. J. 1969. *Causation and the Types of Necessity.* New York: Dover.

Dunbar, G. S., ed. 2001. *Geography: Discipline, Profession and Subject since 1870: An International Survey.* Dordrecht, Netherlands: Kluwer Academic Publishers.

Fischer, E., Campbell, R. D., and Miller, E. S., eds. 1969. *A Question of Place: The Development of Geographic Thought.* 2nd ed. Arlington, Va.: Beatty.

Freeman, T. W., Oughton, M., and Pinchemel, P. (eds). 1977. *Geographers: Biobibliographical Studies.* Vol. 1. London: Mansell.

Glacken, C. J. 1967. *Traces on the Rhodian Shore, Nature and Culture in Western Thought from Ancient Times to the End of the Eighteenth Century*. Berkeley and Los Angeles: University of California Press.

Harley, J. B. and D. Woodward, eds. 1987. *The History of Cartography*. Vol. 1. Chicago: University of Chicago Press.

James, P. E. 1956. "Geography." *Encyclopaedia Britannica*. 14th ed. 10:139–153. Chicago: Encyclopaedia Britannica.

Kish, G. 1978. *A Source Book in Geography*. Cambridge, Mass.: Harvard University Press.

Krauss, R. M. 1968. "Language as a Symbolic Process in Communication." *American Scientist* 56:265–278.

Livingstone, D. N. 1992. *The Geographical Tradition*. Oxford: Blackwell.

Ravenstein, E. G. October 1891. "The Field of Geography." *Proceedings of the Royal Geographical Society* 13, no. 10.

PART 1

CLASSICAL

The strands of classical geography are closely interwoven with those of all other fields of learning, for in those days very little was known and a diligent worker could master the greater part of the world's knowledge. Any one person could contribute to a great variety of what we would now identify as separate fields of study. Most of the ancient Greek scholars are called philosophers or historians, yet a student such as Herodotus wrote both history and geography, and he is also claimed by the anthropologists as the father of ethnography. These were the times of universal scholarship, which continued until the last quarter of the nineteenth century. This period produced such notable figures as Edmund Halley and Francis Galton in Great Britain, the political philosopher Montesquieu in France, Alexander von Humboldt and Carl Ritter in Germany, Thomas Jefferson and Benjamin Franklin (whose enormous versatility is well known) in America, and Mikhail V. Lomonosov in Russia. These were the days when a curious scholar felt no need to recognize disciplinary boundary lines. He or she was free to trespass universally. Thus it was during most of the long history of ideas.

2

THE BEGINNINGS OF CLASSICAL GEOGRAPHY

The fundamental concepts and basic elements used by the Pythagoreans in their theory are odd in comparison to those used by the physicists, since they are not observable; they are arithmetical entities which (except in the case of astronomy) are not a part of our changing world. Despite this, the Pythagoreans are in fact interested only in natural phenomena. They talk about how the universe was created, and always check their theories against observation when discussing the behavior and interactions of objects in the universe, using their particular set of fundamentals solely for theoretical purposes to account for physical reality. Thus in essence they are just like the other natural philosophers, since they agree that the only real things are those that can be perceived with the senses and are part of this world of ours. Their modes of explanation and their fundamental concepts, however, give them direct access to realms which transcend the senses, and to which their way of thinking is really more adapted than to physics. For example, they give no hint how motion and change can possibly be deducted from the only underlying concepts they assume in their theory—from boundedness and unboundedness, or evenness and oddness. Nor do they indicate, since they do not provide for motion and change in their theories, how any physical process can take place—and in particular how the heavenly bodies can move in the way they do.

—*Aristotle's* Metaphysica, *Book 1*

G eography as a field of learning in the Western world had its beginnings among the scholars of ancient Greece. The study of the earth as the home of man excited curiosity outside of Greece, although we might not gain this impression by reading many of the histories of geography written by Europeans. Much attention was given to geographical study in ancient China, and Chinese explorers did as much to "discover" Europe as the Europeans did to reach the "Far East." But Chinese scholarship did not enter the stream of Western thought. The Greeks, like all innovative people, were great borrowers, and much of what they put together in logical and useful order originated from the much older civilizations with which they were in contact, including Egypt, Sumeria, Babylonia, Assyria, and Phoenicia. Greek scholars provided a framework of concepts and a model of scholarly method that guided Western thinking for many centuries. Some Greek concepts had the effect of retarding Western scholarship, so that it can be said that European science could not emerge until the influence of Aristotle had been overcome. But many of the basic procedures of scholarship still in use were first developed by the Greeks.

THE ROOTS OF GREEK SCHOLARSHIP

The Greeks were indebted to the world's earliest scholars in many ways. Egypt has been called the cradle of science because of the very early development of methods of observation, measurement, and generalization in that country. The Egyptian priests had to have a sound working knowledge of mathematics, astronomy, and geometry for the practical purposes of public administration. They developed ways to measure land areas and to identify field boundaries obliterated by the Nile floods so that they could collect taxes. They learned how to fix a north-south line so that their monuments and public buildings could be properly oriented. They invented the art of writing, and they found out how to manufacture something on which to write—papyrus, made from reeds that grew in the marshy Nile Delta.

The civilization of Mesopotamia also contributed to scholarship. The world's earliest mathematicians, who lived in Sumeria, had grasped the basic principles of algebra—although the algebraic symbols we use were not invented until the sixteenth century, some 3000 years later. Without symbols of the kind we use, the Sumerians understood and used such principles as

$$(a + b)^2 = a^2 + 2ab + b^2$$

They also had enough knowledge of algebraic methods to be able to find the square root of any number.

The people of Egypt and Mesopotamia also developed a kind of mathematics based on multiples of 6 and 60—a sexagesimal system. Both Egyptians and Sumerians at first believed that there were 360 days in a year. The Egyptians discovered their error and compensated for it by declaring a five-day holiday period each year. They made additional adjustments every fourth year. The Sumerians divided the year into 12 months, each with 30 days. They also divided the circle of the zodiac into 360 parts. The idea that there are 360 degrees in a circle is a very ancient one.

The priests of these early civilizations also collected a large number of observations regarding the position and movement of celestial bodies. The Babylonians and Assyrians, seeking the meaning of all these observations, developed ideas regarding the influence of the moon and the stars on human affairs—a body of concepts that we call astrology.

The Phoenicians, whose homeland was in modern Lebanon, were among the earliest merchant explorers and navigators. Their voyages went far beyond the limits of the known world, but as merchants they were not anxious to report on what they had found. In a valley near modern Beirut there is an ore body in which copper and tin are naturally combined. The Phoenicians made and sold bronze. But although copper was plentiful in the Mediterranean region, tin was scarce. The Phoenicians made regular voyages to Great Britain to find tin. They also sold cedar logs from the mountains of Lebanon. One of the oldest known pieces of writing is a bill of lading describing the cargo of cedar logs carried by 40 ships that sailed from the Phoenician port of Byblos for Egypt some 3000 years before Christ (Casson, 1959:5). The Phoenicians established trading posts all around the shores of the Mediterranean, including the city of Carthage (near present-day Tunis) (Boyce, 1977).

The Phoenicians, too, developed the world's first phonetic alphabet. It was made up entirely of consonants, like the modern Semitic alphabet. The Greeks added the short vowels to the Phoenician alphabet.

GREEK GEOGRAPHY

Homer

The Greek geographers credited Homer with being the father of geography. This poet, whose existence is not known for certain, was the compiler of the long epic poem, the *Iliad*, which describes episodes of the Trojan War sometimes between 1280 and 1180 B.C. This monumental poem, which is the earliest major literary work of Greek history, was probably put together during the ninth century B.C. A second great epic poem, the *Odyssey*, was also credited to Homer. Whereas the *Iliad* is primarily historical, the *Odyssey* is a geographical account of the fringes of the known world. It records the efforts of Odysseus to return home to Ithaca after the fall of Troy (see Fig. 1). Blown off course by a storm, he spends 20 years wandering in distant places. Many historians of geography and others have attempted to identify the places described in the *Odyssey* and offer evidence to suggest that the poet was indeed describing the Strait of Messina, or an island off the coast of Africa, or other well-known localities. One passage describes a land of almost continuous sunshine, where a shepherd setting out with his flock at daybreak hails another shepherd returning with his flock in the evening. Then, later, Odysseus comes to a land of continuous darkness, shrouded in mist. A Greek poet could not have imagined these scenes. Somehow word of the nature of the world in the far north during the long summer days and the continuous winter darkness had filtered back to Greece, to be woven with other geographical threads into the world's first adventure story.

The Greek sailors of the eighth century B.C. had no way of identifying directions at sea except by reference to the winds and associated weather types. In Homer's time they distinguished four directions: *Boreas* was the north wind—strong, cool, with clear skies; *Eurus* was the east wind—warm and gentle; *Notus* was the south wind on the front of an advancing storm—wet and sometimes violent; and *Zephyrus* was the west wind—balmy but with gale force (Bunbury, 1883:1:36). Much later, in the second century B.C., the Athenians built a tower identifying eight wind directions (Schamp, 1955–56) with sculpture illustrating the weather types associated with each. The tower still stands in the midst of a Roman market at the base of the Acropolis (Fig. 2).

The names *Europe* and *Asia* do not appear in Homer as the names of land masses. But at some later time the name *Europe* was applied to the shore of the Aegean Sea toward the setting sun, and *Asia* was applied to the shore toward the rising sun. The origin of these names is not certain (Bunbury, 1883:1:38; Ninck, 1945:15–23; Tozer, 1897/1964:69).

Thales, Anaximander, and Hecataeus

One of the earliest centers of Greek learning was the town of Miletus in Ionia on the eastern side of the Aegean Sea near the mouth of the Meander River (now the Menderes). Miletus became a major center of commerce and attracted Phoenician and Greek ships from all around the Mediterranean and the Black Sea. The sailors and merchants brought to Miletus a wealth of information concerning what things were like beyond the margins of Greek horizons: information about Europe north of the Black Sea, or about strange countries in Asia to the east, or about what could

Figure 1 Troy

be found to the south of Egypt. Between 770 and 570 B.C. Miletus established some 80 Greek colonies around the shores of the Euxine (Black Sea) and along the Mediterranean shores to the west. In Miletus at this time there was not only a flow of geographical information, but also a group of thoughtful people to speculate about how all this miscellaneous information could be assembled in some kind of meaningful arrangement. To Miletus also came reports on Egyptian geometry, Sumerian algebra, and Assyrian astronomy.

Figure 2 Tower of the Winds

The first Greek scholar to be concerned about the measurement and location of things on the face of the earth was Thales, who lived in the seventh and sixth centuries B.C.[1] Thales was a practical businessman who at one time was able to corner the supply of olive oil to make a large profit for himself. But he was also a genius

---

[1]Additional biographical data on persons mentioned in the text can be found in the Index of Names.

who is credited with a great variety of innovations and is often likened to Benjamin Franklin in the breadth of his contributions and the fertility of his imagination. On a trip to Egypt Thales observed the priests at work measuring angles and base lines and computing areas. Thales returned to Miletus with his head full of mathematical and geometrical regularities that went far beyond the practical utility of trigonometry. Six geometric propositions are credited to him: (1) The circle is divided into two equal parts by its diameter. (2) The angles at either end of the base of an isosceles triangle are equal. (3) When two parallel lines are crossed diagonally by a straight line, the opposite angles are equal. (4) The angle in a semicircle is a right angle. (5) The sides of similar triangles are proportional. (6) Two triangles are congruent if they have two angles and a side respectively equal (Sarton, 1952/1964:171). In the sixth century B.C. no one had ever stated these as general propositions before. But Thales' most important contribution was his recognition that the solution of practical problems of measurement was less of an intellectual accomplishment than the rational generalization of the specific solutions.

Thales also made contributions in astronomy and reported on the magnetism of the lodestone. He speculated about the meaning of this fascinating universe and concluded that the material was made up of water in various forms. The earth he visualized as a disc floating in water. He was trying to offer an explanation of the universe in terms that could be checked by new observation, in sharp contrast to the traditional explanations in terms of manlike deities or astrological influences.

A younger contemporary of Thales in Miletus was Anaximander. He is credited with introducing a Babylonian instrument known as the *gnomon* into the Greek world (see also Heidel, 1937:57–58). This is simply a pole set vertically above a flat surface on which the varying position of the sun could be measured by the length and direction of the shadow cast by the vertical pole. Today we call this instrument a sundial. From the gnomon it was possible to make a variety of observations. Noon could be established by noting when the shadow was shortest; the noon shadow provided an exact north–south line, or meridian (from *merides*, meaning noon). The noon shadow varied from season to season, being shortest at the summer solstice and longest at the winter solstice. By observing the direction of the shadow at sunrise and sunset, it was possible to establish the time of the equinox, for at that time the sunrise and sunset shadows were colinear but opposite.

Anaximander is reported by later Greek historians to have been the first ever to draw a map of the world to scale. To be sure, the Sumerians had drawn pictorial "maps" of some of their cities as early as 2700 B.C., but a true map must show distance and directions to scale. Anaximander's map had Greece in the center, and the other parts of Europe and Asia known to the Greeks were plotted around it. The map was circular and was bounded all the way around by the ocean. A copy of the map was supposed to have been cast in bronze and transported to Sparta in the effort to convince the Spartans that they should join in the war against the Persians. But the Spartans said the map proved that Persia was too far away to worry about.

The scholars who were seeking to explain their observations of the face of the earth and of the relative positions of the celestial bodies found difficulty in understanding how the sun could set in the west and yet get back to the east by the next morning. If the earth were a disc floating in water, how could the sun go under the water? Anaximander suggested that somewhere to the north there must be some very high mountains behind which the sun made the trip back again to the east. The shadow cast by these mountains would account for the night.

Anaximander was also one of the earliest philosophers to provide us with an example of how a word can be used to symbolize something that is not known and not observed. He did not actually reject the idea of Thales that water was the prime substance from which all observable features of the earth were made. But he used the word *apeiron* to symbolize this prime substance. Apeiron, which could not be experienced through the senses, nevertheless became a concept—a specific mental image that by the process of deduction could become a real substance. This thought process is possible for us because we use words to symbolize abstractions. It is still with us in the twentieth century, providing a semantic trap for the unwary who confuse observable reality with the reality of word symbols.

Thales and Anaximander can be recognized as the originators of the mathematical tradition in the study of geography. To Hecataeus goes the credit of originating the literary tradition. Hecataeus, who was born at about the time of the deaths of Thales and Anaximander and who died about 475 B.C., was the first to collect and classify the information brought to Miletus not only from the known world of the Greeks but also from the shadowy world beyond the Greek horizons. One of the two prose works credited to him is the *Ges periodos*, or *Description of the Earth*, only fragments of which survive. But one fragment contains a kind of subtitle in which there is the first record of a "new geography." He says that he has written these things in his book because he believes they are true. "The narrations of the Greeks are many and in my opinion, foolish." Hecataeus divided his work into two parts, each dealing with one of the regional divisions of the earth. One book dealt with Europe, and the other with the rest of the world—Asia and Libya. He followed what was apparently already tradition in separating Europe from Asia along the Hellespont, the Euxine, the Caucasus Mountains, and the Caspian Sea, which he thought was connected with the surrounding ocean (Fig. 3).

Hecataeus was not a theorist. His reaction to the speculations of his predecessors is similar to the reactions of countless generations to follow. He felt that discussions of whether water or apeiron should be accepted as the prime substance, or whether there even was such a prime substance, were futile. Before trying to solve the enigma of the universe, he insisted, we should take stock of what is around us and put the accumulated knowledge about the world together in usable form. The contrast in the approaches of these scholars of Miletus more than 24 centuries ago illustrates the apparent dichotomy between those who seek to formulate generalizations and those who seek to describe unique things.

Herodotus

A century later the ideas of Hecataeus were ridiculed by another large-minded scholar, Herodotus (484–425 B.C.). His great work, which was written while he was residing in Italy, is a history of the Greek struggle with the barbarians and ends with the revolt of the Ionians against the Persians and with the Greek capture of the Hellespont (480–479 B.C.). But his world history meant writing a world geography; the two tasks were inextricably linked. Included were descriptions of the places he

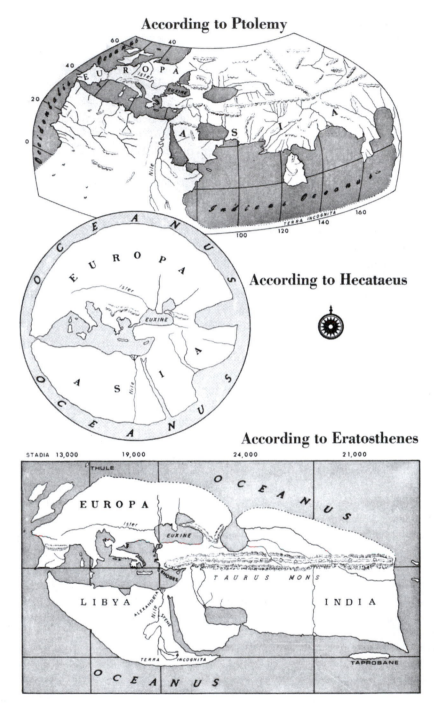

Figure 3 The world according to Ptolemy, according to Hecataeus, and according to Eratosthenes

had visited and the people whose customs he had observed and recorded. In the fifth century B.C. no one was concerned about identifying himself as a member of a separate profession. There were no historians, or geographers, or astronomers, and no professional societies to join. There were no academic departments. Herodotus is usually described as the first great historian, and his work was the first master-piece of Greek prose. But Herodotus is identified as a historian chiefly because there are more historians than there are geographers, for a very large part of his work is readily identified as geography. In fact, Herodotus is credited with the idea that all history must be treated geographically and all geography must be treated histori-cally. It is true that the notion of geography as the handmaiden of history came from Herodotus. Geography provides the physical background, the stage setting, in relation to which historical events take on meaning. Herodotus provides some excel-lent examples of what today we would call both human geography and historical geography—that is, the re-creation of past geographies and the tracing of geographical change through time. But Herodotus is also identified as the father of ethnography because of his vivid portrayal of the cultural traits of people strange to the Greeks.

The contributions of Herodotus to geography were based in part on his own personal observations and interviews during many years of travel. Westward he knew the Mediterranean shores as far as southern Italy, where he resided during the latter part of his life. He went through the straits into the Euxine (Black Sea), reaching the mouth of the Ister (Danube) and traveling for several days northward across the Russian steppes along the valley of the Don. He went eastward over much of the territory of the Persian Empire, visiting Susa and Babylon. Toward the south he visited Egypt many times and went up the Nile as far as the first cataract near Elephantine (Aswan).

In his discussion of Egypt, he takes issue with the tradition of dividing Asia (the eastern side of the Mediterranean) from Libya (the southern side) along the Nile River, as Hecataeus had done. The Nile Valley, he insists, has been built by mud brought down from Ethiopia. This mud is dark colored and easily worked with the plow—quite unlike the light-colored clays of Syria or the red sands of Libya. He insists that Egypt is occupied by Egyptians and that they are not divided into Asians and Libyans along the river. Libya, he says, begins to the west of Egypt, and he was familiar with coastal Africa as far west as Carthage. His is one of the earliest discussions of the properties of regional boundaries and contains many of the arguments used over and over again by later generations.

Herodotus was well aware of some of the physical processes at work on the earth. He used the methods of historical geography to support the hypothesis that the Nile mud, deposited in the Mediterranean, had built the delta. He reconstructed the ancient shoreline and showed that many former seaports were now far inland. The process of delta building, said Herodotus, can be observed in many places—notably in the alluvial plain of the Meander River at Miletus. He also pointed out that the wind blows from cold places to places that are warmer. In the fifth century B.C. it was a significant accomplishment to explain how deltas are formed or to grasp the connection between temperatures and wind directions.

Not all the explanations suggested by Herodotus can be supported in the light of modern knowledge, but even where he was in error he supported his hypotheses with logic. Like all Greek geographers, Herodotus was fascinated with the regularity of the summer floods of the Nile. In this river the water would rise suddenly in mid-May, reach its highest flood stage in September, and then decrease in volume, reaching its lowest stage in April or early May. Since all other rivers known to the Greeks, including the Tigris and the Euphrates, flooded from November to May and reached their lowest stages in summer, students of geography were faced with a challenging problem: What caused this distinctive characteristic of the Nile?

First, Herodotus reviewed the explanations offered by other puzzled scholars and refuted them. For example, he rejected the idea that the strong north wind of winter (the Etesian wind) blowing up the Nile caused the water to back up because floods and low water come whether or not the wind is blowing and no such effect can be observed in other rivers up which the wind is blowing. He rejected the suggestion that the Nile floods were caused by melting snow in Ethiopia, because Ethiopia is closer to the equater than Egypt. Egypt never has any snow, so how could there be snow in Ethiopia?

His own hypothesis was ingenious and illuminates the use of logic in Greek thinking. Like all the Greek scholars, Herodotus accepted as a fundamental principle that the world must be arranged symmetrically. The Ister (Danube), he believed, had its headwaters close to the western coast of Europe and flowed eastward before turning southward through the Euxine, the Hellespont, and the Aegean to reach the Mediterranean. The Nile, in accordance with the principle of symmetry, must follow a similar course, rising close to the western coast of what was then called Libya and flowing eastward before turning toward the Mediterranean to flow northward through Egypt. In winter, he continued, the cold north winds make the sun move along a more southerly course, passing directly along the valley of the upper Nile. The intense heat under the overhead sun draws up the river water, leaving the river with greatly decreased volume in winter. But during the summer, when the sun returns to its course "through the middle of the heavens," the volume of water rises again because the lower Nile crosses the sun's path at right angles, and much less of its water is evaporated. Since this explanation was in agreement with both concepts (symmetry) and direct observations (time of flooding), it was generally accepted by scholars. Here is a case in which accepted concepts structured Herodotus' percepts (a notion developed in Chapter 1).

Herodotus also disagreed with earlier writers who raised doubts about the existence of an ocean all around the margins of the world. Some had reported that there was no ocean to the south of Libya. But Herodotus, in talking with the Egyptian priests, had learned of a Phoenician expedition sent out by King Necho (who ruled Egypt from 610 to 594 B.C.) to sail around the southern end of Libya. The Phoenician ships, it was reported, sailed southward from the Red Sea along the east coast of the continent. They replenished their food supply by stopping from time to time to plant grain and remaining long enough to harvest the crop. It took three years to sail around the southern end of Libya; then northward along the western side; and, finally, to reenter the Mediterranean through the Pillars of Hercules (Gibraltar). This expedition proved that the land is entirely surrounded by water. Then he reported a circumstance that to him "appears incredible, but others may believe," that, while the expedition was near the southernmost part of Libya sailing toward the west, the sun was on their right hand. This observation led many scholars after Herodotus to discount the story of the circumnavigation of Libya. The reality of the Phoenician expedition is now generally accepted, and the circumstance that caused ancient scholars to doubt the story leads modern scholars to find the account plausible. There is also the possibility that some of the Phoenician ships, being caught in the west-flowing equatorial current south of the equator in the Atlantic Ocean, were carried across the relatively narrow ocean to the northeast of Brazil.

One difficulty with the interpretation of these ancient writings involves the things that are omitted. Considering the wealth of detail Herodotus provides concerning some places and some events, it is remarkable that he made no mention of another Phoenician voyage. This was the expedition led by Hanno in about 470 B.C. The expedition was sent out by Carthage to establish trading posts and colonies along the Atlantic coast of Libya south of the Pillars of Hercules. Hanno's descriptions of the things he saw are detailed enough so that the voyage can be charted with confidence. After passing through the strait, he turned southward (Fig. 4). Near the present port of Safi in Morocco he passed a lagoon where elephants were feeding. Farther south on an island in the bay of Rio de Oro on which Villa Cisneros is now located, he established a base that he named Cerne. This remained a Phoenician trading post for many years. From Cerne Hanno led two expeditions farther to the south. On the second of the two voyages he reached Sherbo Island, south of the present site of Freetown in Sierra Leone, almost seven degrees from the equator. Here the explorers came upon "wild men and women with hairy bodies," who, they were told, were called gorillas. They were unable to catch any of the men, but they did catch three women. These they killed and skinned and brought the skins back to Carthage. The record of the expedition was preserved on a bronze plaque in a temple at Carthage.

Plato and Aristotle

The two great Greek philosophers, Plato (428–348 B.C.) and Aristotle (384–322 B.C.), both made important contributions to the development of geographical ideas. Plato, who was a master of deductive reasoning, insisted that the observable things on the earth were only poor copies of *ideas*, or perfect predicates from which observable things had degenerated or were in process of degenerating (Popper, 1945/1962:18–34). At one time, he observed, Attica in Greece (the ancient territory of which Athens was the central city) possessed a very productive soil, capable of supporting the inhabitants in comfort. There were forests on the mountains that not only provided feed for animals but also held the rainwater from pouring down the slopes in floods during heavy rains. "The water was not lost, as it is today, by running off a barren ground to the sea. . . . What now remains, compared with what then existed, is like the skeleton of a sick man, all the fat and soft earth having been wasted away, and only the bare framework of the land being left" (Glacken, 1967:121). Arguing from the general theory to the particular situation in Attica, Plato used this as an example of the degeneration of things from their original perfect state. If Plato had argued from the particular to the general, he might have realized that humans make changes in the land they occupy and that soil erosion and land destruction are parts of cultural history and are repeated in many places. The idea of the individual as

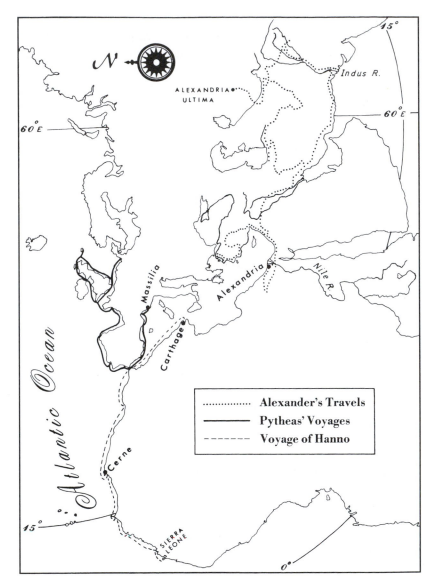

Figure 4 Greek exploration, 470–310 B.C.

an agent of change on the face of the earth was not formulated for many centuries after Plato. As Glacken points out, Plato missed the chance to change the whole history of speculation concerning man–land relations by identifying the individual as destructive agent.

Plato also related the story of Atlantis in the *Timaeus* and the *Critias*. The Greek world, he said, was about to be invaded in the year 9000 B.C. by a highly civilized people who lived somewhere to the west. The Greek armies, after a fierce struggle, were victorious, but just as the invaders were defeated, their homeland suffered a disastrous earthquake and sank beneath the sea. It is possible, he reported, to sail over the sunken city of Atlantis if one is careful to avoid the shallow places. Explorers and popular writers have been searching for Atlantis ever since, even imagining it to have been a land bridge between Africa and America.

Was the earth round or flat? The great majority of the people living in those times did not question the evidence of their senses that the earth was flat; a few philosophers began to think of the earth as a ball on purely theoretical grounds. All the Greek thinkers accepted the idea that symmetry of form was one of the attributes of perfection, and the most completely symmetrical form was a sphere. Therefore, they argued, the earth, which was created in perfect form as the home of man, must be spherical. Pythagoras, who lived in the sixth century B.C., may have been the earliest philosopher to hold this view. At any rate, he worked out some of the mathematical laws for the circular motions of celestial bodies. His pupil, Parmenides, applied these laws to observations made from the surface of a round earth. But Plato, who lived a century after Parmenides, seems to have been the first philosopher to announce the concept of a round earth located in the center of the universe with the celestial bodies in circular motions around it. Whether it was Plato's original concept or whether it was suggested to Plato by Socrates, whom he quotes, cannot be determined. Eudoxus of Cnidus, a contemporary of Plato, developed the theory of zones of climate based on increasing slope (*klima*) away from the sun on a spherical surface. All these formulations were deductions from pure theory—the theory that all observable things were created in perfect form and that the most perfect form was a sphere. But it was Aristotle who first looked for evidence to support the concept.

Aristotle was 17 when he joined Plato's Academy near Athens. At this time (367 B.C.) Eudoxus was acting head during the temporary absence of the master. Aristotle remained at the Academy until Plato's death, at which time Aristotle was 38. During the next 12 years of his life he spent his time traveling widely throughout Greece and around the shores of the Aegean. In the year 335 B.C., when Aristotle was 49, he returned to Athens and founded his own school, which he named the Lyceum (Sarton, 1952/1964:492). By this time he was convinced that the best way to build theory was to observe facts and the best way to test a theory was to confront it with observations. Whereas Plato built theory by intuition and reasoned from the general to the particular, Aristotle built theory by reasoning from the particular to the general. These two ways of thinking about things are known, respectively, as deduction and induction.

Aristotle recognized that observations made through the senses can never provide explanations. Our senses, he said, can tell us that fire is hot but cannot tell us why it is hot. Aristotle formulated four fundamental principles of scientific explanation— that is, of answering the question: "What makes this thing the way it is?" (Aristotle, trans. Gershenson and Greenberg, 1963:2:13). One way is to describe its nature, to tell its essential characteristics. A second way is to specify the kind of matter, the substance, of which it is composed. A third way is to tell what caused the process through which the thing became as it is. And a fourth way, which is complementary to the third, is to tell the purpose the thing fulfills. Unlike Plato, Aristotle assumed that things were in process of physical change leading to a final perfect state. This model for scientific explanation constituted the Western world's first paradigm for the guidance of scholars.

With regard to the matter, or basic substance, of which all material things are made, Aristotle followed Empedocles (490–430 B.C.), who a century earlier had improved on the single-substance idea of Thales (water) by postulating the existence of four basic substances: earth, water, fire, and air. All material objects on the earth are made up of these basic elements in varying proportions. Aristotle added a fifth substance, aether, which did not occur on the earth but was the material from which celestial bodies were made.

Aristotle pointed out that to create the material objects on the earth and in the heavens, some kind of process of change had to take place. First, there had to be empty space. The philosophers of that time recognized the existence of two kinds of space: celestial space and earth space—that is, space on the surface of the earth. There was some speculation concerning interior space within the earth, but there was little knowledge to guide these speculations. Aristotle, modifying the ideas of Empedocles, developed the theory of natural places. Everything had its natural place in the universe, and, if removed from this place, it would seek to return. Earth space was the natural place of earth and water, and, if raised above the surface of the earth, these substances and things composed of them would fall back to the surface. Air and fire, on the other hand, had their natural places in celestial space, and, therefore, they tended to rise. Aether had its natural place in the celestial bodies far out from the earth.

Aristotle agreed in part with the teaching of Plato, derived from Pythagoras and Parmenides, that all things are patterned after numbers. The basic regularities of the universe are those of geometry and mathematics. He complains, however, that "nowadays everybody thinks that science is mathematics, and that it is only necessary to study mathematics in order to understand everything else" (Aristotle, trans. Gershenson and Greenberg, 1963:2:51). Aristotle insists that mathematics can be used to explain the process of change that makes things as they are, but it cannot answer the fourth question concerning the purposes or ideal states. Aristotle was the first teleologist in that he believed everything was changing in accordance with a preexisting pattern or plan, just as a carpenter building a house knows in advance what the house will be like when it is finished. All things, said Aristotle, are not deteriorating from an ideal state but rather are developing toward an ideal state.

Aristotle accepted Plato's concept of a spherical earth and began to seek an explanation of it and to test the concept with observations. His explanation was derived from the theory of natural places. When the solid matter of which the earth is made falls toward a central point, it must form a ball (Sarton, 1952/1964:510). Aristotle was the first scholar to recognize the significance of the observed fact that when the shadow of the earth crosses the moon during an eclipse, the edge of the shadow is circular. He also recognized that the height of various stars above the horizon increases as one travels toward the north, which could only occur if the observer were traveling over the curved surface of a sphere. Strangely, he never realized that additional support for the concept of a round earth could be gained by noting the disappearance of a ship beyond the horizon—hull first. He must have had ample opportunity to observe this fact.

Aristotle's attempt at scientific explanation did not include anything about controlled experiments or the verification of premises, but only the use of logic to formulate and give support for theory. Some of his logical explanations seemed so unassailable in the fourth century B.C. and have been so universally accepted for many generations since that time that his influence on the history of Western ideas has been enormous. It has been pointed out that modern science could not appear until Aristotle had been abandoned. Here we note a very common sequence of events in the history of ideas: The formulation of a new concept is enormously stimulating and results in an increase in the quality and quantity of observations, but the continued acceptance of the concept proves an obstacle to the progress of scholarship among succeeding generations.

An example of this in the field of geography was Aristotle's concept of the varying habitability of the earth with differences of latitude (Glacken, 1956). That habitability was a function of distance from the equator was a notion that seemed to accord with observed facts for people living around the shores of the Mediterranean. If the earth is a sphere and the sun is circulating about it, the parts of the earth where the sun is most directly overhead must be much hotter than places farther away from the sun. The Greeks were familiar with the excessively high temperatures experienced in Libya along the southern side of the Mediterranean. In modern times the world's record for high temperature observed in a standard thermometer shelter (136.4°F) is held by a place in present-day Libya about 25 miles south of the Mediterranean shore and more than 32°N of the equator. If the air gets that hot at this latitude, the Greeks reasoned, it must be very much hotter close to the equator. The people living in the northern part of Libya had black skins, and the Greeks assumed that they had burned black by exposure to the sun. At the equator, then, all life must be impossible because any living thing would be burned in the intense heat. Aristotle reasoned that the parts of the earth close to the equator, the torrid zone, were uninhabitable; that the parts of the earth far away from the equator, the frigid zone, were constantly frozen and also were uninhabitable; and that the temperate zone in between constituted the habitable part of the earth. The *ekumene*, the inhabited part of the earth, was in the temperate zone, but much of it, said Aristotle, was not inhabited because of the ocean. Aristotle also postulated the existence of a south temperate zone, which could not be reached from Greece because of the intense heat of the torrid zone. Many scholars in antiquity accepted Aristotle's south temperate zone but doubted that it could be habitable because in the *antipodes* people would have to hang upside down. This notion of habitability as a function of latitude has had a long history and, in fact, is still widely accepted, especially by nongeographers.

Alexander the Great

Aristotle had many pupils, and he instilled in all of them a desire to test theory by direct observation. He taught them to "go and see" for themselves whether any one theory could or could not be accepted. His greatest pupil was Alexander, who became the king of Macedonia at the age of 20. Alexander studied with Aristotle for only three years (343–340 B.C.), between the ages of 13 and 16, yet no one applied the master's teaching more effectively. When Alexander became king, he started on a career of military conquest. A primary objective in Alexander's conquests was to expand Greek geographic horizons—they were in the nature of armed explorations.

Alexander's conquests pushed Greek knowledge of the earth far to the east (Fig. 3). After conquering the barbarian tribes living north of the Ister (Danube), in 334 B.C. he crossed the Hellespont into Asia. His first marches were close to the coast, where he could be supplied by ship. But then he grew bolder and invaded the central part of present-day Turkey, then part of the Persian Empire. Thence he continued southward along the eastern side of the Mediterranean to Egypt, where he established his rule. He founded the city of Alexandria in 332 B.C., which was destined to become one of the great commercial and intellectual centers of the ancient world. After some exploratory excursions to oases in the Libyan desert west of the Nile, he turned again toward the east, crossing into the heart of the Persian Empire (modern Iran) by way of Babylon and Persepolis. He pushed northward as far as the central Asian market town of Samarkand, then on eastward as far as—and across —the Indus River. Believing that he was only a short distance from the eastern limit of the ekumene, he wanted to march on farther, but his troops mutinied and insisted on returning to Greece. Alexander died in Babylon in 323 B.C., and the empire he had built and ruled with compassion collapsed in strife.

Few teachers have had the lessons they taught applied so well by their students. Alexander's staff included writers to describe the lands they crossed and astronomers to take observations of the height of the bright star Canopus to fix latitude, or distance north of the equator. There were trained pacers, whose duty was to measure distances on the march. As a result, he sent back to the Greek world a wealth of new observations concerning what it was like beyond the Greek horizons and how far and in what direction it was necessary to travel to reach these strange places. At the time of his death he was planning to send out two additional expeditions to find answers to two geographical questions. One was to follow the shores of the Caspian Sea to settle the question whether this sea was connected to the open ocean, as some maps showed. The other expedition was to sail southward along the Red Sea from Egypt to find out whether Libya truly was surrounded by water on the south and whether human beings could survive in the intense heat of the equatorial regions. With his death, plans for both expeditions were abandoned. Alexander accomplished more than any other man of antiquity to extend geographical knowledge.

Pytheas

While Alexander was extending Greek geographic horizons to the east, another Greek explorer was voyaging far to the northwest of the Greek world in western and northern Europe (Fig. 4). This was Pytheas, who is assumed to have made his remarkable voyage sometime between 330 and 300 B.C. Unfortunately, his original report did not survive, and he is known to historians of geography only because of references made to his work by other writers (Bunbury, 1883:1:589–601; Ninck, 1945:218–226; Sarton, 1952/1964:523–525; Tozer, 1897/1964:152–164).

The following account of his voyage is generally accepted today. Pytheas was a native of the Greek colony of Massilia (modern Marseilles), which was engaged at that time in a bitter rivalry with the Phoenicians of Carthage for control of the profitable trade in tin and amber. Whether Pytheas was sent out by Massilia to penetrate the screen of secrecy imposed by the Phoenicians or whether he financed his own voyage to satisfy his curiosity about what lay beyond the Greek horizons is not certain. At any rate, he set out by ship from Massilia along the coast to the Pillars of Hercules and then managed to slip by the Phoenician naval base at Gades (Cadiz). He sailed along the coast of France to the English Channel and then around Great Britain.

Pytheas reported things that were so contrary to Greek experience that the geographic scholars of his day discredited him and treated his important information as pure fantasy. He told about the customs of the people of Britain with detail that could scarcely have been invented. He described the drinking of mead (fermented honey), the use of barns to thresh grain in wet weather, and the change in the character of agriculture from south to north in Great Britain. He also described a sea so full of ice that it could be traversed neither on foot nor in a boat—which exactly describes what the polar explorers call ice sludge. He was also the first Greek to tell about ocean tides (the tides on the Mediterranean are too small to be noticed), and he showed that the tides were related to the phases of the moon.

How much farther north he went is a puzzle, although he reported on the existence of a place called Thule, six days sailing north of Great Britain. It is probable that he sailed along the eastern shore of the North Sea, perhaps as far as modern Denmark. He is quoted as having been to a place where the length of the longest day was between 17 and 19 hours, which would place him 61°N, in the northernmost of the Shetland Islands. He is also quoted as reporting that at Thule the sun remained above the horizon during the whole of the longest day, which would put this place well to the north in Norway or possibly Iceland. It is certain, however, that before he left Massilia he observed the angle of the sun's shadow on a gnomon, and the latitude derived from this measurement was almost exactly correct (43°05′N, instead of 43°18′N).

The details that Pytheas is said to have recorded (and that led the scholars of antiquity to discredit him) led modern scholars to believe that he was indeed reporting his observations correctly. Nowadays Pytheas is accorded his due place among the great explorers of all times.[2]

---

[2]There are many other important contributors to the development of Greek geographical ideas in classical antiquity. There was Hippocrates, the fifth-century B.C. physician, who, among many other writings attributed to him, produced the world's first medical geography. In his book *On Airs, Waters, and Places,* he was the first to present the concept of environmental influence on human character. There was Theophrastus, who succeeded Aristotle as head of the Lyceum and was its director for 35 years. He wrote a prodigious number of books on a wide range of subjects, including meteorology, petrography, ethics, and religion. He is known as the father of plant geography because he described and classified more than 500 species of cultivated plants. There was Dicaearchus, who measured the heights of certain Greek mountains with a primitive theodolite and concluded that the highest mountains were only a slight roughness on the surface of the earth—this, in the late fourth century B.C. Aristarchus of Samos proposed the hypothesis that the sun was the center of the universe and that the earth and planets were revolving around it. He explained day and night in terms of a rotating earth. The inquiring student who wishes to know more about the many other Greek and Roman geographers should consult the following: Bunbury, 1883; Burton, 1932/1969; Heidel, 1937; Ninck, 1945; Sarton, 1952/1964, 1959; Thomson, 1965; or Tozer, 1897/1964.

Eratosthenes, the Father of Geography

Eratosthenes is often identified as the father of geography because, among other contributions, he was the first to coin the word. But, as we have observed, many major contributors to geographical ideas are not identified as geographers. In many ways, however, Eratosthenes set a stamp on the study of the earth as the home of man that still persists.

Eratosthenes was born in the Greek colony of Cyrene in Libya. In Cyrene and later in Athens he received a broad education, including philology and rhetoric as well as mathematics and philosophy. He probably attended both the Academy and the Lyceum. In about 244 B.C. he accepted an invitation from the king of Egypt to become the royal tutor and was also named as "alpha fellow" at the museum in Alexandria. With the death of the chief librarian of the museum in about 234 B.C., he was appointed to that coveted post, which was among the most prestigious in the Greek scholarly world. He remained as chief librarian until his death in about 192 B.C., at the age of 80.

George Sarton gives some interesting sidelights on the attitude of the Greek scholars toward the chief librarian at Alexandria. Eratosthenes had two nicknames: he was called *beta*, which suggests that although he was the senior (alpha) fellow he was still only a second-rate scholar; and he was called *pentathlos*, a name given to athletes who performed well in five different games. Sarton points out that at this time there was a growing specialism in Greek scholarship of a kind that did not again appear until the seventeenth century after Christ. Specialists in a restricted field of study—then as now—are inclined to look with scorn on anyone whose scholarship is based on breadth rather than depth. Here is what Sarton has to say about this very human situation:

> The first nickname, *beta*, shows that the scientists and scholars of that age were already very jealous of one another and all too ready to deflate those whose superiority they misunderstood or resented. Now the professional mathematicians might consider him as not good enough in their field and be displeased with the abundance and variety of his nonmathematical interests. As to the men of letters and philologists they could not appreciate his geographic purposes. Eratosthenes might be second-rate in many endeavors, but he was absolutely first-rate in geodesy and geography and is to this day one of the greatest geographers of all ages. This his critics could not even guess, and therefore they pooh-poohed him. There was among them a man of genius but as he was working in a new field they were too stupid to recognize him. As usual in such cases, they proved not his second-rateness but only their own (Sarton, 1959/1965:101–102).

Eratosthenes' geography was written in three books, of which only fragments survive. In the first he described the form and nature of the earth and changes in its surface; the second dealt with mathematics and measurement; the third described countries, peoples, and politics. These books described the ekumene, the inhabited earth, in which he accepted both the major divisions of Europe, Asia, Libya, and the five zones—a torrid zone, two temperate, and two frigid zones. He improved on Aristotle by giving the mathematical boundaries of these zones. The torrid zone he thought was 48 degrees of the whole circumference. (Twenty-four degrees north and south was calculated as the location of the tropics.) The frigid zones extended 24 degrees from each pole. The temperate zone was between the tropic and the polar circles. In his book he was one of the few who accepted the reports from Pytheas: He extended the ekumene from Thule, near the Arctic Circle, to Taprobane (Ceylon) in the Indian Ocean. The ekumene, he reported, also extended from the Atlantic Ocean to the Bay of Bengal, which he assumed was the eastern limit of habitable land.

He also prepared a world map (Fig. 3). He made use of a frame of north-south and east-west lines, but these were not spaced regularly. Rather, he used the meridian of Alexandria, which he extended southward through Syene and northward through Rhodes and Byzantium, as the prime meridian; and he used the latitude of the Pillars of Hercules, which he thought also passed through Rhodes (Sarton, 1959/1965:106–108). His map of the known world was plotted in relation to these lines.

Yet Eratosthenes is perhaps best known for his calculation of the circumference of the earth. Apparently, he was able to do this because he was the first scholar with the imagination to appreciate the significance of two separate observations of the position of the sun at the time of the summer solstice. One observation came from near Syene (Aswan). On an island in the Nile, just below the first cataract and opposite Syene, there was a deep well, and at the bottom of the well at the summer solstice the image of the sun was reflected in the water. The existence of this well had been known for a long time, and no doubt tourists in ancient times traveled up the Nile to witness this strange occurrence each year. This meant, of course, that on that date the sun was directly overhead. The second observation was made outside the museum in Alexandria, where there was a tall obelisk. Using the obelisk as a gnomon, Eratosthenes measured the length of the shadow at the solstice. He was thus able to measure the angle between the vertical obelisk and the rays of the sun. With these data in mind, Eratosthenes made use of the well-known theorem of Thales, which states that when a diagonal line crosses two parallel lines, the opposite angles are equal. The parallel lines were given by the parallel rays of the sun (Fig. 5). The rays of the sun at Syene, which were vertical, could be extended to the center of the earth (SC). Also, the obelisk, which was vertical at Alexandria, could be extended to the center of the earth (OC). Then the angle between the sun's rays and the vertical obelisk at Alexandria (BOC) must be the same as the opposite angle at the center of the earth (OCS). The next question was this: How much of the whole circumference of a circle is subtended by the angle OCS? Eratosthenes measured this as one-fiftieth of the whole circumference. It was then only necessary to fill in the distance between Syene and Alexandria, which the Egyptians said was the equivalent of about 500 miles, and then multiply this distance by 50. Eratosthenes, therefore, concluded that the whole earth was about 25,000 miles in circumference. (Actually the circumference measured through the poles is 24,860 miles.)[3]

---

[3]Eratosthenes gave his linear measurements in stades. The value of the stade is not known exactly, but it ranges around 10 stades to the modern mile. So his estimate of 5000 stades was approximately 500 miles.

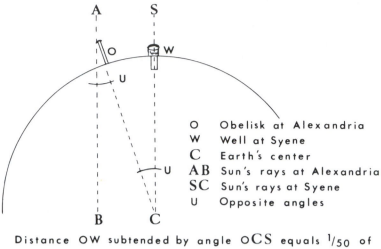

O — Obelisk at Alexandria
W — Well at Syene
C — Earth's center
AB — Sun's rays at Alexandria
SC — Sun's rays at Syene
U — Opposite angles

Distance OW subtended by angle OCS equals ¹/₅₀ of the circumference of a circle

Figure 5 Calculation of the Earth's circumference (by Eratosthenes)

The measurements in those times were far from exact. Eratosthenes assumed that Alexandria was due north of Syene, whereas, in fact, it is about longitude 3°W of Syene. The length of the road between Syene and Alexandria, which the Egyptians said was the equivalent of 500 miles, is actually 453 miles. And Syene is actually at latitude 24°5′N, a little to the north of the tropic. But all these errors canceled out so that the resulting calculation was amazingly close to the correct figure (Thomson, 1965:159–161).

Hipparchus

When Eratosthenes died, he was succeeded as chief librarian (at Alexandria) by Hipparchus. The dates of Hipparchus's birth and death are unknown, but it is certain that he was working at the library in 140 B.C. Hipparchus was more of a mathematician and astronomer than he was a geographer, but he did show, in theory at least, the way to establish the exact position of every point on the earth's surface. He was the first to divide the circle into 360 degrees, based on Assyrian arithmetic; for all nations the circle continues to be so divided. Hipparchus defined a grid of latitude and longitude lines, as Eudoxus had done for celestial space. The equator, he pointed out, was a great circle (one that divides the earth into two equal parts), and the meridians that were drawn converging on the poles were also great circles. The parallels, on the other hand, became shorter and shorter as they approached the poles. Since the earth makes one complete revolution in 24 hours and there are 360 meridians drawn from equator to poles, each hour the earth turns through 15 degrees of longitude.

Hipparchus hoped that geography could be made more exact through the plotting of locations in this theoretical grid. The Greeks did know how to make fairly good measurements of latitude by using the gnomon, but very few such observations had, in fact, been made.[4] Longitude, however, remained a matter of guesswork. There was no way to measure time, especially at sea. Hipparchus suggested that the local times of the start of an eclipse at different places could be compared. The time differences would provide a measure of longitude, but no such system of coordinated observations was even attempted for many centuries after his time. Already in the second century after Christ, geographical studies had become too technical and too mathematical for the use of general readers or of others who wished to find information about particular countries. Polybius, the Greek historian, saw geography as an essential support for the study of history, as Herodotus had suggested.

Hipparchus was also the first to wrestle with the problem of showing the curved surface of the earth on a flat surface. It cannot be done because a spherical surface cannot be made to lie flat without cutting or stretching it. He devised two kinds of projections, however, so that the distortion of the spherical surface on a map could be carried out mathematically. He told how to make a stereographic projection by laying a flat parchment tangent to the earth and extending the latitude and longitude lines from a point opposite the point of tangency. The orthographic projection is similarly produced but by projecting the lines from a point in infinity. On the stereographic map the central portion is too small in relation to the periphery; on the orthographic projection the central portion is too large. These two projections, it should be noted, can only show a hemisphere, not the whole earth.

Posidonius

Another important Greek historian and geographer, who lived shortly before the time of Christ, was Posidonius. Two of his contributions to geographical ideas must be described: one, a wrong idea that persisted for centuries; the other, a correct idea that was overlooked.

It was Posidonius who estimated the circumference of the earth and arrived at a much smaller figure than that of Eratosthenes. Because he felt no confidence in the work of Eratosthenes, he undertook to make his own measurements. He observed the height above the horizon of Canopus (a star of the first magnitude) at Rhodes and Alexandria, which he assumed to be on the same meridian. He then estimated the distance between them based on average sailing time for ships. The figure he arrived at for the circumference of the earth was 18,000 miles. He also greatly overestimated the west-to-east distance from the westernmost part of Europe to the eastern end of the ekumene, then thought to be occupied by India. He declared, therefore, that a ship sailing westward across the Atlantic from Western Europe would reach the east coast of India after a voyage of only 7000 miles. As we will see, when Columbus argued before the scholars at the Spanish court that he could sail west to India, he used the smaller circumference estimated by Posidonius, whereas the scholars favored the larger circumference calculated by Eratosthenes.

---

[4]Hipparchus invented an instrument that was easier to use than the gnomon. This was the astrolabe. A circular dial was marked off into 360 parts, and a rotating arm was fixed at its center. Hanging on the rigging of a ship, the astrolabe made possible the measurement of latitude at sea by observing the angle of the polestar.

Posidonius, however, was right about another matter. He refused to follow Aristotle in believing that the equatorial part of the torrid zone was uninhabitable because of heat. The highest temperatures and the driest deserts, he insisted, were located in the temperate zone near the tropics, and the temperatures near the equator were much less extreme. He arrived at this conclusion—amazing in the first century B.C.—on purely theoretical grounds, for he had no access to credible reports from anyone who had crossed the Sahara, including Hanno's voyage along the African west coast. The sun, he pointed out, pauses longest near the tropics and is overhead for a much shorter time at the equator. The interesting point is that Posidonius's incorrect estimate of the circumference of the earth was widely accepted by those who followed him, while his correct belief concerning the habitability of the equatorial regions was overlooked.

Strabo

A very large part of what scholars think they know about ancient geography came from Strabo. Most of the books written by earlier scholars have disappeared entirely or survive only in fragments. Much of the history of geographical ideas in ancient Greece and Rome must be pieced together from surviving cross-references. But Strabo's monumental work on geography was found almost intact, with only a very few minor parts missing; fortunately, the first part of Strabo's writing is a review of what other geographers since Homer had accomplished. Strabo's work was another example of what has become almost standard practice—the proclamation by an author that what he has produced is, indeed, the "new geography." Here is the way Strabo outlined his task:

> Accordingly, just as the man who measures the earth gets his principles from the astronomer and the astronomer his from the physicist, so, too, the geographer must in the same way take his point of departure from the man who has measured the earth as a whole, having confidence in him and in those in whom he, in his turn, had confidence, and then explain, in the first instance, our inhabited world—its size, shape, and character, and its relations to the earth as a whole; for this is the peculiar task of the geographer. Then, secondly, he must discuss in a fitting manner the several parts of the inhabited world, both land and sea, noting in passing wherein the subject has been treated inadequately by those of our predecessors whom we have believed to be the best authorities on these matters [Strabo, trans. Jones, 1917:429–431].

Strabo was born in Amasia (in what is today central Turkey, some 50 miles south of the Black Sea coast) in about the year 64 B.C. He died in A.D. 20. His family was sufficiently well-to-do that Strabo received a good education and was able to travel widely in the Greek world. He lived for several years in Rome and also worked in the library at Alexandria. His travels took him no farther west than Italy and no farther east than the borders of Armenia. He had sailed on the Black Sea, and, while he was at Alexandria, he made a trip up the Nile (in 24 B.C.) as far as Philae, a short distance above the first cataract. He wrote two major works after he returned to Amasia. One was a history from the fall of Carthage to the death of Caesar, of which only a few fragments have been found. But his *Geography*—almost all of the 17 books—did survive. Strabo was an elderly man when he began to write his *Geography* and perhaps died before it was completed.

Strabo's *Geography* is compiled from the writings of his predecessors. He defends Homer's knowledge of geography at great length but then discards Herodotus as a "fable-monger." He also discards Hanno's voyage along the western side of Africa and Pytheas's exploration of northwest Europe. He accepts Aristotle's zones of habitability, as defined by Eratosthenes, and then goes on to assert that the limit of possible human life toward the equator is at latitude 12°30'N—on what basis he does not say. He also places the northern limit of the habitable earth, where cold is the limiting factor, only 400 miles north of the Black Sea. No one can really be civilized if he lives north of the Alps in Europe because it is necessary to huddle around fires just to keep alive. He accepts the calculation of the earth's circumference made by Posidonius. On the other hand, Strabo gives a correct explanation of the floods of the Nile, attributing them to the heavy summer rains in Ethiopia.

Strabo wrote for a specific group of readers: the educated statesman and the military commanders. His purpose was to provide a text for the information of Roman administrators and military commanders, and his work constitutes the world's first administrator's handbook. He is very critical of those geographers who try to copy Aristotle in the search for explanations; rather, he wants to provide an accurate description of the parts of the ekumene. The rest of the world does not interest him at all. He recognizes the geographer's need for a sound mathematical basis, and he derives this chiefly from Hipparchus and Posidonius. The major part of his work is devoted to detailed descriptions of the various parts of the known world. After two books of introductory material, including the discussion of his sources, Strabo devotes eight books to Europe, six books to Asia, and one book to what we would today call Africa. Most of this book deals with Egypt and Ethiopia; after completing this coverage he then says, "Now let me describe Libya, which is the only part left for the completion of my Geography as a whole" (Strabo, trans. Jones 1917:155). Clearly, he is not much interested in any abstract argument about the placing of regional boundaries, but he accepts without discussion the idea of Herodotus that Libya begins west of the Nile Valley.

Many centuries passed before Strabo's *Geography* was read. When Pliny the Elder wrote his encyclopedia of geography in A.D. 77, based on the reading of some 2000 volumes, he did not even mention Strabo. The administrators who might have benefited from the work never saw it. However, by the sixth century after Christ, Strabo's *Geography* had been "discovered" and had become a classic, as indeed it remained for many centuries thereafter.

ROMAN GEOGRAPHY

Unlike the Greeks, the Romans produced little that was new in the field of geography. Writing shortly before the time of Christ, one Marcus Terentius Varro wrote a compendium of geography that would scarcely have merited the adjective "new" had he not set forth a theory of culture stages that remained almost unchallenged until the nineteenth century. Varro describes human culture as progressing through a regular sequence. Originally, we derived our food from the things that the virgin earth produced spontaneously. From this original state we advanced through a stage of pastoral nomadism, then through an agricultural stage, and finally to the stage of contemporary (first century B.C.) culture (Glacken, 1956:72–73). Varro's stages were generally accepted until the nineteenth century, when Alexander von Humboldt pointed out that there had been no pastoral stage in the Americas and that the theory of stages could not be applied everywhere.

There were other compendia of descriptive writings. Pomponius Mela, writing in A.D. 43, produced such a work, and he was widely quoted in the much larger encyclopedic collection of Pliny the Elder. In addition, extensive sailing directions were published for the guidance of ship captains that described the coastlines and ports with considerable detail and accuracy, such as the *Periplus of Scylax* for the shores of the Mediterranean and the *Periplus of Arrian* for the shores of the Euxine (Black Sea) (Bunbury, 1883:2:384). The most complete work of this kind was an anonymous one that offered a guide for navigators and traders that covered the Red Sea, the east coast of Africa as far as Zanzibar (more than 6 degrees south of the equator), and the northern side of the Indian Ocean as far as the southern end of the Malabar coast in India. This was the famous *Periplus of the Erythraean Sea*, which Bunbury dates as some 10 years after the death of Pliny in A.D. 79. (Bunbury, 1883:2:443–479). The merchants and sailors of the first century after Christ, who had not read Aristotle or Strabo, were happily not conscious of the horrible fate that would come to those who ventured within 12 degrees of the equator or of the impossibility of maintaining life in this central part of the torrid zone. At Zanzibar they carried on a flourishing trade with the inhabitants of the African mainland.

Ptolemy

Ancient geography really came to an end with the monumental work of Ptolemy (Claudius Ptolemaus), who lived in the second century after Christ. Nothing is known of his life except that he worked at the library in Alexandria between A.D. 127 and 150. He is the author of the great work on classical astronomy—the *Almagest*—which long remained the standard reference work on the movements of celestial bodies. His concept of the universe agreed with that of Aristotle: The earth was a sphere that remained stationary in the center while the celestial bodies moved around it in circular courses. This remained accepted doctrine until the time of Copernicus in the seventeenth century.

After completing the *Almagest*, Ptolemy undertook the preparation of a *Guide to Geography*. His teacher, Marinus of Tyre, had already started a collection of data regarding place locations on the basis of which the maps of the known world were to be revised. By this time, in the second century, much new information had been collected by the far-ranging Roman merchants and armies. Ptolemy went on with the work Marinus had started. He adopted the grid of latitude and longitude lines developed by Hipparchus, based on the division of the circle into 360 parts. Every place, therefore, could be given a precise location in mathematical terms. Ptolemy's guide contains some six volumes of tables and forms the world's first geographical gazetteer, on the basis of which he revised the world map. The difficulty is that in spite of the appearance of precision, the work really was a monumental collection of errors. In those days latitude could be determined only approximately, and most voyagers failed to make use of the few instruments available. There was no way to measure longitude. Therefore, each listing of latitude and longitude was, in fact, only a selection among estimates. Furthermore, Ptolemy followed Marinus in taking as his prime meridian a north-south line through the westernmost known islands in the Atlantic—either the Canaries or the Madeira Islands. Therefore, starting in the west with his estimates of longitude, the error accumulated toward the east. Ptolemy not only accepted the smaller estimate of the earth's circumference by Posidonius, but he also increased the error in the eastward extension of the land area. Using the authoritative work of Ptolemy, Columbus estimated that Asia must lie very close to Europe on the west.

The *Guide to Geography* consisted of eight volumes. The first was a discussion of map projections together with a few corrections of the data from Marinus based on actual astronomical observations that he had carried out himself. Books 2 through 7 contained tables of latitude and longitude. The eighth book contained maps of different parts of the world based on the gazetteer (Fig. 3). Ptolemy repeated the commonly accepted idea that the parts of the earth near the equator were uninhabitable because of heat. He also indicated on his maps that the Indian Ocean was enclosed by land on the south, an idea he probably took from Hipparchus, but it is not known where Hipparchus found this information. This *terra australis incognita* was not cleared from the maps until the voyages of Captain James Cook in the eighteenth century proved that such a southern land area did not exist.

Ptolemy's *Geography* was translated into Latin in 1409 and consequently became well known in Western Europe. Ptolemy was considered the authority on matters he had addressed. Many editions of his work were therefore published in the sixteenth century (Marshall, 1972). It was only with the development of science and more travel that his work was rendered obsolete and the last of the ancient geographers ceased to be an authority. Curiously, it was not until 1932 that his famous book was translated into English. This was accomplished by Edward L. Stevenson, an American authority on the history of cartography.

With the death of Ptolemy, the geographic horizons that had been widened both physically and intellectually by the Greeks closed in again. It was many centuries before the effort to describe and explain the face of the earth as the home of humankind again attracted the attention of scholars.

REFERENCES: CHAPTER 2

Aristotle. *Metaphysica*. Trans. D. E. Gershenson and D. A. Greenberg, 1963. Vol. 2, *The Natural Philosopher*. Pp. 5–55. New York: Blaisdell.

Aujac, Germaine. 1978. "Eratosthenes c. 275–c. 195 BC," *Geographers: Biobibliographical Studies*. 2; 30–43. London: Mansell.

Boyce, R. R. 1977. *The Trade of Tyre: Anomaly of the Ancient World*. Seattle: Seattle Pacific College.

Bunbury, E. H. 1883. *A History of Ancient Geography among the Greeks and Romans from the Earliest Ages Till the Fall of the Roman Empire*. 2 vols. London: John Murray.

Burton, Harry E. 1932. *The Discovery of the Ancient World*. reprinted 1969. Freeport, New York: Books for Libraries Press.

Büttner, Manfred, ed. 1979. *Wandlungen im geographischen Denken von Aristoteles bis Kant*. Ferdinand Schöningh.

Casson, L. 1959. *The Ancient Mariners. . . .* New York: Macmillan.

Casson, Lionel. 1989. *The Periplus Maris Erythraei*. Princeton, N.J.: Princeton University Press.

Clarke, Katherine. 1999. *Between Geography and History: Hellenistic Constructions of the Roman World*. Oxford: Clarendon Press.

Dueck, Daniela. 2000. *Strabo of Amasia: A Greek Man of Letters in Augustan Rome*. London: Routlege.

Gallois, Lucien. 1901. "L'évolution de la géographie." Congrès national des Sociétés françaises de géographie, 21st session, Paris, August 20–24, 1900, *Comptes rendus*. Published by the Société de géographie. Pp. 108–119.

Glacken, C. J. 1956. "Changing Ideas of the Habitable World." In W. L. Thomas, ed., *Man's Role in Changing the Face of the Earth*. Pp. 70–92. Chicago: University of Chicago Press.

———. 1967. *Traces on the Rhodian Shore, Nature and Culture in Western Thought from Ancient Times to the End of the Eighteenth Century*. Berkeley and Los Angeles: University of California Press.

Harley, J. B. and David Woodward. 1987. *The History of Cartography*. Vol. 1. Chicago: University of Chicago Press.

Heidel, W. A. 1937. *The Frame of Ancient Greek Maps*. New York: American Geographical Society.

James, Preston E. 1972. "Geography." In *Encyclopaedia Britannica*. Vol. 10, pp. 144–160. Chicago: Encyclopaedia Britannica, Inc.

Keltie, John Scott, and Howarth, O. J. R. 1913. *History of Geography*. ("A History of the Sciences.") New York: G. P. Putnam's Sons.

Leaf, Walter. 1912. *Troy—A Study in Homeric Geography*. London: Macmillan.

Marshall, D. W. 1972. "A List of Manuscript Editions of Ptolemy's *Geographia*." *Geography and Map Division: Special Libraries Association*. Bulletin. 87, 17–38.

May, Joseph A. 1982. "On Orientations and Reorientations in the History of Western Geography." In David Wood, ed. *Rethinking Geographical Inquiry*, ("Geographical Monographs," 11). Pp. 31–72. Toronto: York University, Atkinson College, Department of Geography.

Ninck, M. 1945. *Die Entdeckung von Europa durch die Griechen*. Basel: Benno Schwabe.

Popper, K. R. 1945/1962. *The Open Society and Its Enemies*. New York: Harper & Row.

Ptolemy. 2000. *Ptolemy's Geography: An Annotated Translation of the Theoretical Chapters*. Trans. J. L. Berggren and Alexander Jones. Princeton, N.J.: Princeton University Press.

Romm, J. S. 1992. *The Edges of the Earth in Ancient Thought: Geography, Exploration, and Fiction*. Princeton: Princeton University Press.

Sarton, G. 1952. *A History of Science, Ancient Science Through the Golden Age of Greece*. Cambridge, Mass.: Harvard University Press (reprinted New York: John Wiley & Sons, 1964).

———. 1959. *A History of Science, Hellenistic Science and Culture in the Last Three Centuries B.C.* Cambridge, Mass.: Harvard University Press (reprinted New York: John Wiley & Sons, 1965).

Schamp, H. 1955–56. "Die Turm der Winde in Athen und die Luftkörperklimatologie." *Die Erde* 7–8:119–128.

Strabo. *The Geography of Strabo*. Trans. H. L. Jones, 1917. New York: G. P. Putnam's Sons.

Thomson, J. O. 1965. *History of Ancient Geography*. New York: Biblo & Tannen.

Tozer, H. F. 1897. *A History of Ancient Geography*. Cambridge: Cambridge University Press (reprinted New York: Biblo & Tannen, 1964).

3

GEOGRAPHY IN THE MIDDLE AGES

Afterwards they went to the country of Paris, to King Fransis (i.e., Philippe IV, le Bel). And the king sent out a large company of men to meet them, and they brought them into the city with great honor and ceremony. Now the territories of the French king were in extent more than a month's journey. And the king of France assigned to Rabban Sauma a place wherein to dwell, and three days later sent one of his amirs to him and summoned him to his presence. And when he had come the king stood up before him and paid him honor, and said unto him, "Why hast thou come? And who sent thee?" And Rabban Sauma said unto him, "King Arghun and the Catholics of the East have sent me concerning the matter of Jerusalem."
—From a Chinese account of the visit of the Nestorian Christian Rabban Bar Sauma of Peking to the French king in 1287 (The Great Chinese Travelers, *ed. J. Mirsky*)

During the fifth century after Christ, the Roman world with its system of centralized administration fell apart. For the Greeks the geographic horizons (the limits of the area that was known at least to scholars and merchants) had been extended from the Indus River to the Atlantic and from the Russian steppes north of the Black Sea to Ethiopia. For the Romans the geographic horizons included the vast area brought under Roman jurisdiction. But now geographic horizons closed in again until many of those who lived in Christian Europe after the fifth century were really familiar only with their immediate surroundings. The worlds beyond were peopled with fantastic creatures conjured up by imaginations unfettered by facts. Only in the shelter of monasteries were the flickering flames of the intellectual life preserved.

But this is not a complete picture of the medieval period, which extended from the fifth to the fifteenth centuries. In Christian Europe, although the word *geography* disappeared from the ordinary vocabulary, the study and writing of geography did not entirely cease (Tillman, 1971). Little by little, curiosity concerning other possible worlds that might lie beyond the horizon again prompted some adventuresome people to travel and explore. The Crusades, organized to wrest the Holy Land from the control of the Muslims, took many people out of their localities and then brought some back again to tell of the strange people and landscapes that had been seen. From the thirteenth century on, there were extended travels by missionaries and merchants that reached all the way to China.

Who was discovering whom? Although the geographic horizons had closed in around the communities of Christian Europe, this was a period of greatly widened horizons for the Muslims. Muslim conquests, which started with the conquest of Palestine and Syria in 632, carried followers of Islam eastward to the islands of Southeast Asia, westward to the Atlantic and into southern Europe, and southward across the Sahara. Muslim missionaries and merchants traveled far beyond the limits of Muslim control. Furthermore, Muslim scholars in the great centers of learning were busily engaged in translating the works of the Greek writers into Arabic. Through Arabic, much Greek learning eventually became known to the Latin world of the Christians.

Meanwhile, in the far north of Europe, Norsemen were sailing across the stormy North Atlantic to Iceland, Greenland, and the continent of North America. Since the Norsemen did not write books, news of these discoveries was a long time getting back to the rest of the world.

Of the greatest importance were the accomplishments of the Chinese. Europe and India were "discovered" by Chinese missionaries long before the Christian travelers reached the Orient. According to Joseph Needham (Needham, 1963:117), in the period between the second century before Christ and the fifteenth century after Christ, the Chinese culture "was the most efficient in the world in applying knowledge of nature to useful purposes." The study of geography in China, as a part of a wider scholarly tradition, was well advanced beyond anything known in Christian Europe at this time.

GEOGRAPHY IN THE CHRISTIAN WORLD

The scholars who gathered together in monasteries in Christian Europe were not studying the earth as observers or experimenters. Rather, they were compilers of information from documentary sources and commentators, whose primary effort was to reconcile the geographic ideas recorded in documents with the authority of the Scriptures, especially the Book of Genesis. In the early medieval period, European scholars could work only with Latin documents; only in the latter part of this period did a few of them master the Arabic language. Greek materials remained unknown, except in translation, until the early Renaissance.

John K. Wright, in his masterly study of the geographical ideas available to the Christian scholars of this period, points out the kinds of information that could be found in Latin (Wright, 1925:88–126). Roman geographers, such as Pomponius Mela and Pliny the Elder, were widely used sources. Both of these writers, as we have seen, compiled their books from Greek sources, and through them the medieval scholars had a kind of secondhand and quite incomplete access to Greek concepts (Kimble, 1938). Two medieval scholars—Martianus Capella and Ambrosius Theodosius Macrobius—provided translations of Plato as early as the fifth century. Through the writings of Capella and Macrobius, the medieval Christian scholars had access to the concept of a spherical earth. Although some, like Cosmas, conceived of the world as a round disc rather than as a sphere, other scholars accepted the idea of a spherical earth as demonstrated beyond dispute.

Ptolemy became the major authority in the medieval Christian world for matters pertaining to astrology and astronomy. His work dealing with the effect of the positions of the celestial bodies on human affairs—the *Quadripartitum*—was translated from Arabic into Latin by Plato of Tivoli in 1138; his *Almagest*—the great work on astronomy—was made available in Latin by Gerard of Cremona in 1175 (Kimble, 1938:75–76). As a result, Ptolemy's geocentric model of the celestial universe remained the accepted model for many centuries, and most of the ideas still used by astrologers can be traced back to him.

The geographical ideas of Aristotle were first made available in Christian Europe by translation from Arabic in the twelfth century. The first medieval writer to make use of Aristotle was Albertus Magnus (Tillman, 1971), whose book on the nature of places combined astrology with environmental determinism. The Greek theory of equating habitability with latitude became strongly implanted in medieval writings. Albertus even went beyond the Greeks: From them he accepted the idea that people who live too close to the limits of the habitable earth turned black, but then he insisted that if black-skinned people should move into the temperate latitudes they would gradually turn white (Glacken, 1967:265–271).

There was no really good way to evaluate the conflicting ideas that these trans-lations from the Arabic made available. Furthermore, it was almost impossible to trace the sources of the ideas since in those times it was standard practice to include in one's own writings whole passages taken verbatim from earlier writers without any kind of credit. Isidore of Seville, who compiled a sort of geographic encyclo-pedia during the seventh century, took long passages from Solinus,[1] who, in turn, had taken them from Pliny. When the medieval scholars did seek explanations for natural events, the kinds of events with which they were concerned were spectacu-lar ones, such as earthquakes, volcanic eruptions, or floods. No hypotheses from the Greeks had been presented in Latin concerning the slower and less obvious natural processes, such as the erosion of mountains or the building of deltas. In the absence of a background of theory, these slower processes were not perceived.

Another characteristic of this period in Christian Europe was the deterioration of mapping. The once fairly accurate delineations of the better-known coastlines were lost, and instead maps became pure fancy. This was the period of the so-called T-O maps. The inhabited world was represented by a circular figure surrounded by the ocean. The figure was "oriented" toward the east (Wright, 1925:66–68). In the midst of the land area was a T-shaped arrangement of waterbodies. The stem of the T represented the Mediterranean. The top of the T represented the Aegean and Black Seas on the one hand, and the Nile River and Red Sea on the other. The three divisions—Europe, Asia, and Africa—were accepted as standard. The center of the inhabited world, just above the center of the T, was Jerusalem. At the far east, beyond the limit of the inhabited world, was paradise (Fig. 6).

Medieval Christian Travelers

Meanwhile, some Christians outside the monasteries did travel and make obser-vations, but they had no knowledge of the existence of theoretical concepts regard-ing the nature of the earth as the home of humankind. In A.D. 326 Helena, the mother of the emperor Constantine, made one of the earliest pilgrimages from Rome to the Holy Land. Silvia of Aquitaine, a Roman lady, was one of the earliest woman travellers. She traveled overland to Jerusalem and then on to Egypt, Arabia, and

---

[1]Solinus was the first to describe these seas as "mediterranean" (in the midst of the land), and Isidore was the first to use the descriptive term as a proper name (Wright, 1925:307).

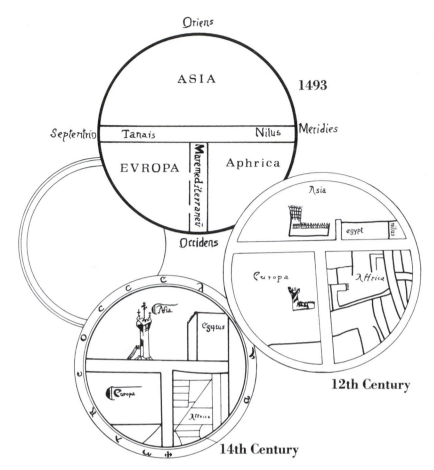

Figure 6 T-O maps

Mesopotamia; eventually, she wrote an account of her travels. As the number of pilgrims increased, itineraries were compiled to guide them on the routes to Jerusalem (Beazley, 1897–1906/1949).

By the eleventh century, the passage of pilgrims overland through what is today Turkey and Syria had become more and more difficult and dangerous. As a result, the Christians of Europe organized a series of military invasions of the Holy Land. Between 1096 and 1270 there were eight separate Crusades, each with the objective of recapturing the Holy Sepulcher at Jerusalem from the Muslims. Some went by sea, some by land; one crusade was even successful in occupying Jerusalem for a short period before the Muslims drove out the invaders. After the eighth formal Crusade, there were other military invasions of Muslim-held territory, one of which in 1365 sacked Alexandria and burned the famous library, where Eratosthenes and other Greek geographers had worked.[2]

---

[2]It is believed, however, that the collection of manuscripts that constituted the major record of Greek geography had long since been ruined due to lack of care even before the destruction of the library by the Christian invaders.

The Crusades had a major impact not only on Christian Europe but also on the Muslims. From almost all parts of Europe, men had been recruited for the war against the infidel and had made the trip to the Holy Land. Nobility, adventurers, pilgrims, soldiers, rogues, peasants, merchants, all joined in the Crusades. Belief in obtaining Jerusalem and economic advantage were doubtless the strongest motives that uprooted so many persons. But the net gain from the undertaking was the enlarging experience thrust upon every European who previously had rarely traveled much beyond his own village. When men returned to Europe, they not only brought with them many new kinds of machines—such as the windmills, later adopted by the Dutch for pumping water—but also exciting stories about strange people and strange landscapes beyond the geographic horizons. The result was a great stimulation of interest in the description of unfamiliar places. For people who knew nothing about geographic theory, popular description and travel became, in essence, geography. Meanwhile, the Muslims, who at first were notably tolerant of people of other faiths, reacted to the violence of the crusaders by becoming aggressively intolerant of unbelievers. One result was the closing of the routes across North Africa–Southwest Asia by which the merchants of Venice and Genoa could make contact with the traders of the East.

Marco Polo

In spite of the blocking of the eastern sea routes, Christian Europe did, in fact, make contact by land with the centers of Chinese culture by following a route to the north of the main Muslim strongholds. This may have been the "silk route." The route was followed both by missionaries sent out from Rome and by merchants. This was at a time near the end of the Crusades, the rise of Venice as a wealthy city-state, Mongol conquest ranging from China to Europe, and papal missions to the Great Khans.

The most celebrated of the travelers to China were the Polo brothers and the son of one of them, Marco Polo. In 1271, when Marco was 17, he started out from Venice with his father and uncle to make the long journey to China (Fig. 7). The Polo brothers had already visited China on a trip that lasted from 1260 to 1269, and the Great Khan, the Mongol emperor of China, had invited them to return. The return journey to China took 4 years, and the Polos remained there for 17 years. Marco served the Khan as ambassador to various parts of China and in various other official capacities, as a result of which he was able to gain intimate knowledge of Chinese culture. In fact, the Polos were so useful to the Khan that he was reluctant to permit their departure. Finally, in 1292 the Khan provided the Polos with a fleet of 14 large ships, some so large that they required a crew of more than 100 sailors. Along with the Polos there were some 600 other passengers. The fleet set sail from a port in southern China, probably the modern Lungch'i, and took three months to reach Java and Sumatra, where it was held up for five months. The expedition then continued to Ceylon and southern India and thence along the west coast to the ancient port of Hormuz on the Persian Gulf. Of the 600 passengers, only 18 survived the voyage. Most of the ships were lost. But the Polos finally returned safely to Venice in 1295 after an absence of 25 years.

Marco Polo, while being held a prisoner in Genoa some years later, dictated his book of travels to a fellow prisoner (Polo, 1930). The manuscript, completed in 1298,

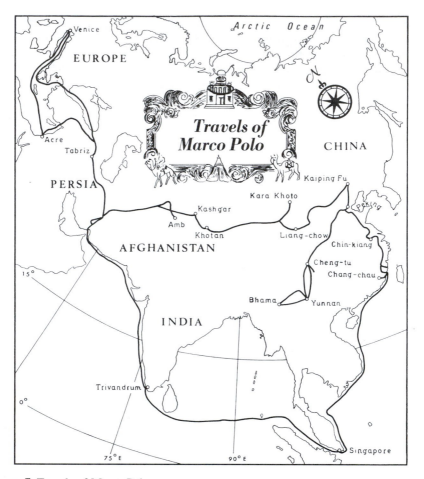

Figure 7 Travels of Marco Polo

was copied frequently for nearly two centuries. In 1477 it was published as a book; since that time it has been translated into several languages and reprinted. Friar Francesco Pipino translated the manuscript into Latin and printed it at Antwerp in 1485. The work itself was an early form of regional geography, and probably the first description of a large part of Asia by a European. His descriptions of life in China and of the perils encountered on the route to and from China were vivid—so vivid, in fact, that they were commonly regarded as the products of a heated imagination. In addition to descriptions of the places he actually visited, he included reports on Cipangu, or Japan, and on the island of Madagascar, which, he said, was near the southern limit of the habitable earth. Since Madagascar is well south of the equator, here was abundant evidence that the torrid zone was not torrid and was actually inhabited.

It is important, however, that Marco Polo was not a geographer and had no knowledge of the existence of such a field of learning. Nor was he aware of the major disputes then going on (1) among those who believed in a torrid zone that was not habitable and those who disagreed with this notion or (2) among those who accepted the smaller estimate of the earth's circumference derived from Posidonius, Marinus, and Ptolemy and those who preferred the larger figure of Eratosthenes. Nor was Marco Polo aware that the Greek geographers thought that the eastern end of the ekumene was near the mouth of the Ganges. Nor was he aware that Ptolemy had said that the Indian Ocean was enclosed by land to the south. It is doubtful if Marco ever thought of measuring the latitude and certainly not the longitude of the places he visited, but he did report that to reach a place required a journey of a certain number of days in such and such a direction. He makes no comments concerning previous geographic ideas. Today we can see that his book must stand among the great records of geographic exploration, yet in medieval Europe it seemed much like many other books of the time, filled with wild but interesting stories. Columbus had a copy of it marked with notations in his own writing (Thomas, 1994).

Brighter Occasions in Medieval Scholarship

Toward the end of the medieval period of Christian Europe, a few scholars began to insist on the need to confront authority with reason. If God gave us the gift of reason, they insisted, there could be no excuse for refusing the use of it. William of Conches, who died in about 1150, was one of the earliest to portray a universe governed by law rather than by the unpredictable acts of a divine authority (Kimble, 1938:79). He presented some remarkably modern ideas concerning the heating of the atmosphere from below and the formation of clouds by the cooling of air. Robert Grosseteste, the bishop of Lincoln, was one of the earliest Christian scholars to master the Arabic language and who, therefore, had access to a much wider range of geographic materials than those written in Latin. As the teacher of Roger Bacon, he was at least in part responsible for refuting the notion of a torrid zone that was uninhabitable through his access to Arabic reports on an inhabited east coast of Africa extending at least as far as 20°S (Wright, 1925:163–165).

Cardinal Pierre d'Ailly, writing in the early fifteenth century, is one of the later medieval scholars who had a major influence on the age that followed. Although he derived his material chiefly from Latin sources, his book *Tractatus de imago mundi* did represent a kind of summary of the work of the period. In a second edition of his book in 1414, he is one of the first to make use of the Latin translation of Ptolemy's *Geography* (published in 1409). He repeats the different opinions concerning the habitability of the torrid zone but without taking any stand on the matter. But he does dispute Ptolemy's idea of an enclosed Indian Ocean. He quotes numerous reports that indicate the existence of an open ocean around southern Africa. This notion had great influence on the Portuguese geographers and navigators, who soon thereafter began to seek a way to India that could avoid the Arabic territory. D'Ailly also accepts the smaller estimate of the earth's circumference, and he was among the first to insist that India could be reached by sailing west, which was influential in building up Columbus's determination to do just this (Kimble, 1938:208–211). It is important, too, that the invention of movable type in about the middle of the fifteenth century made possible the publication of such books in large editions. The works of Pierre d'Ailly were popular, as were all the geographic writings then available in manuscript form.

Another Christian geographer whose writings belong in this period was Pope Pius II (Aeneas Silvius). While he was pope between 1458 and 1464, he wrote a book on Europe and Asia in which he suggests the possibility that the torrid zone is inhabited. He also agrees with Pierre d'Ailly that there is abundant evidence that the Indian Ocean is not enclosed on the south, as reported by Ptolemy.

Navigation and Cartography

During this period several important advances were made in the arts of navigation at sea. Some of these new skills were first developed, as we will see, at the court of Palermo in Sicily, where the Norman king Roger II and the Muslim geographer Edrisi were beginning to learn how to navigate away from the land. Being in Sicily meant that he was close to contemporary activity and at the hub of Mediterranean navigation, and having traveled in Africa, lands of the eastern Mediterranean and Europe meant experience of travel. The first mention of the magnetic compass occurs in Christian Europe in the writings of Alexander Neckam in about 1187. The earliest Arabic reference to a compass was in 1230. Yet there is some evidence that this instrument was in use much earlier and was perhaps independently invented by the Vikings (Kimble, 1938:223). It was certainly in wide use by the fifteenth century and had become indispensable for long sea voyages. In this period, too, the astrolabe had been improved, and it came into common use as an aid to navigation by making possible a more accurate fix on the altitude of the polestar (Kimble, 1938:223–225).

The late fourteenth century also witnessed a notable improvement in the art of mapping. The Portolano charts, of which the earliest is from about 1300, became standard equipment for sea captains. The term *Portolano* means handy or easily available. Instead of a grid of latitude and longitude, the Portolano charts were covered by a network of overlapping lines radiating from several centers in different parts of the chart (Fig. 8). The radiating lines conform to the eight or 16 principal directions of the compass, each corresponding to a wind direction (Bagrow and Skelton, 1964:62–66). Sailors laid out compass courses along these lines (Taylor, 1957:112–114). With the lines to indicate directions from key points, the coastlines, especially around the Mediterranean, were drawn with considerable accuracy.[3]

The famous Catalan map of the world, made in 1375, incorporates the material from numerous Portolano charts. It also includes the west coast of Africa to the south of Cape Bojador, which had not been reached by European sailors. It also shows East and Southeast Asia based on reports by Marco Polo. This was the first map ever to give a proper outline to Ceylon and the Indian peninsula (Kimble, 1938:193; Harley and Woodward, 1987). But the scholars in the monasteries, who

---

[3] With the use of the magnetic compass, it became the usual practice to draw maps with north at the top. When a map is laid on the ground so that directions on the map correspond with compass directions, the map is still said to be "oriented." In the year 800 Charlemagne, hoping to end the confusion about directions, decreed that henceforth there would be only four cardinal directions: north, east, south, and west. North was indicated with the *fleur de lis*, as it still is.

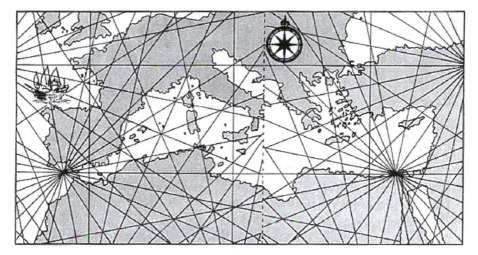

Figure 8 A Portolano chart of the Mediterranean (after Juan de la Cosa, 1500)

were still gathering their information from written documents, had little connection with the mapmakers, whose purposes were strictly practical rather than theoretical. The mapmakers were working for the merchants, and perhaps they were not even aware of the differences of opinion regarding such questions as the habitability of the torrid zone or even the existence of a torrid zone.

GEOGRAPHY IN THE MUSLIM WORLD

One of the events of far-reaching importance in the medieval period was the spread of the Muslims. Muhammad, the prophet who died in 632, was the founder of the religion of Islam, whose followers are known as Muslims. These Arabic-speaking people from Arabia were previously grouped in small, isolated tribes and had no feeling of unity. They were given a common purpose, if not complete unity, by the teaching of the prophet and by the holy book—the Koran. This was the first book written in the Arabic language. It not only provided a religious orientation, but it also gave detailed prescriptions concerning all aspects of life—how to govern, how to carry on commercial transactions (including a prohibition against the payment of interest on loans), how to organize family life, and many other matters. The Koran describes the world in detail, providing explanations for natural phenomena that all "true believers" accept without question.

The followers of Islam embarked on a conquest of the world outside of Arabia. In 641 they conquered Persia, and in 642 they took control of Egypt. The Muslims swept westward across the Sahara, and by 732 all of the Great Desert was under Muslim control. They crossed through the Iberian Peninsula into France and were defeated and turned back only in the battle of Tours (732). For some centuries the Muslims ruled most of southern Spain and Portugal. Muslim rule was also extended eastward into India and eventually to some of the islands of Southeast Asia.

Baghdad

In 762 the Muslims founded the city of Baghdad near the ruins of Babylon, and for more than a century Baghdad was the center of the intellectual world. With the patronage of the caliph Harun al-Rashid, a project was started for the translation of the works of the Greek philosophers and scholars into Arabic. The project was continued under Caliph al-Mamun (813–833), who employed learned men of all faiths to make the translations. Books were collected from all available sources, and the translators were paid the weight of their books in gold (Ahmad, 1947:5). From Baghdad, therefore, a flood of new ideas from varied sources began to spread throughout the Muslim world. Eventually, the innovations were brought into Christian Europe as a result of Latin translations from the Arabic. Among other innovations was the use of the decimal system in arithmetic, which was brought into Baghdad from the Hindus, who had adopted it from the Chinese.

Al-Mamun directed his scholars to recalculate the circumference of the earth. They made use of the same method devised by Eratosthenes some 10 centuries before. On the level plain of the Euphrates, they established a north-south line and fixed the latitude at either end by observations of the stars. They then measured the distance between the fixed points and decided that the length of a degree was $56^{2}/3$ Arabic miles. The scholars made several other measurements, one near Palmyra in Syria, and arrived at almost the same results. These values were much too small, owing to errors in the linear measurements (Wright, 1925:395).

Muslim Contributions to Climatology and Geomorphology

The Arabic geographical writings in the period between 800 and 1400 were based on a much greater variety of sources than were those of Christian scholars in the same period. The Muslims had access not only to their translations from Greek but also to the reports of their own travelers. As a result, they had a much more accurate knowledge about the world than the Christian scholars had. One of the earliest of the great Arab travelers was ibn-Haukal, who spent the last 30 years of his life between 943 and 973 visiting some of the most remote parts of Africa and Asia. On his voyage along the African east coast to a point some 20 degrees south of the equator, he observed that considerable numbers of people were living in those latitudes that the Greeks thought to be uninhabitable. Yet the Greek theory persisted and keeps appearing in different form again and again, even in modern times.

The Arabic scholars made some important observations regarding climate. In 921 al-Balkhi gathered the observations of climate features made by Arab travelers in the world's first climate atlas—the *Kitab al-Ashkal*. Al-Masudi, who died about 956, had gone south as far as modern-day Mozambique and wrote a very good description of the monsoons. He described the evaporation of moisture from water surfaces and the condensation of the moisture in the form of clouds—this, in the tenth century. In 985 al-Maqdisi offered a new division of the world into 14 climatic regions in *The Best Divisions for the Study of Climate*. He recognized that climate varied not only by latitude but also by position east and west. He also presented the idea that the Southern Hemisphere was mostly open ocean and that most of the world's land area was in the Northern Hemisphere (Scholten, 1980).

Two Arab geographers offered important observations regarding the processes shaping the world's landforms. Al-Biruni wrote his great geography of India (*Kitab al-Hind*) in 1030 (Siddiqi, 1991). In this book he recognized the significance of the rounded stones he found in the alluvial deposits south of the Himalayas. The stones became rounded, he pointed out, as they were rolled along in the torrential mountain streams. Furthermore, he recognized that the alluvial material dropped close to the mountains was relatively coarse in texture and that alluvium became finer in texture farther away from the mountains. He quotes the Hindus as believing that the tides are caused by the moon. He also includes the interesting observation that toward the South Pole night ceases to exist—which suggests that some explorers had voyaged to the far south before the eleventh century.

The other contributor to a knowledge of landforms was Avicenna, or ibn-Sina, who observed how mountain streams in central Asia cut down the mountain to form valleys. He formulated the idea that mountains were being constantly worn down by streams and that the highest peaks occurred where the rocks were especially resistant to erosion. Mountains are raised up, he pointed out, and are immediately exposed to this process of wearing down, a process that goes on slowly but steadily. Eight more centuries would pass before James Hutton presented similar ideas concerning the process of erosion; he had never heard of Avicenna and could not read Arabic. Avicenna also noted the presence of fossils in the rocks in high mountains, which he interpreted as examples of nature's effort to create living plants or animals that had ended in failure.

Edrisi and Palermo

The most extensive corrections of the erroneous ideas handed down from Ptolemy were made by the Muslim geographer Edrisi, or al-Idrisi. Educated at the University of Cordoba in Spain, Edrisi was one of the scholars who Roger II of Sicily brought to Palermo. King Roger dispatched observers to many parts of the world where Edrisi said there were uncertainties concerning the actual arrangement of mountains, rivers, or coastlines. The observers brought back much new information to Palermo. As a result, Edrisi was able to write a "new geography" that was really new. In 1154 he completed a book with the title *Amusement for Him Who Desires to Travel Around the World*. He corrects the idea of an enclosed Indian Ocean and the idea of the Caspian Sea as a gulf of the world ocean. He also corrects on maps then available the courses of numerous rivers, including the Danube and the Niger, and the position of several major mountain ranges. As Kimble points out, it is strange that such an important book was not translated into Latin until 1619, at which time the translator did not even know the name of the author (Ahmad, 1947:39; Kimble, 1938:59).

Other important innovations were made at Palermo. Improvements were made in the methods of navigation, including the wide use of coast charts, which were the forerunners of the Portolano charts of the fourteenth century. It is said that the sailors of Genoa learned the arts of navigation from the Sicilians and that the Genoèse passed on this knowledge to the Portuguese in the fifteenth century. The first steps leading to the Age of Exploration were taken in Sicily in the eleventh and twelfth centuries (Wright, 1925:81).

Ibn-Batuta

One of the great travelers of all time was the Muslim ibn-Batuta. He was born at Tangier in 1304 to a family whose members had traditionally served as judges. In 1325, at the age of 21, he set out to make the usual pilgrimage to Mecca, where he proposed to complete his studies of the law. But on the way across North Africa and through Egypt he found himself fascinated more by the people and lands he passed through than by the law. After reaching Mecca he decided to devote himself to travel, and, in his comings and goings through Muslim territory, he carefully avoided following the same route twice. His travels took him to many parts of Arabia never before visited by one person. He sailed along the Red Sea, visited Ethiopia, and then continued southward along the coast of East Africa as far as Kilwa, nearly 10°S of the equator. At Kilwa he learned of the Arab trading post at Sofala in Mozambique, south of the modern port of Beira and more than 20°S of the equator. Ibn-Batuta confirmed what ibn-Haukal had implied—that the torrid zone in East Africa was not torrid and that it was occupied by a numerous native population that justified the establishment of Arab trading posts.

After returning to Mecca, ibn-Batuta set out again to visit Baghdad and Persia and the land around the Black Sea. He traveled in the Russian steppes and thence eventually to Bukhara and Samarkand. Then he crossed the mountains through Afghanistan into India. He served the Mongol emperor in Delhi for several years, traveling widely in India during this time. The emperor appointed him ambassador to China. But delays kept him from reaching China for several more years, during which time he visited the Maldive Islands, Ceylon, Sumatra, and, eventually, China. In 1350 he returned to Fez, the capital of Morocco. But his travels did not end. He made a trip into Spain and then crossed the Sahara to Timbuktu on the Niger River, gathering important information about the Muslim Negro tribes living in that part of the world. In 1353 he settled in Fez, where at the Sultan's command he dictated a lengthy account of his travels (ibn-Batuta, 1958). During some 30 years he covered a linear distance of about 75,000 miles, which in the fourteenth century was a world record. Unfortunately, his book, written in Arabic, made little impact on the Christian world. Even today, when some of our schools teach children about the intense heat of the torrid zone, reference could be made to ibn-Batuta, who, six centuries ago, pointed out that the climate along the equator was less extreme than the climate in the so-called temperate zone in North Africa (Siddiqi, 1992).

Ibn-Khaldun

The last of the Muslim scholars to make a major contribution to geography was the great Arabic historian ibn-Khaldun. Like ibn-Batuta, he was born on the Mediterranean coast of northwest Africa (1332–1406). He lived most of his life in the cities of what is today Algeria and Tunisia and also for a time in the Muslim part of Spain. He spent his later years in Egypt. In 1377, when he was 45, he completed a voluminous introduction to his world history—known as the *Muqaddimah*. This work begins with a discussion of man's physical environment and its influence and with man's characteristics that are related to his culture or way of living rather than to the environment. He discusses various stages of social organization, identifying the desert nomad as the most primitive and purest, and he suggests that the sedentary city dweller is dependent on luxuries and becomes morally soft. He discusses the forms of government, describing a sequence of stages that mark the rise of a dynasty to power, followed by its decline through corruption to its fall. Ibn-Khaldun discusses cities and their proper location. Finally, he discusses the various ways of making a living—commerce, the crafts, the sciences—all of which are shown as both conditions and consequences of urban life. Many of the ideas he develops in his effort to provide a theoretical model of national growth and national decay were ideas that appeared later in nineteenth-century Europe. Yet only recently have the writings of ibn-Khaldun appeared in English (ibn-Khaldun, 1958).

Although ibn-Kaldun, according to Kimble, "may be considered to have discovered—as he himself claimed—the true scope and nature of geographic inquiry," the fact remains that his knowledge of the physical earth is based largely on Greek theory, and his ideas about environmental influence are not highly sophisticated (Kimble, 1938:180). He accepts the traditional seven zones of climate running parallel to the equator. Strangely for an Arabic scholar, he repeats the idea concerning an uninhabitable zone along the equator and an uninhabitable polar zone. He repeats the old idea that people turn black when they live too close to the sun and that when black people move to the temperate zone they gradually turn white or produce white children. The physical environment impressed its characteristics on people in many subtle ways. Such naïve environmentalism is somewhat modified by the recognition of different cultural traditions.

It can be said that ibn-Khaldun was the first Muslim scholar to turn his attention specifically to man-environment relations. He has been proclaimed the greatest historical thinker of Islam who made application of geographical matters to the history of the Muslims (Dickinson and Howarth, 1933).

GEOGRAPHY IN THE SCANDINAVIAN WORLD

The Scandinavians had never heard of Aristotle, or Strabo, or Ptolemy, or Isidore, or ibn-Khaldun. They had no idea that the lands they inhabited were uninhabitable. Among the Scandinavians, the Swedes sent exploring parties far to the east into what is now the central part of Russia. In the ninth century Othar of Helgoland sailed in a Viking ship around the northern tip of Norway and far eastward into the White Sea.

But the greatest accomplishment of the Scandinavians from Norway—the Vikings—was the crossing of the North Atlantic Ocean to the American mainland. In 874 the Vikings reached Iceland and established a settlement there that grew and prospered. In 930 the world's first parliament was organized in Iceland.

Among the people of the Iceland colony was one especially violent and disturbing person named Eric the Red. In 982 he and his family and retainers were banished. Having learned of the existence of land farther west, Eric set sail across the stormy North Atlantic and came upon the southern part of Greenland. In fact, the name, Greenland, that he gave to this new land was perhaps one of the world's earliest examples of real estate promotion, for there was nothing very green about it. Nevertheless, Eric's colony attracted additional settlers from Iceland. Regular voyages were made back and forth between Greenland, Iceland, and the mainland of Norway.

About the year 1000 Eric's son, Leif Ericson, returning to Greenland from a trip to Norway, encountered a severe storm that blew him far off his course. When the skies cleared, he found himself off a strange coast that extended as far as he could see to the north and south. He landed and found fine stands of timber as well as vines on which wild grapes grew. Returning to Greenland, he described this new land farther to the west.

In 1004 Thorvald traveled to the new land, and in 1024 a man named Karlsefni organized an expedition to take another look at this new land. He set sail with a crew of 160 people—men and women—together with cattle and food supplies. There is no doubt that he reached the coast of North America. The large bay with a strong current of water coming out of it was probably the Saint Lawrence estuary, and somewhere along its shore the party made a landing and spent the winter. Here the first European child was born in the Americas. The next summer the party sailed southward, certainly as far as Nova Scotia, probably as far as Cape Cod, and possibly even as far south as Chesapeake Bay. They liked the land they found, but the Indians proved warlike. Their attacks were such a nuisance that eventually the Vikings gave up the effort to settle on the strange shore and sailed back to Greenland. The story was passed on by word of mouth as *The Saga of Eric the Red*. To this day efforts are still being made to identify the places where Karlsefni and his people landed. It is quite possible that there were other expeditions even before the eleventh century, but geographical scholars in the European world heard only rumors of such voyages for several centuries (Cassidy, 1968; Morison, 1963, 1971; Sauer, 1968; Sykes, 1961; Thórdarson, 1924).

GEOGRAPHY IN THE CHINESE WORLD

During all the time that the study of the earth as the home of man was being pursued in ancient Greece and Rome and later in Christian Europe, among the Muslims, and by the remote Scandinavians, there was another major center of geographic study in the world. This was China. Essentially the European and Chinese worlds remained isolated, each discovering the other step by step. Yet there are certain fascinating parallels in the concepts and methods of study that seem to require the existence of contacts, however indirect and remote.

Students steeped in Western history must keep in mind that from about the second century before Christ until at least the fifteenth century after Christ, the people of China enjoyed the highest standard of living of any people on earth (Needham, 1963:117). The Chinese mathematicians had discovered the use of zero and had developed the decimal system, which was vastly superior to the sexigesimal system of Mesopotamia and Egypt. The decimal system was introduced into Baghdad in about 800 from the Hindus, but it is generally believed that the Hindus derived their decimal system from the Chinese.

The Chinese philosophers had a basically different attitude toward the natural world than that held by the Greeks. To the Chinese the individual is not separate from nature—he or she is a part of nature. There is no law-giving deity who created the universe for human use in accordance with a preconceived plan. Death in China is not followed by life in a new paradise or punishment in a hell; rather, the individual hopes to be absorbed in the all-pervading universe of which he or she is an inseparable part. Confucianism developed a way of life that was highly effective in minimizing the frictions among individuals, but it remained relatively indifferent to the development of scientific knowledge.

Joseph Needham repeats the following story to illustrate this attitude:

> When Confucius was travelling in the east, he came upon two boys who were disputing, and he asked them why. One said "I believe that the rising sun is nearer to us and that the midday sun is further away." The other said "On the contrary, I believe that the rising sun and setting sun is further away from us, and that at midday it is nearest." The first replied, "The rising sun is as big as a chariot-roof, while at midday the sun is no bigger than a plate. That which is large must be near us, while that which is small must be further away." But the second said, "At dawn the sun is cool but at midday it burns, and the hotter it gets the nearer it must be to us." Confucius was unable to solve their problem. So the two boys laughed him to scorn saying "Why do people pretend that you are so learned?" (Needham and Ling, 1959:225–226).

Consider what Socrates might have done in this situation, and a very fundamental difference in cultural attitude becomes clear. But this does not mean that there was no interest in finding out what it was like beyond the horizon or in developing methods of recording what was discovered. In fact, the record of geographical work in China is impressive, but it is concerned more with observable things and processes and less with the formulation of theory.

Geographical Work

Chinese geographical work was based on the development of methods for making accurate observations and for using these in constructing useful inventories. For example, there are weather records that date back to thirteen centuries before Christ. The oldest piece of geographical writing is a survey of the resources and products of the nine provinces into which ancient China was divided in the fifth century B.C. For each province the nature of the soil, the kinds of products, and the waterways that provide routes of transportation are described (Needham and Ling, 1959:500). In the second century B.C. Chinese engineers were making accurate measurements of the silt carried by the rivers. In A.D. 2 the Chinese carried out the world's first census of population. Other technical inventions included the making of paper, the printing of books, the use of rain and snow gauges to measure precipitation, and the use of the magnetic compass for navigation.

There was also a record of progress in the understanding of processes. By the fourth century B.C. the nature of the hydrological cycle was understood. At about the same time that Plato was observing the effects of forest clearing in Attica, the Chinese philosopher Mencius (Meng-tzu), who lived two centuries after Confucius, was pointing out that forests once cleared from mountain slopes could not reseed themselves as long as the slopes were grazed by cattle or goats (Glacken, 1956:70).

The Chinese had also learned much about the work of running water in wearing down mountains and forming alluvial plains. At about the same time that Avicenna was writing down his ideas about the erosion of mountains, the Chinese scholar Shen Kuo was presenting the same idea (in 1070) (Xixian, 1987). Referring to a rugged mountain range with jagged peaks and steep-sided valleys, he wrote:

> Considering the reasons for these shapes, I think that (for centuries) the mountain torrents have rushed down, carrying away all sand and earth, thus leaving the hard rocks standing alone.
>
> . . . Standing at the bottom of the ravines and looking upwards, the cliff face seems perpendicular, but when you are on top, the other tops seem on a level with where you are standing. Similar formations are found right up to the highest summits.
>
> . . . Now the Great River (i.e., the Yellow River) . . . (and certain others) are all muddy, silt-bearing rivers. In the west of Shensi and Shansi the waters run through gorges as deep as a hundred feet. Naturally mud and silt will be carried eastwards by the streams year after year, and in this way the substance of the whole continent must have been laid down. These principles must certainly be true (Needham and Ling, 1959:603–604).

Chinese geographical writings, according to Needham, were of eight major kinds: (1) studies of people, which we might classify as human geography; (2) descriptions of the regions of China; (3) descriptions of foreign countries; (4) accounts of travels; (5) books about the Chinese rivers; (6) descriptions of the Chinese coast, of special value to ship captains; (7) local topographies, including special descriptions of areas tributary to and controlled by walled cities, or famous mountains, or certain cities and palaces; and (8) geographical encyclopedias. Considerable attention was given to the origin of and changes in Chinese place-names (Needham and Ling, 1959:508; Renzhi, 1962).

Chinese Exploration

The discovery of the rest of the world by Chinese travelers is an aspect of the history of geography that is often overlooked in Western writings. Travels beyond the far Chinese horizons were undertaken by orders of an emperor or by missionaries and traders (Fig. 9).

The earliest record of Chinese travels is a book of uncertain age that was probably composed sometime between the fifth century and the third century B.C. It was found in the tomb of a man who ruled a part of the Wei Ho Valley in about 245 B.C. The books found in this tomb were written on strips of white silk pasted on bamboo slips and because of their bad condition were recopied in the late third century B.C. The travel books are known as *The Travels of Emperor Mu*, who ruled from 1001 to 945 B.C. Emperor Mu, it is recorded, had the ambition to travel all around the world and to leave the marks of his chariot wheels on every land. Like the *Odyssey* of Homer, this is a story of high adventure, surely embroidered by the imagination of the writer but with details that could scarcely have been invented by fancy. The emperor traveled in forested mountains, encountered snow, and engaged in hunting expeditions. On his return he crossed a wide desert where water was so scarce that he had to drink the blood of a horse. There can be no doubt that at a very early date the Chinese travelers had gone far beyond the original culture hearth in the Wei Ho Valley (Mirsky, 1964:3–10).

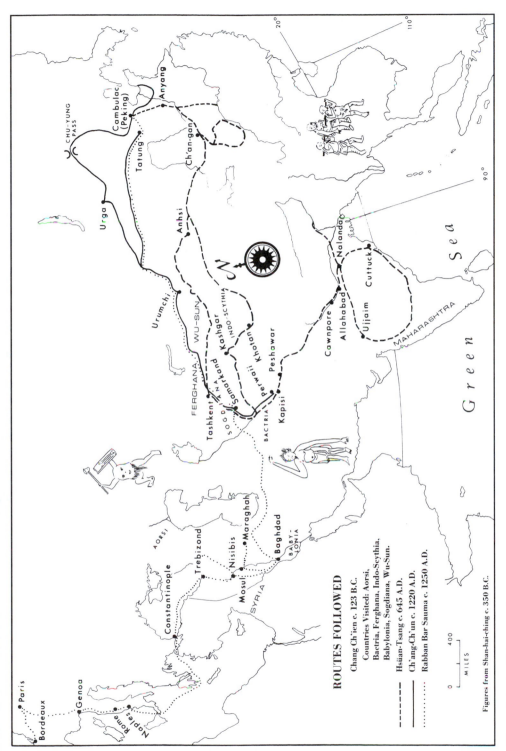

ROUTES FOLLOWED

Chang Ch'ien c. 123 B.C.
Countries Visited: Aorsi,
Bactria, Ferghana, Indo-Scythia,
Babylonia, Sogdiana, Wu-Sun.

-------- Hsüan-Tsang c. 645 A.D.
-------- Ch'ang-Ch'un c. 1220 A.D.
·········· Rabban Bar Sauma c. 1250 A.D.

Figures from Shan-hai-ching c. 350 B.C.

Figure 9 Early Chinese explorers

The discovery of the Mediterranean civilizations is credited to the traveller Chang Ch'ien in 128 B.C. (Mirsky, 1964:13–25; Needham, 1963; Sykes, 1961:21; Thomson, 1965:177–178). The book he wrote describes the land route across inner Asia to Bukhara and thence to Persia and the Mediterranean shore. Over this route traders traveled regularly and had probably made contacts with the West long before the West was discovered officially. The Chinese products that were carried westward included peaches, almonds, plums, apricots, silk, and, eventually, silkworms. The Mediterranean products that were carried back to China included alfalfa, wheat, and the grapevine.

There were a number of other Chinese travelers whose records are complete enough to ensure their place in history. One of the most distinguished was the Buddhist monk Hsüan-Tsang (Sykes, 1961:24–30). In the seventh century after Christ he was able to cross the high, windswept plateaus of Tibet and the world's highest mountains on the way to India. After studying in the centers of the Buddhist faith for several years, he returned to China carrying on the backs of animals a large collection of Buddhist relics and manuscripts. He was the Chinese discoverer of India (Mirsky, 1964). In that same century another Buddhist monk, I-Ching, reached India by sea, stopping first for eight months in Sumatra in 671. When he returned to China, he carried with him more than 10,000 rolls of Sanskrit Buddhist texts, which he undertook to translate into Chinese (Sykes, 1961:30). Several centuries later another Chinese traveler crossed the deserts of inner Asia, encountered endless difficulties, and finally made contact with the Mongol leader Genghis Khan in Samarkand in 1220. And there was the Nestorian Christian monk Rabban Bar Sauma, who in 1287–88 made a pilgrimage to Rome. Finding the pope dead and the new pope not yet selected, he went on through Genoa to Paris and Bordeaux, meeting the kings of France and England. Imagine the amazement of people in thirteenth-century France when they found themselves being discovered by a Christian from China. In 1288 he returned to Rome and received the blessing of the new pope and thereafter journeyed back to Peking. This was several decades before the similar journeys of the Polo brothers to China. And in 1296 the Chinese traveler Chou Ta-kuan visited Cambodia and wrote a detailed account of the strange customs of the Cambodians.

Other Chinese explorers also went by sea. There was never any recorded effort to go far out into the Pacific, although Chinese expeditions did reach Japan and Taiwan. By the thirteenth century Chinese merchants were sailing their junks to Java and Malaya and even as far as India. Marco Polo found them in the Persian Gulf port of Hormuz. But the major work of exploration in this direction was carried out by Cheng Ho, a Chinese admiral, between 1405 and 1433. He led seven separate expeditions, each including a fleet of vessels. His voyages opened regular trade routes to Java, Sumatra, Malaya, Ceylon, and the west coast of India. He also reached the Persian Gulf, the Red Sea, and the east coast of Africa south of the equator. In the east he went as far as Taiwan. On his last expedition, from 1431 to 1433, Cheng Ho returned with ambassadors to China from more than 10 countries. It is even possible that he sent one of his ships to the north coast of Australia (Hsieh, 1968).

Cartography

The Chinese were also expert in the making of maps. The engineer Chang Heng in the second century after Christ was probably the first to introduce the grid system into China, although the maps he made have not survived. There is a reference to him, however, in which he is identified as one who "cast a network [of coordinates] about heaven and earth, and reckoned on the basis of it" (Needham and Ling, 1959:538).

The father of Chinese cartography was Phei Hsiu, who was appointed minister of public works by the Chinese emperor in A.D. 267. He produced a map of the politically organized part of China on 18 rolls of silk. To make the surveys on which the map was based, he measured several base lines and then located places distant from the base lines by the intersection of lines of sight, just as the Egyptians had done long before. He made use of a grid of east-west and north-south lines crossing at right angles to provide the frame on which to plot the rivers, the coastline, the mountain ranges, the cities, and other features. Did he, or Change Heng, or someone much earlier get the idea of using triangulation to locate places and of using a grid of lines from contacts with the Greeks and, perhaps, through them with the Egyptians? It is entirely possible, although no such transmission has ever been established. And it is by no means impossible that many of these methods were devised much earlier in China and diffused westward, as the decimal form of arithmetic certainly was.

Two beautiful examples of Chinese maps were carved in stone in A.D. 1137 based on data that probably had been surveyed before 1100 (Needham and Ling, 1959:547–549). One, "A Map of China and the Barbarian Countries," extended from the Great Wall of China north of Peking southward to the island of Hainan and westward to the mountains of inner Asia. The other, entitled "The Map of the Tracks of Yü the Great," covers essentially the same area but is even more accurate in showing the courses of the great rivers and the coastline from the Gulf of Chihli, north of the Shantung Peninsula, to the island of Hainan. Neither of these maps shows Taiwan. Both maps—like other Chinese maps—were drawn with north at the top. The geographers who made them are not known (Harley and Woodward, 1987; Edson, 1997).

THE MEDIEVAL WORLD IN RETROSPECT

The period from the fifth to the fifteenth centuries was remarkable. Historians of geography studying the matter find disagreement. G. Kimble (1968) entitles a chapter concerning geography in the Middle Ages "The Dark Ages of Geography"; N. Lozovsky (2000) argues that the nature of medieval geography has not been understood and that the earth was to be studied not as an object in itself but as a step toward better understanding of the Creator. She argues that the problem is that we have not understood the nature of prescientific medieval geography, and that it was an integral part of the medieval period.

Yet the essential point remains: In various parts of the world new data were being collected by direct observation, but the lack of close contacts among different peoples, in large part resulting from the barrier of language, meant that geographic knowledge gained by one group was diffused only slowly to others. No rumor of the Viking voyages reached Christian Europe. But Christian Europe did have a variety of documents that contradicted each other. Those who could read Arabic were aware that people were living along the equator and beyond it and that people with white skins had visited these regions without turning black. But there were also reputable authorities, ranging from Aristotle to ibn-Khaldun, who insisted that one would turn black at these very low latitudes. The basic Greek theory, which equated habitability with latitude, seemed firmly established, in spite of the evidence of direct observation. But who could be sure which documents were sober and reliable descriptions of actual conditions and which ones were purely imaginative?

In 1410 two important books appeared. One was the *Imago mundi* of Pierre d'Ailly, which summarized much of the writing of Christian Europe in this period. The other was the Latin translation from Arabic of Ptolemy's *Geography*, which included a large number of errors. Ptolemy did present one important idea: that maps should be based on the precise location by latitude and longitude of specific points. The trouble was that none of the places he located were correctly placed; he gave the weight of authority to several major errors, including the enclosed Indian Ocean.

Starting with Edrisi in Sicily, there had been definite improvement in the arts of navigation, including the wide adoption of the magnetic compass. The Genoese made use of the compass when they rediscovered some of the groups of islands in the Atlantic off Africa, which had been known to the Phoenicians and then lost.

The stage was set for the next step in the discovery of the world. Scholars knew that the earth was a sphere, and some of them insisted that no part of it was beyond human reach. The estimates of the circumference differed considerably, and those who wanted to believe that India was only a short distance west of Spain could find ample authority (Goldstein, 1965). If one could believe Marco Polo, China and Japan were well east of India, which would place them even closer to Spain. Pierre d'Ailly said you could sail westward to reach the East. It was the Florentine physician and scholar Paolo Toscanelli who sent the following remarks to Columbus in 1474:

On another occasion I spoke with you about a shorter sea route to the lands of spices than that which you take for Guinea. And now the Most Serene King requests of me some statement, whereby that route might become understandable and comprehensible, even to men of slight education.

Although I know that this can be done in a spherical form like that of the earth, I have nevertheless decided, in order to gain clarity and to save trouble, to represent (that route) in the manner that charts of navigation do.

Accordingly I am sending His Majesty a chart done with my own hands in which are designated your shores and islands from which you should begin to sail ever westwards, and the lands you should touch at and how much you should deviate from the pole or from the equator and after what distance, that is, after how many miles, you should reach the most fertile lands of all spices and gems, and you must not be surprised that I call the regions in which the spices are found "western" although they are usually called "eastern," for those who sail in the other hemisphere always find these regions in the west. But if you should go over- land and by the higher routes we should come upon these places in the east.

The straight lines, therefore, drawn vertically in the chart, indicate distance from east to west; but those drawn horizontally, indicate spaces from south to north.

From the city of Lisbon westward in a straight line to the very noble and splendid city of Quinsay 26 spaces are indicated on the chart, each of which covers 250 miles. (The city) is 100 miles in circumference and has 10 bridges. Its name means City of Heaven; and many marvelous tales are told of it and of the multitude of its handicrafts and treasures. It (China) has an area of approximately one third of the entire globe. The city is in the province of Katay, in which is the royal residence of the country.

But from the island of Antilia, known to you, to the far-famed island of Cippangu (Japan), there are 10 spaces. The island is very rich in gold, pearls, and gems; they roof the temples and royal houses with solid gold. So there is not a great space to be traversed over unknown waters. More details should, perhaps, be set forth with greater clarity but the diligent reader will be able from this to infer the rest by himself (Morison, 1963:12–14).

REFERENCES: CHAPTER 3

Ahmad, N. 1947. *Muslim Contributions to Geography*. Lahore: Muhammad Ashraf.

————. 1980. *Muslims and the Science of Geography*. Dacca: University Press.

Bagrow, L., and Skelton, R. A. 1964. *History of Cartography*. Cambridge, Mass.: Harvard University Press. The original book by Leo Bagrow, *Geschichte der Kartographie* (Berlin: Safari-Verlag, 1951), was translated into English by D. L. Paisey in 1960. The present book was revised and enlarged by R. A. Skelton.

Beazley, C. R. 1949. *The Dawn of Modern Geography*. 3 vols. New York: Peter Smith (original publication, London: John Murray 1897–1906).

Cassidy, V. H. 1968. *The Sea around Them: The Atlantic Ocean, A.D. 1250*. Baton Rouge: Louisiana State University Press.

Edson, E. 1997. *Mapping Time and Space: How Medieval Mapmakers Viewed Their World*. London: British Library.

Glacken, C. J. 1956. "The Changing Ideas of the Habitable World." In W. L. Thomas, ed., *Man's Role in Changing the Face of the Earth*. Pp. 70–92. Chicago: University of Chicago Press.

————. 1967. *Traces on the Rhodian Shore, Nature and Culture in Western Thought from Ancient Times to the End of the Eighteenth Century*. Berkeley and Los Angeles: University of California Press.

Goldstein, T. 1965. "Geography in Fifteenth Century Florence." In John Parker, ed., *Merchants and Scholars: Essays in the History of Exploration and Trade*. Pp. 9–32. Minneapolis: University of Minnesota Press.

Harley, J. B. and Woodward, D. eds. 1987. *The History of Cartography, Vol. I. Cartography in Prehistoric, Ancient, and Medieval Europe and the Mediterranean*. Chicago: University of Chicago Press.

Hsieh, Chiao-min. 1968. "The Chinese Exploration of the Ocean—A Study in Historical Geography." *Chinese Culture* (Taiwan) 9:123–131.

Ibn-Batuta. *The Travels of Ibn-Battuta, A.D. 1325–1354*. Trans. C. Defrémery and B. R. Sanguinetti, 1958. Cambridge: Cambridge University Press.

Ibn-Khaldun. *The Muqaddimah*. Trans. Franz Rosenthal, 1958. New York: Pantheon Books.

Kimble, G. H. T. 1938. *Geography in the Middle Ages*. London: Methuen.

Lindberg, D. ed. 1978. *Science in the Middle Ages*. Chicago: University of Chicago Press.

Lozovsky, Natalia. 2000. *"The Earth is Our Book": Geographical Knowledge in the Latin West ca. 400–1000*. Ann Arbor: University of Michigan Press.

Mirsky, J., ed. 1964. *The Great Chinese Travelers*. New York: Pantheon Books.

Morison, S. E., trans. and ed. 1963. *Journals and Other Documents on the Life and Voyages of Christopher Columbus*. New York: Heritage Press.

————. 1971. *The European Discovery of America, the Northern Voyages*. New York: Oxford University Press.

————. 1974. *The Southern Voyages*. New York: Oxford University Press.

Needham, J. 1963. "Poverties and Triumphs of the Chinese Scientific Tradition." In A. C. Combie, ed., *Scientific Change*. Pp. 117–153. New York: Basic Books.

Needham, J., and Ling, W. 1959. *Science and Civilization in China*. Vol. 3, *Mathematics and the Sciences of the Heavens and the Earth*. Cambridge: At the University Press.

Nunn, G. E. 1924. *The Geographical Conceptions of Columbus*. New York: American Geographical Society, Research Series No. 14.

Polo, M. *The Travels of Marco Polo* (revised from Marsden's translation, edited and with an introduction by Manuel Komroff). New York: Liveright, 1930.

Said, M. M., ed. 1979. *Al-Biruni Commemorative Volume*. Karachi: Times Press.

Sauer, C. O. 1968. *Northern Mists*. Berkeley and Los Angeles: University of California Press.

Scholten, Arnhild. 1980. "Al-Muqaddasi, c. 945–c. 988." *Geographers: Biobibliographical Studies* 4:1–6.

Siddiqi, Akhtar H. 1991. "Al-Biruni, A.D. 973–1054." *Geographers: Biobibliographical Studies* 13:1–9.

————. 1992. "Ibn Battuta, 1304–1378." *Geographers: Biobibliographical Studies* 14:1–12.

Sykes, P. 1961. *A History of Exploration from the Earliest Times to the Present Day*. New York: Harper Bros.

Taylor, E. G. R. 1957. *The Haven-Finding Art: A History of Navigation from Odysseus to Captain Cook*. New York: Abelard-Schuman.

Thomas, W. L. 1994. "Marco Polo, 1254–1324." *Geographers: Biobibliographical Studies* 15:75–89. London: Mansell.

Thomson, J. O. 1965. *History of Ancient Geography*. New York: Biblo & Tannen.

Thórdarson, Matthias. 1924. *The Vinland Voyages*. New York: American Geographical Research Series, no. 18.

Tillman, J. P. 1971. *An Appraisal of the Geographical Works of Albertus Magnus and His Contributions to Geographical Thought*. Ann Arbor: University of Michigan, Department of Geography, Publication, no. 4.

Wright, J. K. 1925. *The Geographical Lore at the Time of the Crusades. . . .* New York: American Geographical Society, Research Series, no. 15.

Xixian, Yu. 1987. "Shen Kuo, 1033–1097." *Geographers: Biobibliographical Studies* 11:133–137.

4

The Age of Exploration

Terra Incognita: these words stir the imagination. Through the ages men have been drawn to unknown regions by Siren voices, echoes of which ring in our ears today when on modern maps we see spaces labelled "unexplored," rivers shown by broken lines, islands marked "existence doubtful. . . ."

The Sirens, of course, sing of different things to different folk. Some they tempt with material rewards: gold, furs, ivory, petroleum, land to settle and exploit. Some they allure with the prospect of scientific discovery. Others they call to adventure or escape. Geographers they invite more especially to map the configuration of their domain and the distribution of the various phenomena that it contains, and set the perplexing riddle of putting together the parts to form a coherent conception of the whole. But upon all alike who hear their call they lay a poetic spell.

—John K. Wright, from his presidential address to the
Association of American Geographers, 1946

The sudden increase in exploring activity in Europe in the fifteenth century was a major turning point in world history. In China exploration supported by the emperor ceased after the seventh expedition of Cheng Ho (1431–33); and there was no great Muslim traveler after ibn-Batuta in the fourteenth century. But in Europe exploration for the first time was planned and supported by governments or by merchant companies and, for the first time, was directed to the open oceans.

What were the motives? In this case the challenge of the unknown—the urge to find out what it is like somewhere else—was subordinate to two other compelling objectives. One was the zeal to spread the Christian faith; the other was the critical need to replenish the European supplies of precious metals and spices. The two motives were conveniently blended and were never permitted to interfere with each other.

The zeal to spread the Christian faith was in part a response to the long series of conflicts with the spreading Muslims. The greater part of Spain and Portugal had been under Muslim rule since the eighth century. But the Portuguese had freed themselves of Muslim rule by the middle of the thirteenth century, thus forming the world's first example of a nation-state—that is, a politically organized territory occupied by people with a unified national consciousness. From 1391 to 1492 the Spaniards waged continuous warfare with the Muslims, and the last Muslim stronghold (Granada) was taken by the Christians just before Columbus set sail for America. Young people, for that whole century, were brought up in an atmosphere of religious warfare, and, when the last Muslim foothold was pried loose, there was urgent need to find other infidels to kill or convert. The native peoples of America provided what was needed.

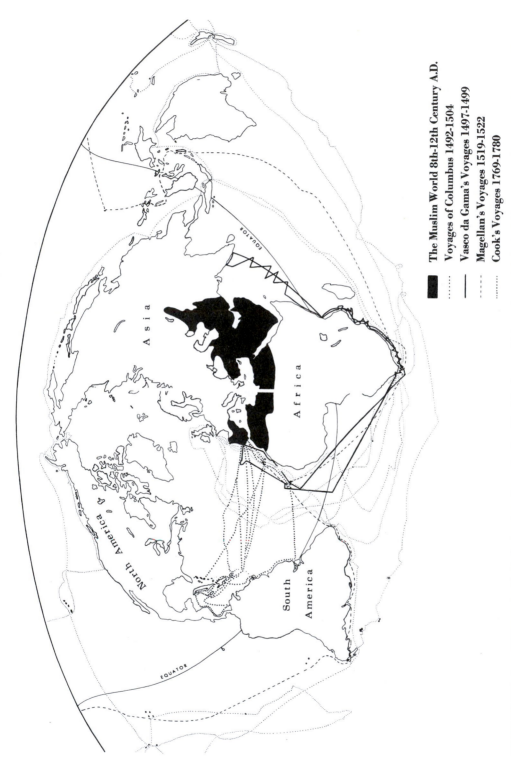

Figure 10 The Age of Discovery, Bartholomew's polar projection (the Muslim world from the eighth to twelfth centuries; Columbus; Vasco da Gama; Magellan; Cook)

The amazing legend of Prester John should not be overlooked. In 1170 a letter was sent to the pope in Rome and to the emperor in Byzantium asking for help in warfare against the infidels. The letter purported to come from a powerful and wealthy Christian ruler located somewhere in the East. To find and bring aid to Prester John became a very real motive for sending out exploring expeditions. The difficulty was to be sure where to look for him. In Marco Polo's time Prester John was associated with the Nestorian Christians in China, perhaps a reflection of the visit of Bar Sauma to Rome and Paris in 1287–88. By 1340 his location had been shifted from Asia to Ethiopia, and one of the objectives of the Portuguese voyages of the fifteenth century was to make contact with Prester John. An ambassador was actually dispatched to represent the Portuguese king. Other Portuguese visited Abyssinia, but the truth concerning Prester John was not revealed until the eighteenth-century English traveler Bruce revealed the hoax.

The need for precious metals and spices in Europe and the desire for personal wealth by the explorers was never far from the surface. Europeans were increasing their trading activities, and to provide financial support for trade there was urgent need for a supply of gold and silver or valuable gems. The Europeans also needed spices. In those days the supply of sugar for sweetening was not adequate; spices were used instead. Without refrigeration meat spoiled quickly unless it was dried or salted—but dried or salted meat is almost unpalatable unless spices are used in cooking. The spices included cloves and nutmegs, both of which came from the Spice Islands (the Moluccas) in what is today Indonesia. The trade in precious metals, gems, and spices had been under the control of the merchants of Genoa and Venice, but the Arabs blocked direct contact between the Europeans and the Asian sources. The Arabs, in fact, had thrown a screen between Europe and Asia and between Europe and Africa by their control of the great dry land area of North Africa–Southwest Asia (Fig. 10). The Arabs were in a position to control the supply of these products to Europe and to collect the larger share of this profitable trade.

PRINCE HENRY AND THE PORTUGUESE VOYAGES

The first initiative toward wider exploration came from Portugal. The individual leader was Prince Henry (called the Navigator), the third son of the Portuguese king. In 1415 Prince Henry commanded a Portuguese force that attacked and captured the Muslim stronghold on the southern side of the Strait of Gibraltar at Ceuta. This was the first time a European power took possession of territory outside of Europe—in a sense the occupation of this portion of Africa began the period of European overseas colonization. At Ceuta Prince Henry learned from Muslim prisoners that the gold, ivory, ostrich feathers, and slaves sold in the Arab markets of North Africa had been brought across the great desert by caravan from Africa south of the Sahara. Why would it not be possible to reach Guinea by sea and thus capture some of this profitable trade for Portugal?

In the year 2000 a book was published by Peter Russell, former Director of Portuguese Studies at the University of Oxford. It is the product of a lifetime of research concerning the Henrican legend. In short, this work sweeps away the myth that has been retold in countless numbers of books and articles: a piece of history has been rewritten. Henry, born in Portugal, was the leading anti-Muslim force in North Africa, which ran close to obsession (Koelsch, 2001). He remained very largely tied to the land of Portugal and only departed to raid Moors. The much published notion of the Institute at Sagres, the palace, and Henry's home, is now discarded. Scholars or cartographers, occasionally part of his court, remained with him in Lagos or Lisbon, and not Sagres. Henry did found a chair at the University of Lisbon in theology, which was his keenest interest. He did, however, have a lifelong interest in marine cartography and was recognized in this area as an authority. He was overly concerned to join with the (mythical) forces of Prester John dismissing the Muslims en route; concerned for the acquisition of worldly wealth; and vitally concerned to record the facts of coastlines, customs, and countries. These activities led Henry into a form of organized exploration, which seems to have been the first of its kind.

At Lagos, Prince Henry's naval architects were busy designing new and better ships. The width, or beam, of the older ships had been about half the length from bow to stern, but the new designs made the beam only a third to a quarter of the length. The new Portuguese caravels were lateen rigged with two or even three masts. The ships, with high castles fore and aft and of about 200 tons burden, were slow but seaworthy.

Navigation aids included the magnetic needle, which was mounted on cardboard— as taught by the sailors of Genoa and Venice, who had probably been previously instructed by the geographers of Palermo. Improved astrolabes were made of brass, permitting the more accurate measurement of star altitudes at sea, and tables were prepared to give the altitudes of various stars in different seasons and different latitudes. Portolano charts were drawn on sheepskin.

Prince Henry's captains were able to gain experience and confidence by sailing to the three groups of islands in the Atlantic off North Africa and Europe. The easiest voyages were to the Canary Islands (Ptolemy's Fortunate Islands, which he used for the 0° meridian on his maps). These were less than 100 miles off the African coast. The Madeira Islands offered more of a challenge, requiring a voyage of some 400 miles into the open ocean. After some unsuccessful efforts, the Madeira Islands were reached in 1420. (They had been reached by some English ships in 1370.) The Azores were still more of a challenge, requiring 1000 miles of sailing. To be sure, the Azores were shown on the Portolano chart of 1351, but the bearings shown on these charts were not accurate, and at first the ships missed them. Then Prince Henry made the necessary corrections, and in 1432 the islands were "discovered."

Cape Bojador

Meanwhile, Prince Henry sent his more experienced captains southward along the African coast. The first probe in that direction was sent out in 1418 but soon turned back because the crews were afraid of what might lie closer to the equator. In spite of repeated efforts it took 16 years for the Portuguese ships to get south of latitude 26°7′N. At this latitude, a little south of the Canary Islands, there is a low, sandy promontory on the African coast known as Cape Bojador. Crews were super-stitious and did not wish to sail south beyond this point. The Portolano chart of 1351 certainly showed nothing unusual about Bojador, which was, after all, just a minor cape. In 1434 Gil Eannes tried a new method, perhaps suggested by Prince

Henry. From the Canary Islands he sailed boldly out to sea, far out of sight of land. When he had sailed south of the latitude of Bojador, he turned eastward. At last the barrier of Bojador had been passed. The very next year the Portuguese ships sailed 390 miles south of Bojador.

By 1441 Prince Henry's ships had sailed far enough south to reach the southern zone of transition between the desert and the wetter country beyond. South of Cabo Blanco, in the area today included in Mauritania, the Portguese explorers captured a man and a woman and then a group of 10 men and women. They also found some gold. When they returned to Portugal they created a sensation. All at once there were hundreds of volunteers who wanted to go south. Between 1444 and 1448 some 40 ships sailed along the African coast, and 900 Africans were brought back to be sold as slaves. The business of exploring was forgotten in the rush to profit from the capture and sale of slaves. In those years the African slave trade began in earnest.

Around Africa to India and Beyond

After a decade or so, however, Prince Henry was able to get his sailors to return to business. Now he realized that there was a bigger prize beyond if ships could sail all the way around Africa to India. The Guinea coast was explored in 1455 and 1456, and the Cape Verde Islands were visited. Although Prince Henry died in 1460, the work he had started was carried on, after a hiatus, and new voyages were sent southward. In 1473 a Portuguese ship crossed the equator without burning up. A few years later the Portuguese went ashore and set up stone monuments to indicate their claim to the African coast. Monuments set up at the mouth of the Congo were reported still standing in the twentieth century. In 1486–87 Bartholomew Dias sailed southward from the equator, meeting headwinds and a north-flowing current. To avoid stormy weather, he sailed far out westward away from the land, and only when the weather improved did he again turn eastward. But after sailing for a longer time than he thought he should, he turned toward the north hoping to find land. He came upon the coast of southern Africa at Algoa Bay (Port Elizabeth). Returning, he passed Cape Agulhas, the southernmost point of Africa, and then the Cape of Good Hope.

The great voyage of Vasco da Gama took place between 1497 and 1499 (Fig. 10). Like Dias, he avoided the strong north-flowing Benguela current and the headwinds along the coast south of the equator. He made a wide circle out into the Atlantic before turning eastward along the latitude of Cape Agulhas. He then followed the east coast northward to Mozambique, where, for the first time in these waters, the Portuguese came into contact with the Arabs. With the aid of an Indian pilot, Vasco da Gama sailed all the way across the Indian Ocean to Calicut in 23 days, thereby completing the "end run" around the Arab-held lands. But for some reason Vasco da Gama was not informed about the monsoons, which had been known and used for navigation for many centuries. On the way to Calicut in late April and May, the northeast monsoon was breaking up, and he had favorable southwest winds. But when he tried to return in August, he encountered steady headwinds against which he had to tack. It took him three months to beat back to the African coast, during which time his crews developed scurvy. So many died that he had to destroy one of his three remaining ships. When he finally returned to Lisbon he had sailed 24,000 miles in more than two years; of the 170 men who started with him, only 44 returned.

In the sixteenth century the Portuguese voyages continued and were extended farther and farther to the east. In 1510 the Portuguese took Goa and made this port into a major trading center on the west coast of India. In 1511 they established a base at Malacca on the strait between the Malay Peninsula and Sumatra. By 1542 the Portuguese even reached Japan. In 1557 they leased Macao from the Chinese as a base from which to control the profitable trade in a wide variety of new products. In 1590 they reached Taiwan and gave it the Portuguese name, Formosa.

CHRISTOPHER COLUMBUS

Christopher Columbus was born of Spanish parents in Genoa in 1451. He was fascinated by ships and by the problems of navigation; he himself said he started navigating at the age of 14. As a young man he served on ships in the eastern Mediterranean, and in 1476 he arrived in Portugal. In that same year he sailed on an English ship, which probably went as far as Iceland. In 1478 he was married and lived for a few years on the islands of Madeira, but he never ceased to increase his knowledge of the ocean and was training himself to be one of the most highly skilled sailors of his time. After the death of his wife, he went to sea again for the Portuguese, sailing several times after 1482 to the Guinea coast (Morison, 1942, 1963; Phillips, 1992; Phillips and Phillips, 1993).

The Geographical Ideas of Columbus

At an early age Columbus had started to think about the possibility of sailing westward to Asia. There were at least five books that he read with enough care to write marginal notes in them. One of these was the *Imago mundi* of Cardinal Pierre d'Ailly, in which the conclusion was reached that China was little more than 3000 miles west of the Canary Islands. He read books by Aeneas Silvius (Pope Pius II), who reported on the Greek and Roman concepts regarding the earth. He also read Latin translations of Ptolemy's *Geography*. He was thoroughly familiar with the first Latin edition of Marco Polo and also with the popular favorite supposed to have been written by the fictitious Sir John Mandeville.

Columbus had no doubts that the earth was round. In those days no educated person, and no person with experience in navigation, had any illusions about the earth being flat. Columbus faced two problems, both of them related to the distance he would have to sail westward to reach Asia. The first was that if the whole circumference of the earth were divided into 360 parts, as Hipparchus had done and as had been common practice since his time, then what was the length of one part, or 1°? The second problem was what authority to believe about the eastward extension of the known lands of the earth.

Regarding the first of these problems, Columbus had a wide range of calculations from which to choose (Nunn, 1924). Eratosthenes had shown how the circumference of the earth could be calculated. Just three measurements were needed: the altitude of the sun on any given day at two places on the same meridian and the distance between the two places. Eratosthenes had come very close to reaching the correct figure; Posidonius, using a different set of principles, had reached a much smaller figure. Marinus of Tyre favored the smaller figure (Fischer, 1975). Ptolemy reported both, but himself followed Marinus. During the ninth century Muslim geographers had made several new sets of observations in Mesopotamia, and from these data they estimated that the length of a minute was about 1480 meters. The actual length of a minute along a great circle is about 1829 meters.

Of course, Columbus accepted the smallest estimate of the distance around the world. Furthermore, his belief was strengthened by a calculation carried out by Prince Henry's navigators. The north-south line they used extended from Lisbon to a place in Guinea near the modern port of Conakry. In those days, when the precise measurement of longitude was impossible, this was accepted as a north-south line. (Conakry is actually 4°33' west of the longitude of Lisbon.) The distance along this line was estimated on the basis of many voyages, and the estimate was actually quite close to the correct distance. But there was real trouble with the latitudes at either end. Ptolemy placed Lisbon at 40°15'N, and the Muslim geographers at 42°40'N (actually it is 38°42'N). The southern end of the line was placed at 1°5'N. (Actually, it is 9°30'N.) Based on the data accepted by the Portuguese, the length of the degree was almost exactly the same as that calculated by the Muslims of Baghdad—about 56.66 Italian nautical miles. Columbus used this confirmation of the size of the earth to convince himself that reaching Asia would not be problematic. Columbus reckoned that the east coast of Asia was located just about where the east coast of Mexico is actually located. Columbus had received additional support for his estimates in correspondence with Toscanelli.

From Ideas to Action

Convincing the scholars in the Spanish and Portuguese courts that Asia was not very far west of Europe was another matter. King João II of Portugal, already committed to finding a way around Africa, rejected Columbus's proposal in 1484. In Spain Columbus did not fare much better. When he finally secured an audience with Ferdinand and Isabella, they appointed a royal commission to study the matter, a time-honored way of postponing a difficult decision. The commission did not report until 1490, at which time they rejected Columbus's plan because they had no confidence in his estimates either of the earth's circumference or of the east-west extension of Europe and Asia. They accepted a figure of 70 Italian nautical miles for the length of a degree, and, therefore, they figured that any ships attempting the long westward voyage would surely be lost. Isabella, however, was impressed with Columbus's unshakable faith that he could make the voyage. In spite of the adverse reports, she agreed to make ships and supplies available to him (Davies, 1967:341).

The story of Columbus's four voyages is too well known to be repeated here (Fig. 10). He came upon land just about where he expected to find it; although he did not find any advanced civilization, such as that reported by Marco Polo, he did remain convinced that he had found Asia. His belief was strengthened when he found the southern coast of Cuba and the coast of Central America trending toward the southwest, just as the coast of Asia was shown on Ptolemy's map. When he heard from the Indians of Central America that there were sources of gold only a short distance west and that there was another great ocean beyond, he was sure it must be the Indian Ocean. Furthermore, noting the great flow of water along the northern coast of South America, he thought that all this water had to go somewhere. There had to be a strait connecting the Caribbean with the Indian Ocean. Of course, he did not know of the Gulf Stream. Instead of turning northward to reach the latitudes where Marco Polo had reported the existence of a Chinese civilization, he continued his exploration toward the southwest. One reason was that he expected to reach the opening into the Indian Ocean. But there was also another reason, as Carl Sauer points out. It was commonly believed at this time that gold was produced by the heat of the tropical sun, and thus more gold would be found closer to the equator (Sauer, 1966:23–24).

Columbus was one of the earliest explorers to make use of the wind systems of the Atlantic (Burstyn, 1971). From his previous experience he knew of the presence of easterly winds in the low latitudes and of westerlies in the higher latitudes. He sailed westward on his first voyage along the latitude of the Canaries with the wind at his back. But when it was time to return to Spain, he sailed northward to the latitude of the Azores; then with the westerlies behind him, he returned to Europe.

The Treaty of Tordesillas

According to the analysis by Arthur Davies (Davies, 1967), Columbus was not only a skilled sailor but also a skilled diplomat. To gain possession of the lands he discovered, he arranged to have them confirmed by the pope, and he negotiated a treaty with King João II of Portugal. This was the Treaty of Tordesillas signed by Spain and Portugal in 1494. By this treaty, the world was divided between these two countries, and the dividing line was to be drawn 270 leagues west of the Azores, or 370 leagues west of the Cape Verde Islands (Fig. 11). Portugal was to have the exclusive right to the lands east of that line, Spain to the west of it. This gave Portugal a free hand in the Indian Ocean, and it gave Columbus a free hand in the lands he had found west of the Atlantic (Thrower, 1996:69).

Several problems resulted from this treaty. Since longitude could only be estimated, no one really knew where the dividing lines should be drawn. Soon it was clear that part of eastern South America was in the Portuguese hemisphere. The line passed through the Guianas if leagues were estimated on the basis of the longer measurement of a degree, or it passed east of the mouth of the Amazon if the shorter measure were used. Actually, the agreed-upon line is approximately along 48°W of Greenwich. On the other side of the world, the boundary would be approximately along 132°E of Greenwich. The Philippines, which became Spanish, lie well within the Portuguese hemisphere; the Spice Islands (the Moluccas) are on the border.

LOOKING FOR ASIA

Columbus died in 1506 still believing that the lands he had discovered were a part of eastern Asia. By that time, however, almost all the geographers and explorers were convinced that the world was really much larger than Columbus had reckoned and that a new continent lay between them and Asia. This belief was confirmed when Pedro Alvares Cabral, sailing on a Portuguese voyage to India and setting a course southwestward from the Cape Verde Islands, came on land within the

Figure 11 The Treaty of Tordesillas

Portuguese hemisphere, which he promptly claimed for Portugal. In 1501–1502 Amerigo Vespucci, sailing for Portugal, explored the coast of Brazil probably as far as La Plata River. He spent a month exploring the disputed area on the border between the Portuguese and Spanish hemispheres, which is now called Uruguay. Perhaps because he was the first to announce that the land he was seeing was in fact a new continent and not a part of Asia, a German cartographer wrote the name America on a map published in 1507.

Within two decades explorers had sketched in the whole eastern coast of the newly discovered continent. Greenland and Labrador had been rediscovered by John Cabot, and Cabot's son, Sebastian, sailed into Hudson Bay. The Cabots were born in Genoa and Venice, but they sailed for England. Other voyages sent out by France included one by Jacques Cartier, which proved that the Saint Lawrence was a river, and one by Giovanni da Verrazano, who was the first European to sail into New York Harbor (in 1524).

Magellan

It was the Portuguese explorer Fernão de Magalhães, or Magellan, who was the first to reach eastern Asia by sailing west. Like Columbus, he tried first to interest the Portuguese king in supporting such a voyage, but the king refused to have any-thing to do with the project. In Spain, however, he found greater interest, and in 1518 the Spanish king agreed to support the voyage. The next year Magellan set sail with a fleet of five ships, all in poor condition (Fig. 10). Looking for an open-ing through the Americas, he examined the Brazilian coast carefully, entering every bay to be sure that it was not the strait they were seeking. The expedition spent the winter (from March to August 1520) in southern Patagonia, during which time some of his men mutinied. When the blustery winter winds subsided, the expedition started southward again; on October 21, 1520, they found the entrance to the strait that bears Magellan's name. It took 38 days for the ships to pass through the 360-mile strait. Only an exceptional period of light winds made the passage possible at all because normally the westerly winds are too strong, even in summer. Leaving the strait, Magellan sailed in a northwesterly course for 98 days before he reached the island of Guam. It is amazing that in that long voyage through an ocean dotted with islands, he came upon only two small and desolate rocks (Saint Paul's Island and Shark Island). His men were sick with scurvy, and his supply of food and water was exhausted to the point that the explorers had to eat rats, oxhides, and sawdust and drink putrid water before they reached Guam. After replenishing supplies and regaining health, they started again westward, reaching the Philippines on April 7, 1521. Magellan, on a previous voyage for Portugal, had reached the Spice Islands (the Moluccas), which are east of the Philippines. Therefore, at this point he had become the first man to sail all the way around the earth.[1]

Magellan was killed in a fight between Philippine natives on April 27, 1521. Thereafter one of his remaining ships, the *Vittoria*, commanded by Juan Sebastian del Cano, continued the voyage across the Indian Ocean after taking on a load of cloves at the Moluccas. They rounded the southern end of Africa and then sailed northward, reaching Seville on July 30, 1522. Of the original five ships, only the *Vittoria* returned to Spain. But the whole cost of the expedition was more than covered by the sale of the cloves.

Outlining the Continents

Between 1521, when Magellan was killed by natives in the Philippines, and 1779, when James Cook was killed by the Hawaiians, the exploration of the world was directed primarily at fixing the outlines of the land and water on the world map. At first there was no single answer to the question about what to observe. Everything was novel, and the sober descriptions of new lands could scarcely be distinguished from the imaginative writings of those who followed Sir John Mandeville. But the outlines of the map could be filled in. Gradually, as the Age of Exploration continued, scholars began to try to digest the flood of new information and to formulate scientific generalizations. In the next chapter we will consider the impact of exploration on the shape of geographic ideas.

---

[1]The second voyage around the world was led by Sir Francis Drake, 1577–80 (Bawlf, 2003; Thrower, 1984).

FIVE PROBLEMS

Meanwhile, the explorers themselves faced five major problems, all of which were solved in the period between Magellan and Cook. These were: (1) the problem of maintaining health on long sea voyages; (2) the problem of perfecting the accuracy of navigation; (3) the problem of making precise determinations of longitude; (4) the problem of showing the whole spherical surface of the earth, or large parts of it, on flat pieces of paper; and (5) the problem of eliminating from the world map the many erroneous concepts of Ptolemy.

Maintaining Health

Scurvy was the disease that posed the greatest health problem on long voyages away from land. The problem did not appear as long as navigation was close to the shore, where fresh food could be taken on board frequently. But when the Portuguese learned how to sail far out to sea, scurvy began to appear among the crews. The first expedition to suffer seriously from scurvy was that of Vasco da Gama when he was returning across the Indian Ocean from Calicut to Malindi in Africa in 1498. No one knew why the men died.

Scurvy is a disease of dietary deficiency and is easily avoided or cured when adequate amounts of vitamin C are supplied by fresh vegetables, such as potatoes, cabbages, onions, carrots, and turnips or by such fruits as lemons or limes. When these fresh foods are lacking, the skin becomes sallow, gums are tender, and the victim is miserable with muscular aches. Eventually, the gums become like putty, teeth fall out, and massive hemorrhages occur. When Magellan sailed for 98 days from the Strait of Magellan to Guam, most of the members of his crew had scurvy and either died or were too weak to do any work. There are many stories of ships drifting helplessly at sea with no one able to handle the sails (Penrose, 1952:203–207).

Apparently, the British were the first to discover a remedy. In 1601 a British captain on a voyage to the Indian Ocean had to stop in southern Africa because so many of his men had scurvy. Perhaps by chance he served them some fresh lemon juice, and miraculously the symptoms of scurvy began to disappear. Yet not until 1607 did another British captain try the same remedy, with encouraging results. Reports of this way of controlling scurvy were not widely read or believed because even in the late nineteenth century explorers in the Arctic were still suffering from the ravages of scurvy. But the first voyage on which fresh fruit and vegetables were regularly supplied for the purpose of controlling the disease was the second voyage of Captain James Cook to the Pacific, 1772–75. On this voyage the method of controlling scurvy was demonstrated beyond doubt, although it was still many decades before it was widely adopted and still longer before the function of vitamin C was understood.

Perfecting the Accuracy of Navigation

There remained the ancient, unsolved problem of measuring distance and direction and fixing position on the surface of the earth. The Greeks learned how to do this in theory, but they never possessed instruments accurate enough to carry out the observations they knew were needed. When long sea voyages were undertaken, the need for precise determinations of distance and direction became critical.

The magnetic compass proved to be one of the best ways for determining direction, especially when skies were cloudy. The Chinese are credited with the invention of this device, long before the time of Marco Polo, but the first mention of the compass in European writings was in 1180. A magnetized needle was first attached to a card mounted on a pivot during the thirteenth century, but not until the eighteenth century was an improved compass for use at sea made available. Columbus was the first explorer to mention the fact that the needle does not always point north and that it varies from north according to longitude. In fact, there was a period when it was believed longitude could be identified by noting the variation of the compass needle. In 1699–1700 Edmund Halley, the British astronomer, made a long ocean voyage for the purpose of plotting the magnetic variation on a map. His use of lines to show equal magnetic variation resulted in the first isarithmic map ever made. He proved clearly that the variation was not the same for each longitude.

The determination of latitude depends on observations of the height of certain stars or of the sun above the horizon. The astrolabe that Hipparchus had invented had been improved by expert workmanship, and for centuries it was the only way to determine latitude at sea. In the sixteenth century the cross-staff was invented. This is a pole about 3 feet in length that the observer points toward the horizon. A rod mounted vertically on the staff can slide back and forth. The observer places his eye at the end of the staff and moves the vertical piece until the sun or star is exactly at the tip of the rod. There were two significant difficulties with this device. The observer had to look at two things at once—the horizon and the sun or star. And if he were using the sun's altitude, he had to look directly at it. In 1594 the polar explorer John Davis invented the back-staff, in which the sun's altitude is measured by its shadow. This was later improved by using a mirror instead of a shadow and was known as a quadrant. John Hadley's octant was invented in 1731. The sextant, still in use, was a still later improvement based on the principle first adopted by Davis.

Another problem was the measurement of speed at sea. Sailors voyaging within the narrow confines of the Mediterranean developed skills at estimating a ship's speed, but on the open ocean the errors of such estimates became intolerable. At one time a navigator would throw some object overboard near the bow of the ship and then measure the time required for the object to reach the stern. The time interval could be estimated by the speed of pacing or by observing the sand in an hourglass. British sailors are credited with the invention of the log method. A thin but strong cord is attached to a heavy log, and the log is thrown into the water. The log remains stationary, causing the line to play out at a speed that can be measured by the time interval between knots on the line. The log was thrown out whenever there was a change in wind direction or of the ship's course (Taylor, 1957:201).

Determining Longitude

None of these methods of observing distance and direction were accurate enough to permit a precise determination of longitude, and until longitude could be determined, navigation remained dangerously inaccurate. The measurement of longitude required some way of keeping accurate time at sea. The Greeks knew that the observation of the time of an eclipse at two places could be used to check estimates of longitude. But eclipses did not always occur when they were needed. As early as 1522 the principle was clearly understood that a dependable timepiece would solve the longitude problem. The difficulty was that the clocks of the sixteenth century required frequent winding and would gain or lose up to 15 minutes a day. Furthermore, when clocks were carried into climates with different temperatures, the metal parts would expand or contract, thus making measurements of time unreliable. In 1657 Christian Huygens invented the pendulum clock, which could keep time with great accuracy, but the clock would not function satisfactorily at sea.

The need for an accurate clock was underlined in 1707 when an English fleet was wrecked on the Scilly Islands because the navigator was far off in his estimate of longitude. The great loss of life on this occasion brought the need to the attention of the public. In 1714 the British Parliament offered a reward of £20,000 to any person or persons who could devise a way to measure time at sea with sufficient accuracy. To receive the award a clock had to be built that, on a voyage from England to the West Indies and back, would not gain or lose more than two minutes. For many years no clocks were offered.

At this time in England there was an expert clockmaker named John Harrison, whose excellent pendulum clocks were well known. In 1729 he decided to try for the reward, and he went to work on a clock suitable for use at sea (Quill, 1966). One of his earliest models did so well on a voyage to Lisbon that the Board of Longitudes of the British government gave him a grant to help continue the work. Not until 1761, however, did Harrison complete clock number four, which he felt could meet the requirements. The clock was tested on a voyage to Jamaica and back and lost less than two minutes on the trip. Another trial was ordered, and this time the clock was mounted on a movable bracket. On a trip to Barbados and back it lost only 15 seconds in 156 days. The Board of Longitudes still held back the prize because Harrison, they insisted, had not told anyone else how to make such a clock. Another clockmaker was commissioned to make a duplicate to Harrison's specifications. It was this clock that was carried to the Pacific on the second voyage of Captain Cook (1772–75), and it proved to be so accurate that for the first time in the long effort to gain useful knowledge about the earth, an explorer could tell exactly where he was. Harrison, then 82 years old, was finally given the reward in 1775, a year before his death. By this time French and Swiss clockmakers were almost ready with their models. After this date navigators could measure longitude with the same precison with which they could measure latitude (Taylor, 1957:260–263; Sobel, 1995; Andrewes, 1996).

Filling the Need for New Kinds of Maps

Most of the cartographers of the fifteenth century lived either in Venice or Genoa because it was from these two places that Europeans departed on voyages to the eastern Mediterranean to pick up cargoes of valuable items from the east. The fifteenth-century geographers generally started with Ptolemy, and the new maps they produced began the job of correcting the work of the ancient cartographers.

One of the earliest corrections was made in 1459 by a monk named Fra Mauro who lived near Venice. His map showed the Indian Ocean open to the south, thereby breaking with the Ptolemaic tradition of enclosing that ocean. But how did Fra Mauro find this out in 1459? Bartholomew Dias did not sail around southern Africa until 1486–87. The Venetians must have learned about the Indian Ocean from Arab traders, and Fra Mauro believed the reports he received. Presumably, his map gave new confidence to the Portuguese voyagers just at the time of the death of Prince Henry. Incidentally, Fra Mauro's map, like most of the maps of this time, was "oriented" toward the south. But since the maps were usually read by laying them flat on tables rather than hanging them on walls, it apparently did not become customary to speak of "going up south."

The first cartographer to produce a globe showing the whole round earth was Martin Behaim of Nuremberg (Bagrow and Skelton, 1964:103–109; Kimble, 1933). He was among those who advised King João II of Portugal regarding the arts of navigation and the most likely directions in which exploration might bring important results. It is thought that Behaim was the first to make a brass astrolabe to replace the wooden ones in use earlier. He may have sailed southward to Guinea on one of the Portuguese expeditions. In 1490, however, he went back to Nuremberg. There, with the help of a painter named Jörge Glockendon, he produced the world's first globe. He did show the coastline of South Africa in accordance with the reports from Bartholomew Dias, but the other Portuguese discoveries—along the Guinea coast, for example—were inaccurately portrayed. The important thing about Behaim's globe, however, is that he showed the east coast of Asia just about where the east coast of the Americas is located. He scattered the Atlantic with mythical islands, some not far to the west of Europe. His globe was based on the smaller estimate of the earth's circumference, which also had been adopted by Columbus. Yet Columbus could never have known of Behaim's globe. Presumably, Behaim gained his concept of the earth from those who had been influenced by Toscanelli, just as Columbus had.

Many cartographers produced maps during the early period of exploration, making use of the new data. Juan de la Cosa in 1500 drew a map using the observations from the first three voyages of Columbus and also from John Cabot's voyage to North America. But the first world map that shows America as a separate continent and not the eastern part of Asia was drawn by Martin Waldseemüller in 1507. He also made use of the name America for the first time, either because he thought that Amerigo Vespucci had reached the new continent before Columbus or because Amerigo was the first explorer definitely to identify the newly discovered lands as a separate continent. As a result of this decision by Waldseemüller, the new continent was not named after the European who first reported seeing it. Although Waldseemüller's map was labeled *Carta Marina*, it was no more useful for navigation than other maps of this period that made use of the Portolano principle of design. Explorers had already found that when they followed any of the straight lines on these maps for long voyages, they did not arrive at expected destinations.

A search began for new kinds of projections that would make it possible to show the curved surface of the earth on flat paper or parchment (Bagrow and Skelton, 1964). Peter Apian in 1530 produced a heart-shaped map of the earth (Fig. 12) in which both lines of latitude and longitude are curved. On this projection, however, both distance and direction are distorted. Apian was the teacher of Gerhard Kremer,

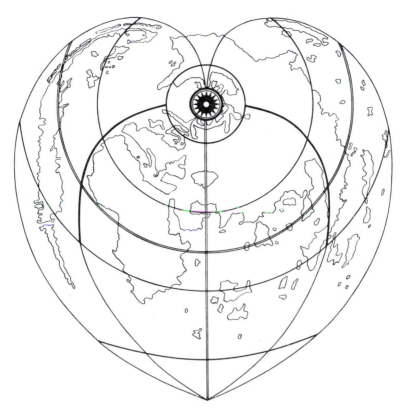

Figure 12 Apian's heart-shaped world, 1530

who adopted the Latin name Gerardus Mercator. Mercator in 1538 made a world map by joining two heart-shaped projections, one for each hemisphere (Fig. 13).

Mercator is famous for the world map he produced in 1569, which became the only map used for navigation in the low and middle latitudes (Crane, 2003). Mercator's distorted continental shapes became the most widely known of all portrayals of the world as a whole. What was needed was a projection on which a compass bearing could be drawn as a straight line so that navigators could plot their courses without having to draw curves. Of course, a straight line on a flat map is not the shortest distance between two points unless it follows the equator or a meridian. To follow other great circles, navigators draw a series of short, straight rhumb lines, approximating the curve of the great circle. The nature of a rhumb line on a flat map had been demonstrated by Pedro Nuñes (curator of maps and charts for the king of Portugal) in 1534. But sailors not trained in mathematics found difficulty in believing that a straight line was not the shortest distance between two points on the surface of a sphere. On Mercator's projection the lines of latitude and longitude are straight lines and cross each other at right angles. This means that the east-west distance between meridians increases with increasing latitude until the poles are stretched out into lines. But if the parallels of latitude are drawn with an even spacing, as they are on a globe, straight lines plotted on them are not true

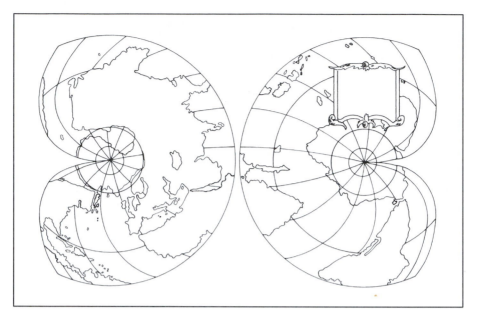

Figure 13 Outline of Mercator's world map, 1538

compass bearings. Mercator's solution was to spread the parallels of latitude in the same proportion as the spread of the meridians. The scale of the resulting map varies from latitude to latitude, increasing progressively away from the equator. The actual amount of the spread was determined by trigonometric tables.

The map that Mercator published in 1569 was, as E. G. R. Taylor puts it, essentially a scholar's map (Taylor, 1957:222). The coastlines were not drawn with care but were generalized from the written reports of the explorers (Fig. 14). Furthermore, Mercator did not explain the method used in making his projection. One gets the impression that here was a man with a theoretical solution to a problem who was not at all concerned with practical applications. The preparation of an accurate world map based on his projection he left to others.[2]

As a result, Mercator's map was not widely adopted as navigators continued to make frustrating use of the Portolanos, often finding that at the end of a long voyage they had reached the wrong place. It remained for an English geographer to explain Mercator in words understandable to people who might make use of his projection. Edward Wright in 1599 produced the trigonometric tables that made it possible for other people to reproduce Mercator's projection. But how to explain this projection to sailors not trained in mathematics? Wright devised a simple, if incorrect, explanation. Think of a spherical balloon, he said, tangent at the equator with the inside surface of a hollow cylinder. The balloon is marked off with lines

---

[2]Mercator was not the first to make use of this principle of projection. In 1511 Erhard Etzlaub of Nuremberg prepared a map of Europe and Africa as far as the equator on which he spread the lines of latitude and longitude proportionally. It is not clear whether Mercator knew of this (Bagrow and Skelton, 1964:148–150).

of latitude and longitude. Now blow up the balloon. As it stretches it comes in contact with the inside of the cylinder, the lines of longitude lie as straight lines against the cylinder, and the lines of latitude are stretched apart in proportion. Of course, the high latitudes will pop out of the ends of the cylinder, and the poles can never be shown at all no matter how far the balloon is stretched. This was a mechanical, nonmathematical demonstration of how such a projection could be made that satisfied the mathematically illiterate—which is most people (Brown, 1960:95). By 1630 Mercator's projection had supplanted all others for use in navigation in low and middle latitudes. It is still the only projection useful for this purpose.

Mercator and his friend Abraham Ortelius of Antwerp both conceived the possibility of preparing the world map in sections and binding them together in atlas form. The Ortelius atlas—*Theatrum orbis terrarum*—was the first to be published, in 1570. Thereafter a large number of atlases were prepared. Mercator started publishing sheets of his atlas in 1585, but the whole work did not appear until after his death (Bagrow and Skelton, 1964:179–189). Amsterdam became a major center for the publication of atlases and wall maps, both of which became very popular, especially in the seventeenth century. King Charles II of England in 1660 commissioned a set of some 40 wall maps, each 6 feet in height, on which the new world revealed by the explorers could be viewed. The Amsterdam publishers also began to make globes for popular use. In France the first producer of world atlases was Nicholas Sanson d'Abbeville, who in the seventeenth century founded a "dynasty of cartographers" that produced maps and atlases for more than a century (Bagrow and Skelton, 1964:185).

During the seventeenth and eighteenth centuries the mapmakers only slowly revised their maps to include the newest information and to acknowledge the existence of gaps in their information (Babcock, 1922; Crone, 1950; Tooley, 1949). The mapmaker who is credited with being the first to eliminate the drawings of weird creatures that traditionally occupied the blank spaces on the maps was Jean Baptiste Bourguignon d'Anville. Jonathan Swift, the author of *Gulliver's Travels* (1726), had pointed out this practice in a well-known jingle:

So geographers, in Afric maps
With savage pictures fill their gaps,
And o'er uninhabitable downs
Place elephants for want of towns.

It was d'Anville in 1761 who first left out the elephants and showed blank spaces where blank places belonged. Even today some map publishers cover their maps with a uniform spread of pictorial symbols, believing, no doubt, that balance and symmetry are more important than the careful portrayal of an asymmetrical world.

Eliminating Erroneous Ideas from the World Map

The fifth problem that was solved between 1521 and 1779 was the removal from maps of the many erroneous ideas inherited from Ptolemy. Ptolemy had become the almost undisputed authority in Christian Europe during the Middle Ages after his *Geography* had been translated into Latin. Each of the numerous errors in his work had to be eliminated, despite the strong resistance of those who treasured the stability of tradition in an insecure age.

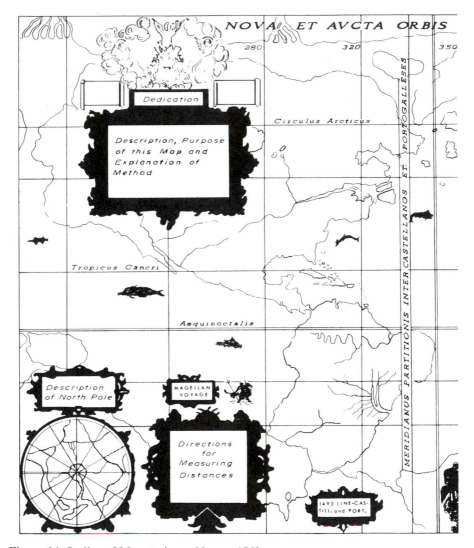

Figure 14 Outline of Mercator's world map, 1569

One of Ptolemy's most influential errors had to do with the great southland, which he called the *terra australis incognita*, the unknown southern land. On his world map (Fig. 3) this unknown land encloses the Indian Ocean on the south. The Arabs knew that the Indian Ocean was open to the south, and Fra Mauro believed them. But there were many others who either did not know about the Arabic writings or discredited them. The Portuguese navigators had to find out for themselves things that certainly had been found out by others—even by the Phoenicians of the seventh century B.C. Still, the Portuguese voyages around southern Africa did not eliminate the idea of a great unknown land mass in the Southern Hemisphere. When Magellan

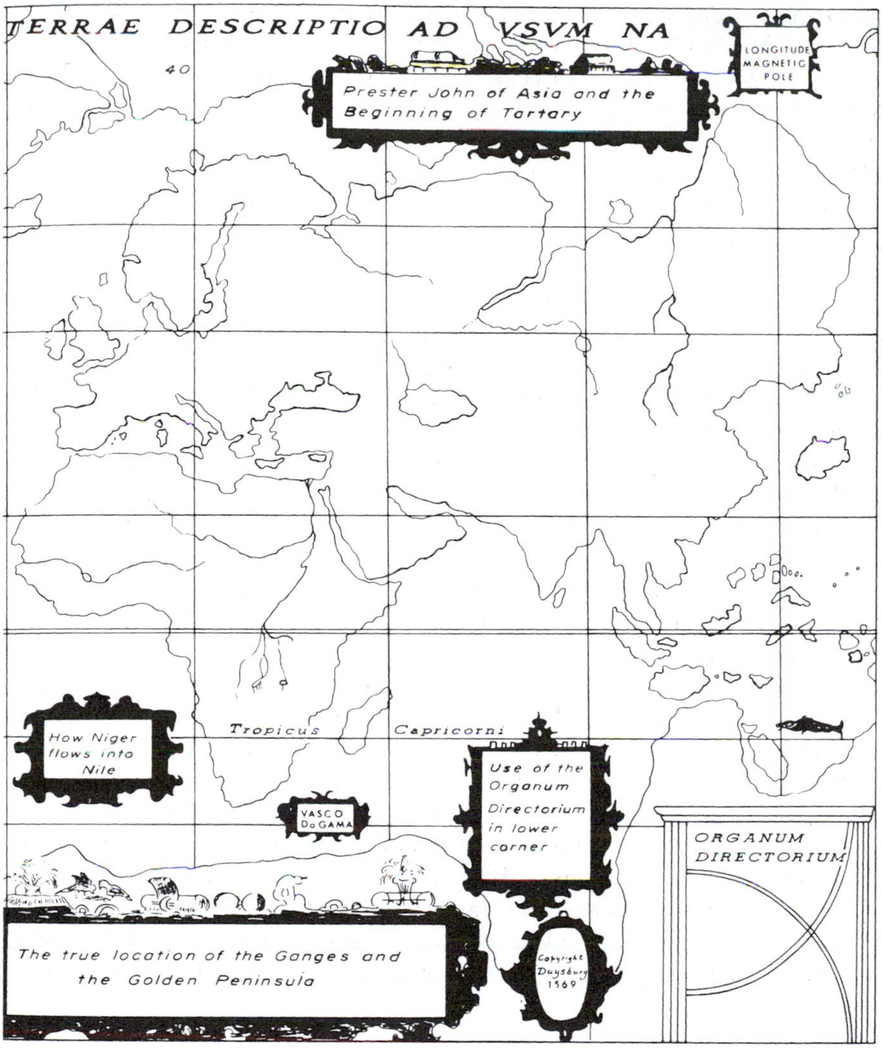
TERRAE DESCRIPTIO AD VSVM NA
40
LONGITUDE MAGNETIC POLE
Prester John of Asia and the Beginning of Tartary
How Niger flows into Nile
Tropicus Capricorni
VASCO Da GAMA
Use of the Organum Directorium in lower corner
ORGANUM DIRECTORIUM
The true location of the Ganges and the Golden Peninsula
Copyright Duysburg 1569

passed through the strait near the southern end of South America, he was certain that Tierra del Fuego was a part of Ptolemy's southland. Later mapmakers extended the southland all the way across to Tasmania. Step by step, the explorers sailed into higher and higher latitudes of the Southern Hemisphere without encountering the unknown land.

Captain Cook was the explorer who finally drew the outlines of the Pacific Ocean and eliminated Ptolemy's southland (Fig. 10). On his first voyage (1768–71) he sailed to Tahiti to carry out observations on the transit of Venus across the sun. The observations he made did not prove very useful, but he did sail southward from Tahiti to about latitude 40°S, and, finding no land, he turned westward to New Zealand. He proved that New Zealand was not a part of any southland, and then he went on to make a careful chart of the east coast of Australia. He passed through the Torres Strait between Australia and New Guinea, rediscovering a passage that a Spanish explorer had reported in 1606.

Cook's second voyage is remarkable for several reasons. He was able to keep his crew from contracting scurvy through the regular consumption of fresh fruits and vegetables. He had a chronometer that permitted him to fix his position accurately at all times. He was also accompanied by such trained observers as the Forsters, father and son, of whom we shall hear more later. This was the first of the exploring expeditions to start gathering information about matters other than just coastlines. But it was also the voyage on which the southland was finally eliminated. Cook left England in 1772, and from December 1772 to March 1773 he sailed in the far south of the Atlantic and Indian oceans. From November 1773 to February 1774 he sailed in the far south of the Pacific Ocean. His farthest point south was 71°10'S. Still there was no land; although Cook expressed the belief that there was an ice-covered land farther south, he was unable to reach it. From February to October in 1774 he sailed through the tropical South Pacific, discovering and locating numerous islands that Magellan had missed. When he returned to England in 1775, he brought the first authoritative report to reach Europe regarding the vast expanse of water in the Southern Hemisphere.

Cook's third voyage ended in disaster, but not before he had completed the outlines of the Pacific Ocean. In 1776 he sailed into the North Pacific, reaching the Hawaiian Islands in 1778. He followed the coast of North America northward through the Aleutian Islands and through the Bering Strait into the Arctic Ocean, where he was stopped by ice floes at latitude 70°44'N. He then returned to Hawaii for the winter. After sailing away from the islands in early 1779, he was forced to return to repair a mast on one of his ships. He was killed in a fight with some of the native islanders. Thereafter his second in command again took the ships to the far north to prove that there was no Northwest Passage. The ships returned to England in 1780. The world map had been essentially completed, at least in its major outlines.[3]

THE SHAPE OF THE EARTH

With the general dimensions of the earth established, the problem of the shape of the earth became increasingly important. Columbus had thought that it was pear-shaped and that it was necessary to climb toward the equator. In 1687 Isaac Newton and Christian Huygens arrived at the conclusion mathematically that the earth must be flattened at the poles and that it must bulge at the equator. But this concept was challenged by Jacques Cassini in 1720. We need to become acquainted with the Cassini family.

_____

[3] For accounts of the explorers and the routes they followed see Brown, 1979; Friis, 1967; Hakluyt, 1965; Hanson, 1967; Parker, 1965; Parks, 1928; Penrose, 1952; Rogers, 1962; Sykes, 1961; Taylor, 1957.

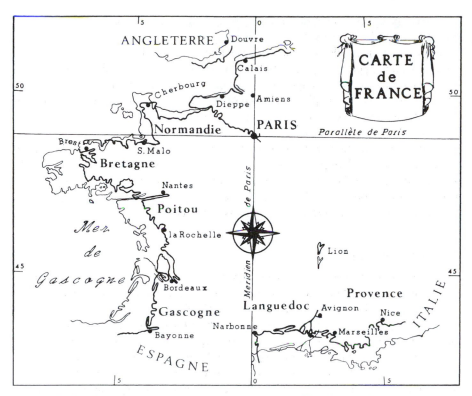

Figure 15 The Cassini survey changes the position of the French coastline

There were four generations of Cassinis who were in charge of the astronomical observatory in Paris and who carried out the first major topographic survey of a large country—France. Giovanni Domenico Cassini, an Italian, became the director of the observatory in 1667 and began to carry out a series of measurements to fix certain locations within France. His son, Jacques Cassini, who became director in 1712, in 1713 carried out a measurement of the arc of the meridian extending from Dunkerque through Paris to Perpignan. His son, César-François Cassini de Thury, used the meridian surveyed by his father to start a network of triangulation on which to fix the position of the sheets of the topographic survey. He started the survey in 1744 but died before it was completed. The work was finally carried through to completion by his son Jacques Dominique Cassini (Fig. 15) (Crone, 1950).

In 1720 the first Jacques Cassini, who made the measurement of the Paris meridian, published a paper entitled *De la grandeur et de la figure de la terre*. His measurement did not show any indication that the earth was not a regular sphere and was, therefore, a direct challenge to the mathematical concepts of Newton and Huygens. The French Academy decided to settle the controversy by carrying out measurements of the arc of the meridian at different latitudes. From 1735 to 1748 a French expedition was working in the high Andes of Ecuador, which was joined by the French scientist Charles Marie de La Condamine together with Louis Godin and Pierre Bouguer. Sometimes the surveyors who had set up their instruments on the tops of the high peaks were forced to wait for weeks for the rare clear days when observations could be made. Eventually, however, the job was done, and de La Condamine returned by way of the Amazon, carrying out the first scientific exploration of that river. Meanwhile, another French expedition, headed by Pierre de Maupertuis, went to Lapland. In spite of the hardships imposed by the very low temperatures they encountered, the survey of a meridian was completed in 1736–37. These two surveys proved that the earth was flattened at the poles and did bulge at the equator and that Cassini was wrong. These measurements were accepted as correct until the modern period of satellite technology, which shows the earth to be dented and bulging like an old baseball, with a major bulge south of the equator. The deviations from the perfect sphere revealed by remote sensing could not have been detected by the eighteenth-century instruments.

RETROSPECT

During the period between the voyages of Magellan and Cook, explorers were finding out more and more about the earth. Bartholomew Diaz rounded the Cape of Good Hope (1487) followed by Vasco da Gama (1497). Meanwhile four Columbian voyages to the West were made (1492, 1493–96, 1498–1500, 1502–4); John Cabot, Gaspar Vortereal, Jacques Cartier, and Amerigo Vespucci began to comprehend something of the shape of Greenland, Newfoundland, the east coast of North America, and the Caribbean. Then Balboa visited the Pacific Ocean. All these were soon followed by many others, hence the term "the age of exploration." Explorers were finding out more and more about the earth. Expeditions had penetrated the Arctic, seeking either a Northwest Passage or a Northeast Passage between Europe and the Orient. Edmund Halley had plotted the trade wind zones of both hemispheres. He had also mapped the monsoons of Asia and had set forth the hypothesis that these seasonally shifting winds were the result of the differential heating of land and water. A vast amount of knowledge flowed into Europe, and the impact of all this on intellectual life was enormous.

At first the explorers tried to describe everything they found in the strange lands. But, having no concepts regarding the new phenomena they found, an attempt was made to describe things by analogy. Unfamiliar animals were described by reference to familiar ones. For example, when a fifteenth-century Florentine traveler first saw a giraffe he described it as "almost like an ostrich save that its chest has no feathers but has very fine white wool. . . . It has a horse's feet and bird's legs. . . . It has horns like a ram" (Hale, 1966:164). Only later did the explorers begin to describe things in terms of dimensions, texture, or color, and, finally, in terms of abstract categories. Stories of strange creatures gradually gave way to sober accounts that could be, and were, verified. After the first contacts with unfamiliar places when observers reported on anything—especially on strange and unusual things— geographers began to ask questions about what should be observed; how they should be observed; and, especially, how the things observed could be related to some generalization of the empirical perceptions. This, in turn, led to new and sharper perceptions. In the next chapter we will examine the impact of all this on the geo- graphic ideas inherited from centuries past.

REFERENCES: CHAPTER 4

Andrewes, J. H. 1996. *The Quest for Longitude*. Cambridge, Mass.: Collection of Historical Scientific Instruments.

Babcock, W. H. 1922. *Legendary Islands of the Atlantic, a Study in Medieval Geography*. New York: American Geographical Society, Research Series No. 8.

Bagrow, L. and Skelton, R. A. 1964. *History of Cartography*. Cambridge, Mass.: Harvard University Press (see the reference in Chapter 3).

Bawlf, S. 2003. *The Secret Voyage of Sir Francis Drake 1577–1580*. New York: Walker and Company.

Beazley, C. R. 1895. *Prince Henry the Navigator*. New York: G. P. Putnam's Sons.

Brown, L. A. 1960. *Map Making: The Art that Became a Science*. Boston: Little, Brown.

Brown, L. A. 1979. *The Story of Maps*. New York: Dover Publications.

Burstyn, Harold. 1971. "Theories of Winds and Ocean Currents from the Discoveries to the End of the Seventeenth Century." *Terrae Incognitae* 3:7–31.

Butzer, K. D., ed. 1992. "The Americas Before and After 1492: Current Geographical Research." *Annals of the Association of American Geographers* 82:343–565.

Cormack, L. B. 1997. *Charting an Empire. Geography at the Universities, 1580–1620*. Chicago: University of Chicago Press.

Crane, N. 2003. *The Man Who Mapped the Planet*. New York: H. Holt.

Crone, G. R. 1950. *Maps and Their Makers, an Introduction to the History of Cartography*. New York: Capricorn Books.

Dampier, W. C. 1987. *A History of Science and Its Relation with Philosophy and Religion*. Cambridge: Cambridge University Press.

Davies, A. 1967. "Columbus Divides the World." *Geographical Journal* 133:337–344.

Debenham, F. 1960. *Discovery and Exploration—An Atlas—History of Man's Wanderings*. New York: Doubleday & Co.

Friis, H. R., ed. 1967. *The Pacific Basin, a History of Its Geographical Exploration*. New York: American Geographical Society, Special Publication No. 38.

Gilbert, E. W. 1972. *British Pioneers in Geography*. Newton Abbot: David and Charles.

Hakluyt, R. 1965. *Hakluyt's Voyages*. Ed. I. R. Blacker. New York: Viking Press.

Hale, J. R. 1966. *Age of Exploration*. New York: Time Inc.

Hanson, E. P., ed. 1967. *South from the Spanish Main*. New York: Delacorte Press.

Henige, D. 1992. *In Search of Columbus: The Sources for the First Voyage*. Tucson: University of Arizona Press.

Kelley, J. E. Jr. 1977. "The Oldest Portolan Chart in the New World." *Terrae Incognitae* 9:22–48.

Kimble, G. H. T. 1933. "Some Notes on Mediaeval Cartography with Special Reference to M. Behaim's Globe." *Scottish Geographical Magazine* 49:91–98.

———. 1938. *Geography in the Middle Ages*. London: Methuen. Reissued 1968, London: Russell and Russell.

Koelsch, W. A. 2001. *Prince Henry "The Navigator": A Life*. Review of . . . Association of Pacific Coast Geographers Yearbook 63:145–151.

Lumley, S. 1998. "Ferdinand Magellan (Fernao de Magalhaes), c. 1480–1521." *Geographers: Biobibliographical Studies* 18:53–66.

Morison, S. E. 1942. *Admiral of the Ocean Sea, A Life of Christopher Columbus*. New York: Little, Brown.

———. trans. and ed. 1963. *Journals and Other Documents on the Life and Voyages of Christopher Columbus*. New York: Heritage Press.

Nunn, G. E. 1924. *The Geographical Conceptions of Columbus, a Critical Consideration of Four Problems*. New York: American Geographical Society, Research Series No. 14.

Oliveira Martins, J. P. 1914. *The Golden Age of Prince Henry the Navigator*. Trans. J. J. Abraham and W. E. Reynolds. London: Chapman & Hall (Portuguese title: *Os Filhos de D, João I*, Lisbon, 1901).

Parker, J., ed. 1965. *Merchants and Scholars, Essays in the History of Exploration and Trade*. Minneapolis: University of Minnesota Press.

Parks, G. B. 1928. *Richard Hakluyt and the English Voyages*. New York: American Geographical Society, Special Publication No. 10.

Penrose, B. 1952. *Travel and Discovery in the Rennaissance, 1420–1620*. Cambridge, Mass.: Harvard University Press.

Phillips, W. D. and C. R. 1992. *The Worlds of Christopher Columbus*. Cambridge: Cambridge University Press.

Pigafetta, A. 1975. *Magellan's Voyage: A Narrative Account of the First Navigation*. Translated and edited by R. A. Skelton. London: The Folio Society.

Quill, H. 1966. *John Harrison, the Man Who Found Longitude*. London: Pall Mall.

Rogers, F. M. 1962. *The Quest for the Eastern Christians: Travels and Rumor in the Age of Discovery*. Minneapolis: University of Minnesota Press.

Russell, P. 2000. *Prince Henry "The Navigator": A Life*. New Haven, Conn.: Yale University Press.

Rutherford, J. and P. H. Armstrong. 2000. "James Cook, R. N. 1728–1779." *Geographers: Biobibliographical Studies* 20:9–23.

Sauer, C. O. 1966. *The Early Spanish Main*. Berkeley and Los Angeles: University of California Press.

Skelton, R. A. 1969. Captain James Cook, After Two Hundred Years. A commemorative address before the Hakluyt Society. London: The British Museum.

Spate, O. H. K. 1979. *The Pacific Since Magellan. The Spanish Lake*. Canberra: Australian National University Press.

Stokes, E. 1970. "European Discovery of New Zealand Before 1642, a Review of the Evidence." *The New Zealand Journal of History* 4:3–29.

Sykes, P. 1961. *A History of Exploration from the Earliest Times to the Present Day*. New York: Harper Bros.

Taylor, E. G. R. 1957. *The Haven-Finding Art: A History of Navigation from Odysseus to Captain Cook*. New York: Abelard-Schuman.

Thrower, N. J. W. 1996. *Maps and Civilization: Cartography in Culture and Society*. Chicago: University of Chicago Press.

———. ed. 1984. *Sir Francis Drake and the Famous Voyage. 1577–1580: Essays Commemorating the Quadricentennial of Drake's Circumnavigation of the Earth*. Berkeley, Calif.: University of California Press.

Tooley, R. V. 1949. *Maps and Map-Makers*. New York: Crown.

Wright, J. K. 1925. *The Geographical Lore of the Time of the Crusades*. New York: American Geographical Society. Research series no. 15. Reprinted with a new introduction by C. J. Glacken, 1965. New York: Dover Publications.

5

THE IMPACT OF DISCOVERIES

Furthermore, men avidly made use of the accumulating voyages and travels which had been and were being published. What is the state of nature, the primitive stage in mankind's development? What are primitive peoples like? What influences, according to the travelers, determine the character of far-off peoples? These questions were asked in the seventeenth century also, but its thinkers still leaned heavily on the classical writers. From the middle of the eighteenth to the early part of the nineteenth century, in the writings of Buffon, Montesquieu, Herder, and Malthus . . . one sees what a refreshing and inexhaustible well these voyages and travels had become. . . .

They built on the past, but the world became a richer place for their departures. Even today their questions suggest our questions, but it would be a miracle if they did more than that. For they lived in a world which resembled the past more than what was to come, at least as far as problems of human culture and the natural environment are concerned.

—*C. J. Glacken,* Traces on the Rhodian Shore

What happened to geographical ideas in Europe as a result of the unprecedented flood of new information coming from the voyages of discovery? What kind of meaning could be discerned in this rapidly expanding new picture of an unfamiliar world? The cosmographers, pursuing their ancient goal of increasing their knowledge of the ordered universe, the cosmos, assumed that the coherence and harmony that they observed was the creation of a divine being (Tuan, 1968). Traditionally, the divine being was conceived in terms of human experience: Like a human being he must have a conscious purpose and a plan of action. Answers to questions about the universe were given as logical deductions from the basic and generally accepted theory that the world was created specifically as the human habitat. But from this assumption two quite different interpretations of the individual's place in the universe are possible: Humans may be seen as creatures of their habitat—their activities and even their physical character controlled by their natural (nonhuman) environment, including the astrological influences that were seen as guiding human affairs, or humans may be seen as creatures commanded by the divine plan to conquer and control their natural environment, and in a sense to complete the work of creation out of the raw materials of their natural surroundings. These are two quite different interpretations, and each is logically derived from the concept of a planned creation.

From the early fifteenth century on, as new information poured into the European intellectual world, the cosmographers struggled with these questions. At first they followed the traditional patterns, seeking evidence of the divine plan. As in the medieval period preceding the fifteenth century, most of the scholars accepted almost literally the account of creation presented in Genesis; they went to great trouble to make the new findings about the earth fit the Scriptures. But these methods and traditions became more and more difficult to sustain. The period we are examining was one when men struggled to shake loose from the weight of old beliefs and sought to find new answers to old questions about order, harmony, and meaning.

There was still very little specialization in the kinds of problems scholars investigated or wrote about. The development of special branches of knowledge came later. But, of course, there were some whose approach was primarily mathematical, and there were others whose approach was literary. There were some who asked questions and looked for answers, and others who were fully occupied—and satisfied—with the task of establishing the validity of the facts that were reported by the explorers and of putting together a general picture of what things were like.

NEW CONCEPT OF ORDER IN CELESTIAL SPACE

A few brilliant flashes of genius during these centuries illuminated the road ahead. One of these had to do with the question of whether the earth was fixed in the center of the universe with the sun and stars revolving around it or whether the sun was the center of the universe with the earth revolving about it. The idea of a heliocentric universe was not new: it had been offered as a hypothesis in the third century B.C. by Aristarchus of Samos. But Ptolemy had accepted the geocentric universe, and he was the great authority who could not be easily questioned. Nevertheless, by the fifteenth and sixteenth centuries there were some who doubted Ptolemy. One of these was a Polish scholar, Nicolaus Copernicus, who, between 1497 and 1529, carried out numerous observations of the movements of the planets, the moon, and the stars. The Ptolemaic system was too complicated, but when he calculated the movements of these bodies with the sun as a fixed center, the results were more satisfactory. In 1543 he published his great work, *De revolutionibus orbium coelestium*, in which he presented the picture of a heliocentric universe. He still followed Ptolemy in describing the movements of the planets around the sun as circular.

Three other scholars were of primary importance in breaking the Ptolemaic tradition concerning the planets and the laws of motion. In 1618 Johannes Kepler, a German astronomer, recognizing that the planetary motions were elliptical rather than circular, presented his work on the laws of motion. In 1623 Galileo presented the proof that Copernicus was right about the heliocentric universe and thereby brought a storm of abuse from those who thought that the findings of science might undermine the authority of the church. In 1632 his book comparing the universe conceived by Ptolemy with that conceived by Copernicus was acclaimed by scholars throughout Europe not only for the brilliance of its ideas but also for the literary quality of the presentation. Galileo for the first time formulated the concept of a universal mathematical order, a universe that could be described in terms of mathematical law rather than in the verbal and logical terms of Aristotle. Finally, in 1686 Isaac Newton presented his laws of gravitation. With Copernicus, Kepler, Galileo, and Newton—in about a century and a half—the seeds of the scientific revolution had been planted. With this revolution came the beginnings of specialization, which would end the work of the cosmographers.

THE ACCOUNTS OF THE VOYAGES OF DISCOVERY

There was still the real possibility, in sixteenth-century Europe, that a considerable proportion of the written accounts of the voyages of discovery might be lost—as so many similar documents in ancient times had been lost. Although the printing of books in Europe had begun in the middle of the fifteenth century, there were not enough printing establishments to meet the demand. Three geographers and publishers in the sixteenth and seventeenth centuries started what became important collections of such reports. The earliest was a Venetian named Giovanni Battista Ramusio, whose three-volume work *Navigationi et viaggi* was published between 1550 and 1559. It contains an edition of Marco Polo and a little-known Portuguese report, *All the Kingdoms, Cities, and Nations from the Red Sea to China*, written in 1535. The collection was completed by the publisher after Ramusio's death in 1557.

The most widely known of these collections of voyages is the one started by Richard Hakluyt in England (Gilbert, 1972). The first small edition of Hakluyt's voyages appeared in 1598; the full three-volume collection came out between 1598 and 1600 (Parks, 1928). Hakluyt was a lecturer at Oxford after 1574 and introduced the "new geography" of the late sixteenth century in that university. He was also adviser to many navigators of his time. His recommendation to the polar explorers was to stop trying to find a way from Europe to eastern Asia by the Northeast Passage and to concentrate on the search for a Northwest Passage, north of North America. He also urged Britons to establish colonies along the eastern side of North America.

Hakluyt was influential in getting the third collection of reports on voyages started. This was done by the German publisher, Théodore de Bry. The de Bry collection included some 25 volumes published between 1590 and 1634.

THE CHANGING IMAGE OF THE EARTH

The traditional way of writing about the earth as the home of humankind, which was followed by the early cosmographers of the sixteenth and early seventeenth centuries, was inherited from Strabo. The idea was to bring together all that was known about the different parts of the world. Strabo was very critical of his sources and indicated clearly that he had confidence in some and little or no confidence in others. His judgments were made on the basis of then current preconceptions and established dogma, and in the long run they did not always prove to be right—as, for example, when he rejected Pytheas as a storyteller. As new information came in during the Age of Exploration and as earlier writers were shown to be in error, there was ample opportunity for writers to offer their versions of the new geography. The only new items were the reports on what the explorers had found; the methods of observation, the ideas or concepts that guided the observations, and the kinds of questions asked remained the same.

Not all the writers of this period were as careful to be critical of their sources as Strabo had been. In 1490, almost 20 years after the first Portuguese ship crossed the equator into the Southern Hemisphere without burning up, an Italian writer published a compendium in which he described the torrid zone as sterile and uninhabitable, almost in the same terms used by Aristotle and Ptolemy. He gave a demonstration of the method of thinking deductively from astrological principles when he insisted that there could be no land in the Southern Hemisphere because the heads of the animals in the zodiac were all pointed toward the north (Kimble, 1938:219). And there were whole books purporting to describe new lands discovered by travelers that were, in fact, wholly fiction, like the famous *Travels of Sir John Mandeville*. This book, which was pieced together by an unknown author from earlier fourteenth-century travel accounts (with the trivial and dull passages removed and with the addition of descriptions of the regions traversed and of the strange customs and physical appearance of the inhabitants) was a widely popular but wholly false literary production of the early Renaissance (Fig. 16). Even in the seventeenth and eighteenth centuries Mandeville made good reading and was not always recognized as fiction.[1]

Sebastian Münster

The first important compendium of geography published after the early voyages of discovery was the one by the German cosmographer Sebastian Münster. Münster, who had previously established his reputation as a classical scholar by translating several geographical works from Latin into German, spent 18 years (and made use of the services of some 120 writers and artists) in preparing his *Cosmographia universalis*, published in 1544. The work is strictly in the Strabo tradition; in fact, Münster was known to his contemporaries as "the German Strabo" (Glacken, 1967:363–366).

The *Cosmography* is in six books. The first presents a general picture of the universe in Ptolemaic terms, to which the author adds a discussion of the dispersal of humankind over the earth after the Flood. The other five books deal with the major divisions of the earth. Of these the best ones have to do with southern and western Europe and with Germany. He divides the Old World into the traditional three parts—Europe, Asia, and Africa—a practice whose origin was previously discussed in Chapter 2. Europe he separates from Asia along the Don River; and the line between Asia and Africa is placed along the Nile—presumably he never read Herodotus on this subject. His treatment of Asia and of the Americas is curiously uninformed, considering that the German mapmakers had access to the new information from the explorers and this information was surely available in Germany while Münster was working. He enlivens the weird stories of America and Africa with woodcuts of men with heads in their chests or with only one leg—creatures that come right out of Sir John Mandeville.

Münster's *Cosmography* was regarded as the authoritative work on world geography for more than a century. New editions with few changes were published in 1545, 1546, and 1548. In 1550 there was a new edition that included many additions and corrections. After his death (from the plague in Basel) in 1552 there were numerous later editions that included much supplementary material (Büttner and Burmeister, 1979).

---

[1]The Dover edition, published in 1964 and edited by A. W. Pollard, gives a brief summary of the history of this famous work (Pollard, 1964). Otherwise see Mandeville, 2001.

Figure 16 Drawings adapted from Sir John Mandeville

Cluverius and Carpenter

Writers of books on "universal geography" in the seventeenth century had better access to the new materials than Sebastian Münster had. In 1616 Cluverius (Philipp Clüver) published a book on the historical geography of Germany, of which Carl

Sauer remarked that he skillfully united "knowledge of the classics with knowledge of the land" (Sauer, 1941:11). A similar work on Italy was published posthumously in 1624. Also in 1624 his six-volume compendium of geography appeared.[2] In this work he started with the traditional picture of the universe as portrayed by Ptolemy, showing that he had no knowledge of the work of Copernicus, which had been published 79 years before the death of Cluverius. The other five volumes of the compendium, however, are much more up to date. His work on Europe was especially well done (Partsch, 1891).

The first geographer to attempt a compendium of geography in English was the British geographer Nathanael Carpenter. He was at Oxford during the period after 1609, when Cluverius was a frequent visitor, and must have received many ideas from this source. He also reports numerous examples of how human character is determined by climate, ideas that he derived from the writings of the French scholar Jean Bodin. Both Carpenter and Bodin accepted the old Greek theory that the habitability of a place was a function of its latitude.

Varenius

The European scholar who profoundly influenced the content and scope of geography for more than a century was Bernhardus Varenius (Bernhard Varen). Varenius set forth, more clearly than anyone had done before his time, the relationship between geographical writings that describe the characteristics of particular places and those that describe the general and universal laws or principles that apply to all places. The first he called *special geography* and the second *general geography* (Büttner, 1978). This intellectual problem of the relation between the specific and the general became of major importance in the early part of the seventeenth century as a result of the flood of new information about specific places and the effort to generalize this information. Cluverius made no attempt to connect the general ideas in his first book with the specific observations about places in the other books; Carpenter did very little with descriptions of particular places. Varenius saw that there was a need to demonstrate the close interconnection between these two viewpoints: Special geography is of great practical importance for government and for commerce, but it leaves out the fundamentals of this field of study; general geography provides these fundamentals, but to be of maximum utility they must be applied. Therefore, special geography and general geography did not suggest a dichotomy but rather two mutually interdependent parts of a whole. It is indeed unfortunate that Varenius lived only 28 years and was not able to amplify these ideas.[3]

---

[2]*Introductionis in universam geographiam* (Leiden); translated into French in 1639 and into German in 1678.

[3]J. N. L. Baker points out that Bartholomaus Keckermann, a German geographer, used the terms *general geography* and *special geography* in lectures at Danzig in 1603 and in a book published in 1610. Since Varenius made use of Keckermann's work, it is likely that he adopted this distinction from the older writer and then provided a clear demonstration of the relation between these two points of view (Baker, 1955b:56, 1963:113).

Bernhard Varen was born near Hamburg in Germany in 1622. In 1640 he entered the university at Hamburg, where he studied philosophy, mathematics, and physics (Dickinson and Howarth, 1933:100). After three years he went to Königsberg to study medicine, and shortly thereafter he went to Leiden. In 1647, however, he took a position as a private tutor in a family living in Amsterdam. In this busy commercial city he came into close contact with merchants who were sending their ships to trade in the most distant parts of the earth, including Japan, where the Dutch had a trading post on an island in the harbor at Nagasaki. Responding to the practical need of the Dutch traders to know much more about the people with whom they were doing business, Varenius in 1649 published a book entitled *Descriptio regni Iaponiae et Siam*. This work contains five parts: (1) a description of Japan, compiled from information available in Amsterdam; (2) a translation into Latin of a description of Siam by J. Schouten; (3) an essay on religions of Japan; (4) some excerpts from the writings of Leo Africanus on religion in Africa; and (5) a short essay on government. This was a book aimed at providing the merchants of Amsterdam with useful knowledge about places and people.

At this point Varenius saw that descriptions of particular places could have no standing as contributions to science if they were not related to a coherent structure of general concepts. He began work on his *Geographia generalis* in the fall of 1649 and completed the work in the spring of 1650. In it are references to a series of books on geography that he proposed to write, but in that same year he died in Leiden.

The *General Geography* of Varenius was the standard text in this field for more than a century. Four Latin editions were published in Amsterdam (1650, 1664, 1671, and 1672). Isaac Newton was so impressed with the work that he edited two more Latin editions published at Cambridge University in 1672 and 1681. In 1693 there was an English translation by Richard Blome. Another Latin edition was published at Cambridge in 1712, with comments and corrections by J. Jurin. The edition of 1712 was also translated into English by Dugdale and Shaw, and this translation went through four editions between 1736 and 1765. One of the Latin editions, edited by Isaac Newton, was used as a text at Harvard in the early years of the eighteenth century (Warntz, 1964:117; Warntz, 1989).

Unlike the writers of the earlier cosmographies, Varenius was careful to include the most recent ideas. He accepted the heliocentric universe of Copernicus, Kepler, and Galileo. On this basis, he was the first to note the difference in the amount of heat received from the sun in the equatorial regions as compared with the higher latitudes. The sun's heat, he suggested, thins the air close to the equator, and, therefore, the cold, heavy air of the polar regions must flow toward the equator. This was the first step toward explaining the world's wind systems (Peschel, 1865:396). Varenius had studied Newtonian physics, and he had studied Nathaniel Carpenter's *Geography Delineated in Two Books* of 1625 and 1635, which were based on the work of Keckermann. In *Geography Now and Then*, W. Warntz provides evidence for the widespread use of the *Geographia Generalis* in American colleges during what he refers to as America's first cycle of academic geography (Warntz, 1981).

Varenius's conception of the field of geography was far in advance of his time. He insisted on the practical importance of the kind of knowledge included in special geography, but he also insisted that special geography only becomes intelligible when the specific features are explained in terms of abstract concepts or universal laws. Geography, he wrote, focuses attention on the surface of the earth, where it examines such things as climate, surface features, water, forests and deserts, minerals, animals, and the human inhabitants. The human properties of a place include "a description of the inhabitants, their appearances, arts, commerce, culture, language, government, religion, cities, famous places, and famous men" (Dickinson and Howarth, 1933:101). In general geography, he continued, most things can be proved by mathematical or astronomical laws, but in special geography, with the exception of celestial features (climate), things must be proved by experience—in other words, by direct observation through the senses.

QUESTIONS AND HYPOTHESES

The mounting mass of new observations and the startling new theories concerning the motions and behavior of physical bodies led to new speculation concerning the origin of the earth. But the methods of thinking had not yet broken free from the traditions inherited from ancient Greece and from the accounts of the Creation in the Bible. Even brilliant innovators, such as Galileo and Newton, were as much concerned to demonstrate that their theories did not really depart from the Scriptures as they were to demonstrate that they were supported by observations.

During the second half of the seventeenth century, some groups of scholars took important first steps in the use of the scientific method. In England there were men such as Robert Boyle, discoverer of the law concerning the behavior of a given volume of gas with changes in temperature and pressure; Robert Hooke, developer of laws concerning the elasticity of solid bodies; John Flamsteed, first director of the Royal Observatory at Greenwich (established in 1675) (Willmoth, 1997; Clark and Clark, 2001); and Edmund Halley, astronomer and author of the first scientific explanation of the world's wind systems. These men were making use of the methods of experimental science in arriving at hypotheses concerning the operation of natural processes. In 1663 some of these innovators formed the Royal Society of London "for the improvement of natural knowledge." Amsterdam was a center of intellectual freedom, where observations brought back by the Dutch merchants and explorers were stimulating new kinds of inquiry.

The Origin of the Continents and Oceans

Much speculation was aroused concerning the origin of the earth and of its surface features as a result of the publication of a two-volume work by Thomas Burnet in 1681 entitled *Telluris theoria sacra* (Taylor, 1948). Burnet was a clergyman whose hypothesis regarding the origin of the earth and its present condition came entirely from his own unfettered imagination. When God created the earth, said Burnet, and set it spinning on its axis, the earth became egg-shaped. Since its axis was then perpendicular to the plane of the ecliptic (its orbit around the sun), there were no seasons but only a perpetual spring at the latitude of England. The surface of earth was smooth. But people like Methuselah lived so long with so much leisure time that they were able to develop much evil among men. In anger, God decreed the destruction of the earth. The surface began to crack open and break up into ugly mountains and valleys. Then a flood of water covered the whole earth, released from deep inside where the water had been confined. The shock to the earth knocked its axis away from the perpendicular so that thereafter there were seasons. The surface was broken into continents, mountains, and deep depressions into which the water drained off to form the oceans.

The appearance of Burnet's *Sacred Theory of the Earth* set off a controversy that lasted for decades. Several new theories regarding the origin of the earth were offered. In 1695 John Woodward presented the idea that the Flood that God sent in anger dissolved the rock material of the earth, and this material was later deposited in the form of layers, or strata, some of which contain the fossil remains of vegetable and animal life. William Whiston, who was much impressed by the observations made by Edmund Halley in 1682 regarding the comet that bears his name, developed the theory that the earth itself was made from the debris of a comet. Furthermore, the near approach of a second comet was the cause of the Flood, of the elliptical (rather than circular) orbit of the earth around the sun, and of the continents and ocean basins on the earth's surface. The comet raised tides in the rock crust of the earth similar to the tides raised by the moon on opposite sides of the earth. The tidal crests were represented by the continents, and the troughs by the basins of the Atlantic and Pacific Oceans. Whiston supported his theory by impressive mathematical equations showing how a comet could produce such tides in the rock crust of the earth. However, his calculations left out so many things that he also was immediately attacked. The theologians based their attack on the Bible: How could the sun already be in existence before the earth started moving around it, when Genesis clearly states that God created light on the fourth day?

In Germany Abraham Gottlob Werner taught that the material dissolved in the waters of the flood was deposited over the earth to form a series of layers like the skin of an onion. All the rocks of the earth's surface, he said, were formed in this way. Critics asked how he explained the disappearance of all that water, but the failure to find a plausible explanation for the disappearance of the Flood did not make Werner abandon his theory, which was almost universally accepted well into the nineteenth century. In fact, the Swiss scholar Horace Bénédict de Saussure (who was one of the first men to climb Mont Blanc on August 8, 1786) supported Werner by suggesting that the Alpine valleys had been cut by the violent torrents when the Flood drained off into the present ocean basins.

On the Origin of Landforms

A great difference of opinion persisted throughout the period from the fifteenth to the eighteenth centuries and indeed on into the nineteenth century concerning the origin of the landforms of the earth's surface. Were the landforms created by divine purpose? Were the hills really everlasting as the Bible insists? Or were the mountains and valleys and ocean basins formed by cataclysms of nature, as Burnet believed? Or were the landforms carved out by the slow processes of erosion? The various theories (some of which were inherited from the Greek and Roman geographers) that explain the surface features of the earth as resulting from violent catastrophes, such as earthquakes or volcanic eruptions, are included under the general term *catastrophism*. Theories based on the idea that all the processes of change observable today also operated in the past and can account for all the world's surface features are included under the general term *uniformitarianism*.

In spite of the widespread belief in the principle of catastrophism, a number of students of the earth's surface features rejected the whole idea of convulsions of nature. Of course, neither the Arabic student Avicenna nor the Chinese student Shen Kuo was known in Europe. One of the first Europeans to ridicule the idea of a universal flood was Leonardo da Vinci, who argued that running water could level off the heights of land until the earth was a perfect sphere. Bernard Palissy in France used a soil augur to observe the nature of the soil and suggested that rivers could easily wash the soil away if it were not held in place by forests. During the seventeenth century John Ray, who is famous for his pioneer work in classifying plants and animals, also argued that water running down the slopes could slowly wash away the mountains.

During the seventeenth century increasing numbers of students examined the rocks of the earth's crust and developed new ideas about the way rock structure is reflected in the forms of surface features. In 1719 John Strachey showed how landforms reflect the underlying rock structure, and his work was given support by such field observers as Johann Gottlob Lehmann, who published a study of the rocks and landforms of the Harz Mountains and the Erzgebirge in central Europe in 1756. Georg Christian Füchsel made a similar study of the Thüringerwald in Germany in 1762. In 1777 Simon Pallas published geological maps to show that many mountain ranges had granite cores. His expedition to Siberia from 1768 to 1774 brought back many observations of the relation between rock structure and landforms. Meanwhile, the Italian scholar Giovanni Arduino in 1760 offered a classification of the rocks of the earth in which he used four major categories: primitive, secondary, tertiary, and volcanic. He explained that unconsolidated alluvium might cover all four.

Furthermore, some observers developed basic ideas concerning the mechanics of river flow and valley development. Domenico Guglielmini, who died in 1710, studied the laws of river flow, and in 1786 the French scholar Louis Gabriel Comte de Buat worked out the mathematical equation to describe how the flowing water of a river can establish an equilibrium between velocity and the load of alluvium being transported. Thus, he established what later came to be known as a graded valley—one that slopes just enough to maintain the flow of water. All this was presented in 1786.

Uniformitarianism had its first great supporter during the eighteenth century in James Hutton, the Scottish geologist (Dean, 1992). Hutton provided the first comprehensive treatment of the origin of landforms by processes that can be observed today. The processes that shape the surface of the earth, he wrote, indicated a world of continuing change with no vestige of a beginning and no prospect of an end. Brilliant as his ideas were, Hutton wrote obscure prose, and it was left to the writings of John Playfair to communicate Hutton's ideas to the scholarly world. In 1802 Playfair published his book *Illustrations of the Huttonian Theory of the Earth* (Playfair, 1802/1956), which set off a major dispute between the supporters of Werner and those who began to accept the ideas of uniformitarianism (Chorley, Dunn, and Beckinsale, 1964:3–94).

It is important to note that nothing promotes progress more rapidly than presenting a hypothesis that meets with a highly critical reception. This leads scholars to make new observations and to formulate new hypotheses. In 1704 Edmund Halley wrote that "the detection of error is the first and surest step toward the discovery of truth" (Taylor, 1948:112; Gould, 1987).

Methods of Classifying Plants and Animals

No less productive of scientific progress were the efforts to find more useful methods of classifying plants and animals. By the seventeenth century the voyages of discovery had flooded Europe with descriptions of the lands and organisms of previously unknown parts of the earth and with a steadily mounting volume of collections of plants and animals. These descriptions had to be organized and classified. If the suggested order was a manifestation of a divine plan, so much the better. One of the earliest scholars to offer a new way to classify organic forms was the English student John Ray, who graduated from Cambridge University in the middle of the seventeenth century. In a work published in 1682 he outlined a way of classifying plants, and this work was followed by a suggested classification of fishes and other animals. His great work on the classification of organic life was published in 1691, *The Wisdom of God Manifested in the Works of Creation* (Cooper, 1999).

John Ray was a major influence on the Swedish botanist Carolus Linnaeus (Carl von Linné), who first suggested the systems of classification based on classes, orders, genera, and species, with increasing attention to detail. His categories of plants, published between 1735 and 1753, provided a simple and useful scheme of putting the new plant specimens in some kind of order.

Lamarck, however, was the first to show the need for a system of classification for plants and animals based on their natural characteristics (Pietro, 1988). In a paper on invertebrate animals in 1801, he first set forth the idea that animals could develop new organs and new characteristics in response to their needs and that organs so developed would be inherited. In his *Histoire naturelle des animaux sans vertèbres*, published in Paris between 1815 and 1822, he further developed the concept of the evolution of animals and of their adaptation to environment. In his lectures on zoology, he used the example of the giraffe, an animal that developed a long neck and long front legs as a result—so he believed—of stretching to browse high on the trees. Lamarck challenged the widely held dogma that all plants and animals were created in their present forms and had not changed since the creation. Lamarck formulated the first concept of evolutionary change, to which Darwin later added the mechanism of natural selection, rather than need or use, as the cause of evolution.

Can Man Be Studied Scientifically?

In this period, too, the first steps were taken toward producing a scientific study of population. As early as 1662 William Petty and John Graunt suggested the kinds of statistical studies that could be made if adequate data were available. Graunt showed that certain statistical regularities made prediction of births and deaths possible on the basis of probability. Petty is described as the father of political economy because he blazed new paths in the study of population and economy, even where statistical data were lacking (Glacken, 1967:398–399; Roncaglia, 1985). The first mortality rates were calculated for 1687–91 for the city of Breslau by the astronomer Edmund Halley (Peschel, 1865:685). Halley showed how life insurance rates could be calculated on the basis of probability.

Stimulated by these pioneer statistical studies, it was the German scholar J. P. Süssmilch who first demonstrated the existence of certain statistical regularities in population data. The sexes, he found, tended to remain more or less balanced, and birth and death rates could be forecast. Süssmilch, a Prussian clergyman, pointed to the evidence of God's planning in his book, *The Divine Ordinance Manifested in the Human Race Through Birth, Death, and Propagation* (1741). Thereafter governments undertook the collection of population data, and statistical studies based on these data were rapidly improved. However, it was a century later before Lambert Quetelet published his work *On the Social System and the Laws Which Govern It* (1848), in which he showed that numerical information concerning individuals tends to group around averages in accordance with the theory of probability.

The Influence of Environment on History and Government

The enlargement of geographic horizons during the Age of Exploration also produced much speculation regarding the influence of the natural environment on human behavior. Although there was not yet any clear separation of the world of scholarship into disciplines, nevertheless the scholars who speculated about the origin of the earth and its surface features were not the same as those who speculated about the effect of the physical earth on man. The latter were mostly historians or students of government. At the start these observers of environmental influence were only repeating the concepts set forth centuries earlier by the Greeks. They included a large amount of astrology inherited from the Greeks through Ptolemy and from the writers of the Middle Ages.

One of the earliest of these observers was Jean Bodin, a French political philosopher who lived in the sixteenth century. Bodin's major work, published in 1566, lay in the search for universal principles of law. Anarchy, he wrote 10 years later, was the supreme catastrophe, and he investigated various ways in which political order could be established and maintained. Accepting the Greek concept of climatic zones, he sought to ascertain the influence of the planets on the behavior of the earth's inhabitants. The people of the southern parts of the world, influenced by the planet Saturn, he said, were given to lives of religious contemplation. The people of the northern regions, influenced by Mars, became warlike and excellent in the use of mechanical devices. The people of the middle regions, influenced by Jupiter, were able to achieve a civilized way of living under a rule of law. Nathanael Carpenter, who published the first geographical work in English in 1625, included much material from Bodin. Carpenter helped sustain the concept of the three zones of climate and of the effect of living in these zones on the character of the people. By this time there were numerous reports on the equatorial climate that supported the idea first developed by Posidonius that temperatures were not as high near the equator as they were beyond the tropics. But these reports were not read or, if they were read, they were not believed.

During the centuries after Bodin and Carpenter, scholars continued to seek examples of the influence of climate on human character and behavior. The Abbé de Bos, writing in 1719, found that weather had a definite effect on suicide and crime rates in Paris and Rome (Glacken, 1967:556–558). Suicides are most common either just before the beginning of winter or just after winter, when the wind comes from the northeast. Most of the crimes in Rome are committed in the two hottest months of summer. He also observed that works of art are produced only in the zone between latitudes 25° and 52°N.

One of the most influential eighteenth-century writers was the French political philosopher Montesquieu.[4] One of the major themes in his work on laws had to do with the influence of climate on politics. The sterility of the ground in Attica, he wrote, resulted in the establishment of a popular form of government in Athens, whereas the fertility of the soil around Sparta was reflected in the establishment there of an aristocratic government. People develop different characteristics in cold climates than in hot ones, but by making use of proper laws the effects of climate can be minimized. K. M. Kriesel points out that a careful reading of Montesquieu's work demonstrates that he recognized the importance of other factors than climate alone—such factors as religion, the maxims of government, precedents, and customs. In any one country as some of these factors act with stronger force, the others are weakened. Kriesel, therefore, describes Montesquieu as a possibilist rather than an environmental determinist (Kriesel, 1968).

Nevertheless, Montesquieu was very persuasive in his discussion of the effect of differences of climate on behavior. His famous experiment with a sheep's tongue and the conclusions he drew from it illustrate his method of thought:

> I have observed the outermost part of a sheep's tongue, where, to the naked eye, it seems covered with papillae. On these papillae I have discerned through a microscope small hairs, or a kind of down; between the papillae were pyramids shaped towards the ends like pincers. Very likely these pyramids are the principal organ of taste.
>
> I caused the half of this tongue to be frozen, and observing it with the naked eye I found the papillae considerably diminished: even some rows of them were sunk into their sheath. The outermost part I examined with the microscope, and perceived no pyramids. In proportion as the frost went off, the papillae seemed to the naked eye to rise, and with the microscope the miliary glands began to appear.
>
> This observation confirms what I have been saying, that in cold countries the nervous glands are less expanded: they sink deeper into their sheaths, or they are sheltered from the action of external objects; consequently they have not such lively sensations.
>
> In cold countries they have very little sensibility for pleasure; in temperate countries, they have more; in warm countries, their sensibility is exquisite. As climates are distinguished by degrees of latitude, we might distinguish them also in some measure by those of sensibility. I have been at the opera in England and in Italy, where I have seen the same pieces and the same performers; and yet

---

[4]Charles Louis de Secondat, baron de la Brède et de Montesquieu. His great work was *De l' esprit des lois . . .* (Paris, 1748).

the same music produces such different effects on the two nations: one is so cold and phlegmatic, and the other so lively and enraptured, that it seems almost inconceivable [quoted from *De l' esprit des lois* in Glacken, 1967:569].

Montesquieu made a number of mistakes that can be recognized today. He had no understanding of the method of confronting theory with observations or with controlled experiments. And in spite of the existence of up-to-date geographic concepts and information in France during the eighteenth century, he followed the ancient Greeks in dividing the world into Europe, Asia, and Africa (but adding the Americas, which the ancients did not know about). Europe, he said, has a variety of climates but no extremes; Asia, on the other hand, is either very hot or very cold and has no temperate climates. In other words, he drew up generalizations about climatic conditions by continents rather than by zones of latitude. Yet he made the theory of climatic influence so plausible that these ideas persisted long after his time. Indeed, they are still to be found firmly embedded in some school curricula. The literary quality of his writing and the importance of his ideas about government gave him enormous prestige in the scholarly world.

The Beginnings of Natural History

All these efforts were new in the sense that they offered new hypotheses, or new methods of classification, or new ways to make use of mathematics. But one of the newest of the new ways of viewing the earth came with the ground-breaking work of Count Buffon.[5] Buffon was director of the Jardin du Roi, the botanical garden in Paris, between 1739 and 1788. There he had access to a vast number of specimens of plants and animals and descriptions written by travelers and explorers from all over the world. His *Histoire naturelle* was written with the aid of many collaborators. It represents one of the first works resulting from the reports of voyages of discovery in which attention was turned from the oddities and marvels to a search for regularities and for the laws governing processes of change. His approach was nonmathematical and was not based on deductive reasoning: This was a strictly inductive study aimed at finding some kind of order in the flood of new information (Roger and Williams, 1997).

Buffon accepted the idea of a divinely created earth, but he rejected the notion that the complete and final plan of the creation was in the mind of the creator from the beginning. Man, he says, is one of the animals but differs from the other animals because with his mind he can remember experiences and learn from them. Humans are commanded to conquer the earth and transform it. Wild nature, said Buffon, is ugly: Man changes the face of the earth in the process of developing a civilization. Buffon was the first to focus attention on man as an agent of change, a focus that Plato failed to develop in spite of his observations on man-made changes in Attica and that had escaped the attention of other writers since Plato, except for the Chinese philosopher Mencius.

---

[5]Georges Louis Leclerc, comte de Buffon. His great work was the *Histoire naturelle, générale et particulière*, 44 vols. (Paris, 1749–1804) (Glacken, 1960; Glacken, 1967:655–685, 720–721).

Buffon developed the idea of a cooling earth. The warmth at the earth's surface, he believed, came from the interior, for the heat received from the sun was not enough to counteract the loss of heat from the cooling earth. On the other hand, Buffon gathered evidence to support the notion that when forests are cleared the sun's heat can establish an equilibrium at the earth's surface between incoming and outgoing heat. As evidence of the beneficial results of forest clearing, he pointed out that although Quebec and Paris are both at about the same latitude, Paris, in a cleared area, is much warmer than forested Quebec.

Climate, Buffon believed, affects the people who are exposed to it. But the examples he gave showed how vague was his concept of actual distributions in the world and how completely he missed discovering the regularities of geographic patterns over the earth. He made use of the traditional regional generalizations inherited, as we have seen, from the earliest Greek writers: White people occupy Europe, black people are in Africa, yellow people are in Asia, and red people in America. In spite of the writings of Marco Polo and the reports of the Dutch voyages of the sixteenth and seventeenth centuries, Buffon did not conceive of a similarity of climate in similar continental positions. In America, he continued, there are no black people because the tropical regions are not as extremely hot as those of Africa. This he explained by pointing to the easterly winds that cross the Atlantic and sweep unobstructed across the American tropics, cooling the air and bringing heavy rainfall.

On the other hand, Buffon insisted that humans could adjust to any climate on earth. They were not compelled to react to any climate the way uncivilized native people would react. With proper clothing they could protect their skin color. With a typical European viewpoint he insisted that humans could adjust more easily to life in cold climates than in hot ones.

Buffon developed the curious notion that nature in America was relatively weak compared with the natural conditions of other continents. The forests, he insisted, were less dense, the animals smaller, the potentialities for human settlement poorer. This notion provoked strong reactions from Americans. Thomas Jefferson, who made numerous contributions to the study of geography, visited Buffon in Paris to protest such an interpretation. But he was only able to get Buffon to acknowledge his error when he had a friend in Maine ship the skeleton and hide of a bull moose to the French scholar. Jefferson's geographical study *Notes on the State of Virginia* was one of the results of this dispute (Jefferson, 1787).

Scientific Travelers

When trained scientists were included in voyages of discovery and could make their own observations, knowledge of the earth increased rapidly. The earliest of these scientific travelers was the British astronomer Edmund Halley. But Halley was interested in many things besides astronomy: His great genius consisted in making some kind of order out of complex data (Thrower, 1990). It was he who made the first mortality tables (for the city of Breslau in 1693). He was the originator of many graphic methods for showing the geographical distribution of physical features of the earth (Tooley, 1949:54–55). His maps and discussion of the trade winds of the Atlantic, published in 1686, provided the first illustration of wind directions and wind shifts.[6] On this same chart he was the first to map the equatorial westerlies of the Gulf of Guinea. In 1698–1700 he was on the first voyage undertaken for purely scientific purposes. On the basis of observations from all around the world, he prepared the first map of magnetic variations, using lines of equal variation to show the pattern on a world scale (isogonic lines). These maps were published in 1701–1702 (Thrower, 1969).

Two pioneer field observers were Johann Reinhold Forster and his son Georg F. Forster (Hoare, 1976). They accompanied Captain James Cook on his second voyage, during which he sailed far to the south in the Indian and Pacific Oceans. On many islands of the South Pacific where the expedition stopped, the Forsters carried out botanical observations and made collections. Georg Forster was the first to identify the pattern of temperatures on the eastern and western sides of continents at the same latitude and to point out the climatic similarity of western Europe and western North America. The Forsters, who derived their ideas of physical geography from Buffon and their ideas about the classification of plants from Linnaeus, had the great advantage over their predecessors in that they could make their own observations and were no longer dependent on descriptions of what other observers had seen. They were very critical of ideas about the influence of climate on man, supporting their criticism with careful observations of the Dutch in South Africa and of the Polynesians. Both had important influence on the scholars who followed them. Of special significance was the influence of Georg Forster in the early life of Alexander von Humboldt.

Among the scientific travelers of this period mention should be made of Major James Rennell, onetime surveyor-general of India (1767–77), student of ancient geography, and one of the founders of the study of oceanography (Markham, 1895). At the age of 20, in 1762, he embarked on a voyage that took him through the Strait of Malacca and on to the islands today included in Indonesia. In his journal he gives the results of his observations of the people, the climate, and the coasts of the islands. In 1764 he was appointed to carry out surveys of the East India Company's lands in Bengal, and in 1767 he succeeded to the post of surveyor-general of India. His *Atlas of Bengal* went through numerous editions between 1779 and 1788 and remained the standard work on Bengal until 1850. His professional papers include descriptions of the Ganges and Brahmaputra Rivers and an account of how the alluvium brought by these rivers has built the huge delta at the head of the Bay of Bengal. He also collected a wealth of information regarding the currents of the Indian Ocean and the Atlantic Ocean. He carried out the world's first systematic observation of ocean water, and two years after his death, in 1832, his daughter brought out the book he had almost finished, *An Investigation of the Currents of the Atlantic Ocean and of Those Which Prevail Between the Indian Ocean and the Atlantic Ocean*. This work provided the first comprehensive view of the movements of ocean water in the Atlantic (Baker, 1963).

---

[6]Halley explained the deflection of these winds toward the west by the apparent westward movement of the sun. The first explanation in terms of the earth's rotation was given by John Hadley in 1735.

Problems of Population and Food Supply

Toward the end of the eighteenth century an increasing number of scholars (like the Forsters) began to seek new answers to old questions concerning humans and their universe. Thomas Robert Malthus was one of these scholars. His father followed the ideas common in earlier generations concerning the design of the earth by divine will and concerning the perfectibility of human society as seen by Jean Jacques Rousseau, Condorcet, and others. The son took issue with the father, insisting that the construction of a truly happy society would always be hindered by the tendency of population to increase faster than the food supply. Population tends to increase geometrically, he said, whereas the food supply can only be increased arithmetically. And population always increases until it reaches the limits of subsistence, after which it is checked by war, famine, and pestilence.[7] A population close to the limits of subsistence suffers widespread misery. Malthus published his first essay on population in 1798, and in 1803 he issued a second edition, much enlarged and amplified with examples from various parts of the world. A sixth edition of his essay appeared in 1826.

Malthus's work on population and food supply is recognized as one of the brilliant achievements of this period, when new concepts were being formulated to accommodate the new knowledge about the earth and humankind. One phrase in his essay on population referred to the "struggle for existence." Decades later both Charles Darwin and Alfred Russell Wallace realized independently that this was the essence of the process of natural selection among organisms. Malthus, in seeking an explanation of the impossibility of increasing agricultural production to keep pace with population growth, became the first to formulate the economic law of diminishing returns from inputs of capital and labor.

NEW PERSPECTIVES

By the middle of the eighteenth century, the vastly increased volume of new information about the world required a search for new perspectives and new ways of organizing what had been called cosmographies. This was the time when universal geographies began to appear, replacing the older type of descriptive geography. In some cases the "new geographies" were really innovative, as when Halley made use of isogonic lines to reveal the pattern of magnetic variation in the world, or when Philippe Buache, in 1737, used lines of equal depth (isobaths) to bring out the shape of the English Channel (Dainville, 1970). But in many cases only the facts were new; the method they followed was in the tradition of Sebastian Münster.

Buache and Büsching

Among the long list of authors of world geographies in the eighteenth century, two were of special importance and had considerable impact on later writers. These were

---

[7]Glacken points out that the Italian scholar Giovanni Botero in 1588 wrote that "the population of a city, or of the whole earth, will increase to the number permitted by the food supply" (Glacken, 1967:373).

the French geographer Philippe Buache and the German philosopher and writer of geography Anton Friedrich Büsching. Buache is best known for his concept of an earth marked off into major basins bordered by continuous ranges of mountains. On the land these are drainage basins, and the mountains form the drainage divides between different river systems. The basins continue under the oceans, and here the mountains form strings of islands or submerged sand banks.[8] (LaGarde, 1985).

The persistence of his idea that drainage basins are bordered by mountains is surprising in view of the easily accessible examples in Europe of rivers that rise in one basin and flow through mountains to other basins. Perhaps Buache's concept would not have gained such wide support had it not been for the persuasive writing of the German geographer Johann Christoph Gatterer, who identified the drainage basins as natural regions and used them as the frame of organization for geographical texts. From Gatterer the idea was picked up by several authors in Great Britain. The river basin was widely used as the framework for identifying what we would now call systems of interrelated elements.

It is not so well known that Buache was the first geographer to identify the existence of a land hemisphere. He pointed out that a hemisphere with Paris at its center contains the greater part of the world's land. He presented this idea in 1746, presumably guessing that Ptolemy's *terra australis incognita* did not really exist and that Captain Cook (who would not enter the Royal Navy until 1755) would eventually prove that the land hemisphere concept was correct (Beythein, 1898; Wagner, 1922:268). He was also a keen believer in the existence of a Northwest Passage.

Another new geography was offered by the German philosopher Anton Friedrich Büsching, whose *Neue Erdbeschreibung* was published in six volumes in 1792. This was a discussion of Europe, organized in the traditional manner by political divisions. In many ways it was strictly in the Münster tradition, although the information it contained was up to date and reliable. Büsching did present two new concepts. He was the first to make use of population density as a geographic element (Peschel, 1865:xv; Büttner and Jäkel, 1982). He was also ahead of his time when he pointed out that the transportation of goods by water could free man from dependence on local resources. This was the first suggestion of the principle of economic interdependence among countries, and it was written before the invention of the steam engine. Büsching died in 1793 before he could complete the other parts of his proposed world geography. He was one of the best-known geographers of his time. His thinking exerted an influence on the early studies of Ritter. His influence on Russian geography will be discussed later.

Malte-Brun

Perhaps the most important of the universal geographies of this period was the one written by Conrad Malte-Brun.[9] Malte-Brun, a Dane born Malthe Conrad Bruun, was banished from Denmark in 1800 for liberal activities, whereupon he went to

---

[8]*Essai de géographie physique . . . (1752); Considérations géographiques et physiques sur les nouvelles découvertes au nord de la grande mer* (1753) (Baker, 1955a).

[9]*Précis de la géographie universelle . . .* (Paris, 1810–29); English translation by J. G. Percival, 3 vols. (Boston: Samuel Walker, 1847).

Paris and changed the form of his name. His universal geography was published in eight volumes between 1810 and 1829. He begins with a discussion of the history of geography and then proceeds in Volume 2 to an outline of geographic concepts, including the shape of the earth, the types of projections, and the astronomical relations. He reviews the various theories regarding the origin of the earth and the controversy between the catastrophists and the uniformitarianists, and then observes that the only way to treat physical geography in a useful way is to remain purely descriptive.[10] He believes that it is absurd to suggest that human characteristics are determined by climate. His treatment of the wind systems includes the latest observations from Cook's voyages.

In the second volume of the *Précis*, Malte-Brun devotes much space to a discussion of the land hemisphere concept. The center of the land hemisphere he places in France, to the west of Paris. In a footnote he refers to Père Chrysologue de Gy,[11] whose celestial planispheres were exhibited in Paris in 1778. One of his planispheres was centered on Paris and showed the land hemisphere. Malte-Brun makes no reference to Buache (Godlewska, 1999:89–111).

This was an influential book and one that seems to have influenced universal geographies that followed. The *Précis* was essentially a compilation, attempting to catalog all that was known about the earth. The approach was descriptive and not one larded with theoretical findings. It was orderly, well written, of a popular genre in its time, and subsequently republished until late in the 1800's.

Kant

Another scholar who contributed to the development of geographical ideas in the eighteenth century was the German philosopher Immanuel Kant (Büttner and Hoheisal, 1980; Kuehn, 2001). He lectured on a wide range of subjects at the University of Königsberg (in east Prussia, now Kaliningrad) from 1755 until 1797, seven years before his death in 1804. In 1770 he was appointed professor of logic and metaphysics at that university. In his famous *Critique of Pure Reason* (1781), he rejected the teleological idea of final causes and insisted that explanations must be sought in what is chronologically antecedent. In this view Kant was opposed to such thinkers as Linnaeus, Leibniz, Süssmilch, Büsching, and Herder, but supported and amplified the ideas of Buffon, Maupertuis, Hume, and Goethe.

Kant is also important in the history of geographical ideas because of the lectures he gave in physical geography at Königsberg between 1756 and 1797.[12]

---

[10]"Il vaut mieux revenir à la marche purement descriptive de la géographie physique, la seule méthode vraiment scientifique et instructive." Ibid., 1:495.

[11]Chrysologue de Gy (Noël André), *Théorie de la surface actuelle de la terre* (Paris, 1806).

[12]Kant never published these lectures, but several versions were published on the basis of students' notes. R. Hartshorne points out that for over a century there was uncertainty about the authenticity of the several versions but that the question was largely answered by the careful studies of Erich Adickes, who first published on Kant's ideas about physical geography in 1911 (Hartshorne, 1939:38–39). Erich Adickes is the most reliable authority on what Kant really said (Adickes, 1924–25).

Physical geography, as the term was commonly used in Kant's time, included not only the features of the earth produced by natural processes, but also the races of humankind and the changes on the face of the earth resulting from human action. Kant found that knowledge about the earth as the home of man was a necessary support for his philosophical studies, but he also found the subject inadequately covered (May, 1970). "He devoted a great deal of attention to the assembly and organization of materials from a wide variety of sources and also to the consideration of a number of specific problems—for example, the deflection of wind direction resulting from the earth's rotation" (Hartshorne, 1939:38).

It was Kant's custom in the first lecture of the series each year to comment on the place of geography among the fields of learning. He pointed out that things can be grouped or classified in two different ways for the purpose of studying them. Things that are similar because of similar origin, regardless of where or when they occur, can be grouped together in what Kant called a *logical classification*. Grouping things of diverse character and origin together because they occur at the same time or in the same place is called a *physical classification* of knowledge. A description or classification of things in terms of time is history; a description or classification of things in terms of area is geography (Adickes, 1925:2:394).

Hartshorne points out that Kant did not offer these ideas as something new. Rather, this classification of knowledge was generally accepted and seemed obvious and beyond dispute. Even by the end of the eighteenth century, there was still no great focus of attention on the definition of fields of study or the separation of learning into disciplines. Like Strabo many centuries before, Kant was simply stating his purpose, not presenting an argument. In his lectures Kant followed Büsching in organizing his materials by political units. Kant saw man and his works in intimate association with the physical surroundings, and he also recognized human action as one of the principal agencies of change on the face of the earth. But he made no distinction between human and natural processes.[13] His largest impact on the history of geography was provided by his notions of the essence of the tasks confronted and methods to be adopted rather than the acquisition and display of new data.

---

[13]J. A. May's study of Kant's concept of geography makes it clear that the distinction between geography as the study of things arranged in earth-space and history as the study of things arranged in succession or sequence is a secondary division of his whole classification of the sciences. Physics, said Kant, is a theoretical science, whereas geography "provides systematic knowledge of nature [the world that is external to man]. This implies that geography studies the relations among particular and concrete things rather than among abstract and general characteristics of things, and that it concentrates upon the differentiations rather than upon the similarities of nature." But Kant did not call such differentiations unique. Geography is an empirical science, seeking to present a "system of nature," and is a law-finding discipline. These ideas that Kant presented are analyzed in May 1970, especially pp. 147–151.

REFERENCES: CHAPTER 5

Adickes, E. 1924–1925. *Kant als Naturforscher*. 2 vols. Berlin: W. de Gruyter.

Baker, J. N. L. 1955a. "Geography and Its History." *Advancement of Science* 12:188–198.

———. 1955b. "The Geography of Bernhard Varenius." *Transactions and Papers, Institute of British Geographers* 21:51–60.

———. 1963. "Major James Rennel, 1742–1830, and His Place in the History of Geography." In *The History of Geography*. Pp. 130–157. New York: Barnes & Noble.

Berget, A. 1913. "La répartition des terres et des mers et la position du pole continental de la terre." *Revue de géographie* 7:1–36.

Beythien, H. 1898. *Eine neue Bestimmung des Pols der Landhalbkugel*. Kiel: Lipsius & Tischer.

Bowen, M. 1981. *Empiricism and Geographical Thought: From Francis Bacon to Alexander von Humboldt*. Cambridge: Cambridge University Press.

Burstyn, H. L. 1971. "Theories of Winds and Ocean Currents from the Discoveries to the End of the Seventeenth Century." *Terrae Incognitae, The Journal for the History of Discoveries* 3:7–31.

Büttner, M. 1978. "Bartholomaus Keckermann, 1572–1609." *Geographers: Biobibliographical Studies* 2:73–79.

Büttner, M. and K. Hoheisal. 1980. "Immanuel Kant, 1724–1804." *Geographers: Biobibliographical Studies* 4:55–67.

Büttner, M. and R. Jakel. 1982. "Anton Friedrich Büsching 1724–1793." *Geographers: Biobibliographical Studies* 6:7–15.

Chorley, R. J., Dunn, A. J. and Beckinsale, R. P. 1964. *A History of the Study of Landforms, or the Development of Geomorphology*. Vol. 1: *Geomorphology Before Davis*. London: Methuen.

Clark, D. H. and S. P. H. Clark. 2001. *Newton's Tyranny: The Suppressed Scientific Discoveries of Stephen Gray and John Flamsteed*. New York: W. H. Freeman and Co.

Cooper, N. S. 1999. *John Ray and His Successors*. Braintree: John Ray Trust.

Dainville, F. de. 1970. "From the Depths to the Heights: Concerning the Marine Origins of the Cartographic Expression of Terrestrial Relief by Numbers and Contour Lines." *Surveying and Mapping* 30:389–403. (Translated from the French by A. H. Robinson and M. Carlier.)

Dean, D. R. 1992. *James Hutton and the History of Geology*. Ithaca: Cornell University Press.

Dickinson, R. E. and Howarth, O. J. R. 1933. *The Making of Geography*. Oxford: Clarendon Press.

Glacken, C. J. 1960. "Count Buffon on Cultural Changes of the Physical Environment." *Annals AAG* 50:1–21.

———. 1967. *Traces on the Rhodian Shore, Nature and Culture in Western Thought from Ancient Times to the End of the Eighteenth Century*. Berkeley and Los Angeles: University of California Press.

Godlewska, A. M. C. 1999. *Geography Unbound: French Geographic Science from Cassini to Humboldt*. Chicago: University of Chicago Press.

Gould, S. J. 1987. *Time's Arrow, Time's Cycle*. Cambridge: Harvard University Press.

Hartshorne, R. 1939. *The Nature of Geography, a Critical Survey of Current Thought in the Light of the Past*. Lancaster, Pa.: Association of American Geographers.

Hoare, M. E. 1976. *The Tactless Philosopher: Johann Reinhold Forster (1729–1798)*. Melbourne, Australia: The Hawthorn Press.

Jefferson, T. 1787. *Notes on the State of Virginia*. London: John Stockdale.

Kelley, J. E. Jr. 1977. "The Oldest Portolan Chart in the New World." *Terrae Incognitae, The Journal for the History of Discoveries* 9:22–48.

Kimble, G. H. T. 1938. *Geography in the Middle Ages*. London: Methuen.

Kriesel, K. M. 1968. "Montesquieu: Possibilistic Political Geographer." *Annals AAG* 58:557–574.

Kuehn, M. 2001. *Kant; a Biography*. New York: Cambridge University Press.

LaGarde, L. 1985. "Philippe Buache, 1700–1773." *Geographers: Biobibliographical Studies* 9:21–27.

Mandeville, J. 2001. *The Book of John Mandeville*. Tempe: Arizona Center for Medieval and Renaissance Studies.

Markham, C. R. 1895. *Major James Rennell and the Rise of Modern English Geography*. New York: Macmillan and Co.

May, J. A. 1970. *Kant's Concept of Geography and Its Relation to Recent Geographical Thought*. Toronto: University of Toronto, Department of Geography, Research Paper No. 4.

Parks, G. B. 1928. *Richard Hakluyt and the English Voyages*. New York: American Geographical Society, Special Publication No. 10.

Partsch, J. 1891. "Philipp Clüver, der Begründer der historischen Länderkunde, ein Beitrag zur Geschichte der geographischen Wissenschaft." *Geographische Abhandlung* 5(2). (47 pages.)

Peschel, O. 1865. *Geschichte der Erdkunde bis auf A.v. Humboldt und Carl Ritter*. Munich: J. G. Cotta.

Pietro, C. 1988. *The Age of Lamarck: Evolutionary Theories in France, 1790–1830*. Berkeley: University of California Press.

Playfair, J. 1802. *Illustrations of the Huttonian Theory of the Earth* (reprinted New York: Dover, 1956).

Pollard, A. W., ed. 1964. *The Travels of Sir John Mandeville*. New York: Dover.

Roger, J. and L. P. Williams. 1997. *Buffon: A Life in Natural History*. Ithaca: Cornell University Press.

Roncaglia, A. 1985. *Petty: The Origins of Political Economy*. Armonk, N.Y.: M. E. Sharpe.

Sauer, C. O. 1941. "Foreword to Historical Geography." *Annals, Association of American Geographers* 31:1–24.

Taylor, E. G. R. 1948. "The English Worldmakers of the Seventeenth Century and Their Influence on the Earth Sciences." *Geographical Review* 38:109–112.

———. 1950. "The Origin of Continents and Oceans, A Seventeenth Century Controversy." *Geographical Journal* 116:193–198.

Thrower, N. J. W. 1969. "Edmund Halley as a Thematic Geo-Cartographer." *Annals AAG* 59:652–676.

———. 1990. *Standing on the Shoulders of Giants: A Longer View of Newton and Halley*. Berkeley: University of California Press.

Tooley, R. V. 1949. *Maps and Map Makers*. New York: Crown.

Tuan, Yi-Fu. 1968. *The Hydrologic Cycle and the Wisdom of God: A Theme in Geoteleology*. Toronto: University of Toronto, Department of Geography, Research Publications.

Wagner, H. 1920–1922. *Lehrbuch der Geographie*, 10th ed. Hanover: Hahnsche Buchhandlung. (Part I, 1920; Part 2, 1921; Part 3, 1922.)

Warntz, W. 1964. *Geography Now and Then, Some Notes on the History of Academic Geography in the United States*. New York: American Geographical Society, Research Series No. 25.

Warntz, W. 1981. "*Geographia Generalis* and the Earliest Development of American Academic Geography." In B. Blouet, *The Origins of Academic Geography in the United States*. Hamden, Conn.: Archon Books. 245–263.

Warntz, W. 1989. "Newton, the Newtonians and the *Geographia Generalis Varenii*." *Annals of the Association of American Geographers* 79(2):165–191.

Willmoth, F. 1997. *Flamsteed's Stars: New Perspectives on the Life and Work of the First Astronomer Royal*. Woodbridge, Suffolk: Boydell Press.

6

An End and a Beginning

Alexander von Humboldt and Carl Ritter

> *The fear of sacrificing the free enjoyment of nature, under the influence of scientific reasoning, is often associated with an apprehension that every mind may not be capable of grasping the truths of the philosophy of nature. It is certainly true that in the midst of the universal fluctuation of phenomena and vital forces—in that inextricable network of organisms by turns developed and destroyed—each step that we make in the more intimate knowledge of nature, leads us to the entrance of new labryinths; but the excitement produced by a presentiment of discovery, the vague intuition of the mysteries to be unfolded, and the multiplicity of paths before us, all tend to stimulate the exercise of thought in every stage of knowledge. The discovery of each separate law of nature leads to the establishment of some other more general law, or at least indicates to the intelligent observer its existence.*
>
> —*Alexander von Humboldt,* Cosmos

The two great masters of German geography—Alexander von Humboldt (1769–1859) and Carl Ritter (1779–1859)—loom large across the pages of the history of science (see Figs. 17 and 18). Both lived and worked in Berlin for more than 30 years, and both died in the same year. They were acquainted, but not intimately so. Never before or since have geographers enjoyed positions of such prestige, not only among scholars but also among educated people all around the world.

Many writers refer to Humboldt and Ritter as the founders of modern geography, but there are also good reasons for thinking of them as bringing the period of classical geography to an end. Using the large volume of new information resulting from the voyages of exploration, Humboldt and Ritter, each in his own way, produced massive syntheses. Although these syntheses made use of the new concepts and methods of study developed during the preceding two centuries, they nevertheless sought to present universal knowledge, just as Strabo had done and as had been attempted during the Age of Exploration of Münster, Varenius, Büsching, and others. But since 1859 the volume of recorded observations about the world and the individual's place in it has increased many thousands of times. In the nineteenth century the Age of Specialization came into being. No longer could any one scholar hope to embrace universal knowledge. The classical period had come to an end (Hartshorne, 1939:48–84).

Figure 17 Alexander von Humboldt

ALEXANDER VON HUMBOLDT

Alexander von Humboldt was born into the Prussian landowning aristocracy. His father, an officer in the Prussian Army, died when Alexander was 10 years old. He and his older brother Wilhelm were brought up by their mother, described as "a very aloof and self-contained woman who provided for the education of her sons but gave them no intimacy or warmth. The sons were expected to show her respect and follow her directions" (Kellner, 1963:6). Alexander came to dislike the cold, constrained atmosphere of his home and lavished his affection on his brother and later on his brother's children. Alexander never married (Botting, 1973; Beck and Hein, 1987).

The brothers were educated at first by tutors, from whom they received an excellent grounding in classical languages and mathematics. Alexander had little interest in science but instead decided to undertake a career in the army. This desire was opposed by his mother, who insisted that he study economics as preparation for a position in the civil service. However, events outside of his formal schooling combined with an almost insatiable curiosity about a great variety of matters led him toward a career in science. In Berlin his mathematics tutor introduced him to a group of liberals and intellectuals who gathered at the home of the Jewish philosopher Moses Mendelssohn (the grandfather of Felix Mendelssohn, the composer). Jews and gentiles joined in discussions of the social inequities of an aristocratic society and drew up plans to do something about these things. Alexander also met the physician Marcus Herz, a disciple of Immanuel Kant, who organized a series of lectures on scientific subjects, including demonstrations of scientific experiments.

Figure 18 Carl Ritter

When Alexander was ready to attend a university, he was already excited about the various aspects of the physical world. After a short time at a small university at Frankfurt-on-the-Oder, he returned to Berlin to take a course in factory management at his mother's insistence. He also used the time to increase his knowledge of Greek and even began to study botany. In 1789 he went to the University of Göttingen, where he studied physics, philology, and archaeology. In the late eighteenth century, studying a subject meant taking a course of lectures in which a scholar would impart to his students everything that was up to date in the field.

At Göttingen Alexander met Georg Forster, lately returned from his voyage around the world with Captain Cook. From Forster Humboldt became excited about the study of botany. In 1790 these two started on a hiking trip down the Rhine to the Netherlands and thence by ship to England. The notes that Humboldt took on this trip show his interest and his ability to observe carefully such diverse matters as the varying price of wool or the effect of crop rotation on soil. He had a "success experience" in asking questions about the physical earth and human use of it and in finding answers to his questions. Later, Humboldt said that his interest in geography had started with his acquaintance with Georg Forster.

Humboldt then decided to attend the School of Mines at Freiberg in Saxony, where the celebrated scholar Abraham G. Werner was lecturing. Werner was the originator of a widely supported hypothesis that all the rocks of the earth had been formed by precipitation under water and had been deposited in layers. Humboldt attended lectures in physics, chemistry, geology, and mining. In 1792 he was appointed to an administrative post, first as inspector and later as director of mines in the Prussian state of Franconia. But his active mind was always formulating new questions about almost everything that caught his attention. He studied the effect of different rocks on magnetic declination. Underground in the mines, he carried out experiments with subterranean plants he found growing there. His first scientific paper was on the results of these studies, published in 1793 (Humboldt, 1793). He also established a school for the miners and in various ways attempted to improve their living conditions. When he heard of experiments carried out by the Italian scientist Luigi Galvani regarding the electrical and chemical stimulation of muscles, he undertook some experiments of his own, as a result of which he came very close to discovering how to make an electric battery. There seemed almost no limit to the range of his curiosity. He also wanted to travel and see for himself what different parts of the world were like. When he visited Bavaria, Austria, Switzerland, and Italy, he observed the rock structure of the Alps and tested some of the ideas of the Swiss scholar Horace Bénédict de Saussure, who thought the deep Alpine valleys had been cut by the rush of water in the receding flood.

In 1796 Humboldt's mother died, and he came into possession of a small fortune. His part of the family inheritance was an estate on the east bank of the Oder River, known as Ringenwald. Income from this estate freed Humboldt from the necessity of earning a living. It paid for his travels in America and for the expensive publication of his many reports on these travels. In 1797 he resigned his government position and began to plan for traveling.

Humboldt's preparations for studies in the field were unprecedented. In Paris he gathered together an amazing variety of instruments and learned how to make use of them:

> He had been provided with an eight-inch Hadley sextant by Ramsden with a silver circle graduated at twenty-second intervals and a two-inch Troughton sextant, a kind of pocket edition which he called his snuffbox sextant. It was extremely accurate and very convenient to carry in difficult terrain. His barometers and thermometers had been standardized before his departure with those of the Paris observatory. The longitude determinations were made with a Dollond telescope and a chronometer by Berthoud whose rate of variation had been carefully checked. Three different kinds of electrometer, provided with pith spheres, straws, and gold-leaf, allowed him to observe atmospheric electricity. He also possessed a Dollond balance for the measurements of the specific gravity of sea water, an eudometer for the analysis of atmospheric gases, a Leyden jar and the necessary chemicals and glass bottles as well as a cyanometer designed by Saussure. This was an instrument by which the blueness of the sky could be determined through comparison with prepared gradations of blue colours and correlated with the hygrometrically determined humidity. The magnetic measurements were carried out with a Borda magnetometer, a rather cumbersome instrument (Kellner, 1963:62).

Before leaving Paris Humboldt learned from Pierre Simon Laplace how to use the aneroid barometer to determine elevations above sea level.

Several opportunities to join expeditions that were going overseas did not work out. One was an expedition to Egypt that was called off when Napoleon occupied that country. Another was a trip to the Pacific, following Captain Cook's steps. In 1798 Humboldt and a French botanist named Aimé Bonpland decided to go to Marseilles and there to take passage on a ship for Algiers, from which place they

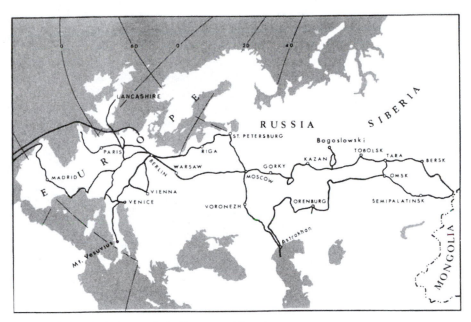

Figure 19 Humboldt's travels in Europe and Russia

intended to travel overland to Egypt. Unfortunately—or fortunately, we might say—these plans also fell through when the ship was wrecked off the coast of Portugal before it ever reached Marseilles. Humboldt and Bonpland then reasoned they might be more successful in getting passage on a ship from a Spanish port, so they set out overland for the city of Madrid, where all such passages would have to be arranged (Fig. 19). On the way to Madrid, Humboldt made daily observations of temperature and altitude. He was the first to make an accurate measurement of the elevation of the Spanish Meseta.

In Madrid Humboldt's position in the Prussian aristocracy gave him access to the ruling aristocracy of Spain. He made a good enough impression on the Spanish chief minister that he and Bonpland were granted permission to visit the Spanish colonies in America—the first such permission granted to any non-Spanish Europeans since the expedition of C. M. de La Condamine to measure the arc of the meridian along the equator in 1735. Humboldt and Bonpland sailed in June 1799.

Humboldt's American Travels

Humboldt's travels in the "equinoctial regions of the new continent" began at Cumaná in Venezuela (Fig. 20). First, the two men went to Caracas and began exploring this long-settled part of the country. One of the first places they investigated was the Basin of Valencia in the midst of which is the Lake of Valencia, some 50 miles southwest of the capital. Humboldt noted that at one time the lake was much deeper and had an outlet to a tributary of the Orinoco but that in 1799 the lake had no outlet. Crops were being grown on the flat lakebed soils from which the lake waters had receded. Why should this event have taken place? The connection between the

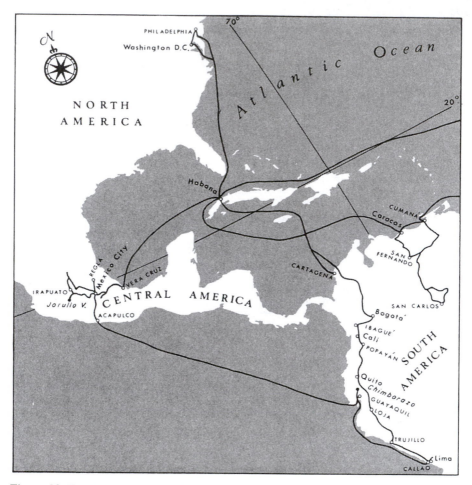

Figure 20 Humboldt's travels in the Americas

removal of forests and the drying up of rivers had been presented by Buffon and others, but Humboldt was the first to test this theory by confronting it with observed facts in a particular place. Here is what he had to say about the Lake of Valencia:

Felling the trees which cover the sides of mountains, provokes in every climate two disasters for future generations: a want of fuel and a scarcity of water. Trees are surrounded by a permanently cool and moist atmosphere due to the evaporation of water vapor from the leaves and their radiation in a cloudless sky. They have an effect on the incidence of springs, not as was long believed by a peculiar attraction for the atmospheric vapor but because they shelter the soil from the direct action of the sun and thereby lessen the evaporation of the rainwater. When forests are destroyed, as they are everywhere in America by the hands of European planters, the springs are reduced in volume or dry up entirely. The river beds, now dry during part of the year, are transformed into torrents whenever there is heavy rainfall in the mountains. Turf and moss disappear with the brushwood from the sides of the hills; the rainwater rushing down no longer meets with any obstructions. Instead of slowly raising the level of the rivers by progressive infiltration, it cuts furrows in the ground, carries down the loosened soil, and produces those sudden inundations which devastate the country. It follows that the destruction of the forests, the lack of springs, and the existence of torrents are closely connected phenomena (Humboldt, 1814–25, Williams translation, 1825:4, 143).

Around the Basin of Valencia, Humboldt observed that the once continuous cover of tropical forest had been entirely removed and the lands were used for agriculture. The Lake of Valencia became a famous example of the application of a concept formulated by earlier writers but without carefully recorded direct observations to support it. Curiously, the idea that forests cause an increase of rainfall persists.

During the year 1800 Humboldt and Bonpland carried out what must be one of the great exploratory efforts in man's continuing urge to see beyond the horizon. They mapped some 1725 miles of the Orinoco River, mostly through uninhabited forests. In small boats and canoes they paddled upstream from the junction of the Orinoco and the Apure. Many years earlier de La Condamine had reported a story told by a Jesuit missionary—Father Manuel Ramón—that the water of the upper Orinoco split into two channels, one of which spilled over to reach a headwater of the Rio Negro and the Amazon. This was the Rio Casiquiare. But Philippe Buache, in compiling maps and reports on various parts of the world in accordance with his theory of continuous mountain chains, rejected de La Condamine's report. He showed a range of mountains along the divide between the Orinoco and the Amazon drainage basins. In 1800 Humboldt surveyed the Casiquiare and confirmed Father Ramón's observation of the bifurcation of the Orinoco. Modern geomorphologists recognize this as an example of river capture that is currently ongoing, in which, over long periods of time, the upper Orinoco will be cut off to become a part of the Amazon drainage. The Orinoco, we would now say, will be beheaded.

The trip up the river and along the Casiquiare imposed severe hardships. The travelers had to subsist largely on bananas and fish and were constantly exposed to the bites of clouds of mosquitoes, ants, and other insects as well as to poisonous snakes, man-eating fishes, and crocodiles. Almost everyone came down with fever, but Humboldt himself seemed immune and retained the necessary vigor to undertake whatever travel was necessary for his observations. With his instruments he established the exact latitudes of places and came very close to correct longitudes. He collected thousands of plant and rock specimens, all of which were transported back to Caracas and then to Cuba. Among his specimens were plants from which the poison curare is extracted. This kind of poison was first reported by Sir Walter Raleigh, but Humboldt brought back to Europe the first specimen. In November 1800 the two men returned to Cumaná and sailed for Cuba.

In 1801 Humboldt and Bonpland arrived in the Colombian port of Cartagena and from there began their exploration of the Andes of Colombia, Ecuador, and Peru. With altitudes established for the first time by the aneroid barometer, with temperatures actually recorded by the use of the thermometers, and with the exact location of every observation fixed by latitude and longitude, Humboldt was able to give the first scientific description of the relations of altitude, air temperature, vegetation, and agriculture in tropical mountains. His description of the vertical zones of the northern Andes is a classic. He also examined the numerous volcanoes of Ecuador, descending again and again into active craters for the purpose of collecting the gases emanating from within the earth. Looking closely at the rocks of the Andes, he decided that Werner was quite wrong about the origin of rocks and that granite and gneiss and other crystalline rocks were of volcanic origin.

Humboldt climbed most of the volcanoes of Ecuador. After the expedition of de La Condamine, it was believed that Mount Chimborazo was the world's highest mountain. Attempting to reach its summit, Humboldt and Bonpland, on June 9, 1802, reached an altitude of 19,286 feet, which was the highest altitude that had been reached by man up to that time. This remained the record for 29 years until in 1831 Humboldt's protégé, Joseph Boussingault, reached an altitude of 19,698 feet on the same mountain. (Chimborazo, 20,561 feet high, was finally conquered by the British mountaineer Edward Whymper in 1880.) Among the high peaks, Humboldt was able to observe and report on the effect of altitude on human beings and to describe the symptoms of mountain sickness, or *soroche*. He explained the feeling of dizziness as resulting from low air pressure. (It is now known to be due to lack of oxygen.)

Humboldt and Bonpland finally reached Lima. Here Humboldt was able to observe the transit of the planet Mercury across the sun. This gave him an exact measurement of the longitude of Lima and made it possible to check his chronometer, which proved to be quite accurate. On the Peruvian coast he investigated the chemical properties of guano, or bird droppings. He sent some samples back to Europe and as a result started the export of guano as a fertilizer. On a sea voyage from Callao in Peru to Guayaquil in Ecuador, he measured the temperature of the ocean water and for the first time described the movement of ocean water, including the upwelling of cold water from below. He named this the Peruvian Current; he always objected to its being called the Humboldt Current because, he said, he had not discovered the current but had only measured its temperature and velocity. Modern oceanographers have agreed to name all currents by geographical names. Therefore, it is officially known today as the Peru Current.

In March 1803 Humboldt and Bonpland sailed from Guayaquil to the Mexican port of Acapulco. The Viceroyalty of New Spain, as Mexico was then called, was at the peak of its prosperity owing to the relaxation of restrictions on trade, to the investment of new capital in mining, and to the presence of an unusually capable group of administrators and ecclesiastical leaders. In 1794 New Spain was the first Latin American country to take a census of population. Humboldt updated the population figures to 1803 by consulting with the parish priests. He also found a rich collection of statistical data on production and trade. Traveling throughout the country, he continued to climb the mountains, measure altitudes, fix locations by latitude and longitude, and investigate the many questions about human-land relations that occurred to his imaginative mind.

In 1804 the travelers sailed to Havana, Cuba. Humboldt now faced a problem that scientific travelers have always faced—how to avoid the loss of notes and specimens painfully collected in the field. He and Bonpland had accumulated a large number of boxes containing notes written on their travels and specimens of plants and rocks of inestimable value. He dispatched the whole lot to Europe by different ships, some of which were destined for Paris and others for London. Almost all his notes and drawings were made in duplicate, which was amply justified when some of the shipments failed to reach their destinations.

Humboldt and Bonpland's visit to the United States was a memorable occasion. They reached Philadelphia in May 1804. There they visited the American Philosophical Society and then started for Washington by way of Baltimore. From June 1 to 13 they were in Washington, where Humboldt had several meetings with Thomas Jefferson, whose interest in geography has already been noted. Humboldt and Jefferson became close friends, and the great German scientist was given access to the White House without special invitation. Humboldt was greatly moved by the liberal ideas so eloquently expressed by the author of the Declaration of Independence, with whom he was thoroughly in accord. He and Bonpland finally set sail from Philadelphia for the return voyage to Bordeaux on June 30, 1804.[1]

In Paris

First, Humboldt returned to Berlin, but especially after the defeat of the Prussians by Napoleon in the battle of Jena in 1806, he found himself isolated from the world of science and scholarship. After a brief visit to Italy to observe an eruption of Vesuvius, he went to Paris on a diplomatic mission but remained there for the next 19 years (Godlewska, 1999).

It was in Paris that Humboldt wrote and published the 30 volumes in which the results of his American field studies were presented. In the French capital he was able to secure assistance from other scholars in putting his 60,000 plant specimens in order—specimens that included many species and genera never before known to Europeans. And, too, in Paris he could work with skilled publishers and engravers. The 30 volumes are included under the general title: *Voyage aux régions équinoxiales du Nouveau Continent* (Humboldt, 1805–1834).[2]

---

[1]For a detailed account of Humboldt's stay in the United States, see Herman R. Friis in Schultze, 1959:142–195. See also Friis, 1963.

[2]The contents of the 30 volumes as originally published are as follows: 1–2 *Plantes équinoxiales* . . . , ed. A. Bonpland (Paris: Levrault et Schoell, 1808–1809) (143 plates). 3–4 *Monographie des mélastomacées* . . . , ed. A. Bonpland (Paris: Librairie grecque-latine-allemande, 1816–23) (120 plates). 5 *Monographie des mimoses et autres plantes légumineuses*, ed. C. S. Kunth (Paris: N. Maze, 1819–24) (60 plates). 6–7 *Révision des graminées* . . . , introduction by C. S. Kunth (Paris: Gide fils, 1829–34) (220 plates). 8–14 *Nova genera et species plantarum* . . . , with A. Bonpland, ed. C. S. Kunth (Paris: Librairie Schoell, 1815–25) (700 plates). 15–16 *Vues des Cordillères et monuments des peuples indigènes de l'Amérique* (Paris: F. Schoell, 1810) (63 plates). 17 *Atlas géographique et physique du Nouveau Continent* . . . (Paris: Dufour, 1814). (32 maps; with a supplement of an additional 7 maps published later). 18 *Examen critique de l'histoire de la géographie du Nouveau Continent et des progrès de l'astronomie nautique aux 15e et 16e siècles* (Paris: Gide, 1814–34.) 19 *Atlas géographique et physique du Royaume de la Nouvelle Espagne* (Paris: F. Schoell, 1811) (20 maps). 20 *Tableau physique des Andes et pays voisins (Géographie des plantes equinoxiales)* (Paris: F. Schoell, 1805). 21–22 *Recueil d'observations astronomiques, d'opérations trigonometriques, et de measures*

The *Relation historique* (Volumes 28–30, of which the fourth volume was never published) made an enormous impact on the scholarly world. It was translated into many European languages, appearing in English in 1825 and in German in 1859–60 (London: trans. H. M. Williams, 1825; Berlin: trans. H. Hauff, 1859–60). In his *Ansichten der Natur* (*Views of Nature*) (Humboldt, 1808) he declared that his purpose was "to win the attention of educated but non-scientific readers for the fascination of the discovery of scientific truths" (Kellner, 1963:75). Later Charles Darwin said that he had read and reread this account of scientific travels and that it had changed the course of his life. There can be no doubt that these volumes stimulated numerous field studies in different parts of the world. Actually, the *Relation historique* (or the *Personal Narrative*, as it was translated into English) dealt with Humboldt's own experiences and hardships very briefly but devoted most of its pages to a sober report on scientific problems that had been investigated and on the results achieved. Yet for a world emerging from the first shock produced by the impact of the discoveries, Humboldt's books were like a fresh breeze because they were filled not only with the excitement of travel in strange places, but also with the reports of careful scientific investigation and the seeking of answers to questions about the interconnections among the phenomena grouped together in rich diversity on the face of the earth. As early as 1805 (Volume 27), he presented a synthesis of his detailed findings as a basis for the study of plant geography.

Another part of his great work that was widely influential was the *Essai politique sur le Royaume de la Nouvelle Espagne* (Volumes 25–26). This was one of the world's first regional economic geographies, dealing with the resources and products of a country in relation to population and political conditions. Humboldt was much impressed with the far greater prosperity he found in New Spain in comparison with the countries of northern South America, and he was curious about the reasons for the difference. His interpretation was based on the theory that the only proper way to increase the general prosperity of a country was to make more effective use of natural resources, of which Mexico seemed to have an abundance. He supported his explanations with a wealth of statistical data he found in New Spain, organized and enriched on the basis of his own observations. One of the

---

barométriques, faites pendant le cours d'un voyage aux régions équinoxiales du Nouveau Continent, ed. J. Oltmanns (Paris: Schoell, Treuttel & Würtz, 1808–1820). 23–24 Recueil d'observations de zoologie et d'anatomie comparée faites dans l'Océan Atlantique, dans l'intérieur du Nouveau Continent, et dans la Mer du Sud pendant les années 1799–1803, in collaboration with Cuvier, Latreille et Valenciennes (Paris: Schoell & Dufour, 1805–1833) (54 plates). 25–26 Essai politique sur le Royaume de la Nouvelle Espagne (Paris: F. Schoell, 1811). (In later editions with a supplement: Essai politique sur l'île de Cuba.) 27 Essai sur la géographie des plantes, accompagné d'un tableau physique des régions équinoxiales . . . (Paris: F. Schoell, 1805) (one large, folded plate). 28–30 Relation historique du voyage aux régions équinoxiales du Nouveau Continent, faites en 1799–1804 par A. de Humboldt et A. Bonpland (Paris: F. Schoell, 1814–25). (The account ends with the first part of the travels in Peru in 1801; a fourth volume was planned but never published.) (All 30 volumes have been republished in facsimile by Theatrum Orbis Terrarum, Amsterdam, 1973.)

numerous digressions that interrupts the main theme of his work is the suggestion that a canal should be dug across the Isthmus of Central America and that the best place to do this would be in Panama.

In later editions of the *Essai politique* (after 1826) he included a supplement dealing with the island of Cuba (*Essai politique sur l'île de Cuba*). In this short essay, he deplored the institution of slavery and outlined a procedure whereby slavery could be eliminated from Cuba without serious disruption of the economy.

During his stay in Paris, Humboldt enjoyed many stimulating meetings with the numerous scholars concentrated there. He became a close personal friend of the French physicist François Arago, pioneer in the study of electromagnetism and in the wave theory of light. Humboldt enjoyed universal acclaim and was generally recognized as second only to Napoleon among famous Europeans. People came to visit him from all over the world, including the future leader of the independence movement in northern South America, Simón Bolívar, then in exile in Spain. Humboldt encouraged and actually aided many young scientists, including Louis Agassiz (the Swiss scholar who developed the hypothesis of universal glaciation and later lectured at Harvard), Justus von Liebig (the German biochemist), Joseph Boussingault (the French geologist), and many others.

In Berlin

In 1827 Humboldt moved back to Berlin. His own personal fortune had been exhausted by the cost of his travels and, especially, by the cost of printing his books. When he was offered a position as chamberlain in the court of the Prussian king, which included a steady income, he accepted. In 1829, at the invitation of the Russian czar, he went to Saint Petersburg and then on by horse and carriage into Siberia as far as the borders of China (Fig. 19), He visited the shores of the Caspian Sea. The whole trip was a triumphal tour, for as his carriage approached a village or town the inhabitants turned out to line the road and give their distinguished visitor an ovation.

On this trip Humboldt was impressed by his observations of temperature. He could see clearly that temperatures varied at the same latitude in accordance with distance from the ocean. On his return to Saint Petersburg, he urged the czar to set up a network of weather stations at which weather data could be recorded regularly and in accordance with standard procedures to make the results comparable. The czar promised to do this; by 1835 the Russian network of recording stations extended all the way from Saint Petersburg to an island off the Alaska mainland. From these stations Humboldt later received the data that permitted the construction of the first world map of average temperatures (Fig. 21). Following the example of Halley and Buache, who had used lines to connect points of equal value, Humboldt for the first time used lines to connect points of equal temperature (isotherms) (Robinson and Wallis, 1967). Noting how the isotherms departed from the lines of latitude, Humboldt developed the concept of continentality: that continental climates are colder in winter and warmer in summer than places near the oceans at the same latitude.

On his Siberian trip Humboldt also observed and described the permanently frozen soil, which is now called *permafrost*. He saw the remains of a mastodon that had been frozen in the ground and preserved in this way. But he did not see any evidence of

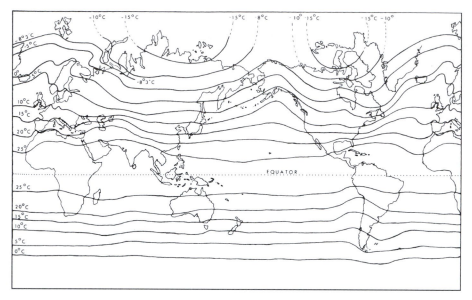

Figure 21 Isotherms of average annual temperature, 1845, according to Humboldt

glaciation, and, therefore, he remained skeptical of the idea of a universal Ice Age then being advanced by the Swiss scholar Louis Agassiz. Humboldt was right— large parts of Siberia were not covered by ice sheets during the Ice Age.

The Kosmos

In the winter of 1827–28 Humboldt offered a series of public lectures at the Royal Academy of Sciences in Berlin. His lectures had drawn such large and enthusiastic audiences that he had to repeat them in a larger room. In these lectures he not only made science interesting to the educated layperson, but he also made it acceptable to the religious leaders of the time. Religions, he insisted, offer three different things to humankind: a lofty moral idealism, which is common to all religions; a geological dream regarding the origin of the earth; and a legend concerning the origins of the religion. He always emphasized the unity and coherence of nature. Although he made clear the wonder of the universe, some of his admirers complained that nowhere in his lectures or in his books did he mention God (Kellner, 1963).

For nearly 50 years Humboldt had been forming in his mind the plan of a book, or a series of books, that would

> give a scientifically accurate picture of the structure of the universe which would attract the general interest of the educated public and communicate some of the excitement of scientific study to the non-scientific mind. Since he saw nature as a whole and man as a part of nature, and therefore all intellectual and artistic activities as having a share in natural history, he linked his main theme to an exposition of its development through the centuries and to the history of landscape painting and descriptive poetry of nature. . . . The book, when it was finally completed, followed fairly faithfully the scheme of the course of lectures he had given in 1828 (Kellner, 1963:199).

He wrote the book, which he called the *Kosmos*, during the last years of his life. The first volume was published in 1845, when he was 76; the fifth volume, published posthumously in 1862, was based on the copious notes he left.[3] Written in superb literary style, the *Kosmos* became the most prestigious scientific work ever produced up to that time. It was an immediate success: The first edition of Volume 1 sold out in two months, and soon the work had been translated into many languages, including almost all the European languages.

The *Kosmos* put together in one unified work all the various interests and discoveries of Humboldt's lifetime. The first volume makes a general presentation of the whole picture of the universe. The second volume starts with a discussion of the portrayal of nature through the ages by landscape painters and by poets and then continues with a history of man's effort to discover and describe the earth since the time of the ancient Egyptians. Humboldt's enormous erudition becomes especially clear in this second volume. The third volume deals with the laws of celestial space, which we would call astronomy. The fourth volume deals with the earth, not only with geophysics but also with man. Here is what Humboldt had to say about man as a part of nature near the end of the first volume:

> The general picture of nature which I have endeavoured to delineate, would be incomplete, if I did not venture to trace a few of the most marked features of the human race, considered with reference to physical gradations—to the geographical distribution of contemporaneous types, to the influence exercised upon man by the forces of nature, and the reciprocal, although weaker, action which he in his turn exercises on these natural forces. Dependent, although in a lesser degree than plants and animals, on the soil, and on the meteorological processes of the atmosphere with which he is surrounded—escaping more readily from the control of natural forces, by activity of mind, and the advance of intellectual cultivation, no less than by his wonderful capacity of adapting himself to all climates—man everywhere becomes most essentially associated with terrestrial life (Humboldt:1: 378: Otté translation:1: 360–361).

Humboldt believed that all the races of man had a common origin and that no race was necessarily inferior to the others: All races, he insists, are equally destined for freedom, individually or in groups.

Humboldt emphasized again and again the need for careful observation of nature in the field and for the careful and precise measurement of observations. Yet he was always seeking to formulate general concepts, or what we would now call abstract models or theory. However, he thought that observation had to come first. In Volume 1 he wrote:

> We are still very far from the time when it will be possible to reduce, by the operation of thought, all that we perceive by the senses to the unity of rational principle. On the other hand the exposition of mutually connected facts does not

---

[3]*Kosmos, Entwurf einer physischen Weltbeschreibung*, 5 vols. (Stuttgart and Tübingen: J. G. Cotta Verlag, Vol. 1, 1845; Vol. 2, 1847; Vol. 3, 1850; Vol. 4, 1858; Vol. 5, 1862). The best English translation is by E. C. Otté, *Cosmos, Sketch of a Physical Description of the Universe* (London: H. G. Bohn, 1849–58). Also: *Cosmos*, vols. 1 and 2. (Baltimore: Johns Hopkins University Press, 1997).

exclude the classification of phenomena according to their rational connections, the generalizing of many specialties in the great mass of observations, or the attempt to discover laws (Humboldt:1: 67–68; Otté translation:1: 58).

During Humboldt's long life, the need for closer attention to the definition of fields of special study became important. By the time of Immanuel Kant, as we have seen, the course of lectures on physical geography started with a definition of the field. It was quite clear that history dealt with problems of chronology and that geography was concerned with problems of areal association and distribution. In addition, Kant's logical classification of knowledge made room for the specialist in the study of particular processes without reference to time or space. That this was a generally accepted division of the world of scholarship and not an invention of Kant is made clear by Humboldt's earliest studies of the subterranean plants in the mines at Freiberg. In the introduction to this monograph (1793:ix–x), Humboldt pointed out that he was not studying plants as such, but rather the plants in relation to their surroundings. Humboldt reprinted his earlier statement in a footnote (in Latin) in the *Kosmos* (1, 486–487). Hartshorne suggests that Humboldt was probably presenting ideas he had gained from his teacher, A. G. Werner (Hartshorne, 1958:100). In the introduction to the *Kosmos*, Humboldt points out that

> the terms physiology, physics, natural history, geology, and geography were commonly used long before clear ideas were entertained of the diversity of objects embraced by these sciences, and consequently of their reciprocal limitation (Humboldt:1: 51; Otté translation:1: 39).

Geography, which Humboldt called *Erdbeschreibung* (earth description), deals with the variety of different kinds of interrelated phenomena that exist together in areas or segments of earth space. This was essentially the same idea as that suggested by Kant, although there is no evidence that Humboldt was quoting Kant.

CARL RITTER

Carl Ritter was born in 1779, 10 years after Humboldt. His father was a physician, and when he died the widow lacked any means of support for her family of five. At this time Carl, the youngest of the family, was only five years old. By the greatest good fortune, in 1784 a German schoolmaster named Christian G. Salzmann was founding a new school to experiment with the radical innovations in educational methods then being proposed. He wanted a child who had never been exposed to the traditional methods and could be trained from the beginning by new pedagogical procedures. The child he selected for this purpose was Carl Ritter.

The Education of a Geographer

In the late eighteenth century in Germany and France the traditional educational methods were being vigorously challenged. It had long been customary to expect children to memorize selected passages from books, often in Latin or Greek, and then repeat them out loud. To repeat the words correctly was enough whether or not the passages were understood. In 1762 Jean Jacques Rousseau in his novel *Emile* outlined a new educational procedure that would end rote learning and would encourage a child to develop inborn potentialities. The Swiss educator Johann Pestalozzi further developed the ideas of Rousseau, insisting on the principles that clear thinking must be based on the careful observation of things and that words could have no meaning unless they were matched with perceptions. This helped establish *Heimatkunde*—the study of the neighborhood—as a fundamental of geography teaching. The Pestalozzi method was reinforced by the notions of Friedrich Fröbel, another of the large-minded pedagogues of the time. And so it came to be that geographical study began at home, and then with a knowledge of facts and principles, extended beyond one's familiar environment. Salzmann was enthusiastic about these new suggestions and established his school at Schnepfenthal in the Thuringerwald to experiment with them.

The teacher who was selected to supervise the young Carl Ritter was a geographer named J. C. F. GutsMuths, whose own special field of interest was in the observation of natural features and who had already made some contributions to the teaching of geography by the new methods (Hartshorne, 1939:50–51). At an early age Ritter was able to observe the close involvement of humans with the natural features of their surroundings. He was encouraged by his teachers to formulate for himself the concept of the unity of man and nature; from the richly varied landscapes of this region of hills and low mountains, he derived the idea of unity in diversity, which became a basic theme of his mature writings. He could only account for such unity as evidence of God's divine plan. Without Rousseau, Pestalozzi, Salzmann, and GutsMuths, this kind of educational experience would not have been available, but because of these men, Ritter received the best possible basic training for a career as a teacher of geography.

At the age of 16, when he was ready to go to a university, Ritter was again fortunate in finding financial support. A wealthy banker, Bethmann Hollweg, agreed to pay for Ritter's university expenses if Ritter, in turn, would agree to tutor the two Hollweg sons. At the University of Halle, with the commitment to become a teacher in mind, he took work with the famous educator Professor Niermayer. Ritter continued his own studies after he started tutoring the Hollweg children at their home in Frankfurt-on-the-Oder. He undertook to learn Latin and Greek and to read widely in geography and history. With his pupils he made frequent field trips around Frankfurt, where in the process of teaching field observation he increased his own competence as an observer. He became a master of the art of landscape sketching, which, even after the age of photography, remains an effective way to preserve field observations for future study. Later he extended the range of his field trips to Switzerland and Italy, during the course of which he met many of the leading scholars of that period. In 1807 he met Humboldt and was deeply impressed by him. In 1811 he published a two-volume textbook on the geography of Europe, making use of previously prepared maps of the geographic features of that continent.[4]

---

[4]These maps, prepared between 1804 and 1806 (*Sechs Karten von Europa mit Erklärendem Text* [Schnepfenthal]), were among the first to make use of hypsometric symbols to describe the shapes of surface features. The earliest such map was a world map by A. Zeune, 1804. See the discussion of the use of the contour method in Joseph Szaflarski, 1959: "A Map of the Tatra Mountains Drawn by George Wahlenberg in 1813 as a Prototype of the Contour-Line Map," *Geografiska Annaler* 41:74–82.

After one of his pupils died, Ritter went to the University of Göttingen to accompany the other Hollweg son. At that university between 1813 and 1816 he studied geography, history, pedagogy, physics, chemistry, mineralogy, and botany.

Ritter as a Teacher and Lecturer

Unlike Humboldt, Ritter held several academic positions during his life. In 1819 he became professor of history at Frankfurt. He held this appointment for only one year, during which time he married. In 1820 he was appointed to the first chair of geography established in Germany—at the University of Berlin; he continued to offer courses of lectures at this university until his death in 1859. During this time he held other positions. He lectured on military history at the Prussian military school and became the director of studies for the Corps of Cadets. He was appointed a member of a scientific commission on geography and history, and he founded the *Gesellschaft für Erdkunde zu Berlin* (the Berlin Geographical Society). He was the private tutor for Prince Albert of Prussia. In addition to these numerous undertakings, he continued to lead field trips each summer to various parts of Europe.

Ritter was a brilliant and influential lecturer. In interesting contrast to his obscure style of writing, his lectures were clear and well organized. He was a master of the art of using the blackboard to illustrate his ideas. After his first two or three lectures at the university, when he found his lecture room empty or with only a few students, his lectures became immensely popular, and his lecture room frequently included 300 to 400 students. Many were the young students whose enthusiasm for geography and for Ritter's interpretation of it was kindled by attendance at his lectures and who went forth to spread the word in other countries. Among his best known disciples were Elisée Reclus of France and Arnold Guyot, who became professor of physical geography and geology at the College of New Jersey (later Princeton) in 1854. In Russia Semenov-Tian-Shanski helped translate some of Ritter's works. In Cracow W. Pol spread the ideas of Ritter, and in Hungary Hunyaldy did likewise.

Ritter's public lectures were also highly successful. Some of his basic ideas concerning the influence of the earth's major features on the course of history were developed in lectures before the Royal Academy of Sciences in Berlin.[5]

Ritter's Geographical Ideas

Ritter emphasized repeatedly that he was teaching a "new scientific geography," in contrast to the traditional "lifeless summary of facts about countries and cities, mingled with all sorts of scientific incongruities" (Bögekamp, 1863:37). His scientific geography was based on the concept of unity in diversity, which he had developed for himself at an early age. His purpose was not just to make an inventory of the

---

[5]Between 1826 and 1850 Ritter gave five lectures of great importance: "The Geographic Position and Horizontal Extension of the Continents," 1826; "Remarks on Form and Numbers as Auxillary in Representing Relations of Geographical Spaces," 1828; "The Historical Element in Geographical Science," 1833; "Nature and History as the Factors of Natural History, or Remarks on the Resources of the Earth," 1836; "The External Features of the Earth in Their Influence on the Course of History." 1850 (Gage, 1863).

things that occupy segments of earth space. Rather, he sought to understand the inter-connections, the causal interrelations, that make the areal associations cohesive. Again and again he used the German word *zusammenhang* (literally, hanging together) to refer to this quality of cohesion among diverse things.

To refer to the new scientific geography he made use of the word *Erdkunde*, or earth science. This he preferred to Humboldt's term *Erdbeschreibung*, or earth description. *Erdkunde* is a German synonym for the Greek word *geography*. There was never any uncertainty in Ritter's mind that he was studying the earth as the home (*wohnort*) of man and, therefore, that he was dealing with the earth's surface. Later, as we will see, some German geographers took the word *Erdkunde* literally and focused their work on the whole body of the earth, not just its surface.

Ritter insisted that geography should be empirical, in the sense that the student should progress from observation to observation in the search for general laws and not from preconceived opinion, to hypothesis, to observation. The student should ask the earth itself for its laws (Ritter, 1822:1:23). By avoiding preconceptions and making his own empirical observations of the surface features of Europe, Ritter was among the first to point out the error in Buache's concept of continuous mountain chains. Ritter urged a more exact scientific foundation for comparative geographical analysis relating to political units and to natural regions. He had a critical attitude toward all attempts to practice a one-sided investigation of physical phenomena: He sought the inclusion of his specialized areas of study—regional geography, ethnology, and history, thereby making of geography a truly comparative under-taking (Linke, 1981).

Ritter's search for unity in diversity led him to make use of the regional approach to geography rather than the systematic study of individual features. Yet he realized the importance of systematic studies and acknowledged his indebted-ness to Humboldt, whose general studies made Ritter's special studies of regions possible. For his larger regional units Ritter made use of the traditional continents and proceeded to formulate generalizations concerning the continents and their human inhabitants. The continued use of continents as major regional entities not only for the teaching of geography but also for the formulation of concepts may have retarded the progress of geographical scholarship. Unfortunately, Ritter's identification of races by skin color and his identification of color by continents has produced only obscurity (Europe for white people; Africa for black people; Asia for yellow people; America for red people).

Ritter's concepts regarding the meaning of the observed geographical patterns on the earth were strongly teleological. Following the philosophers Immanuel Kant and Johann Gottfried von Herder, Ritter saw in all his geographical studies the evidence of God's plan. A Supreme Being, an all-wise Creator, was identified as the author of a plan for building the earth as the home of man, and all through Ritter's writings and lectures are words of praise for the divine creation. Even the arrange-ment of the continents Ritter saw as evidence of God's purpose. Asia, said Ritter, represents the sunrise—here the early civilizations originated. Africa represents the noon—because of the smoothness of outline as well as the uniformity of climate, the inhabitants are induced to slumber and to shun outside contacts. Europe is espe-cially designed to bring out man's greatest accomplishments—because it represents the sunset, or the end of the day, the culmination of human development is found there. But the discovery of America now suggests the approach of a new sunset and a new culmination toward which man continues to strive. The polar regions represent midnight, when land and people are locked in eternal sleep. Ritter enlarges on the concept of the land hemisphere, as suggested by Buache and developed by Malte-Brun, and points out that this, too, is a part of God's plan. Only in this central location among the earth's land areas can a world-conquering civilization arise.

Ritter's teleological ideas were criticized by his contemporaries. It was Julius Fröbel who said that one might, with equal truth, say that grass had been created as feed for cattle (Fröbel, 1831). Ritter replied that among all the creatures on the earth only man could comprehend the existence of a divine plan and so could adjust his life to it and make maximum use of God's gifts (Hartshorne, 1939:62).

The Erdkunde

Ritter, like Humboldt, produced one great work that represented his major scholarly achievement. This was *Die Erdkunde*.[6] The translation of the full German title presents the basic purpose: *The Science of the Earth in Relation to Nature and the History of Mankind; or, General Comparative Geography as the Solid Foundation of the Study of, and Instruction in, the Physical and Historical Sciences.* Before Ritter became the professor of geography at Berlin in 1820, he was still thinking of geography as the basis for the writing of history. The first two volumes of the *Erdkunde* (1817–18) were intended to be followed by a study of history. But when Ritter went to Berlin he decided to devote himself to doing a more thorough piece of work on geography. In 1822 he published a second edition of Volume 1 and in 1832 a second edition of Volume II. But by this time he realized the magnitude of the work he had started. After 1831 he gave up many of his positions so that he could devote himself more fully to the completion of the *Erdkunde*. Between 1832 and 1838 he completed 6 more volumes, and between 1838 and 1859, 11 more. Yet the 19 volumes of the *Erdkunde* Ritter actually finished covered only Africa and a part of Asia.

Unlike Humboldt, Ritter's great work was largely put together on the basis of other people's observations. He said that his field studies in Europe made it possible for him to interpret what other people reported. Fritz Kramer comments on the interesting point that Ritter's descriptions of places he had never seen were vivid and accurate, whereas his descriptions of places he had seen often lacked zest (Kramer, 1959).

In contrast to the clarity of his lectures, Ritter's published works are often obscure. Scholars have struggled to find suitable translations for some of his passages that would make sense in another language and yet not do violence to his ideas.[7]

---

[6]*Die Erdkunde, in Verhältniss zur Natur und zur Geschichte des Menschen, oder allgemeine vergleichende Georgraphie, als sichere Grundlage des Studiums und Unterrichts in physikalischen und historischen Wissenschaften*, 19 parts and 21 volumes. (Berlin: G. Reimer, 1817–18; 1822–59).

[7]For example, see the discussion of the meaning of one of Ritter's frequently quoted statements that geography is the study of *"der irdisch erfüllten Räume der Erdoberfläche"* (from Ritter, 1852) in Hartshorne, 1939:57.

RETROSPECT

So these two great scholars, who died in the same year in Berlin, each in his own way attempted to establish a "new geography." Each tried to embrace the knowledge of humankind concerning the earth as the home of man. Both of them saw the field of geography as dealing with things and events of diverse origin that were interconnected in segments of earth space, as did Kant and others. Both were tireless workers who wrote many books and whose influence on the scholarly world was very great. Both recognized the need for seeking generalizations, and both recognized that little progress toward more advanced theory could be made in their time. But both had confidence that continued geographical study would eventually bring to light the inner meaning of the universe. Humboldt was an agnostic; Ritter once remarked that although the *Kosmos* was a magnificent piece of work, one found in it no single word of praise for the Creator. Ritter saw all of his studies of the earth and man as revealing more and more of God's plan.

Both men shared the fruits of the Enlightenment. Both shared technical and other knowledge gained by journeying in the Pacific, improvement in instruments and transportation facilities, and both lived before Darwin's proclamation of the evolutionary journey of life forms. Yet the two men were fundamentally different in their approach. Humboldt could not look at the world around him without finding innumerable questions demanding answers. He not only described what he saw with care and precision, but he also formulated hypotheses to account for the things he observed—and then he also subjected his hypotheses to the test of new observations. There was an ecologic particularism about his viewpoint. Ritter, on the other hand, viewed the landscape historically and thought regionally. Ritter also had a vision of an ordered and harmonious universe, but instead of asking questions about it, he wanted to communicate to others the meaning he had found. As a teacher he wanted to make clear to his disciples how God's plan was revealed in the harmony of man and nature. Each in his own way was enormously successful, and each enjoyed wide personal prestige.

When Humboldt and Ritter died there was no one to replace them. Classical geography had come to an end—no individual scholar could hope any longer to master the world's knowledge about the earth. The specialization of subject matter disciplines resulted in the development of new technical jargon, new paradigms of scientific behavior. As a result much that had been called geography was partitioned among a variety of logically defined fields. In Germany no one was appointed to fill Ritter's chair. Some years later when geography was reestablished as a university study, the scholars invited to teach it had had no previous training in a field called by that name.

What do Humboldt and Ritter mean to us today? Ritter did influence his disciples to identify a new scientific geography based on the organic unity of man and nature (Guyot, 1860). But his teleology, reflecting the contemporary thinking of such philosophers as Kant and Herder, became outmoded and raised a barrier to the continued acceptance of this kind of new geography. Moreover, Ritter's regional studies deal for the most part with such large areas that the material he included had to be highly generalized. The interconnections he described could not be perceived by direct observation. Today Ritter's *Erdkunde* has chiefly an antiquarian interest.

Humboldt's systematic studies are also outdated, although the methods he used represent important steps in the progress of geography. But Humboldt's regional studies "cannot become obsolete" (Hartshorne, 1939:82), especially his comparative studies of New Spain and Cuba, which provide invaluable material for studies in historical geography. Humboldt dealt with areas that were small enough that he could discuss all the factors relevant to a problem that could be tested by direct observation—for example, the study of the Lake of Valencia basin in Venezuela. These two great intellects of the nineteenth century, each in his own way, dedicated their lives to the geographical undertaking. Their contributions have been incorporated in innumerable ways into a changed conception and corpus of contemporary geography.

REFERENCES: CHAPTER 6

Beck, H. 1959–1961. *Alexander von Humboldt*. Vol. 1 (1959), *Von der Bildungsreise zur Forschungsreise, 1769–1804*. Vol. 2 (1961), *Vom Reisewerk zum "Kosmos," 1804–1859*. Wiesbaden: Franz Steiner.

Beck, Hanno. 1979. *Carl Ritter: Genius of Geography. On His Life and Work*. Bonn-Bad Godesberg: Inter-Nationes.

Beck, Hanno and W. H. Hein. 1987. *Alexander von Humboldt: Life and Work*. Ingelheim am Rhein: C. H. Boehringer Sohn.

Bögekamp, H. 1863. "An Account of Prof. Ritter's Geographical Labors." In W. L. Gage, trans., *Geographical Studies by the Late Professor Carl Ritter of Berlin*. Pp. 33–51. Boston: Gould & Lincoln.

Botting, D. 1973. *Humboldt and the Cosmos*. London: Michael Joseph.

Bowen, M. 1981. *Empiricism and Geographical Thought: From Francis Bacon to Alexander von Humboldt*. Cambridge: Cambridge University Press.

De Terra, Helmut. *Humboldt: The Life and Times of Alexander von Humboldt, 1769–1859*. New York: Alfred A. Knopf, 1955. (Reprinted New York: Octagon Books, 1979.)

Dickinson, R. E. 1969. *The Makers of Modern Geography*. London: Routledge & Kegan Paul.

Downes, Alan. September 1971. "The Bibliographic Dinosaurs of Georgian Geography (1714–1830)." *Geographical Journal* 137, part 3:379–387.

Friis, H. R. 1963. "Baron Alexander von Humboldt's Visit to Washington, D. C., June 1 Through June 13, 1804." *Records of the Columbia Historical Society*. Washington, DC 1–35.

Fröbel, J. 1831. "Einige Blicke auf den jetsigen formellen Zustand der Erdkunde." *Annalen der Erd-, Völker-, und Staatenkunde* 4:493–506.

Gage, W. L., trans. 1863. *Geographical Studies by the Late Professor Carl Ritter of Berlin*. Boston: Gould & Lincoln.

———. 1867. *The Life of Carl Ritter: Late Professor of Geography in the University of Berlin*. New York: C. Scribner.

Godlewska, A. M. C. 1999. *Geography Unbound: French Geographic Science from Cassini to Humboldt*. Chicago: University of Chicago Press.

Guyot, A. H. 1860. "Carl Ritter." *Journal of the American Geographical and Statistical Society* 2:25–63.

Hartshorne, R. 1939. *The Nature of Geography, a Critical Survey of Current Thought in the Light of the Past*. Lancaster, Pa.: Association of American Geographers.

———. 1958. "The Concept of Geography as a Science of Space, from Kant and Humboldt to Hettner." *Annals AAG* 48:97–108.

Humboldt, A. von. 1793. *Florae fribergensis subterraneas exhibens*. Berlin: H. A. Rottman.

————. 1805–1834. *Voyage aux régions équinoxiales du Nouveau Continent*. Paris. (See footnote on pp. 115–116 for titles of the 30 volumes.)

————. 1808. *Ansichten der Natur, mit wissenschaftlichten Erläuterungen*. 2nd ed., 1849. Stuttgart: Cotta.

————. 1814–1825. *Relation historique du voyage au régions équinoxiales du Nouveau Continent* (Vols. 28–30, 1805–1834). English translation by H. M. Williams, 1825. *Personal Narrative of Travels in the Equinoxial Regions of the New Continent During the Years 1799–1804*. 5 vols. Paris, German translation by H. Hauff, 1859–1860. Personal Narrative 1995 ed. J. Wilson, London, New York: Penguin Book. *Alexander von Humboldt's Reise in die Aequinoctial Gegenden des neuen Continents*. 4 vols. Stuttgart.

————. 1845–1862. *Kosmos: Entwurf einer physischen Weltbeschreibung*. 5 vols. Stuttgart: Cotta. (Vol. 1, 1845; Vol. 2, 1847; Vol. 3, 1850; Vol. 4, 1858; Vol. 5, 1862.) English translation by E. C. Otté, London: H. G. Bohn, 1849–1858.

————. 1847. "Geschichte der physikalischen Weltanschauung." In *Kosmos: Entwurf einer physische Weltbeschreibung*. Vol. 2, pp. 135–520. Stuttgart and Tübingen: J. G. Cotta'scher Verlag.

————. 1934. *Alexander von Humboldt's Vorlesungen über physikalische Geographie nebst Prolegomenen über die Stellung der Gestirne. Berline im Winter von 1827 bis 1828*. Berlin: Miron Goldstein.

Humboldt, Alexander von, and Bonpland, Aimé. *Personal Narrative of Travels to the Equinoctial Regions of America during the years 1799–1804*. Translated from French by Thomasina Ross. London: HG Bohn, 1852–53. 3 vols. (Reprinted: New York: Ayer Co., 1969.)

Kellner, L. 1963. *Alexander von Humboldt*. London: Oxford University Press.

Kramer, F. 1959. "A Note on Carl Ritter, 1779–1859." *Geographical Review* 49:406–409.

Lenz, Karl, ed. 1981. *Carl Ritter: Geltung und Deutung*. Berlin: Dietrich Reimer.

Linke, Max. 1981. "Carl Ritter, 1779–1859." *Geographers: Biobibliographical Studies*. Ed. T. W. Freeman. 5. 99–108.

Meyer-Abich, A. 1967. *Alexander von Humboldt in Selbstzeugnissen und Bilddokumenten*. Rowohlt: Kurl Kisenberg.

Ritter, C. 1822–1859. *Die Erdkunde, im Verhältniss zur Natur und zur Geschichte des Menschen; oder allgemeine vergleichende Geographie, als sichere Grundlage des Studiums und Unterrichts in physikalischen und historischen Wissenschaften*. 19 vols. Berlin: G. Reimer.

————. 1852. *Einleitung zur allgemeinen vergleichenden Geographie, und Abhandlungen zur Begründung einer mehr wissenschaftlichen Behandlung der Erdkunde*. Berlin: G. Reimer.

————. 1862. *Allgemeine Erdkunde*. Berlin: G. Reimer.

————. 1880. *Geschichte der Erdkunde und der Entdeckungen. Vorlesungen an der Universität zu Berlin gehalten von Carl Ritter*. Ed. H. A. Daniel. 2nd ed. Berlin: Druck und Verlag von G. Reimer.

Robinson, A. and H. Wallis. 1967. "Humboldt's Map of Isothermal Lines: A Milestone in Thematic Cartography." *Cartographic Journal* 4:119–123.

Schultz, J. H., ed. 1959. *Alexander von Humboldt: Studien zu seiner universalen Geisteshaltung*. Berlin: W. de Gruyter.

Sinnhuber, K. A. 1959. "Carl Ritter, 1779–1859." *Scottish Geographical Magazine* 75:153–163.

Stoddart, D. R. 1985. "Humboldt and the Emergence of Scientific Geography." In P. Alter, ed. *Humboldt*. London: Heinemann.

Troll, C. 1959–1960. "The Work of Alexander von Humboldt and Carl Ritter: A Centennary Address." *Advancement of Science* 64: 441–452.

Tuan, Yi-fu. 1997. *Alexander von Humboldt and His Brother: Portrait of an Ideal Geographer in Our Time*. Los Angeles: University of California, Los Angeles Department of Geography.

PART 2

MODERN

A major innovation in the world of scholarship took place in nineteenth-century Germany. The university as an institution first appeared in medieval Europe when charters were issued by religious or secular authorities giving certain faculties the right to teach. The University of Paris in the twelfth and thirteenth centuries became the chief center (other than Rome) for the teaching of orthodox Christianity. But in 1809 Wilhelm von Humboldt, the brother of Alexander, founded the University of Berlin with the support of King Friedrick Wilhelm III of Prussia. For the first time anywhere, the attachment of either faculty or students to any particular religious creed or school of thought was explicitly repudiated. Hitherto, universities were places where the accepted dogma of state and church was taught to students. After 1809 the university as a free community of scholars began to appear.

Geography as a field of advanced study taught by professionally qualified individuals first appeared in Germany in 1874. Within a few decades geography departments offering graduate training leading to advanced degrees were established not only in Germany, France, and Britain but also all around the world. This was the "new geography," and it was guided for the first time in history by professional geographers. A profession had come into existence that could establish the paradigms of geographical study. We date the modern period in the history of geographical ideas with the establishment of professional staffs in universities.

7

WHAT WAS NEW?

*The labors of Humboldt, of Ritter, of Guyot, and their followers have given the
science of geography a more philosophical, and, at the same time, a more imaginative character than it had received from the hands of their predecessors.
Perhaps the most interesting field of speculation, thrown open to the new school
of the cultivators of this attractive study, is the inquiry: how far external physical conditions, and especially the configuration of the earth's surface, and the
distribution, outline, and relative position of land and water, have influenced
the social life and social progress of man.*

*But it is certain that man has done much to mould the form of the earth's
surface, though we cannot always distinguish between the results of his action
and the effects of purely geological causes; that the destruction of the forests,
the drainage of lakes and marshes, and the operations of rural husbandry and
industrial art have tended to produce great changes in the hygrometric,
thermometric, electric, and chemical condition of the atmosphere, though we
are not yet able to measure the force of the different elements of disturbance,
or to say how far they have been compensated by each other, or by still obscurer
influences; and, finally, that the myriad forms of animal and vegetable life, which
covered the earth when man first entered upon the theater of a nature whose
harmonies he was destined to derange, have been, through his action, greatly
changed in numerical proportion, sometimes much modified in form and
product, and sometimes entirely extirpated.*

— *G. P. Marsh*, Man and Nature, or Physical Geography
as Modified by Human Action

Those scholars who from time immemorial have been seeking more and more
useful knowledge concerning the face of the earth, including our use of it, have
always confronted five basic problems, none of which has yielded to permanent
solution. These problems, as suggested by Fred Lukermann, are: (1) What things
in the universe should humans select to observe and record? (2) What is the best
way to observe them? (3) How can the resulting observations be generalized to
reveal some kind of significant geometric arrangement on the earth? (4) How can
the patterns of arrangement be explained or made plausible? (5) How can the
results be communicated?

The flood of new information that swept over the world of European scholarship as a result of voyages of discovery greatly complicated the search for answers
to these questions. At first, attention was focused on the marvels that were reported,
and writers with vivid imaginations, such as the author of *The Travels of Sir John
Mandeville*, could scarcely be distinguished from sober reporters, such as Marco
Polo. The world revealed by the explorers was full of strange things, and there was

131

no lack of subjects to be observed and recorded. Then, little by little, attention shifted from the marvels to things that formed some kind of pattern with familiar things at home. It became more important to report similarities than differences. Cluverius and Carpenter in the seventeenth century began omitting references to weird creatures and unusual natural phenomena, but not until 1761 did Jean B. B. d'Anville remove the drawings of strange creatures that hitherto had adorned the blank places on maps.

The seventeenth century witnessed the beginning of the scientific revolution that led to the development of more useful ways of generalizing, explaining, and communicating. The effort to provide more exact descriptions of specific things was replaced by the effort to formulate general theory in relation to which specific things could be made significant. In the formulation and testing of theory and in communicating the findings, a step of major importance was the independent development of calculus by Newton and Leibniz. The use of mathematical procedures made the process of reasoning more precise and provided a universal language for the communication of the results. Most of the present fields of science had their roots in the eighteenth century, during which time the acceptable methods of study were being formulated and reliable procedures for verifying hypotheses were being established. No longer could a hypothesis be supported by its plausibility, for scholars were learning that things perceived by the senses were not necessarily the outline of reality. Controlled experimentation began to bring spectacular results, and after the eighteenth century it would no longer be possible for a Montesquieu to draw conclusions from the study of a sheep's tongue.

By the end of the eighteenth century, the ideas expressed by Kant had become generally accepted. As bodies of theory developed and proved useful, special fields of study appeared, each defined in terms of the segment of the universe being investigated. These new fields of study became what Kant called the logical division of knowledge—in contrast to the physical classification of knowledge in terms of time and space. Each logically defined field provided a method for describing and demonstrating the significance of a particular segment of human experience—and of creating new experiences through use of the experimental method.

The last great figure who could claim universal scholarship was Humboldt. No student of the earth before or since has enjoyed such acclaim by his contemporaries. Ritter, too, attempted to embrace the whole of geographical knowledge concerning the earth and man but was less successful.

The world of scholarship underwent a basic change during the nineteenth and twentieth centuries. Not only have the fields of learning—the academic disciplines —been greatly elaborated, but also the total number of scholars has reached unprecedented size.[1] Moreover, the number of recorded facts about the earth has

---

[1]For example, there were 12 professional geographers in Germany in 1880 (Wagner, 1880); in 1921 there were 70 (Joerg, *Geographical Review*, 1922, p. 442). In 1964 the international directory, *Orbis Geographicus*, ed. E. Meynen (Wiesbaden: Franz Steiner), listed 546.

In 2004 geography could be studied at 62 universities: in these alone there were more than 800 professional geographers in Germany.

increased astronomically. A Humboldt could once master a very large part of the available knowledge concerning the earth, but this is no longer possible. The number of books and articles is causing libraries to bulge and research workers to seek new kinds of information retrieval systems. The computer was invented just in time to provide a mechanical means for data storage.

The question is: What is new about all this? Are the basic questions still the same? How much of what went on before the modern period is of any importance today— other than to satisfy the curiosity of historians of science?

THE LOGICAL SYSTEMS

The logical systems have become the familiar divisions of the academic curriculum. In the broadest sense these systems include the physical sciences, the biological sciences, the social sciences, and the humanities.

The Physical Sciences

The physical sciences, which appeared as separate disciplines earlier, are now the most advanced in the building of theory and in the continued testing of theory by controlled experiment. A particular physical process is artificially isolated in a laboratory and can then be observed free from the complications resulting from the presence of a great variety of logically unrelated processes in the total environments of particular places on the earth. General models can then be formulated to describe the observed sequences of events. What the physical scientist is trying to do is to find order in human experience and to describe this order in the simplest possible terms. Ptolemy identified a kind of order in the movements of the celestial bodies, but Copernicus found Ptolemy's picture of celestial order too complicated and with too many motions unexplained. An example of the simplicity of order sought by the physical scientists is the law of gravitation formulated by Newton. This law states that every particle of matter in the universe is attracted to every other particle with a force proportional to the masses of the particles involved and inversely as the square of the distance between them. Albert Einstein showed that this simple statement of the law of gravitation works only with large numbers of particles and must be modified for studies of atomic physics.

Another major achievement of the late eighteenth century was the discovery by the French chemist Antoine Laurent Lavoisier (1743–94) and by the English chemist Henry Cavendish (1731–1810) that Aristotle's four basic substances (air, fire, earth, and water) were not really basic elements, no matter how plausible this might seem. In 1783 Lavoisier announced that water is made up of hydrogen and oxygen. Henry Cavendish in England anticipated Lavoisier by a few years, but his studies of water were not published until 1784–85. He also carried out the first measurements of the composition of the air, anticipating Gay Lussac. These scientific accomplishments required use of the experimental method.

A number of specialized fields of study emerged from the unspecialized cosmography to give attention to particular groups of processes on the face of the earth. The study of celestial bodies is now left to that branch of physics known as astronomy. The study of the interior of the earth is entrusted to the various branches of geophysics, including seismology (the study of earthquakes). The study of things and events on the surface of the earth involves geomorphology, geology, and the even more specialized mineralogy, petrography, and paleontology (where geology overlaps with biology). There is also the study of water on the land, hydrology. Oceanography became a separate discipline after Matthew Fontaine Maury started his collection of observations concerning winds and oceans currents. Climatology became the study of the average state of the atmosphere and meteorology the study of atmospheric processes that produce weather.

The Biological Sciences

The later voyages of discovery in the eighteenth century had a special impact on the development of biology as a separate field of study. Captain James Cook commanded the first expeditions on which scientifically trained people were included (Rutherford and Armstrong, 2000). Among those who sailed with Cook on his first voyage were Sir Joseph Banks and David C. Solander, a pupil of Linnaeus. On the second voyage Cook took with him the two Forsters. Georg Forster was the one who focused Humboldt's interest on botanical observations. Humboldt himself brought back to Europe some 60,000 specimens of plants never before known to Europeans. When Jean Lamarck was appointed professor of zoology at Paris in 1773 he had access to a wealth of new plant and animal collections. The tropical parts of the world, long feared because of the persistence of ideas inherited from Aristotle, attracted much attention after they were vividly described by Forster and Humboldt. It was a copy of Humboldt's narrative that inspired Charles Darwin to turn to the study of plants and animals.

Darwin was greatly influenced by the uniformitarian ideas of Hutton and Sir Charles Lyell. Accepting Lamarck's concept of evolution, Darwin began looking for the processes of change in species that would explain the diversity of organic life on the earth. From December 1831 to October 1836 Darwin sailed around the world on H.M.S. *Beagle*, observing a great variety of physical and biotic processes, much as Humboldt had done. His concept of the stages in the transformation of coral reefs—from fringing reefs, through barrier reefs, to atolls—was published in 1842 (Darwin, 1842; Davis, 1928). It was on this voyage that Darwin formulated his hypothesis concerning the mechanism whereby random variations in plant and animals species would be selectively preserved and by inheritance lead to changes in species.

At about the same time another young scientist was traveling to the tropical parts of the world. This was Alfred Russel Wallace. In 1848 Wallace accompanied Henry W. Bates on a voyage up the Amazon River. Although his collections were lost when his ship burned on the return voyage, Wallace had been fascinated with the problem of how evolutionary changes in organisms could take place. From 1854 to 1862 he explored the Malay Archipelago, and in the course of this exploration he identified a sharp boundary that separated areas with very different kinds of native mammals. The line passed between Borneo and Bali on the west and Lombok and Celebes on the east (in an area now sometimes referred to as Wallacea). The more primitive animals farther east had been protected from competition with more advanced species farther west. The line between the two kinds of fauna was called Wallace's Line.[2] In the process of plotting these geographic differences on a map, Wallace, in a flash of intuition, saw the significance of what he called natural selection. Influenced by the ideas of Malthus regarding the relation of population to food supply, he applied the same idea of the struggle for existence to animals. Wallace promptly dashed off a letter to Charles Darwin in England, presenting his hypothesis in brief form. The letter arrived just as Darwin was preparing to present a major paper to the Royal Society of London on exactly the same hypothesis, which Darwin had now verified by laboratory experiments. The idea was presented in a joint paper by Darwin and Wallace in 1858 entitled *On the Tendency of Species to Form Varieties, and on the Perpetuation of Varieties and Species by Natural Means of Selection.*

Darwin's *Origin of Species* was published in London in 1859, the year in which both Humboldt and Ritter died (Darwin, 1859). He demonstrated that evolutionary change in organisms was not the result of need or use, as Lamarck had thought. The giraffe did not get its long neck by stretching. Rather, the individual giraffes that were born with longer necks were better able to survive than their shorter necked relatives and so could pass on this characteristic to later generations. Darwin's contribution was to throw light on the mechanism whereby evolutionary change could take place. But he also produced clear evidence of the randomness of such variations. If evolutionary changes resulted from random variations, the teleological concept of a divine plan had to be abandoned. In spite of continued resistance by some biologists (such as Louis Agassiz) and the reluctance of Darwin himself to face the full implications of his conclusions, the world of science could never return to previously held beliefs.

The concept of evolutionary change was so stimulating that it was applied by analogy to many other fields beside biology. Applied to the study of landforms, it appeared as the theory of the cycle of erosion. Applied to soils, it was reflected in the concept of mature soils as developed from young or immature soils and parent materials. Applied to the survival of social groups because of the ability to adjust to environmental conditions, it became environmental determinism. As D. R. Stoddart points out, the geographers adopted the notion of evolutionary change as described by cause and effect sequences, but they overlooked the concept of random variations and failed to apply the theory of probabilities (Stoddart, 1966).

The Social Sciences

Among the logical systems aimed at the study of human group behavior, the first to develop as a special field was political economy, or economics, as it was renamed in the twentieth century. Some of the earliest attempts to formulate general theory regarding population and resources are associated with a group of eighteenth-century Scottish scholars at the University of Edinburgh. The group includes such people as David Hume (1711–76), Adam Ferguson (1723–1816), and Adam Smith

---

[2]Others have drawn "lines" pursuant to Wallace. One of these "lines" was suggested by Max Weber (originally based on the distribution of freshwater fish).

(1723–90). Adam Smith's study, *An Enquiry into the Nature and Causes of the Wealth of Nations*, was published in 1776. The real source of a nation's wealth, he said, is its annual labor, its use of productive resources; wealth can only be increased by making its use of resources more effective, by increasing the specialization of labor, and accumulating profit in the form of capital. Money, he pointed out, was not wealth, but only the means of carrying on trade.

Adam Smith's work was followed by the essay on population by Malthus. It was Malthus who put into clear language the previously known law of diminishing returns from investments of capital and labor. David Ricardo in 1817 published a study, *Principles of Political Economy and Taxation*, in which he developed a theory of value. By 1830 political economy was a recognized field of study in most European universities, and in the course of more than a century it has led to the formulation of a large body of theory and methods of study.

During this time, economics has itself been further subdivided into distinct specializations. There is a separation between economics as a pure science and economics as an applied science dealing with public and private problems of policy. Another distinct field is economic history; another is econometrics, based on the application of mathematical procedures to economic problems. Still another field embraces the study of economic theory.

And History?

History, like geography, did not fit easily into what Kant called the logical classification of knowledge. The various substantive fields into which the study of human behavior was partitioned were built around particular conceptual structures suitable for enlarging or testing these concepts. But traditionally history had dealt with whatever kinds of processes were necessary to understand the sequences of events —social processes, political processes, economic processes, military events, and especially the people who made an impact on the course of events. In a modern university, is history to be found among the humanities or among the social sciences? Actually, it may be found in either.

Historians have long been concerned with the question about whether they should seek to identify general laws of human behavior or only to reach a more precise and "correct" knowledge of the sequence of events. Many felt that the verification of unique sequences of events constituted an ample justification for historical scholarship. Others, however, felt impelled to seek universal laws around which to arrange the historical facts. The earlier historians in Europe accepted the common belief that human behavior was a manifestation of the divine plan and that man was in process of development toward the perfect state that God had established as the goal. But during the eighteenth and nineteenth centuries historians gradually shifted away from this interpretation of history. The perfectibility of man, they said, was not demonstrated by empirical evidence but remained only an article of faith. What general laws of behavior, then, could be discerned from the historical record?

As the volume of historical data increased, historians had to find some way to specialize—to reduce the size and complexity of the questions being investigated. The last universal history was published in 1681 by J. B. Bossuet; thereafter historians became specialists in particular countries or cultures and then just certain limited periods within those countries. Or perhaps they became specialists in the biography of a particular person. Specialization had gone so far by World War I that H. G. Wells sensed the need for another universal history in which certain general and repeated historical trends could be noted for all of human-kind. His book, which was criticized by some professional historians, took the chronological record back to the origin of the earth and the evolution of organic life.[3]

The idea that historians should seek to formulate laws and models to explain the course of events became stronger as the teleological interpretations were abandoned. In the early eighteenth century the Italian historian Giambattista Vico identified certain cycles that were repeated again and again in the history of different peoples. He agreed that historical law could not have the precision found in natural law but that at least the broad trends could be discovered.[4] The German scholar Herder presented the idea that to understand any sequence of events three interconnected factors had to be known: time, place, and national character.[5] Many other writers of history formulated general laws to explain the course of events.[6]

In Germany during the nineteenth century, there was a similar discussion con-cerning the objectives of the new field of economics. Gustav F. Von Schmoller insisted that economics was only a branch of history and that the so-called laws then being formulated by economists could really only be applied to situations unique in time and place. On the other hand, Karl Menger said that the only purpose of economics was to identify the general and universal laws of man's economic behavior, not to record unique events. Menger's views came to be accepted as the pattern of eco-nomic scholarship characterized by a nomothetic approach—that is, a field of study in which general laws are identified—rather than an undertaking characterized by an idiographic approach, or one that describes unique situations without reference to general laws. Economic history was left intermediate between history and econ-omics. Generally speaking, the parts of history in which bodies of theory have been formulated tended to develop as separate academic disciplines (such as sociology, anthropology, political science, or economics), whereas those parts of history where general laws seem to be less useful have been cultivated by scholars whose primary objectives were the discovery of new sources of information and the use of more precise methods for verifying the authenticity of the data (Shafer, 1969:1–9, 37–42).

---

[3]H. G. Wells, *The Outline of History, Being a Plain History of Life and Mankind* (New York: Macmillan, 1920). Reprinted 2001. Classic Books (Murrieta, California).

[4]Giambattista Vico, *Scienza nuova* (Rome, 1725).

[5]J. G. von Herder, *Ideen zur Philosophie der Geschichte der Menschheit* (Berlin, 1784–91).

[6]Among these: Etienne Bonnot de Condillac (1714–80); the Marquis de Condorcet (1743–94); Auguste Comte (1798–1857); John Stuart Mill (1806–73); Karl Marx (1818–83); Friedrich Engels (1820–95); Oswald Spengler (1880–1936); Arnold J. Toynbee (1889–1975).

What Happened to Geography?

Essentially, the elements that were included in what Varenius had called general geography were divided up among the separate disciplines, each with its own body of theory and its own methods of connecting observations with theory. Humboldt and Ritter recognized that when the newly emerging disciplines had divided up what used to be general geography, there still remained an area of study not included in these substantive fields. Humboldt asked questions about the earth and man that were not asked by workers in any of these other fields. He wrote regional studies of Mexico and Cuba that were not just descriptions of unique places but that provided explanations in terms of general theory. Ritter in his regional studies of Africa and Asia did not just describe each element as a separate and distinct phenomenon. Rather, he sought the interconnections among things of diverse origin. These interconnections among the physical, biotic, and human features of the face of the earth Ritter identified as evidence of God's plan to lead humankind toward a state of perfection. This harmony of interconnected parts is what Ritter described by the expressive German word *zusammenhang*. Stoddart has suggested that geography became an objective science when Cook entered the Pacific in 1769 and in the next 10 years charted one third of the coastlines of the world. Cook took with him scientists, Harrison's chronometer, and a sturdy vessel. Stoddart regards this as the announcement of a geography that was objective science, which, however, is different from the new geography (Stoddart, 1986).

Meanwhile, the substantive fields of study included in the general categories of physical science, life science, and social science made spectacular progress by isolating the processes each examined and formulating an ideal or abstract model of how each process works in isolation. These sciences moved forward by specifically excluding the disturbing effect of *zusammenhang*. Chemistry and physics could isolate the processes they studied in laboratories. Biology set up experimental programs to test the validity of theory, also in isolation from the total environments of particular places on the earth. The social sciences had more difficulty in isolating the processes they studied, but economics in particular established at least a symbolic isolation by the use of the phrase "other things being equal." The law of diminishing returns, for example, operates in undisturbed form only when the impact of irrelevant interconnections is eliminated. Now "other things" can be made equal by statistical procedures.

Geography was left with three major tasks. One was the continued collection of information about the still unknown or inadequately known parts of the earth and the presentation of this information in useful form. The second was the study of particular places in the world, whether for the purpose of throwing light on the processes at work in them or for the practical needs of government administrators, military commanders, or businessmen who needed clear descriptions of the facts and conditions relevant to particular problems. The third task was the formulation of concepts: empirical generalizations, hypotheses, and perhaps even theory. There never was a time when geographers as a professional group were satisfied to describe unique situations without seeking to illuminate the geography of particular places by reference to generalizations or to seek explanations in terms of models.

The New Cartography

In all these tasks new kinds of maps were needed to go along with the new geography. Pioneering work in making large-scale topographic maps had already been done by the Cassinis in France and by Nicolas Cruquius in the Netherlands, who made use of lines of equal elevation to show landforms in 1728 (Goode, 1927). Improvements in topographic mapping had to await new methods of printing. The use of copper plates had started in 1493, but lithography was not invented until 1800. Electrotyping and photography were developed between 1840 and 1850. Only then could finely engraved details be reproduced with precision.

One of the earliest of the new cartographers was Adolf Stieler. Stieler had received a law degree from the University of Göttingen in 1797, but he had also developed a keen interest in geography and in the problem of representing geography on maps. He attended lectures by Gatterer and even taught geography in a girls' school at Gotha (which is 8 miles east; northeast of Schnepfenthal, now called Waltershausen, where Ritter attended school). At Gotha Stieler found the old German publishing house of Justus Perthes, which he helped to convert into one of the world's leading centers of geographic study and cartography. In 1817 he published the first sheets of the *Stieler Handatlas*; in 1831, when the first edition of the Stieler atlas was complete, it contained 75 maps of the world as a whole and its different parts. Between 1829 and 1836 Justus Perthes published Stieler's map of Germany in 25 sheets.

Another German geographer-cartographer who contributed to the development of cartography and to the spread of German maps and atlases to other countries was Heinrich Berghaus. As a young man he was employed by the Prussian War Ministry in a field survey of Prussia, and from 1821 to 1855 he taught geometry and cartography at a school in Berlin. In nearby Potsdam he established a school of cartography at which several famous mapmakers were trained between 1839 and 1848. Berghaus was one of the many scholars encouraged by Humboldt. Much of the information for his atlas maps came from Humboldt; Berghaus's *Berghaus Physikalischer Atlas*, 1837–48 (revised 1849–52), was intended to supplement the *Kosmos*. He included a great variety of thematic maps covering the latest information on climatology, hydrography, geology, earth magnetism plant geography, zoogeography, anthropogeography, and ethnography—93 maps in all (Beck, 1956). Berghaus also published a number of texts in geography as well as scholarly works. His six-volume *Allgemeine Länder- und Völkerkunde* (Stuttgart, 1837–43), and his five-part *Grundriss der Geographie* (Breslau, 1840–43) were widely read (Hartshorne, 1939:74).

Work on the *Berghaus Physikalischer Atlas* was carried on by Hermann Berghaus, nephew of Heinrich and one of those trained at the school in Potsdam. Hermann moved to Gotha in 1850 and remained there until his death in 1890. He compiled and edited a third edition of the *Berghaus Physikalischer Atlas*, which was published in 1883–91. Also at Gotha were Karl Vogel, who for many years kept successive editions of the *Stieler Handatlas* up to date, and Eric von Sydow, who in the 1830s recognized the need for large wall maps for use in classrooms and who prepared the first series of such maps for Justus Perthes. Of this series the first, on Asia, was published in 1838. Sydow's *Schulatlas* (1847–49) not only made the new information about the earth available in the schools of Germany and elsewhere but also set the standards for the use of blues, browns, and greens on hypsometric maps.[7]

Among the widely known scholars trained by Heinrich Berghaus, who later worked at Gotha, was August Petermann. In 1845, at the age of 23, Petermann went to Edinburgh to assist the Scottish map publisher Alexander Keith Johnston in bringing out an English edition of the *Berghaus Atlas*. He remained in Edinburgh and London until 1854, during which time he was appointed cartographer to the queen and introduced numerous ideas and techniques from Germany into Britain. While he was in London he joined the controversy then raging regarding the existence of an ice-free polar sea. It is ice free, he insisted, because of the warming effect of the Gulf Stream, which flows into the Arctic Ocean. In 1854 he returned to Gotha; the next year he founded the famous geographical periodical *Petermanns Geographische Mitteilungen*, which is still among the leading professional periodicals in the field of geography. Petermann's maps, for which the *Mitteilungen* became famous, set new standards in cartography. In the first 24 volumes and 56 supplements (*ergänzungshefte*) that he edited, he published a total of 850 maps. The topographic map, said Petermann, is the highest achievement of geography since it furnishes the most accurate reproduction of the earth's surface and thereby provides the best basis for all knowledge (*Petermanns Geographische Mitteilungen*, 1878:208).

When Petermann was in Edinburgh, he was assisted by the young John Bartholomew, son of the director of the map-publishing firm of John Bartholomew & Son. When John became head of the firm in 1856, he was the fourth member of the family with the same name to occupy this position.[8] With his son, John George Bartholomew, he introduced the use of layer tints for hypsometric maps into the English-speaking world. These two Bartholomews also recognized the need for having a map-publishing firm closely supported by a geographic research center when they established the Edinburgh Geographic Institute. The next John Bartholomew, who died in 1962, was the editor of the world-famous *Times Survey Atlas of the World* in 1922, which was expanded into a new edition in 1955. Since then many atlases have been produced.

German cartographic ideas also had an important influence in the United States. Daniel Coit Gilman, who was professor of physical and political geography at the Sheffield Scientific School at Yale from 1863 to 1872, was in close touch with developments in Germany, especially with the mapping of statistical information by the Prussian statistical office. When the ninth U.S. census was being planned under the direction of the economist Francis Amasa Walker, Gilman brought the German materials to Walker's attention. The result was the publication in 1874 of the *Statistical Atlas of the United States* under Walker's direction. The atlas was an immediate success, and in succeeding censuses the data on economic production

---

[7]See the *Bulletin of the American Geographical Society* 44 (1912):846–848.

[8]The seventh Bartholomew is now head of the firm; John Bartholomew, 1831–93; John George Bartholomew, 1860–1920; John Bartholomew, 1890–1962; John C. Bartholomew was made head of the firm in 1962. *Reader's Digest* acquired Bartholomew in 1980 and in 1989 became part of Harper Collins Publishers.

and population were effectively presented on maps. The tenth census (1880), which Walker also directed, was published in 22 large volumes together with the atlas, which reviewed the changing geography of the United States since 1790. This census is a major source for the study of American historical geography. Neither Gilman nor Walker was primarily a geographer, yet their influence on the development of the new geography in America was very great.[9] (Short, 2001)

THE LEGACIES OF HUMBOLDT AND RITTER

Humboldt and Ritter left quite different legacies for future generations. Humboldt sought answers to a great variety of specific questions (de Terra, 1955). For example, he attempted to develop a general picture of the distribution of average temperatures in the world in relation to the distribution of continents and oceans. With the assistance of the Russian network of weather stations, he was able to do so. He attempted to define the effect of altitude of tropical mountains on plants, animals, and man. This he did on the basis of personal observation in tropical America. But Humboldt did not leave a school or disciples. The method of asking questions and seeking answers that he so effectively demonstrated was not "rediscovered" until several decades after his death. Since he did not restrict himself to the study of physical, biotic, or cultural processes in isolation, his contributions to the substantive fields are usually considered minimal. With Georg Forster he laid the groundwork for the study of plant geography—but he is not listed by botanists or zoologists as a major contributor to their fields. He was, in fact, a geographer because he asked about the interconnections among things and events of diverse origin, sweeping aside the then-growing barriers among disciplines to get a last majestic view of the cosmos.

Ritter did found a school, in the sense that his enthusiastic teaching aroused a similar enthusiasm among his disciples (Sinnhuber, 1959). Many disciples undertook to continue his plan of the *Erdkunde* in parts of the world he had been unable to complete. There were German studies of Australia and more detailed works dealing with parts of Europe, especially Germany. His most famous disciples, however, were Elisée Reclus and Arnold Guyot.[10]

Elisée Reclus

Elisée Reclus was a French geographer and anarchist who studied briefly under Carl Ritter (Dunbar, 1978). Departing France following revolutionary activism in 1852,

---

[9]Francis Amasa Walker was an economist who taught at Yale from 1873 to 1881. Later he was president of the Massachusetts Institute of Technology. Daniel Coit Gilman became the president of the University of California in 1872 and of the newly organized Johns Hopkins University in 1876. At Johns Hopkins he introduced the German concept of the university as a community of free scholars. Advanced study for the degree of doctor of philosophy was made available, and a faculty was selected on the basis of excellence in scholarship rather than ability to lecture to undergraduates (Wright, 1961).

[10]Another of Ritter's students was the Russian geographer Petr Petrovich Semenov Tyan-Shanski. Semenov's work is discussed in Chapter 11.

he spent the next five years traveling in Great Britain, the United States, and Colombia. He returned to France in 1857, was imprisoned in April 1871, and was banished for 10 years in February 1872. In March 1879 the ban was lifted. By then he had become an ardent supporter of the antimarriage movement and demonstrated his sincerity by permitting his two daughters to live with their partners without either civil or religious sanction. He was identified by the French government as one of the leading promoters of anarchism. Although he lived in France from 1890 to 1894, he maintained his residence in Switzerland, which prevented his arrest. In 1892 he was appointed professor of comparative geography at the University of Brussels in Belgium. Because of his continued activity as a revolutionary anarchist, his appointment was canceled. From 1898 until his death in 1905, he was director of the Institut Géographique, which he had founded at the New University of Brussels.

One reason for his immunity from any penalty other than banishment from France was his standing as a scholar and the resulting efforts of European scholars to give him protection. In 1867–68 he published *La terre*, a two-volume descriptive systematic geography with preponderant emphasis on physical geography, which was accomplished in the Ritterian manner (Reclus, 1867–68). But his major work was the completion of the kind of universal geography that Ritter had started. His 19-volume "new universal geography" was, in a sense, the last echo of the classical period, when one scholar could present all available knowledge about the earth as the home of man (Reclus, 1876–94). Reclus took great care with the accuracy of his sources and with the clarity of his writing. Unlike Ritter's *Erdkunde*, which is well known for its numerous obscure passages, Reclus's work was easy to read and understand. Also, he was one of the few disciples of Ritter who eliminated the teleological element. His standing as a scholar was given further support by numerous other writings, including a detailed description of the history of a stream and a similar work on a mountain (Reclus, 1869, 1880).

Arnold Guyot

Arnold Guyot was born in Switzerland; in 1839 he was a colleague of Louis Agassiz at the University of Neuchâtel. Agassiz turned the attention of his younger colleague to the study of glaciers and the effect of glacial action in producing distinctive kinds of mountain landforms. Guyot also studied with Ritter and became one of his most devoted disciples (Libbey, 1884). In 1848 Guyot came to the United States, where he was invited to deliver a series of lectures in Boston in 1849 outlining the rudiments of what was to become the "new geography." His lectures were published in book form and served to make Ritter's ideas known in America (Guyot, 1849). Guyot attacked the traditional descriptive geography wherein encyclopedic collections of facts were given to students to be memorized (James, 1969). The "new geography" should not only describe but also compare and interpret: "it should rise to the how and wherefore of the phenomena that it describes" (Guyot, 1849:21). When the Massachusetts Board of Education asked Guyot to deliver a series of lectures on the "new geography" and the methods of teaching it, his influence on American schools spread rapidly. For decades his textbooks set the standards for elementary and secondary classes in geography. He taught his pupils to observe their surroundings and to match their perceptions with the word symbols they used to describe them.

When distant regions were studied, Guyot urged that pupils should become better acquainted with these places by the close examination of topographic maps.[11]

Guyot, who held the position of professor of physical geography and geology at the College of New Jersey (Princeton) from 1854 to 1880, remained a vigorous supporter of Ritter's ideas. Even as the widespread acceptance of the concepts of evolution as developed by Darwin, Wallace, and Huxley swept away the philosophical ideas of the teleologists, Guyot's stand remained unshaken. When he retired in 1880, the "new geography" he preached was not only old, but its philosophical basis had been largely discredited. Here is what he had to say about the purpose of physical geography in 1873:

> The Earth, as an individual organization, with definite structure, character, and purpose, is the subject of geographical science. . . . A careful study of physical geography tends to lead the mind to the conclusion that the great geographical constituents of our planet—the solid land, the ocean, and the atmosphere—are mutually dependent and connected by incessant action and reaction upon one another; and hence, that the earth is really a wonderful mechanism all parts of which work together harmoniously to accomplish the purpose assigned to it by an all-wise Creator (Davis, 1924; 165–169; Guyot, 1873).

NEW APPROACHES TO GEOGRAPHY IN AMERICA

Not all the scholars who contributed to knowledge about the earth were Europeans or derived their ideas directly from European sources.[12] Several made notably original contributions to geographic knowledge. Among these were George Perkins

---

[11] In addition to Arnold Guyot, who taught at the College of New Jersey (Princeton) from 1854 to 1880, (Ferrell, 1981) and Daniel Coit Gilman, others who were important in the introduction of European ideas into American geography include: Jedidiah Morse (1761–1826); whose texts, *American Geography* and *American Universal Geography*, were published in 1789 and frequently revised thereafter; they were widely used in schools and read in American homes for many decades (James, 1969:474–475); John Daniel Gross, who was professor of German and geography at Columbia College from 1784 to 1795; John Kemp, who was professor of geography at Columbia College from 1795 to 1812; Louis Agassiz, who was professor of zoology at Harvard from 1848 to 1873 and introduced the ideas of natural history to America. He was strongly opposed to Darwin's concept of evolution through the survival of the fittest; his student, N. S. Shaler, was the teacher of William Morris Davis (Lurie, 1960).

[12] A more complete survey of the history of geographical ideas would include other eighteenth-century Americans. For example, like Adams and Jefferson Benjamin Franklin was a keen observer and, for his time, a careful scientist. His discovery of the nature of lightning and of electricity is well known. He also was the first to measure the temperature of the Gulf Stream. There was Hugh Williamson, who in 1760 observed that the warmer the waters of the Gulf Stream, the colder the weather that might be expected in New England. Lewis Evans, whose *Analysis of a Map of the Middle British Colonies of America* was published by Franklin in 1755, is credited with being the leading geographer of his time (Brown, 1951:192; Davis, 1924:160–162). Either Evans or Franklin first recognized that storms with northeast winds actually came from the southwest.

Marsh, Matthew Fontaine Maury, and the numerous outdoorsmen who participated in the exploration and survey of the American West (Allen, 1997; Colby, 1936; Curti, 1943; Glick, 1974; Wheat, 1958–1962).

George Perkins Marsh

Marsh was described by David Lowenthal as the "versatile Vermonter" (Lowenthal, 1958) and later "prophet of conservation" (Lowenthal, 2000). Few men have excelled in a wider range of fields of interest. After graduating from Dartmouth in 1820 and passing examinations for the bar, Marsh set up a law office in Burlington, Vermont, where for many years he carried on a small practice. In 1844 he was elected to Congress by the Whig party, but when he was defeated for reelection in 1848 he was appointed the U.S. minister to Turkey. In 1861 President Lincoln named him minister plenipotentiary to the Kingdom of Italy, a post that he held until his death in 1882. During all these years he wrote on an amazing variety of scholarly questions. He was a master of the English language and in addition could read some 20 other languages. His books include a grammar of the Icelandic language, a treatise on the habits and uses of the camel (which he recommended for importation to the dry parts of America), and a book on the origin and history of the English language.

Marsh also occupies an important position in the history of geographical ideas. At an early age he began to notice the destructive effects of man's use of the land. In his wide reading, especially of the works of Humboldt, Ritter, Guyot, and Mary Somerville (British writer, 1780–1872), he recognized the stirrings of a geography different from its predecessors, one focusing on the close interconnections between man and his natural surroundings. Marsh, who had never studied geography as such, developed a novel approach to the study of the relations of man to the land: He turned his attention to man's effect on nature, to the modifications of the organic and the inorganic parts of the habitat that resulted from human action. This was the point of view that Plato missed and that Buffon promoted. But Marsh began to seek examples of man's destructive use of land as a result of the damage he had seen done by widespread forest clearing in Vermont. His years of residence in Turkey and Italy gave him an opportunity to observe even more startling examples of the damage done by human action.

Many years of field observation and reading went into the preparation of Marsh's great works (Marsh, 1864, 1874). He had been working on the theme of the modification of nature by human action long before he left for Turkey. In 1847 he gave an address to the Agricultural Society of Rutland, Vermont, on "man's alteration of the landscape, intentional and unintentional, desirable and dangerous" (Rosenkrantz and Koelsch, 1973, pp. 340–368). He describes the objectives of his book, *Man and Nature*, as follows:

> To indicate the character and, approximately, the extent of the changes produced by human action in the physical conditions of the globe we inhabit; to point out the dangers of imprudence and the necessity of caution in all operations which, on a large scale, interfere with the spontaneous arrangements of the organic and of the inorganic worlds; to suggest the possibility and the importance of the

restoration of disturbed harmonies and the material improvement of wasted and exhausted regions; and, incidentally, to illustrate the doctrine that man is, in both kind and degree, a power of a higher order than any of the other forms of animated life, which, like him are nourished at the table of bounteous nature (Marsh, 1864:iii).

Marsh's warning was sounded in a country with seemingly endless resources and at a time when the need for conservation programs had yet to be formulated. In Russia a similar note was sounded in 1901 by Alexander Ivanovich Voeikov, who was especially concerned about the destruction of the grasslands owing to over-grazing and about the supposed modifications of climate that resulted from changes in the cover of vegetation (Voeikov, 1901). In America it was not until 1905 that Nathaniel S. Shaler, professor of geology at Harvard and a student of Louis Agassiz, returned to the theme of man's destructive effect on earth resources. Shaler, however, was especially concerned about the depletion of mineral resources, whereas Marsh paid slight attention to minerals (Shaler, 1905). Marsh had to await rediscovery until the modern period, when the destructive effect of human action was recognized as a major and very practical concern (Thomas, 1956).

Matthew Fontaine Maury

Another American who contributed major new concepts about the earth during the nineteenth century was Matthew Fontaine Maury, a Virginian by birth who was brought up in the backlands of rural Tennessee (Leighly, 1977; Williams, 1963), Maury received an appointment as midshipman in the U.S. Navy and was on the *Vincennes* when it became the first naval ship to be sailed all the way around the world, a voyage that lasted from 1826 to 1830. At an early age Maury had developed an insatiable curiosity concerning all matters that lay beyond his immediate horizon. His voyage around the world left him with many unanswered questions concerning the characteristics of the oceans. In 1839 Maury was appointed director of the Navy Depot of Charts and Instruments (which later became the U.S. Naval Observatory and Hydrographic Office). He devised a blank form for ship's logs on which the captains could enter specific observations of winds and currents and other conditions of the sea. Each observation was located by latitude and longitude; when the logs were sent back to Washington, the data were plotted on maps. Maury also devised new instruments for sounding ocean depths and was, therefore, able to produce the first map of the floor of the North Atlantic Ocean—information of the highest practical value in planning the route of the first transatlantic cable. The wind and current data were plotted on charts that were published along with an explanatory text. On the basis of the new picture of winds and currents his data revealed, Maury was able to advise ships' captains concerning the best routes to follow. His sailing directions cut the trip from New York to Rio de Janeiro by 10 days. The trip from New York to San Francisco, which used to take an average of 183 days, was cut to 135 days. Maury's sailing directions were fully as important as new rigging and new design in making possible the speed records of the clipper ships (Maury, 1851).

Maury was not satisfied just to collect data. He sought to develop a generalized picture of the surface winds of the earth by identifying the prevailing winds and

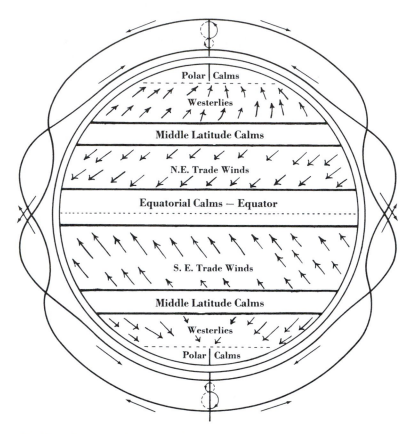

Figure 22 Maury's model of atmospheric circulation

eliminating the temporary and local interruptions. His model of atmospheric circulation is shown in Figure 22 (Maury, 1850:137; 1855:75). Along the equator is a zone of equatorial calms, which became known as the doldrum belt. On either side of the equator as far as about latitude 30° are the trade wind zones, with prevailing winds from the northeast in the Northern Hemisphere and from the southeast in the Southern Hemisphere. At about latitude 30° in each hemisphere is a zone of middle latitude calms, which became known as the horse latitudes. In the middle latitudes, roughly between 30° and 60° in each hemisphere, are the prevailing westerlies. Maury showed the regions around both poles as zones of calms.

It is instructive to note the way Maury solved the problem of generalizing his information about wind directions. He recognized that there would be many interruptions of his simplified scheme owing to irregularities in the distribution of land and water. Major interruptions are the monsoons and the local land and sea breezes. He did not include these interruptions on his model. For example, his information about wind directions in the tropical South Atlantic (Fig. 23) showed that there was a high probability of encountering northeast winds rather than southeast winds along

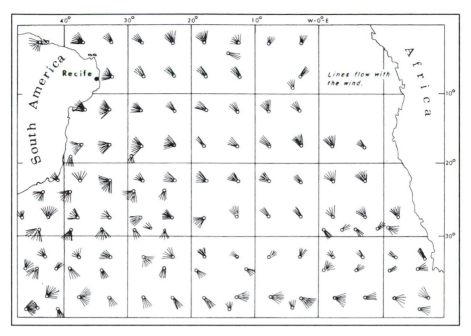

Figure 23 Maury's wind chart of the tropical Atlantic Ocean, 1859

the coast of Brazil south of 10°S. In his sailing directions he was sufficiently realistic to suggest that captains should stay close to the coast on the way to Rio de Janeiro, and it was this recommendation that cut the sailing time from New York. On the other hand, he did not modify his generalized model to take these northeast winds into consideration. His wind zones and belts of calms, running all around the earth, fitted so well into the traditional torrid, temperate, and frigid zones that Maury's model was widely accepted and taught in schools.

Maury also showed the wind directions aloft around the margin of his generalized model. If air moves steadily from latitude 30° toward the equator and if observations of air pressure show that pressure is high in the horse latitudes and low along the equator, then clearly there had to be a return current of air aloft moving in a direction opposite to the surface winds. If the surface winds in the middle latitudes are from the southwest or the northwest, air must be moving in the opposite direction aloft unless the equilibrium of the world's atmosphere were to be destroyed. His vertical section shows winds crossing each other (Fig. 22). Modified later to avoid this difficulty, diagrams showing the atmosphere in vertical section have been widely used in elementary texts.

Having developed his model, Maury proceeded to make deductions from it. For example, the coastal desert of Peru, lying in the zone of the southeast trades, can be explained by the descent of air on the lee side of the Andes. This notion, that the Peruvian desert is a lee coast in the trade wind zone, is still found in many books. Yet it is an established fact, with which Maury was well acquainted, that the surface winds along the Peruvian coast come from the southwest. These interruptions of Maury's model should have suggested the need to develop a different model, yet because of the wide acceptance of Maury's concept, the error contained in the concept was permitted to persist (James, 1964).

Maury himself was very explicit about his attempt to formulate a general picture of the movements of air and water. He said:

> I am wedded to no theories, and do not advocate the doctrines of any particular school. Truth is my object. Therefore, when the explanation which I may have at any time offered touching any facts fails to satisfy further developments, it is given up the moment one is suggested which will account for the new, and equally as well for the old system of facts. In every instance that theory is preferred which is reconcilable with the greatest number of known facts (Maury, 1855; quote from preface to the sixth edition, 1856:xv).

Maury also gave his support to an idea that was being hotly debated during the middle years of the nineteenth century—the concept of an open polar sea. This concept had had a long history and perhaps dated from 1527, when an English merchant suggested using the polar route to reach the Spice Islands (Wright, 1953). It was given strong support when a Dutch geographer in the seventeenth century provided a hypothetical explanation for the existence of a mild climate near the pole. Maury believed that the warm Gulf Stream flowed under colder surface water and emerged again at the surface in the vicinity of the pole. In 1853, when Elisha Kent Kane was preparing for a voyage to the far north, Maury explained his belief in the existence of such an open sea at great length. When Kane's ship was frozen in the ice with no open water in sight, he sent a sledge party farther to the north. At latitude 80°30′N, standing on a high cliff in northern Greenland, the leader of this party reported that "not a speck of ice was to be seen." The vast open sea extended northward beyond the horizon and was moved by the kind of huge swell that only could develop on a large expanse of open water. He provided the perfect demonstration of how a concept, vividly presented, can be so firmly implanted in the mind that evidence supporting it can be perceived and confidently reported. The idea of an open polar sea was not finally abandoned until the voyage of Fridtjof Nansen of the *Fram* (1893–96), who came to within a few degrees of the pole and did not find open water.

Maury sought to extend his collection of wind and current data by international cooperation. Enlarging on a suggestion made by the British, Maury promoted the idea of an international conference to which delegates from all the major maritime powers would be sent. He also wanted to include a system of land observations, but at this same time Joseph Henry, director of the Smithsonian Institution in Washington, was attempting to set up a network of weather-reporting stations in the United States and Canada for the study of storms. In 1835 the Russians had already set up such a network as a result of Humboldt's visit. But Henry did not like the idea of cooperating with Maury. When the international conference was held in Brussels in 1853, the proposed cooperation was restricted to observations at sea. As a result of the conference, the flow of new data into Washington was greatly increased. Humboldt himself gave his support to Maury for establishing a new field of scholarship—the physical geography of the sea.[13] *The Physical Geography of the Sea* (1855) by Maury remained the standard work on the subject for many years. It was translated into six languages and experienced numerous editions. In 1861 the work was expanded and retitled *The Physical Geography of the Sea and Its Meteorology*.

Other American Climatologists

The conflict that developed between Maury and Joseph Henry is one of those sad episodes in the history of geography that have become more and more common during the past century. Maury, as we have seen, lacked an academic background. As a scientist he was self-taught, but he made up for his deficiency by a boundless enthusiasm and by a vast experience. He was a salesman, a promoter who sold the basic notion that knowledge of the ocean and of the atmosphere could only be brought together by widespread cooperation. His observations at sea had been carried out by sailors under proper direction. Why, said Maury, should not a similar system of weather observations be extended to the land and be carried out by farmers? In the 1850s he worked steadily for the establishment of a national weather bureau, even to the extent of securing a leave of absence for the purpose of making a lecture tour to gain popular support.

Joseph Henry was a scientist. As a student of physics he had invented some of the devices required for electric motors and generators. He and Michael Faraday had developed these devices independently and at the same time, but Faraday had published his results sooner. In 1832 Henry had joined the faculty of the College of New Jersey (Princeton), where he taught physics and mathematics and a variety of other subjects, including geology and astronomy. In 1846 he was appointed the first secretary of the new Smithsonian Institution in Washington. Among the projects that Henry undertook at the Smithsonian was the establishment of a network of weather stations, each of which was to report to Washington by telegraph. Henry wanted these reports in order to chart the movements of winter storms. When Guyot set up a series of weather stations in both New York state and Massachusetts, it further strengthened Henry's weather station plan.

Of course, the two proposals—one by Maury and the other by Henry—met head on. Although Maury continued to insist that the proposals were not competitive, the fact is that Maury had access to government funds and Henry did not. Henry continued to oppose Maury's plan, insisting that Maury was not qualified to direct such a program or to make use of the results. Maury's attempted explanations of atmospheric circulation and his endorsement of the open polar sea only made

---

[13] Maury's career as a naval officer and as a geographer came to an abrupt end in 1861 when he resigned his commission to join the Confederacy. Born in Virginia, he felt a strong sense of loyalty to his native state. For the defense of the James River, he invented the first torpedoes to be detonated by electricity. He went to England to purchase supplies and ships for the Confederacy. After the war he went to Mexico briefly to promote the settling of Virginian colonists who wanted to leave the United States. From 1868 to the time of his death in 1873, he was professor of physics at the Virginia Military Institute. His school textbooks in geography were widely adopted (Williams, 1963; Leighly, 1977).

Henry rate him still lower as a scientist. It was the familiar story of the enthusiastic promoter with little capacity for patient scientific work meeting the careful scientist. Henry was probably right: If Maury's plan had been approved by the Congress it would have submerged the Smithsonian project. So Henry with a small group of leading scientists saw to it that Maury's plan did not pass.

An official U.S. Weather Service was established in 1870 and assigned to the Signal Corps of the Army. In 1891 it was transferred to the Department of Agriculture, and in 1940 it was moved from Agriculture to the Department of Commerce.

Meanwhile, several scholars in America were working with climatic data (Leighly, 1949). Samuel Forry made one of the earliest maps of the United States showing the distribution of temperature by making use of Humboldt's isotherms. In 1853 Lorin Blodget used new climatic data to draw a temperature map, but, since he permitted his isotherms to end on the map and to split, his method was something less than expert. In 1857 he published *The Climatology of the United States and of the Temperate Latitudes of the North American Continent*. This was perhaps the most valuable contribution on the subject to this time. His studies of the impact of climate on both agriculture and industrial resources were considered valuable, and his study relating the climatic impact on industry in Pennsylvania might be considered a contribution to location studies. His relevance as a representative of the new geography is also revealed in his engineering work for the Pacific railroad surveys in the 1850s and his later office as secretary of the Philadelphia Board of Trade. In 1854 James H. Coffin prepared a wind map of the Northern Hemisphere, and in 1875 he extended the map to the whole globe. His paper on the laws of atmospheric circulation went far beyond Maury. In 1846 Loomis presented the first synoptic weather map that was helpful in weather prediction. He went beyond this in 1868 with *Treatise on Meteorology* and in 1882 compiled the first rainfall map of the earth (James and Jones, 1954:334–361).

Meanwhile, a major contribution to the understanding of atmospheric circulation had been made by a high school teacher of mathematics in Nashville, Tennessee, William Ferrel. In 1865 he had read Hadley's explanation of the deflection of moving air by the earth's rotation and decided to work out the problem for himself. To Ferrel goes the credit for giving the first mathematical explanation of the way moving bodies must behave on a rotating sphere.[14]

THE GREAT SURVEYS

In addition to the collection of new data on population or the oceans or the world's climates, another kind of geography emerged from studies of geology in

---

[14]William Ferrel, "The Motions of Fluids and Solids Relative to the Earth's Surface," *Mathematical Monthly* 1 (1859):140–148, 210–216, 300–307, 366–373, 397–406; 2 (1859–60):89–97, 339–346, 374–390. There was also a brief report in *American Journal of Science and Arts* 31 (1861):27–50. For a review of the various scholars who have contributed to an understanding of the deflective force of the earth's rotation, including the work of G. G. Coriolis (1835), see C. L. Jordan, *Bulletin of the American Meteorological Society* 47 (1966):401–403; see also 47 (1966):887–891.

the nineteenth century in the form of systematic surveys of unsettled territory. By midcentury the United States was in possession of a vast extent of thinly occupied land stretching roughly from the one hundredth meridian west to the Pacific Ocean. California was occupied, and the discovery of gold had resulted in a rush of new settlers. But what was the land like in between? (Worster, 1985). This country had been penetrated initially by Indians, then by fur trappers, missionaries, and a long list of explorers including Zebulon M. Pike, Stephen H. Long, Lewis Cass, John C. Frémont, William H. Emory, and Isaac Stevens (Carter, 1999).[15]

Lewis and Clark

The Lewis and Clark Expedition of 1803–1806 was also primarily exploratory. The chief purpose was to explore the Missouri River and its tributaries and to find the best route to the headwaters of the Columbia River and thence to the Pacific Ocean. But the president, Thomas Jefferson, who was a keen student of geography, gave Lewis and Clark a careful and detailed directive. In addition to establishing the latitude and longitude of key places along the route, the expedition was to prepare a systematic record of observations concerning the nature of the country and its inhabitants. Jefferson specified that the reports should include observations on the number and characteristics of the Indians, their manner of making a living, their languages, and their relations to neighboring tribes. The reports should also provide information on the soil and face of the country, its vegetation, its animals, its minerals, its climate, including temperature; proportion of rainy, cloudy, and clear days; the occurrence of lightning, hail, snow, and ice; the prevailing wind directions; the dates of the first and last frosts; and the time of the year when particular plants lose their leaves (Coues, 1893/1965). The expedition left Saint Louis in May 1804 and returned in September 1806. The diaries and maps represent the most useful collection of data concerning a previously little-known land area ever assembled up to that date (Dillon, 1965; McLaughlin, 2003).

Ferdinand V. Hayden

The term "Great Surveys" is usually applied to the expeditions specifically organized to map and make inventories of the western territories. There were four such expeditions working more or less simultaneously in the years 1866 to 1879 (Bartlett, 1962). The four leaders were Ferdinand V. Hayden, Clarence King, George M. Wheeler, and John Wesley Powell.

Hayden directed the U.S. Geological and Geographical Survey of the Territories, which was sent out by the Department of the Interior each year between 1869 and 1878. Hayden's men surveyed the mountain country of northern Colorado and

---

[15] Major Emory surveyed and prepared a geographic report on the United States-Mexican border, House Document 135, 34th Cong., 1st Sess., 1857; also *Reports of Explorations and Surveys to Ascertain the Most Practical and Economical Route for a Railroad from the Mississippi River to the Pacific Ocean*, 1853–1855, Senate Document 46, 35th Cong., 2d Sess., 1855. For Isaac Stevens, see Meinig, 1955.

Figure 24 Group of all members of the Hayden Survey. In camp at Red Buttes, junction of the North Platte and Sweetwater Rivers

Figure 25 Geological exploration of the Fortieth Parallel in the King Survey

Figure 26 Lieutenant Wheeler, U.S. Engineer, camp and scientific party near Belmont, Nevada

Figure 27 The Powell party, Sweetwater County, Wyoming, May 1871

Wyoming. He reported on the geysers and hot springs of Yellowstone and the spectacular scenery of the Teton Mountains. Hayden was one of the most effective proponents of the creation of Yellowstone National Park. Hayden's men were the first to report on the cliff dwellings of the Mesa Verde region. His reports, however, were hurried and were less than objective. He was enthusiastic about the possibilities of settlement and even defended the notion that the rainfall would increase with settlement. Nevertheless, his maps and the landscape sketches prepared by William H. Holmes remain major accomplishments. One of his workers was Henry Gannett, who went on to become one of the founders of the Association of American Geographers and one of the earliest population geographers. The *Checklist of United Sates Public Documents, 1787–1901* lists some 50 publications of the Hayden Survey. Emanating from Hayden's survey work were the Timber Culture Act (1873) and the Desert Land Act (1877) stemming from his (erroneous) belief that rainfall followed settlement.

Clarence King

Clarence King, who in 1863 was operating the California Geological Survey, was appointed director of the U.S. Geological Exploration of the Fortieth Parallel. The survey carried out a geological cross section of this western country along latitude 40°N between 1867 and 1872. He identified and named the now-dry lake basins of the Great Basin in Utah and Nevada (Lake Bonneville and Lake Lahontan, respectively). His studies of the Uinta Mountains and the Wasatch Mountains made it possible for him to correlate some of the fossils found there with fossils of the same geologic age in Europe. Although the findings of his survey team were very considerable, he failed to convince the War Department that his survey should be broadened; it was felt that his approach was too narrowly geological.

In 1879 King was named the first director of the U.S. Geological Survey (which in that year replaced the separate surveys of the West). He resigned after one year.

George M. Wheeler

For many years before the Civil War (from 1813 to 1863), the U.S. Army Topographic Engineers had been engaged in making detailed maps of various parts of the United States. But during the war years this function was dropped. After the war the four surveys resumed the kind of work that had once kept the engineers busy in peacetime. General Humphreys of the Army Engineers felt that the surveys, especially those of Hayden and King, were too exclusively focused on geological problems and that any mapping accomplished was only incidental. For this reason he rejected King's bid to secure additional appropriations. But in 1871 he appointed Lieutenant George M. Wheeler, a West Point graduate, to head the U.S. Geographical Surveys West of the One Hundredth Meridian. Wheeler was directed not only to make topographic maps but also to undertake a land classification. The potential use of land was identified in four categories: (1) land suited for agriculture, (2) land suited for timber, (3) land suited for grazing, and (4) arid land. Between 1874, when the land classification work was added to the mapping, and 1879, when all the surveys were combined in the U.S. Geological Survey, some 175,000 square miles were classified. It is interesting that the army wanted not only maps showing landforms but also man-made features, such as mines, farms, roads, dams, and settlements. Wheeler was able to recruit some of the best-qualified field scientists available, including Grove Karl Gilbert, who made some highly important original contributions to the understanding of landforms. In 1875 Gilbert resigned to join the Powell survey.

In 1881 Wheeler was sent to Venice to attend the Third International Geographical Congress. His report on the principal government land and marine surveys of the world is an important record of the progress of topographic mapping since the earliest mapmaking efforts of the Russians in 1720 under Peter the Great (Wheeler, 1885).

John Wesley Powell

The story of John Wesley Powell is so important for the history of geography that it needs to be told in somewhat more detail (Davis, 1927; James, 1979; Pyne, 1980; Worster, 2001). Powell, like the other three leaders of Great Surveys, had not received any formal geographic training. From the enthusiasm of a high school teacher in Ohio, he developed a keen interest in natural history and a capacity to observe natural phenomena out of doors. In 1859 while he was a schoolteacher in Illinois, he became secretary of the Illinois Natural History Society. But when the Civil War began, he volunteered for the army. In the Battle of Shiloh he was wounded and lost an arm, but he continued on active duty and served as engineer officer in the siege of Vicksburg. He left the army as major in 1864.

By this time Powell was fascinated by the unknown country to the west. He sought funds to organize an expedition but was repeatedly turned down in Washington. He did get some funds from several Illinois colleges, and in 1867 and 1868 he was able to travel to the Rocky Mountains and climb Pikes Peak and Longs Peak. Powell became a national figure in 1869 when, with a small party and four boats, he sailed down the Green River and the Colorado River through the Grand Canyon (Darrah, 1951; Stegner, 1954).

In 1870 the U.S. Geographical and Geological Survey of the Rocky Mountain Region was financed by an appropriation from Congress. It continued to examine the mountain country until 1879, when it was combined with three other surveys of the West. In 1871 Powell carried out a second canyon trip, but most of his attention was given to surveying the land, the resources, and the people. He made the first records of Indian customs and language. In 1875 he secured the service of Grove Karl Gilbert, whose *Report on the Geology of the Henry Mountains* (Gilbert, 1878) is a classic in the field of landform studies. Gilbert is credited with the first clear statement of the concept of grade—that is, the equilibrium reached between slope, volume of water, velocity of flow, and load of detritus. Powell himself, during his first trip through the Uinta Mountains along the canyon of the Green River, saw for the first time just how a river could cut down through a mountain that was being raised up across its course. Fortunately, Powell had not read Buache and did not know that mountains were supposed to border drainage basins. He called a river that continues in its course through a rising mountain range an antecedent river.

Gilbert and Powell enjoyed two advantages not shared by the students of land-forms in Europe. In the western part of North America, there are large areas that are arid or semiarid, which means that the forms of the surface are not hidden under a thick cover of vegetation. Furthermore, these field men were not blinded by preconceptions based on earlier influential studies. Powell never had to consider the hypothesis set forth by Charles Lyell that landforms were chiefly the result of marine erosion: In the western part of the United States marine erosion did not seem plausible, and, clearly, the landforms had been sculptured by running water. Of course, the Colorado River had cut its own canyon—the conclusion was inescapable when the canyon was viewed from a boat. Here in the western part of the United States a truly new geography was born.

Powell was also concerned about the human use of this country, but he did not share Hayden's enthusiasm about the possibilities of settlement. He could find no evidence to support the idea that the rainfall would increase if more of the area were settled. In fact, his own observations led him to insist that the only way to maintain the flow of water from the mountains to the bordering lowlands was to preserve the forest cover. Without ever having read Humboldt, he nevertheless predicted that the clearing of the forests on the watersheds would bring disaster. One of his projects was to produce a map showing the relatively small areas that could be supplied with enough water to support irrigated farms.

In 1880 Powell became the second director of the U.S. Geological Survey, following King's resignation. The Bureau of Ethnology had been established the year before to carry on Powell's work with the Indian cultures. But the Geological Survey was specifically directed not to undertake geographical work (Powell, 1885). Why? Does it not seem clear that the Congress must have wanted information regarding the potential uses of land and about the recommended ways of maintaining the flow of water in the rivers? But no, the members of Congress were subjected to pressures from people who did not want Powell's information published. There were people who wanted to sell land to settlers: What of it if the land turned out to be in an area designated as arid by both Wheeler and Powell? There were also grazing interests that presented good arguments in favor of clearing the forests to enlarge the area of pasture. Neither of these groups wanted Powell's maps, and, in spite of vigorous efforts by Powell himself, the Congress directed him to continue the search for mineral resources.[16]

The story of Powell's years as director of the Survey from 1880 to 1894 offers a fascinating study in how the director of a government agency can find ways to carry out the programs he feels are necessary in spite of opposition. In his *Report on the Arid Regions of the United States*, Powell suggested a land classification system and new laws relevant to these lands. He recommended creation of 80-acre irrigation units with water rights attaching to ownership. But opposition developed in the form of politicians and groups with special interests in the West but also from scholars in the universities, whose opposition to Powell reminds us of the very similar treatment of Maury. From the scholars' point of view, the attacks on Powell

---

[16]Powell was also one of a small group of scholars in Washington who founded the Cosmos Club in 1878 (Stegner, 1954:242).

—and men like him—were entirely justified. Here was a man who had never held a teaching post except in a high school many years before. Here was a man with no scholarly reputation—a tough field observer, a man with a brilliant mind, but lacking the patience to do the tedious job of testing hypotheses; a man who would jump ahead to explanations before accumulating the needed supporting evidence. A field of study progresses because both kinds of people contribute to it, but this does not make the apparently inevitable conflicts any easier to live with.

The data and viewpoints developed by the Surveys replaced the conjectural geography of the time. A remarkable collection of publications replaced much of the guesswork and imagination with data drawn from the land. Americans, hitherto, had been too busy exploring their country and had left description and analysis to foreign writers (the first serious work on the geography of America was written by the German Christoph D. Ebeling in seven volumes, 1793–1816). The point is that the Surveys laid a foundation for the beginnings of a science of the land, suggestively encouraged legislation, and inspired a body of thought that was emerging into a new geography.

EXPEDITIONS AND GEOGRAPHICAL SOCIETIES

Gradually, then, the exploratory kind of expedition gave way to a new kind of expedition—a field survey. Of course, exploration continued: In the Arctic and Antarctic both poles were reached on foot (Robert Peary in 1909, Roald Amundsen in 1911, and Robert F. Scott in 1912) before airplanes could fly over them. Mountains were climbed (the summit of Mount Everest was reached by Edmund Hillary and Norgay Tenzing in 1953), and deserts were crossed for the first time by Europeans. A series of famous sea voyages was carried out by specially equipped research vessels: The *Challenger* expedition (1872–1876) traveled 69,000 miles bringing back to England 13,000 animals and plants, 1440 water samples, and hundreds of sea-floor deposits. Other vessels in *Challenger's* wake returned with large numbers of specimens that were sent to *Challenger's* specialists, resulting in 50 published volumes of this activity, beginning in 1880 and reaching completion in 1895.

These and many other scientific expeditions were reported chiefly in the various geographical societies that were founded during the nineteenth century. The prototype of all the geographical societies was the Association for Promoting the Discovery of the Interior Parts of Africa, formed in London in 1788. This and the Palestine Association were merged in 1830 to form the Royal Geographical Society. The Société de géographie was founded in Paris in 1821, and in 1828 the Gesellschaft für Erdkunde zu Berlin was founded. The first geographical societies in America were established in Rio de Janeiro (Instituto Histórico e Geográfico Brasileiro) in 1838 and in Mexico City (Sociedad Mexicana de Geografía) in 1839. The Russian Geographical Society was founded in Saint Petersburg in 1845, and the American Geographical Society of New York[17] was founded in 1851 (Wright, 1951, 1952). In 1888 the National Geographic Society was established in Washington, D.C.,

---

[17]Formed as the American Geographical and Statistical Society; the "and Statistical" was dropped in 1871 to shorten the name.

for the increase and diffusion of geographical knowledge. The first monograph of the society, published in 1896, included a series of papers on geography, one of which was Powell's division of the United States into physiographic regions.

By 1875 there were 28 geographical societies in Europe and one in Cairo (Ginsburg, 1972). The meetings of these societies and the periodicals each of them published were devoted to accounts of scientific expeditions in different parts of the world. The first International Geographical Congress was held in Antwerp in 1871, and congresses have been held approximately at four-year intervals since that time.[18] The International Geographical Union was formed in 1922 to tie all these and other geographical activities together and to coordinate their programs (Martin, 1996; Pinchemel, 1996).

REFERENCES: CHAPTER 7

Allen, J. L. 1997. Vol. 1 *North American Exploration: A New World Disclosed*. Vol. 2 *North American Exploration: A Continent Defined*. Vol. 3 *North American Exploration: A Continent Comprehended*. Lincoln, Neb.: University of Nebraska Press.

Armstrong, P. 1984. "Charles Darwin and the Development of Geography in the Nineteenth Century." *History of Geography Newsletter* No. 4:1–10.

———. 1991. *Darwin's Desolate Islands: A Naturalist in the Falklands, 1833 and 1834*. Chippenham, UK: Picton Publishing.

Bartlett, R. A. 1962. *Great Surveys of the American West*. Norman: University of Oklahoma Press.

Beck, H. 1956. "Heinrich Berghaus und Alexander von Humboldt." *Petermanns Geographische Mitteilungen* 100:4–16.

Bossuet, J. B. 1681. *Discours sur l'histoire universelle*. Paris.

Brown, R. H. 1951. "A Letter to the Reverend Jedidiah Morse, Author of *The American Universal Geography*." *Annals AAG* 41:188–198.

Carter, E. C., ed. 1999. *Surveying the Record: North American Scientific Exploration to 1930*. Philadelphia: American Philosophical Society.

Colby, C. C. 1936. "Changing Currents of Geographic Thought in America." *Annals AAG* 26:1–37.

Commission on History of Geographical Thought, International Geographical Union. 1972. *Geography through a Century of International Congresses*.

Coues, E., ed. 1893. *History of the Expedition under the Command of Lewis and Clark*. New York (republished in 3 vols., New York: Dover, 1965).

Curti, M. 1943. *The Growth of American Thought*. New York: Harper Bros.

Darrah, W. C. 1951. *Powell of the Colorado*. Princeton, N.J.: Princeton University Press.

---

[18]International Geographical Congresses have been held at Antwerp (1871); Paris (1875); Venice (1881); Paris (1889); Bern (1891); London (1895); Berlin (1899); Washington, D.C. (1904); Geneva (1908); Rome (1913); Cairo (1925); Cambridge, England (1928); Paris (1931); Warsaw (1934); Amsterdam (1938); Lisbon (1949); Washington, D.C. (1952); Rio de Janeiro (1956); Stockholm (1960); London (1964); New Delhi (1968); Montreal (1972); Moscow (1976); Tokyo (1980); Paris (1984); Sydney (1988); Washington, D.C. (1992); The Hague (1996); Seoul (2000); and Glasgow (2004).

Darwin, C. R. 1842. *The Structure and Distribution of Coral Reefs*, 2nd ed. 1874. London: John Murray (3rd ed., New York: D. Appleton, 1889).

———. 1859. *On the Origin of Species by Means of Natural Selection, or The Preservation of Favoured Races in the Struggle for Life*. London: John Murray.

Davis, W. M. 1915. "Biographical Memoir of John Wesley Powell, 1834–1902." *National Academy of Sciences, Biographical Memoirs*, vol. 8:11–83.

———. 1924. "The Progress of Geography in the United States." *Annals AAG* 14:159–215.

———. 1927. *Biographical Memoir of Grove Karl Gilbert, 1843–1918. Biographical Memoirs, National Academy of Sciences*. 21, no. 5. 303 pages.

———. 1928. *The Coral Reef Problem*. New York: American Geographical Society, Special Publication No. 9.

de Terra, Helmut. 1955. *Humboldt: The Life and Times of Alexander von Humboldt, 1769–1859*. New York: Knopf.

Dillon, R. 1965. *Meriwether Lewis*. New York: Coward-McCann.

Dunbar, G. S. 1978. *Elisée Reclus: Historian of Nature*. Hamden, Conn.: Shoe String Press.

———. 1981. "Lorin Blodget, 1823–1901." *Geographers: Biobibliographical Studies* 5:9–12.

Ferrell, E. H. 1981. "Arnold Henry Guyot, 1807–1884." *Geographers: Biobibliographical Studies* 5:63–71.

Gilbert, G. K. 1878. *Report on the Geology of the Henry Mountains*. Washington, D.C.: Department of the Interior.

———. 1890. *Lake Bonneville*. Washington, D.C.: U.S. Geological Survey, Monograph No. 1.

Ginsburg, N. S. 1972. "The Mission of a Scholarly Society." *Professional Geographer* 24:1–6.

Glick, T. F., ed. 1974. *The Comparative Reception of Darwinism*. Austin: University of Texas Press.

Goode, J. P. 1927. "The Map as a Record of Progress in Geography." *Annals AAG* 17:1–14.

Guyot, A. 1849. *The Earth and Man: Lectures on Comparative Physical Geography in Its Relation to the History of Mankind*. Boston: Gould & Lincoln.

———. 1873. *Physical Geography*. New York: Scribner, Armstrong & Co.

Hartshorne, R. 1939. *The Nature of Geography, a Critical Survey of Current Thought in the Light of the Past*. Lancaster, Pa.: Association of American Geographers.

James, P. E. 1964. "A New Concept of Atmospheric Circulation." *Journal of Geography* 63:245–250.

———. 1969. "The Significance of Geography in American Education." *Journal of Geography* 68:473–483.

———. 1979. "John Wesley Powell: 1834–1902." In *Geographers: Biobibliographical Studies*. Vol. 3, pp. 117–124. London: Mansell.

James, P. E., and Jones, C. F. eds. 1954. *American Geography, Inventory and Prospect*. Syracuse, N.Y.: Syracuse University Press.

Joerg, W. L. G. 1922. "Recent Geographical Work in Europe." *Geographical Review* 12: pp. 431–484.

Leighly, J. 1938. "Methodological Controversy in Nineteenth Century German Geography." *Annals AAG* 28:238–258.

———. 1949. "Climatology since the Year 1800." *Transactions of the American Geophysical Union* 30:658–672.

———. 1977. "Matthew Fontaine Maury: 1806–1873." In *Geographers: Biobibliographical Studies*. Vol. 1, pp. 59–63. London: Mansell.

Libbey, W., Jr. 1884. "The Life and Scientific Work of Arnold Guyot." *Bulletin of the American Geographical Society* 16:194–221.

Livingstone, D. N. and C. W. J. Withers, eds. 1999. *Geography and Enlightenment.* Chicago: University of Chicago Press.

Lorenz, E. N. 1966. "The Circulation of the Atmosphere." *American Scientist* 54:402–420.

Lowenthal, D. 1958. *George Perkins Marsh, Versatile Vermonter.* New York: Columbia University Press.

———. 2000. *George Perkins Marsh: Prophet of Conservation.* Seattle: University of Washington Press.

Lurie, E. 1960. *Louis Agassiz: A Life in Science.* Chicago: University of Chicago Press.

Marsh, G. P. 1864. *Man and Nature, or Physical Geography as Modified by Human Action.* New York: Charles Scribner (republished, David Lowenthal, ed.; Cambridge, Mass.: Harvard University Press, 1965).

———. 1874. *The Earth as Modified by Human Action*, 2nd ed. 1885. New York: Charles Scribner.

Martin, G. J. 1996. "One Hundred and Twenty-five Years of Geographical Congresses and the Formation of the International Geographical Union: Or from Antwerp to The Hague." *I.G.U. Bulletin* 46:5–26.

———. 1998. "The Emergence and Development of Geographic Thought in New England." *Economic Geography* extra issue. 1–13.

Maury, M. F. 1850. "On the General Circulation of the Atmosphere." *Proceedings of the American Association for the Advancement of Science* 3:126–147.

———. 1851. *Explanations and Sailing Directions to Accompany the Wind and Current Charts.* Washington, D.C.: C. Alexander.

———. 1855. *The Physical Geography of the Sea.* New York: Harper Bros.

Mayhew, R. J. 2000. *Enlightenment Geography: The Political Languages of British Geography, 1650–1850.* New York: St. Martin's Press.

McLaughlin, C. 2003. *Arts of Diplomacy: Lewis and Clark's Indian Collection.* Seattle: University of Washington Press.

Meinig, D. W. 1955. "Isaac Stevens: Practical Geographer and Historian." *Geographical Review* 45:542–558.

Newton, H. A. 1895. "Biographical Memoir of Elias Loomis." *Biographical Memoirs. National Academy of Sciences* 3:213–252.

Pinchemel, P., M. C. Robic, A. M. Briend, and M. Rössler. 1996. *Géographes face au monde: l'Union géographique internationale et les Congres internationaux de géographie.* Paris: L'Harmattan.

Powell, J. W. 1878. *Report on the Lands of the Arid Region of the United States with a More Detailed Account of the Lands of Utah.* Washington, D.C.: 45th Congress, 2nd Session.

———. 1885. "The Organization and Plan of the United States Geological Survey." *American Journal of Science* 29:93–102.

Pyne, S. J. 1980. *Grove Karl Gilbert: A Great Engine of Research.* Austin: University of Texas Press.

Reclus, E. 1867–1868. *La terre, description des phénomènes de la vie du globe.* 2 vols. Paris: Hachette.

———. 1869. *Histoire d' un ruisseau.* Paris: J. Hetzel.

———. 1876–1894. *Nouvelle géographie universelle, la terre et les hommes.* 19 vols. Paris: Hachette. English translation by E. G. Ravenstein and A. H. Keane. *The Earth and Its Inhabitants.* London: 1878–1894.

———. 1880. *Histoire d' une montagne.* Paris: J. Hetzel.

———. 1905–1908. *L'homme et la terre.* Paris: Librairie Universelle.

Rosenkrantz, B. G. and W. A. Koelsch. 1973. *American Habitat: A Historical Perspective.* New York: Free Press.

Rutherford, J. and P. Armstrong. 2000. "James Cook, RN 1728–1779." *Geographers: Bio-bibliographical Studies* 20:9–23.

Shafer, R. J., ed. 1969. *A Guide to Historical Method.* Homewood, Ill.: Dorsey Press.

Shaler, N. S. 1905. "Earth and Man: An Economic Forecast." *International Quarterly* 10:227–239.

———. 1912. *Man and the Earth.* New York: Duffield & Co.

Short, John R. 2001. *Representing the Republic: Mapping the United States, 1600–1900.* London: Reaktion Books.

Sinnhuber, K. A. 1959. "Carl Ritter, 1779–1859." *Scottish Geographical Magazine* 75:152–163.

Stegner, W. 1954. *Beyond the Hundredth Meridian: John Wesley Powell and the Second Opening of the West.* Boston: Houghton Mifflin.

Stoddart, D. R. 1966. "Darwin's Impact on Geography." *Annals AAG* 56:683–698.

———. 1985. "Humboldt and the Emergence of Scientific Geography" in P. Alter, ed. *Humboldt.* London: William Heinemann.

———. 1986. *On Geography and Its History.* Oxford and New York: Basil Blackwell.

Thomas, W. L., ed. 1956. *Man's Role in Changing the Face of the Earth.* Chicago: University of Chicago Press.

Van Orman, R. A. 1984. *The Explorers: Nineteenth-Century Expeditions in Africa and the American West.* Albuquerque: University of New Mexico Press.

Voeikov, A. I. 1901. "De l'influence de l'homme sur la terre." *Annales de géographie* 10:97–114, 193–215.

Wagner, H. 1880. "Bericht über die Entwicklung der Methodik der Erdkunde." *Geographisches Jahrbuch* 8:523–598.

Ward, R. deC. 1914. "Lorin Blodget's *Climatology of the United States,* an Appreciation." *Monthly Weather Review* 42:23–27.

Wheat, C. I. 1958–1962. *Mapping the Trans-mississippi West, 1540–1861.* 5 vols. Menlo Park, California: The Institute of Historical Cartography.

Wheeler, G. M. 1885. *Report upon the Third Geographical Congress and Exhibition at Venice, Italy, 1881. Accompanied by Data Concerning the Principal Land and Marine Surveys of the World.* Washington, D.C.: House Executive Document 270, 48th Congress, 2nd Session.

Williams, F. L. 1963. *Matthew Fontaine Maury, Scientist of the Sea.* New Brunswick, N.J.: Rutgers University Press.

Worster, D. 1985. *Rivers of Empire: Water, Aridity, and the Growth of the American West.* Oxford: Oxford University Press.

———. 2001. *A River Running West: The Life of John Wesley Powell.* New York: Oxford University Press.

Wright, J. K. 1951. "The Field of the Geographical Society." In *Geography in the Twentieth Century.* Pp. 543–565. G. Taylor, ed., New York: Philosophical Library.

———. 1952. *Geography in the Making: The American Geographical Society, 1851–1951.* New York: American Geographical Society.

———. 1953. "The Open Polar Sea." *Geographical Review* 43:338–365.

———. 1961. "Daniel Coit Gilman: Geographer and Historian." *Geographical Review* 51:381–399.

Yochelson, E. L., ed. 1980. "The Scientific Ideas of G. K. Gilbert." *Geological Society of America Special Paper* 183.

8

THE NEW GEOGRAPHY IN GERMANY

If Galileo's famous experiment had merely demonstrated that when he, Galileo, dropped two specific objects of different weight from the Leaning Tower of Pisa they fell together at the same rate of speed, that fact would have found but a small place in scientific knowledge. Its great importance, of course, was that subsequent experiments showed that he had illustrated a universal, a relation that was true regardless of where, when, or by whom the weights were dropped. Few will question that it is an essential function of science to seek for such universals. On the other hand if later experiments had shown that the same results were not obtained elsewhere or at other times, the fact that they did take place on that one occasion, if substantiated as a fact, even though never explained, would represent a bit of scientific knowledge. . . . One might say that it is an axiomatic ideal of science to attain complete knowledge of reality—expressed as completely as possible in terms of universals, but in any case expressed in some way.

—R. Hartshorne, The Nature of Geography

The nineteenth century witnessed a fundamental change in the role of universities. The traditional university, as we have seen, was a place where students could be indoctrinated in the religious and political beliefs of the community and where studies were focused on Greek and Latin classics, on theology and law, or logic and rhetoric. But as the newly identified disciplines came into existence, the universities had to assume responsibility for the advanced training of younger generations of scholars. The members of each professional field set the standards of professional behavior (Kuhn, 1963). This kind of education required courses of advanced study in graduate schools.

The change started in Germany in 1809 with the establishment of the University of Berlin as a free community of scholars. Slowly at first, then more rapidly after the middle of the century, two principles came to be accepted regarding the new functions of universities. First, the students were freed from standard curricula and were permitted to select whatever course of study interested them. Second, the appointments to positions on the faculties were made largely on the basis of scholarly performance; after an appointment was made, the faculty member was granted the right to engage in research and to teach the results of his research free from any restraints except those established by professional opinion.

When these major innovations in the functions of universities began to appear, only a few universities in the world offered a young scholar training in the concepts and methods of a field called geography. Only a very few university teachers were conscious of being geographers, and most of these were pupils of Carl Ritter at Berlin.

Yet geographical research was being carried on. In America, as we have seen, the field men working on the Great Surveys were not trained as geographers. They were geologists, naturalists, engineers, or men trained for the military services. They were curious about the processes that produced the unfamiliar landscapes of the American West, but they were also faced with questions of pressing practical importance. There was an urgent need to know more about the resources and shortcomings of a region that was about to be occupied by waves of new settlers.

In Europe the situation was quite different. There were no nearby unoccupied regions where new pioneer settlement created an obvious need for more knowledge about the earth as the potential home of man, nor were there regions of unfamiliar appearance to challenge the curiosity. Men such as Alexander von Humboldt had to travel to other continents to find such regions. In Germany there were centers of geographic research where the observations of scientific travelers could be collected and where exciting new maps were being devised to present this information in useful form. Geographical societies were organized so that the latest knowledge about the earth could be presented in lectures or in professional periodicals. But advanced instruction in universities was not available. In spite of Wilhelm von Humboldt's effort to improve the quality of instruction in the secondary schools, there were no places where teachers of geography could be trained. The result was a continued decline in the content of secondary-school geography, and even a return to the kind of rote learning against which Pestalozzi and Ritter had waged so vigorous a campaign.

Geography as a field of advanced study in the universities appeared in Germany in the 1870s. From Germany the movement spread rapidly to France, to other European countries, and also to America. In most countries after 1870, it is possible to identify one outstanding pioneer who was chiefly responsible for establishing the scope and methods of the new geography at the university level. But at first the persons who were appointed to university positions in geography had had no previous training in this field. There was no professionally accepted paradigm to serve as a guide to the study of geography. The new appointees had been trained in history, geology, botany, zoology, mathematics, engineering, or journalism. In the absence of any guidelines regarding the field of geography, each new professor felt the need to set forth his own ideas concerning the scope of the field. Each tried to provide a definition of geography that would give it unity and that would establish its position among other academic disciplines. All around the world in the late nineteenth century the question echoed through the academic corridors: What is geography? In this chapter we will review the answers that were given in Germany (Dickinson, 1969:51–185; Fischer, Campbell, and Miller, 1967:81–174; Hartshorne, 1939:84–148; Hettner, 1927; Schelhaas and Hönsch, 2001; Van Valkenburg, 1951).

THE NEW GEOGRAPHY IN GERMANY

After Ritter's death, geography in Germany lacked a focus to give it unity. Those who worked at Gotha were devoted to the careful plotting of information on maps and to the publication of clearer maps of finer design. There was no need at Gotha to ask what geography was—geography was anything that could be put on a map.

Those who were engaged in military careers appreciated the practical need for more accurate and useful information about the earth as the scene of warfare. Those who were soon to be administrators of Germany's new colonial possessions and those who were engaged in doing business away from home also needed more knowledge of geography—not theoretical geography, but compilations of useful facts about places. The former pupils of Carl Ritter, who were giving lectures on geography in some of the German universities, were chiefly concerned to provide a background for the study of history.[1]

Oscar Peschel

Oscar Peschel's appointment as professor of geography at Leipzig in 1871 was the first new professorship created after the death of Ritter. Peschel was then 45 years of age and had already established his reputation as an editor and a writer. He had been the assistant editor of the *Allgemeine Zeitschrift* at Augsburg and from 1854 to 1870 was the editor of *Das Ausland*, a periodical that printed articles about foreign countries and about problems of foreign affairs. He had written on the history of ancient geography, and on this basis the Historical Commission of the Royal Academy of Sciences invited him to write a book on the history of geography to form a part of a series on science in Germany. Peschel's *Geschichte der Erdkunde* was published in 1865 (Peschel, 1865).

His studies of the history of geography led him to take issue with Ritter's method of making comparisons between regions, which Ritter called *vergleichende Erdkunde* (comparative geography). Peschel pointed out that Ritter made his comparisons between whole continents or major parts of continents and that such units were really composite concepts and were not properly comparable. He demonstrated how he would make comparative studies by focusing attention on a particular kind of landform, which he examined on the most detailed maps available to him. For example, he studied the much-indented fjorded coasts, which, he observed, occur on the western sides of continents in higher middle latitudes. He offered the hypothesis that the fjords were fissures in the earth's crust that had been occupied and gouged out by glaciers. His systematic studies of these features together with lakes, islands, valleys, and mountains were published in 1870 (Peschel, 1870).

Peschel is credited with being one of the founders of modern physical geography. Yet he was not unmindful of the importance of showing the relation of the physical features of the earth to human use of the earth. He served for only four

---

[1]Between 1859 and 1871 courses of lectures in geography were being given at three German universities. At Berlin, where Ritter had been professor of geography, he was replaced by a *dozent* (more or less equivalent to an assistant professor). This was Heinrich Kiepert, a classical historian, who was promoted to professor in 1874. At Göttingen there was Johann Eduard Wappäus, who was appointed as *privat dozent* (assistant professor or instructor) in 1838 and became a professor in 1854. At Breslau there was Karl, J. L. Neumann, who had been appointed professor of geography and ancient history in 1856. G. B. Mendelssohn had been at Bonn, but he died in 1857 and was not replaced for two decades (Dickinson, 1969:51–55; Wagner, 1880).

years at Leipzig before he died at the early age of 49. His book on physical geography was published posthumously (Peschel, 1879).

New Professorships in Germany

In 1874 an event of major importance took place in Berlin. The Prussian government decided to establish a chair of geography (to be occupied by a scholar with the rank of professor) in each of the Prussian universities. Prussia was the largest and most influential of the separate political units that came together to form the German Empire in 1871 at the end of the Franco-Prussian War (1870–71). Prussia's action was followed elsewhere in the newly unified Germany.

Why did Prussia take this step in 1874? The answer is not entirely clear. It is reasonable to assume that the Franco-Prussian War resulted (as all wars do) in a popular demand for the teaching of more geography. The officers of the army, many of whom had studied geography under Ritter at the university and the Prussian Military Academy in Berlin, wanted more and better geography taught in the schools and universities. In addition, the new interest in people of German origin who had settled outside Germany after 1848 contributed to a recognition of the need to learn geography. It is also recorded that Hermann Wagner, then a teacher of geography at the gymnasium (secondary school) in Gotha, was greatly concerned about the poor preparation of secondary school teachers of geography. With Alfred Kirchhoff, who had lectured at the military academy in Berlin and had been appointed professor of geography at Halle in 1873, Wagner urged the Prussian government to provide advanced instruction in geography in the universities (Dickinson, 1969:59). When the government took this action, numerous openings for university professorships in geography appeared suddenly, and many of them were filled within a decade. By 1880 Hermann Wagner reported that there were professors of geography in 10 of the Prussian universities, and vacancies in 3 more were about to be filled (Wagner, 1880:591).

Ferdinand von Richthofen

A geologist, Baron Ferdinand von Richthofen, became the leading figure in the introduction of the new geography into the universities of Germany. He was an experienced field observer. As a young man he had carried out geological studies in the Alps and the Carpathians. In 1860 the Prussian government selected him to undertake an expedition to eastern Asia for the study of lands and resources. After working in China, he sailed across the Pacific to California, where he spent six years in active geological studies. Still challenged to find out more about the resources of China, he was able to get the Bank of California to finance field work there with the agreement that he would report his findings to the Chamber of Commerce in Shanghai. In his subsequent survey of China, he was the first to report and map the Chinese coal fields.

Richthofen did much more than locate minerals and fuels. He also sought to formulate hypotheses to explain China's surface features. He noted the presence of a fine, powdery material covering the land on the eastern side of the Gobi, and he was the first to identify this material as wind-blown dust, or *loess*. He also noted that in this part of Asia the loess, as well as the stratified rocks, were deposited over a relatively level or gently rolling surface and that this more-or-less level surface cuts across ancient rocks of varying degrees of resistance to erosion. He concluded that the only force strong enough to level such resistant rocks must be the ocean. Ocean waves working on gradually sinking land could, he believed, produce the great, nearly level plains of the world. His studies of China were published in five volumes over the period 1877 to 1912 (Richthofen, 1877–1912; Kolb, 1983).

Richthofen's answer to the question "What is geography?" was enormously influential inside and outside Germany. His was the pioneer statement of the scope and method of the new geography. His ideas were initially presented in the first volume of his study of China (Fischer, Campbell, and Miller, 1967:84–87; Richthofen, 1877–1912:1, 729–732) and then repeated in his inaugural lecture at Leipzig, delivered on April 27, 1883 (Fischer, Campbell, and Miller, 1967:88–95; Richthofen, 1883). It was the distinctive purpose of geography, he said, to focus attention on the diverse phenomena that occur in interrelations on the face of the earth. To reach useful and reliable conclusions, he believed, a geographical study of any part of the face of the earth must start with a careful description of the physical features and then must move on to an examination of the relationships of other features of the earth's surface to the basic physical framework. He distinguishes the study of the processes creating the surface features (geology) from the description of the surface features themselves as the frame of reference to which other elements of the face of the earth (including the works of man) are to be related. The highest goal of geography is the exploration of human relationships to the physical earth and to the biotic features that are also associated with the physical features. This became the basic pattern for geographical studies, not only in Germany but also in other parts of the world.

Richthofen also thought carefully about two other problems that bothered all the newly appointed professors of geography. The first of these problems was similar to the questions being asked during the same period by historians and economists. Is geography concerned only with describing the unique features of particular regions or is it also concerned with formulating generalizations or theory? The second problem had to do with the relationship of what Varenius had called *general geography* and *special geography*. Richthofen combined the answers to these two questions. In the first instance, he reasoned, the essential observations on which any framework of concepts must be built had to be made in the field in particular areas where the features are unique. This, he said, is *special geography*. Regional study first must be descriptive, but it must also go beyond the description of unique features to seek regularities of occurrence and to formulate hypotheses that explain the observed characteristics. Loess, for example, can be observed, measured, and carefully described, but a regional study must also look for the process by which loess is accumulated and for the consequences of the presence of loess in the cover of plants and in man's use of the region. In this procedure Richthofen was specifically following the lead of Humboldt as that master had carried out field studies in New Spain and Cuba.

Richthofen, being an experienced field observer, had a feeling for what could be learned from direct observation out of doors. The purpose of developing the general concepts regarding the world distribution of phenomena (general or topical geography) is to throw light on the causal interrelations among the diverse things in particular areas. To this process he applied the term *chorology,* or regional study— a term already widely used in Germany (Hartshorne, 1939:92). He also realized that in addition to looking at the world as a whole, it was necessary to examine smaller and smaller segments of the earth's surface. He distinguished the different methods of study in areas of different size, which he named (in order of decreasing size): *Erdteile* (major divisions of the earth), *Länder* (major regions), *Landschaften* (landscapes or small regions), and *Örtlichkeiten* (localities).

Furthermore, Richthofen helped define the word *Erdkunde.* Ritter had preferred this term, which is of German origin, rather than the word *Géographie,* which is of Greek origin. Ritter intended to use *Erdkunde* as a synonym for *Géographie,* but some of the German writers took the word literally to mean the study of the whole earth body. Richthofen insisted that geography (and *Erdkunde*) must refer to a study of the surface of the earth (*Erdoberfläche*), where the lithosphere, hydrosphere, atmosphere, and biosphere are in contact with each other.

Richthofen published another statement of the method of geographic study in 1886 when he accepted the professorship at Berlin.[2] He wrote a systematic outline of the processes shaping the surface of the earth that could serve as a guide to the field study of any region (Richthofen, 1886).

Friedrich Ratzel

Peschel and Richthofen had laid down the guidelines for the systematic study of the earth's physical features and at the same time had defined geography as a unified field of study by organizing the treatment of biotic and cultural features around the fundamental framework of the physical earth. It was Friedrich Ratzel who provided the guidelines for a comparable systematic study of human geography, or, to use the word he coined, *anthropogeographie.*

Ratzel, who was 11 years younger than Richthofen, came to geography from a quite different background (Bassin, 1987a). He completed his graduate study at Heidelberg in 1868, working in zoology, geology, and comparative anatomy. This was just that exciting period when Darwin's revolutionary concepts regarding the process of evolution were sweeping away old ideas. Ratzel's dissertation was perhaps inspired by the work of Ernst Haeckel. At this time he was studying zoology; it would be several more years before he was drawn to geography (Wanklyn, 1961).

---

[2]Richthofen occupied several professorships in German universities. In 1875 he was offered the chair at Berlin and was immediately granted leave of absence to work on the manuscripts of his China study. Then in 1877 he accepted an appointment at Bonn. In 1883 he moved to Leipzig, and then in 1886 he moved to Berlin. For a time he served as rector of the University of Berlin. Richthofen trained many of those who became leaders of the later generations of geographers in Germany and elsewhere. They included: E. Banse, O. Baschin, K. Dove, E. V. Drygalski, K. Futter, M. Groll, E. Hahn, S. Hedin, K. Kretschmer, W. Meinardus, S. Passarge, A. Philippson, A. Rühl, O. Schlüter, G. Schott, R. Sieger, W. Sievers, E. Tiessen, W. Volz, and A. Wegener.

But Ratzel was more interested in the observation of plants and animals out of doors than he was in laboratory studies. To study in the field required travel, and traveling cost money. He had his first taste of field study when he accompanied a French naturalist on a trip to the countries around the Mediterranean. When the Franco-Prussian War started in 1870, Ratzel joined the Prussian Army and was wounded twice during the war.

The unification of Germany in 1871 was, for Ratzel, a very exciting occurrence. His strong sense of national pride drew his attention away from academic studies to field observation of the ways Germans lived and made use of resources. He wanted to continue his traveling, if possible, to seek out and describe the people of German origin now living outside Germany. Some letters he wrote for a newspaper in Cologne so impressed the editor that the young Ratzel was employed as a roving reporter. Payment for articles written for the *Kölnische Zeitung* made it possible for him to extend his travels to more distant places. He visited Hungary and Transylvania, reporting on the German minorities in that part of eastern Europe. In 1872 he crossed the Alps for an extended visit to Italy.

Ratzel's visit to the United States and Mexico in 1874–75 was a turning point in his career (Sauer, 1971). Not only was he impressed by the contributions of people of German origin to American life in the Midwest and the Southwest, but also his attention was focused on the success of other minority groups, such as the Indians, the Africans, and the Chinese in California. He began to formulate some general concepts regarding the geographic patterns resulting from contact between aggressive and expanding human groups and retreating groups. He recognized and described examples of the destructive use of land (to which the Germans attach the expressive term *Raubbau*), which he hoped was a characteristic of land settlement at an early stage and would be remedied as settlement matured. He later published his studies of the United States, based on his own keen observations and a remarkably penetrating study of earlier accounts, especially regarding the settlement of the West. This experience in the interpretation of an extensive region turned his attention specifically toward the study of geography (Speth, 1977; Steinmetzler, 1956:73–74).

Returning to Germany in 1875, he resigned his position with the newspaper and was appointed *privat dozent* in geography at the Technische Hochschule (Institute of Technology) in Munich. He was promoted to *dozent* in 1876 and to professor in 1880. In 1886 he accepted an offer to become the professor of geography at Leipzig, where he remained until his death in 1904.

It was at Munich that Ratzel first began to publish his ideas concerning the systematic study of human geography. In 1882 he published the first volume of *Anthropogeographie*, in which he traced the effects of different physical features on the course of history (Dickinson, 1969:64–72; Hassert, 1905; Ratzel, 1882; Steinmetzler, 1956; Wanklyn, 1961). Meanwhile, some other geographers, notably Kirchhoff, had used the opposite approach to the study of human geography. Instead of describing the influence of the physical earth on human affairs, they focused on man himself. Human societies were studied in relation to the physical features, but the most attention was given to the culture of human groups rather than to the physical earth. This was the method that Ratzel adopted in the second volume of *Anthropogeographie*, the first edition of which was published in 1891 (Hartshorne, 1939:91).

Richthofen was much more influential than Ratzel among later generations of German geographers, but Ratzel had more influence than Richthofen outside Germany, especially in France and the United States. Ratzel, like Ritter, was a brilliant lecturer, and his classes at Leipzig were always full. Sometimes as many as 500 people crowded into his lecture room. One of his most eloquent disciples was the American student Ellen Churchill Semple, who studied at Leipzig in 1891–92 and again in 1895. Here is what she had to say about his ideas:

> Moreover the very fecundity of his ideas often left him no time to test the validity of his principles. He enunciates one brilliant generalization after another. Sometimes he reveals the mind of a seer or poet, throwing out conclusions which are highly suggestive, on the face of them convincing, but which on examination prove untenable, or at best must be set down as unproven or needing qualification. . . . He grew with his work, and his work and its problems grew with him. He took a mountain-top view of things, kept his eyes always on the far horizons, and in the splendid sweep of his scientific conceptions sometimes overlooked the details near at hand. Herein lay his greatness and his limitations (Semple, 1911:v–vi).

Among Ratzel's "brilliant generalizations" that led some of his disciples to carry his ideas further than he himself intended was the application of Darwin's biological concepts to human societies. This analogy suggested that groups of human beings must struggle to survive in particular environments much as plant and animal organisms must do. This is known as social Darwinism.[3] Ratzel, like many other scholars in this period, was greatly impressed with the ideas of Herbert Spencer regarding the similarity of human societies to animal organisms. In his *Political Geography*, Ratzel compares the state to an organism: "Der Staat als Bodenständiger Organismus" ("the state as an organism attached to the land") (Ratzel, 1897:Chap. 1). However, he is very careful to point out that this analogy is not to be taken literally, for there are many ways in which human groups act quite differently from organisms. The comparison is not intended, he says, to be a scientific hypothesis, but only an illuminating remark (Ratzel, 1897; 2nd ed., 1903:13). Nevertheless, having said this, he proceeds to show that a state, like some simple organisms, must either grow or die and can never stand still. When a state extends its borders at the expense of another state this is a reflection of internal strength. Strong states

---

[3]Social Darwinism can be traced to the writings of the English philosopher Herbert Spencer (1820–1903). Spencer pointed out the close resemblance of human societies to animal organisms. In both there are regulative systems (a central nervous system and a system of government); in both there are systems for the production of energy (the digestive system and the economic system); in both there are systems of distribution (veins and arteries and roads and telegraphs) (Spencer, 1876–96). It was Spencer who coined the phrase "survival of the fittest" (Spencer, 1864:1:444).

must have room to grow in order to survive. This is only one small step removed from the concept of *Lebensraum* (living space) that rationalized the right of superior people to enlarge their living space at the expense of inferior neighbors. Ratzel himself never subscribed to the idea of superior and inferior races. But some later geographers made use of the *Lebensraum* concept as a pseudoscientific support for the national policy of the 1930s (Smith, 1980).

Ratzel did not cease to be critical of his own generalizations (not a common trait among scholars). He made special note of cases where cultural differences were more important than differences in the physical character of the land. For example, in a regional geography of Germany (Ratzel, 1898), he pointed out the great contrast in the way people make use of the land in two places that are physically very much alike. The two places are the low mountain regions on either side of the middle Rhine Valley—the Vosges Mountains in France and the Schwarzwald in Germany. The differences to be observed in these two regions are related to the contrast between the French and German cultural traditions.

Ratzel's systematic studies of human geography led him to discard a number of ancient concepts. Ever since the time of Dicaearchus in the fourth century B.C., it had been accepted that man's use of the land had advanced through a succession of stages: first the primitive hunters, fishers, and collectors; then the pastoral nomads; then the agriculturists; and finally the horticulturists, such as the rice farmers of southern and eastern Asia. Humboldt had noted the absence of the stage of pastoral nomadism in America owing to the absence of domestic animals. Ratzel expressed doubt about the possibility, in every case, of distinguishing between herders and agriculturists since many people were a little of both (Ratzel, 1891:2:741).

Ratzel's criticism of the concept of stages was noted by Eduard Hahn, who in 1892 published a new map of the economic systems (*Wirtschaftsformen*) of the world. He distinguished six major kinds of rural economies: (1) hunting and fishing; (2) hoe culture (*hackbau*); (3) plow culture where the land is worked with the plow (*ackerbau*); (4) European-West Asian agriculture, where farming and herding are combined; (5) unmixed herding; and (6) horticulture, as in China or Japan (Hahn, 1892). Hahn provided ample evidence to demonstrate that the three stages of man's use of the land had to be discarded. In fact, there was evidence that in some cases pastoral nomadism developed out of an earlier agricultural way of life. Hahn's work ended a period when man's relation to the physical earth could be described by deduction from a theory and when a new phase of inductive study of the origins of agriculture could be undertaken (Hahn, 1896, 1919; Kramer, 1967).

The Problem of Unity and Diversity

Although Ratzel has provided the guidelines for the systematic study of human geography and scholars such as Hahn had proceeded along the lines Ratzel suggested, many of the newly appointed professors of geography in the German universities were bothered by the diversity of the subject matter they had to command. The question could be put this way: Are the concepts and methods of physical geography so utterly different from those of human geography that these two branches cannot be properly included in a single discipline? And did Richthofen's restriction of geography to the study of the surface of the earth really provide sufficient unity or did it require those who hoped to become geographers to master the concepts and methods of many diverse fields?[4]

Numerous suggestions were made in the search for unity and coherence. At one extreme were those who wanted to exclude human geography entirely. Georg Gerland, who was appointed professor of geography at Strassburg in 1875, concluded that geography should be exactly what the word *Erdkunde* implied: the study of the whole earth body without reference to man. Physical science, said Gerland, can formulate exact laws, but no such laws can ever be formulated to account for the behavior of human groups (Hartshorne, 1939:89). Few of the geographers could accept such a radical break with the tradition, nor did Gerland himself follow his own prescription, for he always included the geography of man in his courses at Strassburg.

The Concept of Chorology

It was Alfred Hettner[5] who elaborated Richthofen's concept of chorology. His earliest methodological statement was published in Volume 1 of the *Geographische Zeitschrift*, a professional periodical that he founded in 1895 and of which he was the editor until 1935. Hartshorne quotes Hettner's first statement of the nature of geography as follows:

> If we compare the different sciences we will find that while in many of them the unity lies in the materials of study, in others it lies in the method of study. Geography belongs in the latter group; its unity is in its method. As history and historical geology consider the development of the human race or of the earth in terms of time, so geography proceeds from the viewpoint of spatial variations (Hartshorne, 1939:97).

Hettner continued to develop his methodological ideas in published papers in the *Geographische Zeitschrift* during the next two decades. His more important

---

[4]Many of those who had been trained in geology were scornful of their colleagues who accepted appointment as geographers. One remark that was commonly repeated among the geologists at the turn of the century was a kind of pun: "Ein Geograph ist einer der die Erdoberfläche oberflächlich studiert" ("A geographer is one who studies the surface of the earth superficially"). Contributed by one-time Dean Edward H. Kraus of the University of Michigan, who studied mineralogy, chemistry, and geology at Munich in 1899–1901.

[5]Alfred Hettner was the first professor in a German university after Ritter who was trained as a geographer. He also studied philosophy. He began his advanced studies with Kirchhoff at Halle, worked with Fischer, Richthofen, and later with Ratzel, and completed his doctorate with Gerland at Strassburg. He wrote a dissertation on the climate of Chile and western Patagonia. Field studies in Germany and in the Andes of Colombia led to major publications in geomorphology. In 1897 he was appointed professor of geography at Tübingen, and he gave his inaugural lecture there in 1898. In 1899, however, he was appointed professor at Heidelberg, where he remained until his retirement in 1928. At Heidelberg he supervised more than 30 doctoral dissertations, and many of his students became leaders of the next generation in Germany—including Wilhelm Credner, Fritz Jaeger, Albert Kolb, Friedrich Metz, Oskar Schmieder, Heinrich Schmitthenner, and Leo Waibel (Dickinson, 1969:113; Plewe, 1982).

Alfred Hettner

Joseph Partsch

Albrecht Penck

Walter Penck

Friedrich Ratzel

Ferdinand von Richthofen

Otto Schlüter

Carl Troll

pronouncements were published in 1895 and 1905 (Hettner, 1895, 1905). In 1927 he brought all his methodological ideas together in one book, together with a history of geographic thought (Hettner, 1927). Hartshorne quotes his concept of chorology as presented in 1927:

> The goal of the chorological point of view is to know the character of regions and places through comprehension of the existence together and interrelations among the different realms of reality and their varied manifestations, and to comprehend the earth surface as a whole in its actual arrangement in continents, larger and smaller regions, and places (Hartshorne, 1959:13).

Geography as a chorological science was not a new concept when it was developed by Hettner. Hartshorne has traced the repeated appearances of this view of the place of geography among the fields of learning in the writings of different scholars since the eighteenth century (Hartshorne, 1958). In his lectures at the University of Königsberg, Immanuel Kant presented this view of geography as axiomatic—the view that geography was the field in which things were considered in their areal context on the face of the earth, much as history was the field in which things were considered in their time context. Humboldt expressed the same view of the field, but there is no evidence that he received the idea from Kant. In 1832 Julius Fröbel pointed out the similarity of the ideas of Kant and Humboldt, but Fröbel's writings were overlooked and only rediscovered many decades later. Richthofen accepted the same idea without finding it necessary to quote any earlier authority. When Hettner presented this view, it had apparently been accepted by so many generations of geographers that he felt no need to support it with references to earlier writings.

Hettner also faced and answered the question of whether geographers should be concerned only with the unique things and events that occur in a spatial context or whether geographers should seek to formulate general concepts. Hettner recognized that the study of geography is both idiographic and nomothetic in nature, as indeed almost all other fields of learning must be. If attention were paid only to universals, there would be a residue of information considered irrelevant because it could not be explained by existing models. If attention were paid only to unique things and events, the full richness of interpretation resulting from deduction from theory would be lost. But why is the question posed as a dichotomy? Geography, like other fields of learning, must deal both with unique things and with universals. Is the astronomer who maps the face of the moon less of an astronomer than one who studies planetary motions? Is the economist who describes the unique factors producing inflation in Brazil less of an economist than the one who formulates or tests the causes of inflation in general? Are not both considered to be valuable members of the profession? Hettner rejected the view that geography could be either idiographic or nomothetic but not both. As Varenius had decided more than two centuries earlier, special geography is concerned with many unique characteristics of places, yet this approach can be effectively pursued only when it is illuminated by general concepts that are necessary even to identify uniqueness. General geography is that aspect of geographic study in which general concepts are formulated and in which every unique element is a challenge that leads to the search for better theory. Since both Hettner and Hartshorne made this view of geography entirely clear (Hartshorne, 1939:382–384; Hettner, 1927:221–224), it is discouraging to find some writers who continue to accuse Hettner and his followers of defining geography as essentially idiographic (Schaefer, 1953), thereby obscuring the underlying continuity of geographic thought.

Hettner's ideas concerning the organization of geographical studies influenced the course of German geography for decades. One result was the continued focus of attention on the theme of man's relation to his physical and biotic surroundings. The traditional outline of a regional study started with location or position and then proceeded through chapters on geology, surface features, climate, vegetation, natural resources, the course of settlement, the distribution of population, the forms of the economy, the routes of circulation, and the political divisions (Dickinson, 1969:122). The outline was based on the notion that this formed a kind of sequence of cause and effect; in dealing with each topic, the relationships with the physical base were discussed, but not the relationships with other topics (Gradmann, 1931a; Hettner, 1907; Spethmann, 1931). In spite of the apparent rigidity of the outline, some highly effective regional studies were produced (Gradmann, 1931b; Philippson, 1904).

Geography as Landschaftskunde

Hettner's thought did not remain unchallenged. Some geographers were uneasy about the identification of geography as a chorological science and, hence, one that, like history, was defined by its method rather than its subject matter and its concepts. Others were also concerned about the overemphasis on the significance of the physical features of an area that resulted from following the schema. By tying everything to the physical features, other important relationships were overlooked —such as the relation of population density to the economy, or the economy and the routes of circulation, or even the relation of all of these to the political units. Furthermore, many of the interrelations observed in regional studies were in the process of change through time; only by examining the past geographies and the changes taking place could the idea of processes, or sequences of events, be introduced to geographical work (Troll, 1950).

As a means of reaching a more balanced treatment of the interrelations of things in particular areas, it was suggested that attention be given to the overall appearance of the landscape. The idea was first set forth by J. Wimmer in *Historische Landschaftskunde* in 1885 (Hartshorne, 1939:218). But the concept of geography as *Landschaftskunde* ("landscape science") was widely adopted in Germany after the inaugural address given at Munich in 1906 by Otto Schlüter[6] (Schlüter, 1906). Schlüter recommended that geographers look first at the things on the surface of the earth that could be perceived through the senses and at the totality

---

[6]Otto Schlüter began his academic training as a student of the German language and of history. At Halle he studied with Kirchhoff, who turned his interests toward geography. In 1895 he went to Berlin to study with Richthofen and to work as his assistant. He went to Munich briefly but was the professor of geography at Halle from 1911 until his retirement in 1951 (Schick, 1982).

of such perceptions—the landscape. He objected to the chorological definition of geography and noted that accepting the landscape as the subject matter of geography would give the field a logical definition, like that of any other academic field except history. The nonmaterial content of an area—such as political organization, religious beliefs, economic institutions, or even the statistical averages of climate—could not be considered primary objects of geographical study, although these could be introduced to explain the observable landscape.

Both Hettner and Schlüter were concerned about the variations in the character of the face of the earth, which became known later as *areal differentiation*. Both recognized that there were distinctly different kinds of areas on the earth that were distinguished from their surroundings and that showed a certain degree of homogeneity within boundaries that could be defined. But Hettner stressed the ways in which the features of a region reflected the basic patterns of the physical earth, whereas Schlüter focused attention on the interrelations of these features that gave the region its distinctive appearance. He used the method of historical geography in analyzing landscapes. First, he identified what he called the *Urlandschaft*—that is, the landscape that existed before major changes were introduced through the activities of man. He would then trace the sequences of change whereby an *Urlandschaft* was transformed into what he called a *Kulturlandschaft*, a landscape created by human culture (Schlüter, 1920, 1928). To trace these changes, he said, was the major task of geography.

When Schlüter was over 40 years of age, he started a major research program to apply his ideas to a specific region. His problem was to reconstruct the *Urlandschaft* of central Europe. He found that the approximate date when the major movement of new settlement into the forests began was A.D. 500. By using the evidence of place names and reading the descriptions written by ancient Greek and Roman geographers, he reconstructed the pattern of forested land and open land for that date. He then traced the process of settlement that created the *Kulturlandschaft*. His three-volume work was published after his death in 1952 (Dickinson, 1969:126–136; Lautensach, 1952).

The German geographers who followed Schlüter in identifying geography as *Landschaftskunde* surrounded the concept of landscape with a kind of mystical significance (Hartshorne, 1939:149–174). Unfortunately, the word *Landschaft* has two distinct meanings. Those geographers who made use of the word in a technical sense were not always careful to distinguish these meanings. *Landschaft* can, and traditionally did, refer to an extent of territory distinguished by a more or less uniform aspect. The word carried this connotation for many centuries before it was used in the fifteenth century by artists to refer to the painting of scenery—the aspect of the face of the earth as seen in perspective but without connotation of areal extent (Schmithüsen, 1963:17).[7] Since then, writers have used the word without being clear about its meaning. Humboldt used *Landschaft* to refer to the visual impression, the

---

[7]Similarly, before A.D. 1000 the word *landscipe* in Old English referred to an extent of territory. The word *landscape* was reintroduced into English from the Dutch (*landschap*) in the early seventeenth century (James, 1934:78–79).

aesthetic appreciation of beauty in scenery. J. Wimmer also used it in this sense in 1885 (Hartshorne, 1939:97), and even in 1953 Ewald Banse used the word with this meaning (Fischer, Campbell, and Miller, 1967:168–174). Meanwhile, the use of the word in the other sense—referring to an area of uniform aspect—never entirely disappeared.

But there have been other difficulties with the concept of *Landschaft*. Schlüter defined the landscape as the total impact of an area on man's senses, including such invisible phenomena as wind and temperature. Schlüter specifically included man as a part of the landscape. But some of his followers insisted on restricting the term to those material objects that could be seen. On the other hand, other students of landscape used the material features of the earth's surface only to indicate the existence of a region and did not hesitate to include in their studies such nonmaterial elements as law, religion, and economic institutions.

Leo Waibel, in a paper discussing the meaning of *Landschaft* (Waibel, 1933), pointed out that the word came into common use in Germany just at the time when geographers were focusing their studies on smaller and smaller areas, for which the word *Landschaft* (with its connotation of a small region) seemed more appropriate than the word for a larger region (*Gebiet*). Furthermore, there was continued emphasis on Richthofen's concept of the interconnections among things associated in area, and the word *Landschaft* seemed to carry the connotation of harmony of related parts.

The uncertainty regarding the exact meaning of this word is responsible for a conceptual error that persists. The sequence of ideas runs as follows: first, it is agreed that *Landschaft* is made up of concrete, observable features; second, it is agreed that the word *Landschaft* is a synonym for a homogeneous area, or region. Combining these meanings, it becomes possible to assert that a region is a concrete reality and not just an intellectual construct (Hartshorne, 1939:263).

In one way or another, the concept of landscape as an area with a more-or-less uniform appearance, the interpretation of which requires study of the nonmaterial as well as the material phenomena in an area, was adopted by most of the geographers of Germany before World War II (Bobek and Schmithüsen, 1949; Krebs, 1923). H. Lautensach, writing in 1952, reported that most German geographers followed Schlüter in identifying the study of landscape as the central purpose of geography (Lautensach, 1952:226). Schmithüsen, in 1963, says that "every landscape is a dynamic structure, a thing-area-time (*Sach-Raum-Zeit*) system of specified quality inside the whole geosphere (*Geosphäre*)." This is what we might describe as a spatial system. It is an open system, not one that is closed, like an organism (Schmithüsen, 1963:13).

The Concept of Chorology Applied to General Geography

In Germany and in other countries as well, it became common to accept the notion that general (or systematic) geography is necessarily analytic and makes use of general concepts, whereas regional geography is necessarily synthetic and deals with unique situations. This error, according to Hettner, can be blamed on Richthofen (Hettner, 1927:400). The concept of chorology, or the examination of the areal associations of things of diverse origin, can be applied to general geography as well as to studies of segments of the whole face of the earth. Before and during the spread of the concept of geography as *Landschaftskunde*, some German scholars were also looking at the whole surface of the earth.[8] One of the leading geographers of the late nineteenth and early twentieth centuries whose contributions were chiefly in the systematic physical aspects of geography was Albrecht Penck (Meynen, 1983).

Penck—who is credited with the first use of the term *geomorphology* to refer to the origin and development of the earth's landforms—showed how the systematic study of physical features can be approached from the chorological point of view.[9] In 1910 Penck suggested the hypothesis that the climate of a region so impresses itself on the observable features of the landscape that a classification of climates can be made even where instrumental records are lacking. He was the first to point out that the effective rainfall of a place is a balance between rainfall, runoff, and evaporation and that evaporation increases with higher temperatures. Although he started with the observable features of the landscape, he did not overlook the nonmaterial factors because of any rigid definition. Nor did he overlook man and the works of man as essential to an understanding of the varying characteristics of the face of the earth (Fischer, Campbell, and Miller, 1967:99–106).

Penck realized that no one geographer, and certainly no class or seminar, could actually see any considerable part of the earth. He realized, therefore, that the compilation of accurate maps, showing at least the major features that are associated in areas, was essential for the adequate study of geography. The maps and atlases that had been produced at Gotha and other places were excellent and informative, but it was not possible to show enough detail on the relatively small scales used in atlases. The large-scale topographic map, showing not only the shape of the landforms but also water bodies, vegetation, the works of man, and other things, was the ideal way to bring the real surface of the earth down to a size that would permit study. But only a very small part of the world was covered by surveys detailed enough to permit the compilation of topographic-scale maps. Penck proposed a compromise. In 1891 (at the International Geographical Congress held in Bern, Switzerland) he

---

[8]The whole surface of the earth is what Richthofen and others called the *Erdoberfläche*. Schmithüsen, in suggesting a number of new technical terms for the purpose of gaining greater precision, offers the word *Geosphäre*, or geosphere, to replace the older word (Schmithüsen, 1963:10).

[9]Albrecht Penck was professor of geography at Vienna from 1885 to 1906. In that year he succeeded Richthofen at Berlin, where he remained until his retirement in 1926. He was the rector of the university in 1917–18. In Vienna Penck collaborated with Eduard Brückner in the identification of four separate ice ages in the Alps—see *Die Alpen im Eiszeitalter*, 3 vols. (1901–1909). He was associated at the university with Eduard Suess, who prepared maps of the major geological regions of the world, outlining the great crystalline shields of ancient rocks; see *Das Antlitz der Erde* (1883–1908), translated by Emmanuel de Margerie as *La face de la terre* (1897–1918); it has also been translated into Italian and English. Also at Vienna was Julius von Hann, the famous climatologist whose work *Handbuch der Klimatologie* (1883, 1897, 1908–1911) has been translated into English by Robert deC. Ward as *Handbook of Climatology* (1903). Penck's son, Walther Penck, was noted for his challenge to the Davis system of geomorphology (Martin, 1974).

suggested that the nations of the world should cooperate in the production of an International Map of the World at a scale 1/1,000,000 (or about 1 inch to 15.8 miles). He proposed that agreement be reached regarding the standards of accuracy, the categories of things to be included, the symbols to be used, and the projection. Not until 1913 was such a conference called together in Paris to reach the agreements Penck had suggested. Work on the "millionth map" started slowly; when the pressing need for such a map was underlined during World War II, it was still far from complete. In 1953 the Central Bureau (which had been established to coordinate efforts in different countries to facilitate the project) was placed under the Economic and Social Council of the United Nations. Another international conference in 1962, attended by delegates from some 40 nations, reviewed the procedures for compiling the map and the symbols to be used to take advantage of the new technology then available for mapmaking (Hard, 1969, 1970).[10] Replacing the traditional sheets for this project were operational navigation charts. These are aeronautical charts that provide topographic information for air navigation. They were adopted in 1992 as the basis for a Digital Chart of the World, sponsored by the U.S. Defense Mapping Agency. These were integrated with UNESCO maps of vegetation and Food and Agriculture Organization soil maps to produce a near approximation to the millionth scheme, which was never completed.

Climate and Landscape

Penck's suggestion concerning the imprint of climate on the landscape inspired a number of studies at different scales to elaborate on the hypothesis. One of those who contributed to *Landschaftskunde*, not only in the detailed investigation of small areas but also at the global scale, was Siegfried Passarge.[11] Passarge's field study of the landscapes of the Kalahari Desert was published in 1904. His interest in the treatment of landforms as a part of a broader geographical study of landscape led him to call for empirical landform description rather than the genetic method proposed by Davis (Passarge, 1919–20). This led to a vigorous reaction by Davis and his followers. Nevertheless, in the contemporary period the tendency is to describe the surface features of an area without attempting what Davis called "explanatory description."

Passarge rejected the description of landscapes as unique. He insisted that a landscape must be viewed as a type, and he gave some examples of what he had in mind. He saw a landscape type as what we would call a spatial system, an assemblage of interrelated elements. His way of studying landscapes led him to focus

---

[10]The United Nations *Report* for 1966 (published 1968) shows that of the 975 sheets then needed to cover the land areas of the earth, Europe, Asia, Australia, and South America were completely covered, Africa almost completely covered, and North America partially covered.

[11]Siegfried Passarge was trained as a medical doctor and a geologist, receiving a degree in both fields. He served as a medical doctor in World War I. His detailed field studies were in the Kalahari Desert, in Algeria, and in Venezuela. From 1908 to his retirement in 1936, he was at the Kolonial Institut in Hamburg (Dickinson, 1969:137–141).

his attention on boundaries to be drawn around the area occupied by a type. In a discussion of the problem of defining the area of a landscape type (*landschaftsraum*), he postulated the existence of forested mountains rising above a grass-covered plain. The difficulty, as he saw it, was that in wetter parts of the plain, the forest extended beyond the mountains. If the landscape is described as *Waldgebirge* (forested mountains), should the boundary of this landscape be drawn around the whole forested area, even where the forest extends onto the plain?[12] That such a problem could exist seems like a good example of a semantic trap that results from the nature of the German language rather than the nature of the German landscape.

When looking at the world on a global scale, Passarge suggested that the best indicator of the extension of landscape types would be the vegetation. His *Landschaftsgürtel der Erde* (*Landscape Zones of the Earth*) is based on major categories of vegetation. On the basis of his world map, it is possible to identify certain regularities in the arrangement of the world's major landscape zones in relation to latitude and position on the continents (Fischer, Campbell, and Miller, 1967:143–154; Passarge, 1923).

Penck's suggestion concerning a classification of climates based on observable features of the landscape was picked up by the Russian-born climatologist Wladimir Köppen, who from 1875 to 1919 was employed as a meteorologist in the German oceanographic office (Deutsche Seewarte) in Hamburg. Between 1884 and 1918 Köppen made several attempts to produce a satisfactory classification of climates. At first he used only temperature distinctions, attempting to select values that would correspond as closely as possible to categories of vegetation. But after the publication of Penck's ideas concerning the relation of rainfall effectiveness to temperature and to seasonal changes, Köppen devised a new system of classification in which he made use, for the first time, of annual variations of temperature and rainfall. He published this new classification in 1918.[13] He continued to revise his definitions (Köppen, 1923), and his final version appeared in 1936 (Köppen, 1936).

---

[12]"Aus einer Steppenplatte erhebt sich ein Waldgebirge, an dessen Fuss aber der Wald über feuchte Teile der Ebene greift. Wo soll man die Grenze ziehen?" (Passarge, 1930:34).

[13]W. Köppen: "Die Wärmezonen der Erde, nach der Dauer der heissen, gemässigten und kalten Zeit, und nach der Wirkung der Wärme auf die organische Welt betrachtet," *Meteorologische Zeitschrift* 1 (1884):215–226; "Versuch einer Klassification der Klimate, vorzugsweise nach ihren Beziehungen zur Pflanzenwelt," *Geographische Zeitschrilft* 5 (1900):593–611; "Klassification der Klimate nach Temperatur, Niederschlag, und Jahreslauf," *Petermanns Geographische Mitteilungen* 64 (1918):193–203, 243–248. See also R. deC. Ward, "A New Classification of Climates," *Geographical Review* 8 (1918):188–191; and P. E. James, "Köppen's Classification of Climates: A Review," *Monthly Weather Review* 50 (1922):69–72. A five-volume work on the climates of the world by various authors was planned but never completed. The first volume appeared in 1930, *Handbuch der Klimatologie*, W. Köppen and R. Geiger, eds. (Berlin: Gebrüder Borntraeger), Köppen's final definitions of his categories of climate are in Volume 1. These same definitions are in P. E. James, *A Geography of Man*, 3rd ed. (Waltham, Mass.: Blaisdell, 1966), pp. 478–487. These definitions differ from those used in *Goode's World Atlas*.

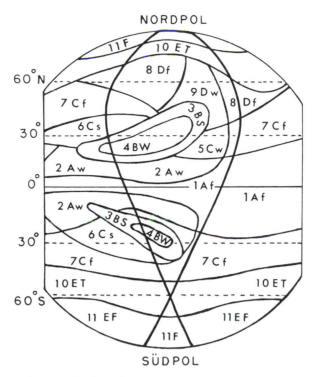

Figure 28 Köppen's generalized continent

Whenever climatic phenomena are plotted on world maps, certain regularities of pattern become evident, as Passarge had suggested in his highly generalized landscape zones. Humboldt had demonstrated this in his maps of temperature: Because continental areas are colder in winter and hotter in summer than places exposed to the open oceans, any winter isotherm in crossing a land mass bends equatorward; any summer isotherm bends poleward. Similarly, deficiencies of moisture throughout the year are found on the west sides of continents between latitudes 20° and 30° in both hemispheres, owing to the cold water offshore. Areas of abundant moisture are found on continental west coasts poleward of about 40° and along all the east coasts. Köppen demonstrated these regularities by drawing a geometric figure for a generalized continent and plotting on it the hypothetical positions of his categories of climate (Fig. 28). The generalized continent eliminates mountains and irregularities of coastline. The hypothetical pattern of climates shows how the climates would be arranged "other things being equal." It represents an imaginative new effort to identify certain of the regularities discovered on the face of the earth at a global scale.

The Geography of the Oceans

Regularities of pattern also exist in the oceans, and the first comprehensive treatment of this subject was by the German ocean geographer Gerhard Schott, who from 1894 to 1933 worked in the Deutsche Seewarte in Hamburg (Schott, 1912, 1935).

Schott's monumental work on the oceans deserves a place beside the similar treatments of the land by Eduard Suess and of the climates by Julius von Hann. The range of subject matter that Schott includes in his two volumes is impressive. He describes with great care the hydrological conditions (motion, temperature, salinity, color, and other elements), the climates over the oceans, the configuration and geological structure of the ocean basins, the marine organisms, the areas that support fishing or whaling industries, the routes of ocean commerce, and the air routes over the oceans. There are several chapters as well that deal with the history of discovery and exploration. Both volumes contain many maps of ocean features.

Schott produced a world map of oceanic regions. These are areas within which characteristic oceanic conditions are to be found, based on associations of hydrological conditions and marine organisms. This map, like the generalized maps of climate and landforms, reveals certain regularities of arrangement on the earth. It is possible to prepare a diagram of a generalized ocean similar to the generalized continent of Köppen (James, 1936:669).

Mention should be made of two other thrusts of German work relating to geography: the work of the Germans in polar exploration from the 1870s onward (Murphy, 2002), and the work of Alfred Wegener, who worked out the scheme of continental drift, arguably fathering plate tectonic theory, which emerged some decades later. It is probable that Wegener's work would have received more attention had not it been published at a time when World War I prevented its proper consideration by an international audience.

German Geography between the Two Wars

Samuel Van Valkenburg describes the period from 1905 to 1914 as the "golden age of German geography" (Van Valkenburg, 1951). This was a period of rapid growth and increasing productivity. The methodological discussions made working in this professional field not less, but more, exciting. Each new proposal regarding the scope and method of geography resulted in new field studies to provide examples. A number of new professional periodicals were founded as the influence of German geography spread throughout the world.

German exploration at both ends of the world, across the Pacific, and in Africa for purposes of colonization and geographical discovery typify the expansive mood of German geography. Industrial expansion, increasing prosperity, and a rising nationalism seemed to encourage the institutionalization of geography. Chairs had been filled in nearly all German universities, and geographical societies had been founded in several towns. Military studies, boundary studies, and geographies of patriotism bespoke a confidence in German geography.

Then came World War I, with death and destruction on all sides. Geography at once became an important subject (as it was with all the combatant nations). Emphasis of the subject shifted from physical to human geography. Economic geography and studies in propaganda came to the fore. Then came defeat, Versailles, reparation, restrictions, and loss of empire to be followed by the raging inflation of 1923 (Schelhaas and Honsch, 2001). Yet even in these difficult times, German geographers continued the tradition of foreign field work. Following the example of Humboldt and Hettner, geographers from this period also journeyed to South

America for field studies, for example, Schmieder, Waibel, Pfeifer, Willhelmy, Troll; others undertook field work in other parts of the world. From the middle 1920s until the mid 1930s, a stream of German geographers were employed in the Berkeley geography department, then under the stewardship of Carl Sauer. Sauer had hoped to establish an exchange plan and institute with Heidelberg University (Martin, 2003), but this did not come to fruition.

Although these were unhappy days for Germany, the country possessed a formidable assemblage of native geographers who were contributing works of enduring value. S. Passarge wrote a three-volume work, *The Bases for the Study of Landscape* (1919–20); A. Hettner wrote *The Surface Features of the Continents* (1921), *The Spread of Culture over the Earth* (1923), and *Geography: Its History, Nature, and Methods* (1927); O. Schlüter produced *Forest, Swamp and Settled Land in Old Prussia Before the Period of the Germanic Order* (1921); L. Waibel contributed *Problems of Agricultural Geography* (1933); and K. Sapper wrote *Geomorphology of the Humid Tropics* (1935). Other outstanding authors of this period included W. Credner, N. Krebs, H. Lautensach, A. Rühl, O. Schmieder, H. Schmitthenner, and C. Troll. Special mention might be made of Robert Gradmann who published the two-volume *Southern Germany*, 1931[14]. The geographers of Germany paid little attention to the innovative study of the functional organization of space in south Germany published in 1933 by Gradmann's student at Erlangen, Walter Christaller (Christaller, 1933). Christaller's contribution to the development of central-place theory and the other German writings on the theory of location during this period will be discussed in the concluding chapter.

Troll does not mention Christaller, but he does point to some other positive advances made by German geographers during this period. There was a conservation movement, to which the geographers contributed field studies as a basis for landscape planning. There were studies of the relation of people to living space that were not politically oriented (Troll, 1949:115). Largely through the efforts of Max Eckert, German cartographers formed the *Deutsche Kartographische Gesellschaft* in 1937 for the purpose of promoting cartography as a separate discipline (Troll, 1949:124). There was increased interest in the status and economic development of places all around the world occupied by Germans who had emigrated from Germany.

Germany was involved in two disastrous wars, and between 1933 and World War II the rise of the Nazi state made studies in objective scholarship increasingly more difficult. The government frequently interfered in academic affairs. Most of the geographers chose to remain discreetly silent on matters of policy, but there were some who came out in support of the "New Germany." Only those who were Jews were actually sent to concentration camps—such as Alfred Philippson, the

---

[14]Robert Gradmann was trained as minister of a Lutheran church, then as a botanist, was then directed toward geographical studies by Karl Sapper at Tübingen, and accomplished the habilitation for his study of grain cultivation in ancient Greece and Roman civilizations. In 1919 he became professor of geography at the University of Erlangen; while there, he completed the famous regional study of southern Germany, one of the outstanding examples of *Landerkunde* (Gradmann, 1931b). He also edited the German research series *Forschungen zur deutschen Landeskunde* (Schroder, 1982).

distinguished author of the study of the Mediterranean region (Philippson, 1904).[15] Hettner's *Geographische Zeitschrift*, which had been edited by Heinrich Schmitthenner after 1935, was suspended in 1943. (Hettner was proclaimed a quarter Jew by the Reich and removed from the editorial post [Plewe 1982].) The discipline of geography was bent to Nazi specification. Sufficient history has been written to reveal this (Rössler 1987, 1989; Sandner 1988, 1994). In 1947 Carl Troll wrote a review of what had been going on in German geography between 1933 and 1945, which was published in the first volume of the new geographical periodical that he established, *Erdkunde* (Troll, 1947). Troll's review was translated in part by Eric Fischer and published in the United States (Troll, 1949).

One of the streams of geographic thought that developed in Germany between the wars was the application of geographic concepts to politics. This was given the German name *Geopolitik* (geopolitics). Geopolitics differs from political geography and is not to be confused with it. According to Karl Haushofer, *Geopolitik* is the art of using geographical knowledge to give support and direction to the policy of a state. Haushofer himself published his *Geopolitik des Pazifischen Ozeans* in 1924, analyzing the significance of location around the Pacific basin with reference to the various threatening conflicts. Haushofer drew his ideas from the Swedish political scientist Rudolf Kjellén (1846–1922), who developed the basic ideas of *Geopolitik* by using Ratzel's analogy of the state as an organism that required room to grow. Haushofer was also influenced by the British geographer H. J. Mackinder. At first there were numerous geographers writing papers for publication in the *Zeitschrift für Geopolitik* dealing with problems in political geography. But as the periodical became the chief vehicle for disseminating geographic writings in support of Nazi policies and as the unscientific nature of the periodical became clear, many of the geographers withdrew their support from the movement. It seems that although Haushofer may have had some influence on Nazi policies, his importance may have been exaggerated. With the collapse of Nazi Germany in 1945, Haushofer was interrogated by the Allies. In 1946 he committed suicide.

Postwar German Geography

In the aftermath of the disaster of World War II, German geography had to be rebuilt. Many universities and libraries were in ruins, and the funds for foreign field research were almost nonexistent. The situation after the war is described in some detail (Smith and Black 1946; Troll, 1949; Van Valkenburg, 1951). Geography as an institution had to be rebuilt. Relationships between geographers similarly had to be rebuilt. By 1949 two Germanys had been created, a West Germany and an East Germany (the Federal Republic of Germany and the German Democratic Republic). The two Germanys would assume differing postures politically, economically, and intellectually.

In 1949–50, immediately following the establishment of West Germany, 24 universities possessed a total of 25 chairs in geography (the University of Hamburg

---

[15]Philippson was not deported to Poland owing to the intervention of Sven Hedin of Sweden, who was moved to act in behalf of Philippson by Carl Troll.

alone had two chairs). It seems that more than 1060 students studied geography as a "major," and 600 as a "minor" in that year. It was this student generation that produced the leaders of German geography from the 1960s to the 1990s.

By the late 1960s new universities were created and existing universities and faculties were expanded. Between 1968 and 1980 10 new departments were founded. The number of chairs in geography increased from 80 in 1970 to 145 in 1980. This is a very visible part of a remarkable economic recovery. Accompanying this advance in geography were hidden difficulties associated with a profusion of subjects, some perhaps marginal to the undertaking, and increasing disparity in the quality of teaching and research by individuals. This increase in the apparatus of the discipline did not necessarily mean concomitant increase in profundity of discipline development.

An occurrence with long-term implication was the elimination after 1970 of many geography teaching positions in the primary and secondary schools. Geography was integrated with social studies (a situation similar to the United States in the 1920s). The reduction of geography from an independent school subject detracted from the image of geography and led to a surplus of geography students looking for employment. This led to more emphasis on applied geography. Most geography departments did not have such a program of studies and, therefore, did not introduce them until the late 1960s, by which time many "applied" vacancies had been filled by nongeographers.

The traditional organizations maintained themselves during these years. These included the umbrella organization responsible for the biennial scientific meeting of German academic geography (founded 1881); the organization for school teachers of geography (1912); and the organization for those who had succeeded with the *Habilitation* (a postdoctoral degree) and who were the dominant force in research and representation on high councils. Those without the *Habilitation* began to feel excluded and began to plan their own organization. This happened in 1969–70. Now there were two groups in university geographical work. With such a division present, it is not surprising that a third group was brought into existence: This group comprised geographers who trained teachers at the universities or teacher training college.

The most active group, however, is the Association of German Applied Geographers, founded in 1950, with a membership of 1300 by 1991. Its relevant problem-solving applications, regional groupings, regional and national conferences, and permanent office space have placed it at the forefront of German geographic organizations. All of these organizations were combined in a central association of German geography. This was the structure that permitted West German geography to rebuild a geographic profession that had been reduced during the Nazi regime. Schools and universities formed closer connections, curricula were revised, and more textbooks were written providing wider choice.

Meanwhile, German geographers in the 1960s went forward with the traditional concept of geography as *Landschaftskunde* but with the use of new and more precise methods of analysis. Hermann Lautensach's study of the Iberian Peninsula offers a new approach to the study of landscapes. Instead of seeking to define landscape regions as outlined by boundaries, Lautensach describes the variations from place to place as forming a continuum. The continuous variation of the landscape throughout the Iberian Peninsula is a function of four variables: latitude, elevation, distance from the ocean, and direction from the nearest coast (Lautensach, 1964). It is this kind of innovative approach that keeps regional geography (*Länderkunde*) as the central interest of the great majority of German geographers (Dickinson, 1969:184–185; Pfeifer, 1965; Schmieder, 1966; Schmithüsen, 1963; Schwind, 1981; Uhlig, 1983; Zimm, 1982).

Some decades ago German geographers reacted against the traditional physical determinism stemming from the teaching of Richthofen and Hettner. The reaction began when Schlüter focused attention on the historical changes of landscape resulting from human action. In the postwar period a kind of "cultural determinism" has replaced the earlier physical determinism. The new emphasis on culture is called *social geography* (*Sozial-geographie*) (Bobek, 1948; Schmithüsen, 1959). The purpose is to interpret the cultural landscape but with the clear recognition that the major force for landscape change is the human group—the "attitudes, objectives, and technical skills" that are parts of human culture. Many collections of papers on the new social geography have been published (Hajdu, 1968; Hartke, 1960; Storkenbaum, 1967, 1969; see also comments on Hajdu's paper in the *Annals AAG*, 59 [1969]:596–599).

Substantively, in postwar German geography several outstanding geographers contributed to both physical and human geography (Bartels and Peucker, 1969; Beck, 1957; Jager, 1972). These workers include particularly H. Bobek, H. Mortenson, H. Schmitthenner, and C. Troll. It is a significant matter when workers in a field of study may so specialize that they find themselves far removed from each other. Physical geography subdivided itself into geomorphology, geophysics, meteorology, climatology, oceanography, and so forth. Cultural geography also found itself the object of subdivision. An interest in the history of German geographers is revealed by Büttner (1978, 1979) and Büttner and Burmeister (1979). These subdivisions have been shared between geography and other disciplines. It is then that the regional concept and undertaking brings workers and their points of view closer to each other. The contributions of a variety of sciences and specializations are integrated for the purpose of understanding areal complexes. Perhaps the two strongest advocates of this, the regional undertaking, have been Carl Troll and Hermann Lautensach (Troll, 1966).

In 1967 there came a break with tradition. Dietrich Bartels presented his habilitation thesis, which included a bibliography providing a large number of English, Scandinavian, and American geographical references. Whereas German geography had for decades been exporting its product, now it was importing ideas. Non-German geography was being read in Germany possibly as never before. Bartels's views concerning the integration of foreign viewpoints were accepted at a meeting of young geographers in October 1967 and summarized in a brochure containing 12 points.

A new scientific revolution made itself apparent at the annual meeting of the German geographers at Kiel in 1969. The younger geographers who could not attain the highest positions were encouraged to be patient. At this meeting the professional association of geography students created a general meeting that was the first of its kind. In the months that followed, a school of social geography was established at Munich, where Wolfgang Hartke and Karl Ganser had led the way since 1966;

Franz Schaffer and Karl Ruppert came to the fore in November 1968. This stage in West German geography was marked by an integration of the traditional with modern geography, and led to a new pluralism in geography. *Geographische Zeitschrift* and *Erdkunde* were the leading research publications, although in addition there were six journals sponsored by geographical societies and more than 50 "series" issued by universities. More recently there has been a strong movement to return to political geography; closer self-examination of geography and of the geographers of the Nazi period has been especially evident at Hamburg University. Significantly stronger ties between West Germany and the United Kingdom and the United States were developed in the 1980s and continued to the present. Even so, the regional tradition continues, and a substantial percentage of geographers still travel abroad to conduct their research.

Facilitating research in regional geography in foreign countries is what Wirth refers to as a particular tradition of German geography (Wirth, 1988). In any case, there has been steady growth in this area since 1945, and this has been facilitated by substantial financial support from the German Research Council. Research tended to anthropogeography in the 1960s and early 1970s, then toward physical geography from the late 1970s. Throughout regional geography of Europe and overseas (with special reference to Africa, Latin America, and the Islamic World) has been undertaken and other foundations have also provided financial assistance such as the Volkswagen Foundation and the Thyssen Foundation. However, special mention must be made of the fact that most university departments have developed their own publication series. Most of these have been established since 1950, and they each have active lists; for example, in the years 1989 and 1990 a total of nearly 300 monographs were published. These departmental series form a very significant part of German geography. Otherwise, there are a number of periodicals of international standing. These are listed in C. D. Harris's *Annotated World List of Selected Current Geographical Serials.*

GEOGRAPHY IN EAST GERMANY

After World War II departments of geography were located in Berlin, Leipzig (closed in 1969 but merged with Halle), Greifswald, Jena (closed in 1969), Rostock (closed in 1969), and the Technical University of Dresden. The institution once known as Justus Perthes has been renamed the VEB Hermann Haack Institute. The Deutsches Institut für Landerkunde at Leipzig has become an institute for geography and geoecology. During the 1950s geography departments were established in teacher training colleges in Potsdam and Dresden. Economic geography was brought into the Leipzig School of Commerce and the Berlin School of Economics. In 1953 the Geographical Society of East Germany (GDR) was founded, and its journal, *Geographische Berichte*, has been published since 1956. *Petermanns Geographische Mitteilungen*, published in Gotha since 1855, remains the most important geographic periodical in Germany.

In the postwar period new directions were assumed in both teaching and in academic research. Specifically, a break was made with the traditions of the past, and a new geography was created based on a material world order. East Germany

(GDR) looked to the east. The Soviet Union dominated this section of Germany, and Germans learned Russian, read Russian books, and took classes in Marxism. There were still those geographers who were functional in the English language, but the younger generation accepted Russian as a second language. The GDR was inherently poorer than West Germany (FRG), and this, combined with its trading partner the USSR rather than Western countries, further accentuated the economic gap between the two Germanies. Hard currency needed for travel in the West was not available; permission to travel West was rarely granted. Geography was highly organized as was much else, and geographical science and party policy were drawn close. The two Germanys had two geographies. At the 19th International Geographical Congress, Stockholm, 1960, two national committees for the two geographies were recognized. After construction of the Berlin Wall in 1961, there came increasing separation of the two Germanys as the cold war deepened.

From a geographic tradition common to the whole of Germany, international circumstances had created two traditions at variance in little more than a 40-year period. The eastern version, rigorous, structured, subservient to the state, and with only modest resources at its command, was the product of the post-1945 period. Notwithstanding difficulties and constraints, geography in the East Germany was nevertheless responsible for considerable teaching, learning, research, and publication. Then came reunification.

Geography and the Reunification of Germany

It was at the Beijing regional conference of the International Geographical Union, 1990, that the national committees of the two Germanys were unified. Now the task was to reconcile the two geographies that had developed with different traditions over the last 40 years. The Geographentag held at Basel (1991) was the venue for that serious discussion. The two Associations of Academic Geographers were combined to form the Association of Geographers at German Universities. And in 1995, when the Geographentag met at Potsdam—the first time the meeting was held in the former East German—a new organization entitled the German Society of Geography replaced the old established umbrella organization that was still a part of "West German" geography. New chairs were created in the east at Erfurt, Frankfurt on the Oder, and Chemnitz. And the German National Atlas went forward as a combined project of the now-unified Germany.

The one-time West Germany experienced less trauma in the unification process. Geographers continued to look to anglophone geography and to import ideas from this source. British geography books were translated into German, and a lecture was established at Heidelberg, addresses for which were provided by geographers from the United States, Canada, and the United Kingdom.

In the early history of German geography, Germany exported ideas and imported students. Now Germany finds itself importing ideas and exporting students. Part of this reversal of fortune may owe to the early start that Germany obtained (e.g., its early chairs were established more than 60 years before the first chair in Canadian geography), an advantage now no longer applicable. The German language is no longer the lingua franca of science that once, arguably, it was. And, too, persons of capacity may come in groupings, and the first three generations of German professional geographers were unusually able. Finally, one must think of difficulties confronting the advancement of geography in Germany: loss of geographers in both world wars (a loss also suffered by the Allies), exclusion from the international congresses pursuant to both world wars[16], the incongruous impress of Nazism on German geography commencing in 1934, and the post-1945 division of the country and its geography until 1990. It is perhaps too soon to decide on the issue of geographic accomplishment since reunification, but opportunity for benefit accruing from concerted effort is now possible.

REFERENCES: CHAPTER 8

Bartels, D. 1968. "Zur wissenschaftstheoretischen Grundlegung einer Geographie des Menschen," *Geographische Zeitschrift, Beihefte, Erdkundliches Wissen* 19, Wiesbaden.

Bartels, D., and Peucker, T. 1969. "German Social Geography Again." *Annals AAG* 59:596–598.

Bassin, M. 1984. "Friedrich Ratzel's Travels in the United States: A Study in the Genesis of His Anthropogeography." *History of Geography Newsletter* 4:11–22.

———. 1987a. "Friedrich Ratzel, 1844–1904." In *Geographers: Biobibliographical Studies*. Vol. 11, pp. 123–132. London: Mansell.

———. 1987b. "Imperialism and the Nation-State in Friedrich Ratzel's Political Geography." *Progress in Human Geography* 11:473–495.

———. 1987c. "Race Contra Space: The Conflict between German Geopolitik and National Socialism." *Political Geography Quarterly* 6:115–134.

Beck, H. 1957. "Geographie und Reisen im 19. Jahrhundert: Prolegomena zu einer allgemeinen Geschicht der Reisen." *Petermanns Geographische Mitteilungen* 101:1–14.

———, ed. 1959. *Gespräche Alexander von Humboldts*. Berlin: Akademie-Verlag, 1959.

———. 1959–1961. *Alexander von Humboldt*. 2 vols. Wiesbaden: Franz Steiner Verlag GmbH. Vol. 1, *Von der Bildungsreise zur Forschungsreise, 1769–1804*; Vol. 2, *Von Reisewerk zum "Kosmos," 1804–1859*.

———. 1973. *Geographie. Europäische Entwicklung in Texten und Erläuterungen*. Orbis Academicus. Problemgeschichte der Wissenschaft in Dokumenten und Darstellungen. Freiburg-München: K. Alper.

———. 1981. "Zur Geschichte der Geographie, der Pädagogik und des Geographieunterrichts." In *Theorie und Geschichte des geographischen Unterrichts* (Geographisch-didaktische Forschungen 8). (W. Sperling, ed.), pp. 61–86. Braunschweig: Westermann.

———. 1982. "Alexander v. Humboldt—der grösste Geograph der neueren Geschichte (1769–1859)." In *Grosse Geographen: Pioniere—Aussenseiter—Gelehrte* by Beck. Pp. 83–102. Berlin: Dietrich Reimer Verlag, 1982.

Black, L. D. 1947. "Further Notes on German Geography." *Geographical Review* 37:147–148.

Bobek, H. 1948. "Stellung and Bedeutung der Sozialgeographie." *Erdkunde* 2:118–125.

Bobek, H., and Schmithüsen, J. 1949. "Die Landschaftsbegriff im logischen System der Geographie," *Erdkunde* 3:112–120.

---

[16]Only once was the International Geographical Congress held in Germany (Berlin, 1899: compare to the UK, four occasions; France, four occasions; and the United States, three occasions.)

Bohm, H., and E. Ehlers. 1996. "Erdkunde—50 Jahrgange, Archiv für Wissenschaftliche Geographie." *Erdkunde* 50:360–379.

Büttner, M. 1978. "Bartholomäus Keckermann: 1572–1609." In *Geographers: Biobibliographical Studies*. Vol. 2, pp. 73–79. London, Mansell.

———. 1979. "Phillip Melanchton: 1497–1560." In *Geographers: Biobibliographical Studies*. Vol. 3, pp. 93–97. London, Mansell.

Büttner, M., and Burmeister, K. H. 1979. "Sebastian Münster: 1488–1552." In *Geographers: Biobibliographical Studies*. Vol. 3, pp. 99–106. London, Mansell.

Christaller, W. 1933. *Die zentralen Orte in Süddeutschland*. Jena: Gustar Fischer Trans. C. W. Baskin, *Central Places in Southern Germany*. Englewood Cliffs., NJ.: Prentice Hall, 1966.

Dickinson, R. E. 1969. *The Makers of Modern Geography*. London: Routledge and Kegan Paul.

Dunbar, G. S. 1983. *The History of Geography: Translations of Some French and German Essays*. Malibu, Calif.: Undena Publications.

Ehlers, E. ed. 1992. German geography 1945–1992: Organizational and Institutional Aspects. In Eckart Ehlers (ed.), *40 Years After: German Geography. Developments, Trends and Prospects 1952–1992*, Bonn. Joint publication: Bonn Deutsche Forschungsgemeinschaft: Tübingen, Institute for Scientific Co-operation.

———. 1998. "Current German Geography in a Globalizing World of Science: A Documentation and Personal Interpretation." *GeoJournal* 45:1–2:57–68.

———. 2004. " 'Once Upon a Time . . .' Interactions of German and American Geography in a Rapidly Changing (Academic) World." *GeoJournal* 59, 1:9–13.

Elkins, T. H. 1990. Human and Regional Geography in the German-Speaking Lands in the First Forty Years of the Twentieth Century: An Outsider's View, in Eckart Ehlers (ed.), Philippson-Gedächtniskolloquium, *Colloquium Geographicum* 20, Bonn, 21–33.

Fahlbusch, M., M. Rössler, and D. Siegrist. 1989. Conservatism, Ideology and Geography in Germany 1920–1950, *Political Geography Quarterly* 8:353–367.

Fischer, E., Campbell, R. D., and Miller, E. S. 1967. *A Question of Place: The Development of Geographic Thought*. Arlington, Va.: Beatty.

Gradmann, R. 1931a. "Das länderkundliche Schema." *Geographische Zeitschrift* 37:540–548.

———. 1931b. *Süd-Deutschland*. 2 vols. Stuttgart: J. Engelhorn.

Haase, G. 1977. "Ziele und Aufgaben der geographischen Landschaftsforschung in der DDR." *Geographische Berichte* 22:1–19.

Hahn, E. 1892. "Die Wirtschaftsformen der Erde." *Petermanns Geographische Mitteilungen* 38:8–12.

———. 1896. *Die Haustiere und ihre Beziehungen zur Wirtschaft des Menschen*. Leipzig: Duncker & Humblot.

———. 1919. *Von der Hacke zum Pflug, Garten, und Feld: Bauern und Hirten in unserer Wirtschaft und Geschichte*. Leipzig: Quelle & Meyer.

Hajdu, J. G. 1968. "Toward a Definition of Post-War German Social Geography." *Annals AAG* 58:397–410.

Hard, G. 1969. "Die Diffusion der 'Idee der Landschaft': Präliminärien zu einer Geschichte der Landschaftsgeographie." *Erdkunde* 23:249–364.

———. 1970. "Was ist eine Landschaft? Etymologie als Denkform in der geographischen Literatur." In D. Bartels, ed., *Wirtschafts- und Sozialgeographie*. Pp. 66–84. Berlin/Cologne: Kiepenheuer and Witsch.

———. 1973. *Die Geographie: Eine wissenschaftstheore tische Einführung*.

Harke, H., ed. 1984. *Alfred Rühl—Ein hervorragender Deutscher Geograph.* (Wissensch-aftliche Beitrage der Martin Luther Universität Halle-Wittenberg Bd. 21.) Oberlungwitz: VEB Kongres-und Werberdruck 9273.

Harke, H. 1988. Alfred Rühl 1882–1935. *Geographers: Biobibliographical Studies* 12:139–47.

Harris, C. D. 1991. "Unification of Germany in 1990." *Geographical Review* 81, 2:173–182.

Hartke, W., ed. 1960. *Denkschrift zur Lage der Geographie.* Wiesbaden: Franz-Steiner.

Hartshorne, R. 1939. *The Nature of Geography, a Critical Survey of Current Thought in the Ljght of the Past.* Lancaster, Pa.: Association of American Geographers.

———. 1958. "The Concept of Geography as a Science of Space, from Kant and Humboldt to Hettner." *Annals AAG* 48:97–108.

Heske, H. 1986. German geographical research in the Nazi period; a content analysis of the major geography journals, 1925–1945. *Political Geography Quarterly* 5:267–81.

———. 1987. Karl Haushofer; his role in German geopolitics and in Nazi politics. *Political Geography Quarterly* 6:135–44.

Hettner, A. 1895. "Geographische Forschung und Bildung." *Geographische Zeitschrift* 1:1–19.

———. 1905. "Das Wesen und die Methoden der Geographie." *Geographische Zeitschrift* 11:549–553.

———. 1907. *Grundzüge der Länderkunde.* Vol. 1, *Europa.* Rev. eds., 1923, 1932. Vol. 2, *Die Aussereuropäische Erdteile.* Rev. eds., 1923, 1926. Leipzig: Teubner.

———. 1927. *Die Geographie—ihre Geschichte, ihr Wesen, und ihre Methoden.* Breslau: Ferdinand Hirt.

Jacobsen, H.-A., ed. 1979. *Karl Haushofer—Leben und Werk.* 2 vols. Boppard: Bolt, Schriften des Bundsarchivs 24.

Jager, H. 1972. "Historical Geography in Germany, Austria and Switzerland." In A. R. H. Baker, ed., *Progress in Historical Geography.* Pp. 45–62. New York: John Wiley & Sons.

James, P. E. 1934. "The Terminology of Regional Description." *Annals AAG* 24:78–92.

———. 1936. "The Geography of the Oceans: A Review of the Work of Gerhard Schott." *Geographical Review* 26:664–669.

Joerg, W. L. G. 1922. "Recent Geographical Work in Europe." *Geographical Review* 12:431–484.

Kolb, A. 1983. "Ferdinand Freiherr von Richthofen, 1833–1905." In *Geographers: Biobibliographical Studies.* Vol. 7, pp. 109–115. London: Mansell.

Köppen, W. 1923. *Die Klimate der Erde, Grundriss der Klimatologie* (revised and enlarged, 1931). Berlin: W. de Gruyter.

———. 1936. "Das geographische System der Klimate." In W. Köppen and R. Geiger, eds., *Handbuch der Klimatologie.* Vol. 1, Part C. Berlin: Gebrüder Borntraeger.

Kost, K. 1988. The Conception of Politics in Political Geography and Geopolitics in Germany until 1945, *Political Geography Quarterly* 8:369–385.

Kramer, F. L. 1967. "Eduard Hahn and the End of the 'Three Stages of Man.'" *Geographical Review* 57:73–89.

Krebs, N. 1923. "Natur- und Kulturlandschaft." *Zeitschrift der Gesellschaft für Erdkunde zu Berlin:* 81–94.

Kuhn, T. S. 1963. "The Function of Dogma in Scientific Research." In A. C. Crombie, ed., *Scientific Change.* Pp. 347–369. New York: Basic Books.

Lautensach, H. 1952. "Otto Schlüter's Bedeutung für die Methodische Entwicklung der Geographie." *Petermanns Geographische Mitteilungen* 96:219–231.

———. 1964. *Iberische Halbinsel.* Munich: Keysersche Verlagsbuchhandlung.

Lichtenberger, E. 1984. "The German-Speaking Countries." R. J. Johnston and P. Claval, eds., *Geography since the Second World War*. Pp. 156–184.

Linke, M. 1991. "Alfred Philippson, 1864–1953." In *Geographers: Biobibliographical Studies*. Vol. 13, pp. 53–61. London: Mansell.

Lüdenmann, H. 1980. "Geographische Forschung in der DDR: Entwicklung und Perspektiven." *Petermanns Geographische Mitteilungen* 124:97–103.

Martin, G. J. 1974. "A Fragment on the Pencks(s)—Davis Conflict." *Special Libraries Association. Geography and Map Division*, Bulletin 98:11–27.

Meynen, E. 1983. "Albrecht Penck, 1858–1945." In *Geographers: Biobibliographical Studies*. Vol. 7, pp. 101–108. London: Mansell.

———. 1992. "Josef Schmithüsen, 1909–1984." In *Geographers: Biobibliographical Studies*. Vol. 14, pp. 93–104. London: Mansell.

Mohs, G. 1986. "Globale Probleme und regionale Entwicklung im Blickfeld der Geographie." *Geographische Berichte* 31:1–12.

Murphy, D. T. 2002. *German Exploration of the Polar World: A History 1870–1940*. Lincoln: University of Nebraska Press.

Passarge, S. 1919–1920. *Die Grundlagen der Landschaftskunde*. 3 vols. Hamburg: L. Friederichsen.

———. 1923. *Die Landschaftsgürtel der Erde*. Breslau: Ferdinand Hirt.

———. 1930. "Wesen und Grenzen der Landschaftskunde." In *Herman Wagner Gedenkschrift, Ergebnisse und Aufgaben der geographischen Forschung* (Ergänzungsheft). *Petermanns Geographische Mitteilungen* 209:29–44.

Paterson, J. H. 1987. "German Geopolitics Reassessed." *Political Geography Quarterly* 6:107–114.

Penck, A. 1916. "Der Krieg und das Studium der Geographie." *Zeitschrift der Gesellschaft für Erdkunde zu Berlin*: 158–176, 222–248.

———. 1925. Deutscher Volks- und Kulturboden. "In *Volk unter Völkern* (K. C. v. Loesch ed.) pp. 62–73. Breslau: Ferdinand Hirt, Bucher des Deutschtums, No. 2.

Peschel, O. 1865. *Geschichte der Erdkunde bis auf A.v. Humboldt und Carl Ritter*. Munich: J. G. Cotta.

———. 1870. *Neue Probleme der vergleichenden Erdkunde als versuch einer Morphologie der Erdoberfläche*. Leipzig: Duncker & Humblot.

———. 1879. *Physische Erdkunde*. Ed. Gustav Leipoldt. Leipzig: Duncker & Humblot.

Pfeifer, G. 1965. "Geographie Heute?" In *Festschrift Leopold G. Scheidl zum 60 Geburstag*. Pp. 78–90. Vienna: Ferdinand Berger & Söhne.

Philippson, A. 1904. *Das Mittelmeergebiet*, 4th ed. 1922. Leipzig: Teubner.

Plewe, E. 1982. "Alfred Hettner, 1859–1941." In *Geographers: Biobibliographical Studies*. Vol. 6, pp. 55–63. London: Mansell.

Ratzel, F. 1882–1891. *Anthropogeographie*. Vol. 1, *Grundzüge der Anwendung der Erdkunde auf die Geschichte*, 2nd ed., 1889; 3rd ed., 1909. Vol. 2, 1891, *Die geographische Verbreitung des Menschen*, 2nd ed., 1912. Stuttgart: J. Engelhorn.

———. 1897. *Politische Geographie, oder die Geographie der Staaten, des Verkehrs, und der Krieges*, 2nd ed., 1903; 3rd ed., 1923. Munich and Berlin: R. Oldenbourg.

———. 1898. *Deutschland, Einführung in die Heimatkunde*. Leipzig: Grunow.

Richthofen, F. von. 1877–1912. *China: Ergebnisse eigener Reisen und darauf gegründte Studien*. 5 vols. Berlin: Dietrich Reimer.

———. 1883. *Aufgaben und Methoden der heutigen Geographie* (Akademische Antrittsrede). Leipzig: Veit.

———. 1886. *Führer für Forschungsreisende*. Berlin: Robert Oppenheim.

Rössler, Mechtild. 1987. Die Institutionalisierung einer neuen Wissenschaft im National-sozialismus. Raumforschung und Raumordnung 1935–1945, *Geographische Zeitschrift* 75:177–194.

———. 1989. Applied Geography and Area Research in Nazi Society: Central Place Theory and Planning, 1933 to 1945, *Environment and Planning D* 7:419–431.

Sandner, G. 1988. "Recent Advances in the History of German Geography 1918–1945. A Progress Report for the Federal Republic of Germany." *Geographische Zeitschrift* 76(2):120–133.

———. 1994. In Search of Identity: German Nationalism and Geography, 1871–1910. In David Hooson (ed.), *Geography and National Identity*. Oxford: Blackwell, 71–91.

Sauer, C. O. 1971. "The Formative Years of Ratzel in the United States." *Annals AAG* 61:245–254.

Schelhaas, B. and I. Honsch. 2001. "History of German Geography: Worldwide Reputation and Strategies of Nationalization and Institutionalization." In G. S. Dunbar, ed., *Geography: Discipline, Profession and Subject since 1870*. Dordrecht, Netherlands: Kluwer, 9–44.

Schick, M. 1982. "Otto Schlüter, 1872–1959." *Geographers: Biobibliographical Studies* 6:115–122.

Schlüter, O. 1906. *Die Ziele der Geographie des Menschen* (Antrittsrede). Munich: R. Oldenbourg.

———. 1920. "Die Erdkunde in ihrem Verhältnis zu den Natur- und Geisteswissens-chaften." *Geographische Anzeiger* 21:145–152, 213–218.

———. 1928. "Die analytische Geographie der Kulturlandschaft erläutert am Biespiel der Brücken." *Zeitschrift der Gesellschaft für Erdkunde zu Berlin* (Sonderband):388–411.

Schmieder, O. 1966. "Die deutsche Geographie in der Welt von Heute." *Geographische Zeitschrift* 54:207–222.

Schmithüsen, J. 1959. "Das System der geographischen Wissenschaft." In *Festschrift Theodor Kraus. . . .* Pp. 1–14. Bad Godesberg.

———. 1963. *Was ist eine Landschaft?* Erdkundliches Wissen, No. 9. Wiesbaden: Franz Steiner.

Schmitthenner, H. 1947. "Alfred Hettner," In Alfred Hettner (ed. Schmitthenner), *Allgemeine Geographie des Menschen*. Vol. 1. Stuttgart: W. Kohlhammer Verlag.

Schott, G. 1912. *Geographie des Atlantischen Ozean*. 4th ed., 1942. Hamburg: C. Boysen.

———. 1935. *Geographie des Indischen und Stillen Ozeans*. Hamburg: C. Boysen.

Schroder, K. H. 1982. "Robert Gradmann, 1865–1950." In *Geographers: Biobibliographical Studies*. Vol. 6, pp. 47–54. London: Mansell.

Schultz, H.-D. 1980. *Die deutschsprachige Geographie von 1800 bis 1970. Ein Beitrage zur Geschichte ihrer Methodologie*. (Abhandlungen des Geographischen Instituts-Anthropogeographie, 29). Berlin: Selbstverlag des Geographischen Instituts der Freien Universität Berlin.

Semple, E. C. 1911. *Influences of Geographic Environment, on the Basis of Ratzel's System of Anthropo-Geography*. New York: Henry Holt.

Smith, T. R., and Black, L. D. 1946. "German Geography: War Work and Present Status." *Geographical Review* 36:398–408.

Smith, W. 1980. Friedrich Ratzel and the Origins of Lebensraum. *German Studies Review* 3:51–68.

———. 1986. *The Ideological Origins of Nazi Imperialism*. New York: Oxford University Press.

Spencer, H. 1864. *Principles of Biology*. 2 vols. New York: D. Appleton.

————. 1876–1896. *The Principles of Sociology*. 3 vols. New York: D. Appleton.

Speth, W. W. 1977. "Carl Ortwin Sauer on Destructive Exploitation." *Biological Conservation* 11:145–160.

Spethmann, H. 1931. *Das länderkundliche Schema in der deutschen Geography*. Berlin: Reimar Hoffing.

Steinmetzler, J. 1956. *Die Anthropogeographie Friedrich Ratzels und ihre ideengeschichtlichen Wurzeln*. Bonn: Geographische Abhandlung.

Storkenbaum, W., ed. 1967. *Zum Gegenstand und zur Methode der Geographie*. Darmstad: Wissenschaftliche Buchgesellschaft.

————. ed. 1969. *Sozialgeographie*. Darmstad: Wissenschaftliche Buchgesellschaft.

Tilley, P. D. 1984. "Carl Troll, 1899–1975." In *Geographers: Biobibliographical Studies*. Vol. 8, pp. 111–124. London: Mansell.

Troll, C. 1947. "Die geographische Wissenschaft in Deutschland in dem Jahren 1933 bis 1945: Eine Kritik und Rechtfertigung." *Erdkunde* 1:3–48.

————. 1949. "Geographical Science in Germany during the Period 1933–1945: A Critique and Justification." Trans. and ed. by E. Fischer. *Annals AAG* 39:100–137.

————. 1950. "Die Geographische Landschaft und (ihre) Erforschung." *Studium Generale* 3:163–181.

————. 1966. "Hermann Lautensach." *Erdkunde* 20:243–252.

Van Valkenburg, S. 1951. "The German School of Geography." In G. Taylor, ed., *Geography in the Twentieth Century*. Pp. 91–115. New York: Philosophical Library.

Wagner, H. 1880. "Bericht über die Entwicklung der Methodik der Erdkunde." *Geographisches Jahrbuch* 8:523–598.

————. 1920. "Geschichte der Methodik der Geographie als Wissenschaft." *Lehrbuch der Geographie* 1:17–25.

Wagner, P. L. October 1989. "A German Critique of Modern Geography." *The Geographical Review* [New York] 79, No. 4:467–474.

Waibel, L. 1933. "Was verstehen wir unter Landschaftskunde?" *Geographische Anzeiger* 34:197–207.

Wanklyn, H. 1961. *Friedrich Ratzel, a Biographical Memoir and Bibliography*. Cambridge: At the University Press.

Wardenga, U. 1991. "Ernst Plewe, 1907–1986." In *Geographers: Biobibliographical Studies*. Vol. 13, pp. 63–71. London: Mansell.

————. 1995. "Nun ist alles, alles anders!" Erster Weltkrieg und Hochschul-geographie, in Ingrid Hönsch and Ute Wardenga (eds.), Kontinuität und Diskontinuität der deutschen Geographie in Umbruchphasen. Studie zur Geschichte der Geographie, *Münstersche Geographische Arbeiten* 39:83–97.

————. 1999. Constructing Regional Knowledge in German Geography: The Central Commission on the Regional Geography of Germany 1882–1941, in Anne Buttimer, Stanley Brunn, and Ute Wardenga (eds.), Text and image. Social construction of regional knowledges, *Beiträge zur Regionalen Geographie* 49:77–84.

West, R. C. 1990. *Pioneers of Modern Geography: Translations Pertaining to German Geographers of the Late Nineteenth and Early Twentieth Centuries*. Baton Rouge: Louisiana State University. Geoscience and Man, vol. 28.

Wirth, E. 1984a. "Dietrich Bartels 1931–1983." *Geographische Zeitschrift* 72:1–22.

————. 1984b. "Geographie als moderne theorieorientierte Sozialwissenschaft." *Erdkunde* 38:73–80.

Zimm, A. 1982. "Darstellung regionaler Strukturen und Prozesse in der Geographie des sozialistischen Auslands." *Petermanns Geographische Mitteilungen* 126, no. 4:217–222.

9

THE NEW GEOGRAPHY IN FRANCE

*Little by little, I have arrived at the idea that our doubts originated in the conflict
between two concepts of geography: the traditional way of looking at it, which
I have called classical, looking more toward the past and toward the delimitation of regions, and a forward looking interpretation which is not yet sure
of its routes but which is playing an increasingly important role in contemporary studies. . . . Classical geography adopts the positions of Ritter. In spite
of appearances, it owes little to Humboldt: from him it has borrowed the
practical methods of symbolization, a taste for precise scientific description,
but no general concepts. Humboldt's cosmologic vision of geography is lost in
the classical period: geography is now concerned with small, local integrations,
and has given up explaining distributions on a global scale. Modern geographers have returned to the ideas and preoccupations of Humboldt.*

—Paul Claval, Essai sur l'évolution de la géographie humaine *(1964)*

The "new geography" that appeared as a professional field in Germany after
1874 spread to other countries of the world with a lag of at least a decade. In
each country the same kinds of philosophical and methodological questions were
faced that had been faced in Germany by scholars who, like their German colleagues,
had not been specifically trained in a field called geography. In each country somewhat different answers to these questions were found, depending on differences
of language, differences of national scholarly traditions, and differences of earlier
experience with geographical study. In each country the new geography was introduced in the universities by one outstanding scholar who became the "grand old
man" of that national school of geography.

The part played by the universities in the formation of geography as a professional field has been fundamental. To be sure, before the professors of geography
were appointed at universities, there were those who, like Humboldt, wrote books
and gave lectures on geography, because there never has been a time when people
were not curious about the world in which they lived. There were cartographic
centers, where geographical information was compiled and transmitted on maps.
But only through the training of younger generations in an accepted program of geographical study can a professional field make its appearance. As Paul Claval points
out, it is through personal contacts with teachers in seminars that all the philosophical
options involved in a methodological statement can be transmitted (Claval, 1964:20).
The development of schools of geography, each with a somewhat distinctive
approach, appeared after chairs were established at the universities and after professional institutions had been provided, such as professional societies, periodicals,

and texts. Only after a professional group comes into existence can scholarly work be progressively developed. This chapter reviews the development of professional geography in France.

THE NEW GEOGRAPHY IN FRANCE

To understand the distinctive character of French geography, it is necessary to recall what scholars had been writing and thinking about during the preceding century and a half. In 1752 Philippe Buache attacked the use of administrative divisions to provide the frame of organization in presenting geographic information. Instead, he proposed that geography should be organized by natural regions and that a river basin was the best kind of natural region. Buache believed that it was self-evident that high ground must form the water partings between different drainage basins, and "high ground" was easily translated into continuous ranges of hills or mountains. In fact, Buache believed that the mountain ranges continued under the oceans, so that the whole surface of the earth was divided into naturally defined compartments. In spite of gradually accumulating observations that refuted Buache's concept, those who accepted his general picture of an earth sharply set off into compartments could easily overlook evidence by just not seeing it. This is another neat example of the way in which a mental image can determine the perception of so-called reality.

French geography had been centered on cartography until the mid-eighteenth century. Owing to the increasing autonomy of cartography, it experienced a crisis of readaptation at the end of the eighteenth and beginning of the nineteenth centuries (Godlewska, 1999). It was during this period that Giraud-Soulavie introduced the modern conception of a natural region based on rock outcrops and climate (Giraud-Soulavie, 1783). His ideas were developed by a government administrator, Charles Coquebert, Baron de Montbret, who taught physical geography at the École de Mines in Paris in 1796–97 (Gallois, 1908:21). When Baron Coquebert was made director of the French statistical office, he proposed a division of the national territory into natural regions and a provision for a concise description of each one. Statistical data would then be compiled by these regions instead of by the traditional administrative divisions. Unfortunately, the work of identifying these natural regions was not completed when Coquebert was assigned to another post. But the seed of interest in regional divisions had been planted, and Buache's hypothesis concerning river basins, like any challenging hypothesis, was beginning to promote the formulation of counterhypotheses.

In 1810 Malte-Brun in his *Précis de la géographie universelle* offered a carefully considered refutation of the idea that river basins are bordered by ranges of mountains or hills. But so strongly implanted was Buache's concept that Malte-Brun's refutation was overlooked. In 1823 a geologist, J. J. d'Omalius d'Halloy, prepared a geological map of France and the Low Countries, accompanied by a description of the relation of landforms to the soils and underlying rocks. Even this clear contradiction of Buache's hypothesis failed to deter those who saw river basins as natural regions.

Meanwhile, the teaching of geography in French schools suffered from a difficulty similar to that of Germany. There were no geographers in the universities and teachers' colleges; therefore, the teachers in the schools had never been given any training in the concepts and methods of the field. There had been a chair of geography at the Sorbonne in Paris since 1809, but it was occupied by historians. The historian who taught geography was in the Faculty of Letters; in addition, in the Faculty of Sciences a geologist offered courses in landforms.

As a result of the Franco-Prussian War of 1870–71, France lost the border provinces of Alsace and Lorraine to Germany and experienced popular demand for better geography teaching in the schools. Donald V. McKay attributes the development of France as a colonial power after 1871 to the influence of the French geographical societies, all of which urged France to bring the benefits of French civilization to the less developed parts of the world. There were two geographical societies in Paris (the Société de géographie at Paris had been founded in 1821) and one each in Bordeaux, Nancy, Lyons, Rochefort, Marseilles, Montpellier, and Douai. They had wide influence on French public opinion (McKay, 1943).

One other institution was to advance the cause of geography in France. The International Geographical Congress first met in Paris in 1875, then again in 1889, 1931, and 1984.[1] To have arranged the second and fourth meetings in the ongoing series of International Geographical Congresses brought attention to Paris and geography.

Paul Vidal de la Blache

Paul Vidal de la Blache led the development of the new geography in France (Freeman, 1967:44–71). Vidal made his way into geography through his study of ancient history and classical literature. He became familiar with Greek geographical writings when he spent a year (1865) at the French School of Archaeology in Athens. In 1866 he graduated from the École normale supérieure in Paris with high honors and completed work for a doctorate in 1872. For the next 26 years he devoted himself to improving the training of teachers of geography and to making up-to-date materials and ideas available to them.[2] During this time the chair of geography at the Sorbonne was occupied by Louis-Auguste Himly, whose major interest was in the changing political boundaries of Europe. In 1898, when Himly retired, Vidal became the first geographer to be appointed to the chair of geography since the chair was established in 1809.

---

[1] This is the only case of the International Geographical Congress being held in the same city on four separate occasions.

[2] Vidal taught geography at the University of Nancy from 1872 to 1877 and then returned as professor to the École normale supérieure. In 1891, with the collaboration of Marcel Dubois, he founded a new professional periodical in which the best writings on geography could be published. In each number of the new *Annales de géographie* he included a bibliography of published materials where additional information and new geographical concepts could be found. This bibliography has since 1923 been published separately as the *Bibliographie géographique internationale*, compiled each year with the collaboration of geographical societies throughout the world and issued by Armand Colin. In 1894 Vidal published the first edition of the *Atlas générale Vidal-Lablache* (revised editions in 1909, 1918, 1922, 1938, and 1951). He was the professor of geography at the Sorbonne from 1898 to the time of his death in 1918 (Dickinson, 1969:208–212).

In his inaugural address at the Sorbonne delivered on February 2, 1899, Vidal followed the custom of presenting his ideas concerning the scope and purpose of geography. There was need, he said, to focus attention on the close relationships between man and his immediate surroundings (milieu) by studying small homogeneous areas. In France such homogeneous areas are popularly recognized as *pays*, such as the *pays de Beauce* around its urban center of Chartres.[3] Vidal presented an effective refutation of the idea of environmental determinism. From Ratzel's second volume of *Anthropogeographie* he formulated the concept of *possibilism*. Nature, he insisted, set limits and offered possibilities for human settlement, but the way man reacts or adjusts to these given conditions depends on his own traditional way of living (Vidal, 1899). Later, Lucien Febvre (1922) published *La terre et l'evolution humaine*, translated into English in 1925 as *A Geographical Introduction to History*. This was the work above all others that helped replace views of geographical determinism with the philosophy of possibilism ("There are no necessities, but everywhere possibilities").

The concept of a way of living (*genre de vie*) has been widely used in French geography. It refers to the inherited traits that members of a human group learn— what we may call a *culture*, borrowing the term from the anthropologists. The *genre de vie* stands for the complex of institutions, traditions, attitudes, purposes, and technical skills of a people. Vidal pointed out that the same environment has different meanings for people with different *genres de vie*: the *genre de vie* is a basic factor in determining which of the various possibilities offered by nature a particular human group will select (Buttimer, 1971:52–57).

Vidal supported the idea of studying small natural regions, but he was very much opposed to the definition of such regions in terms of drainage basins, as Buache had suggested. The use of drainage basins, he pointed out in 1888, would make it impossible even to recognize the existence of one of France's important natural regions, the Massif Central. This is the low mountain region lying west of the Rhône Valley, a region of massive crystalline rocks and radial drainage. If each river basin were identified as a natural region, the Massif Central itself could not be treated as a region. Vidal suggested that one of the major contributions of geographers would be the identification of useful natural regions, or *pays* (Vidal, 1903).

The idea of identifying the regions of France was picked up by one of Vidal's first students, Lucien Gallois,[4] whose *Regions naturelles et noms de pays* was

---

[3] The French and German languages differ in the clarity of their word symbols. The French *pays* is roughly the equivalent of the connotation of *landschaft* as an extent of territory. The connotation of *landschaft* as aspect or scene is translated in French as *paysage*.

[4] Among the French scholars who were contemporaries of Vidal de la Blache, Lucien Gallois was of outstanding importance. Gallois graduated from the École normale supérieure in 1881. He was Vidal's first disciple and worked closely with Vidal for years. From 1898 to 1919 he was editor of the *Annales de géographie*. He was also the editor of the French geographical series *Géographie universelle*, which Vidal had proposed.

Other contemporaries included Elisée Reclus; Emmanuel de Margerie, translator of Eduard Suess's *Das Antlitz der Erde*; and Franz Schrader, the cartographer, who headed the map division of the Librairie Hachette, publishers.

published in 1908 (Gallois, 1908). In this study Gallois provided a useful review of the history of the regional idea in France. Later, after visiting the United States, he became increasingly conscious of the role of big cities and modern transportation in organizing regions (Vidal de la Blache, 1910, 1917), introducing in this way the idea of the polarized region into French geography.

In 1913 Vidal elaborated his ideas about the method of geographical study (Vidal, 1913). In a much-quoted passage he gave clear support to the concept of chorology as the study of things associated in area, mutually interacting, characterizing particular segments of earth space. In this paper he wrote:

> ... that which geography, in exchange for the help it has received from other sciences, can bring to the common treasury, is the capacity not to break apart what nature has assembled, to understand the correspondence and correlation of things, whether in the setting of the whole surface of the earth, or in the regional setting where things are localized (trans. Harrison-Church, 1951:73).[5]

When Vidal died suddenly in 1918 at the age of 73, he was in the process of writing his definitive work, *Human Geography*. From the partially completed manuscripts and notes, Vidal's son-in-law, Emmanuel de Martonne, completed the book, which was published in 1922 (Vidal, 1922). The chapter and section headings in the *Principes de géographie humaine* give an idea of the breadth of Vidal's scholarship:

Introduction: The Sense and Object of Human Geography
Critical examination of the concept of human geography
    The principle of terrestrial unity and the concept of *milieu*
    Man and the *milieu*

Man as a geographic factor

I.   The Distribution of Man over the Globe
    General view
    The formation of areas of population density [population clusters]
    The European agglomeration
    The Mediterranean regions
    Conclusions

II. The Patterns of Civilizations
    The relations of human groups to the *milieux*
    Tools and materials
    Foods
    Building materials
    The human establishments (*habitats*)
    The evolution of civilizations

---

[5]"... ce que la géographie, en échange du secours qu'elle reçoit des autres sciences, peut apporter au trésor commun, c'est l'aptitude à ne pas morceler ce que la nature rassemble, à comprendre la correspondence et la corrélation des faits, soit dans le milieu terrestre qui les enveloppe tous, soit dans les milieux régionaux où ils se localisent" (Vidal, 1913:299).

III. Circulation
   The means of transport
   The route
   The railroads
   The sea

Fragments on Which Vidal Was Working at the Time of His Death
   The origin of races
   The diffusion of inventions (examples: the plow, the wheel, and the draft
   animals)
   Culture regions
   Cities

La Tradition Vidalienne

After the appointment of Vidal de la Blache to the chair of geography at the
Sorbonne, the number of professorships of geography in the French universities
increased rapidly. By 1921 there were departments of geography in almost all of the
16 French universities[6] (Joerg, 1922:438–441). In almost every case the scholars
appointed to teach geography were pupils of Vidal. In no other country, said Joerg,
not even in Germany around Richthofen, has the development of geography been
so centered around one outstanding teacher. These disciples of the master spread
throughout France the point of view and method that came to be known as *la
tradition vidalienne* (Buttimer, 1971).

The disciple who not only elaborated Vidal's ideas about human geography and
spread them throughout France but also transmitted these ideas to other countries
was Jean Brunhes.[7] Brunhes provided a classification of geographic facts that made
Vidal's concepts easier to understand and to transmit in the classroom. Brunhes said
that two world maps were of chief importance in the understanding of human geography: a map of water and a map of population. He divided the essential facts of
human geography into three categories: (1) the facts of the unproductive occupation
of the soil: houses and roads (including rural habitations, urban agglomerations, and
circulation patterns); (2) the facts of plant and animal conquest: the cultivation
of plants and the raising of animals; and (3) the facts of destructive exploitation:
plant and animal devastation and mineral exploitation. He then illustrated the use
of these categories by several small regional studies of sharply defined and distinctive

---

[6]Geography was also represented in 1921 in four universities in Belgium where French
is spoken and in seven Swiss bilingual universities where both French and German
are spoken.

[7]Jean Brunhes had a broad training in history, natural science, law, finance, and
geography. He taught geography at the University of Fribourg in Switzerland from
1896 to 1912. From 1912 until his death in 1930 he held a research professorship at the
Collège de France. In 1910 he published the first edition of his great work, *La Géographie
humaine* (Brunhes, 1910), which went through several revised and enlarged editions in
1912, 1915, and 1934. The second edition was translated into English by I. C. LeCompte
and edited by Isaiah Bowman and R. E. Dodge in 1920; a shortened English translation
was published in 1952. His work had great impact in America because of the translations.

associations of man and the land: the two Saharan oases—Suf in a sandy desert, Mzab in a rocky desert; the Fang people who carry on a destructive exploitation of a small area in present-day Gabon; and the seasonal seminomadism of the inhabitants of a single valley in the Alps (which was replaced in the English edition by Bowman's study of the Peruvian Andes). The last part of his book went "beyond the essential facts" to a discussion of different kinds of broader geographical analyses under the headings of human geography, regional geography, ethnographical geography, social geography, and political and historical geography.

The French school of geography under Vidal's leadership achieved a notable balance between physical and human components. The French geographers of that period were not bothered by an apparent dichotomy between physical geography and human geography the way the Germans were. The reason why this difficulty did not arise in France can be traced back to the work of two of Vidal's earliest disciples: Jean Brunhes, who led in the development of human geography, and Emmanuel de Martonne, who led the way in physical geography[8] (Fig. 29). De Martonne combined the usual training in history and geography with a sound background in geology, geophysics, and biology. To him physical geography was an essential part of the whole geographical study of an area, as he first demonstrated in his regional monograph on the Wallachian Plain and his studies of the Carpathians (de Martonne, 1902, 1917). During all of his professional career, he maintained a regional interest in central Europe and a systematic interest in geomorphology and climatology (de Martonne, 1909, 1927). He became one of the leading physical geographers of the world and the most influential geographer in Europe between the two world wars. He was a strong supporter of Davis and made Davis's ideas known to the French-speaking world. Like Davis, he was a master of the art of landform description, including the drawing of eloquent landscape sketches with pen and ink (de Martonne, 1917:424). His work on the identification of arid regions through the use of an aridity index was a major contribution to the systematic study of climate (de Martonne, 1927).

An important part of *la tradition vidalienne* has been a relative freedom from concern about whether geography is one field or many. This dichotomy has

---

[8]Emmanuel de Martonne, who became Vidal's son-in-law, graduated from the École normale supérieure in 1899. He taught at the University of Rennes from 1899 to 1905 and at Lyons from 1905 to 1909. He was appointed to the Sorbonne in 1909 and remained there until his retirement in 1944. He founded the *Institut de géographie* at the Sorbonne and was its director from 1927 to 1944. This institute was set up in the Faculty of Letters with the close collaboration of the historians. As a result, questions were never raised concerning the teaching of physical geography in a social science faculty. De Martonne became one of the world's leading physical geographers. He was secretary general of the International Geographical Union from 1931 to 1938 and president from 1938 to 1949. His major work in physical geography, *Traité de géographie physique*, was first published in 1909 in one volume of 910 pages and was later expanded and revised (4th ed., 3 vols., 1925–27). He was also the author of the two volumes of *Europe Centrale* in *Géographie universelle*. See "Emmanuel de Martonne, 1873–1955" by Jean Dresch: *Geographers: Biobibliographical Studies* 1988 12:73–81.

Henri Baulig

Paul Vidal de la Blache

Paul Claval

Lucien Gallois

Emmanuel de Margerie

Emmanuel de Martonne

Philippe Pinchemel

Elisée Reclus

Figure 29a E. de Martonne in the field, Romania, 1937

Figure 29b E. de Martonne with Robert Ficheux (from collection of Robert Ficheux)

worried German geographers for a long time, but French geographers ceased being concerned about methodological questions of this sort after 1920. To a person who thinks in French, no real problem is involved in recognizing that from one point of view geography is a unitary field, whereas from another it seems to tie together a variety of fields. C. Vallaux's book *Les sciences géographiques* (Vallaux, 1925) expresses his understanding that geography is both a unitary and an autonomous field of study and also an auxiliary aspect of many fields. Not only does geography have a philosophy of its own, he writes, but also "it is almost, in itself, a philosophy of the world of man" (Vallaux, 1925:viii). So it is that the French geographers can continue to make important contributions to systematic or topical studies and at the same time continue to produce regional monographs (Beaujeu-Garnier, 1976; Claval, 1964; Dickinson, 1969).

But the scope and method of regional studies have changed since the early part of the twentieth century, just as the scope and method of systematic studies have changed.[9] The earliest regional monographs (Blanchard, 1906; de Martonne, 1902; Demangeon, 1905; Vallaux, 1906) followed a more or less standard outline of topics, starting with the surface features and climate, advancing to the organic life in relation to the physical features, and then proceeding to the human inhabitants, looked at both as controlled by the environment and modifying the environment. The French region was a very similar concept to Schlüter's *Kulturlandschaft*. Yet by 1957 Roger Dion, writing on historical geography in the middle of the twentieth century (L'Information géographique, 1957:185), points to the many errors of interpretation that resulted from attempting to demonstrate strict controls. Man's economic life must be seen as in the process of transition and makes no sense at all unless it is viewed in historical perspective. René Musset, discussing recent regional studies in the same monograph (L'Information géographique, 1957:187–196), shows that the original idea of making a "complete" regional study had to be abandoned if only because of the vast increase of information available. For many decades regional studies have been tightly organized around single central themes or single problems, and all materials not relevant to such themes or problems have been omitted. Pierre Deffontaines's study of the Middle Garonne Valley in 1932 is organized in this way around the changing impact of human society on the landscape (Dickinson, 1969:217). Pierre Monbeig's regional monograph on the state of São Paulo in Brazil is focused on the contrast in settlement between the small pioneer farmers and the large coffee planters (Monbeig, 1952). Many decades ago the French geographers, like their colleagues in other countries, began experimenting with different methods of organizing regional studies (Meynier, 1969:113–119). Even Vidal in his last published work recognized the complexity of Alsace-Lorraine and the need to focus attention on the changing significance of this border region between France and Germany (Vidal, 1917).

---

[9]By defining a regional study as one carried out in the manner suggested by Vidal in 1903, it is of course possible to prove that regional study must be dropped. Using the same reasoning, one could insist that systematic studies must be dropped.

Géographie Universelle

Vidal thought that the field study of relatively small regions was the best possible way to train geographers. In the 1950s and 1960s many French geographers still believed that the regional monograph was the best kind of doctoral dissertation. But Vidal also thought that regional studies could serve practical needs. He planned a series of books covering the whole of the earth's land areas that would be carried out on a smaller scale than the French regional monographs and in this way would deal more broadly with the larger regions of the world. Vidal died before the plan could be carried out, but it was both directed and edited after his death by Lucien Gallois. The first volume, *Les îles Britanniques*, was written by Albert Demangeon and published in 1927. Twenty-three volumes were published and included authors M. Zimmerman, E. de Martonne, P. C. D'Almeida, R. Blanchard, F. Grenard, J. Sion, P. Privat-Deschanel, A. Bernard, F. Maurette, H. Baulig, Y. Chataigneau, M. Sorre, and P. Denis. The series was admired in the English-speaking world. I. Bowman sought to have the set translated into English. The lingering impact of the Great Depression prevented this. The series, except the volume on France in three parts, had been completed before the start of World War II. The last volume, on France, was published in 1948 (Vidal and Gallois, 1927–48). The whole series is a monument to the professional work of the first generation of French geographers after Vidal (Martin, 1964). Beautifully printed and illustrated with many detailed maps, the books include information never before available in one series. The parts dealing with economic conditions, population, and political boundaries are now out of date, but they are of inestimable value as the basis for studies in historical geography. The parts dealing with the physical earth and its cover of vegetation are as useful today as when they were written 60 to 70 years ago. Jean Gottmann has written of the volume on *La France physique* by Emmanuel de Martonne:

> De Martonne's physical geography of France . . . gives him the opportunity of presenting a complete picture of the problems and ideas that have been his main concern throughout his life. It is probably the most authoritative and best-written study ever produced on the physical aspects of any large section of Western Europe; it will long remain a classic (Gottmann, 1946:82).

The person of Jean Gottmann, deriving from the Vidalian tradition, and more especially from Albert Demangeon, demands further attention. Gottmann escaped France in the late 1930s, took a post with Princeton and, in 1943, removed to Johns Hopkins University. Five years later, he returned to Paris, then in 1968 accepted the chair of Geography at the University of Oxford. He left behind him a wealth of publication, including, more notably, *A Geography of Europe* (1950), *Virginia at mid-century* (1955), and *Megalopolis* (1961). He remained a French regional geographer at a time when that genre was beginning to recede and he spent most of his years outside France, reducing the opportunity to develop a sustained following "at home." With the publication of *Megalopolis*, the thrust of his work became perhaps better described as spatial analysis (Johnston, 1996) and his work more international than French in character.

French Geography since World War II

Since World War II there has been a great increase in the number of geographers in France (from approximately 70 university teachers in 1955 to more than 1200 at present) and in the range of their interests. As a result, "*la tradition vidalienne*" has been weakened, and disciplinal coherence has been lost as French geography reacts to international contemporary currents of thought. More specifically, in the years 1950 to 1968 study of regional geography declined. The rural areas themselves were modernized, and traditional regional geography was not able to explain a world that was no longer dominated by its past, but increasingly subject to present, ever more complex, functions. However, the regional idea remained of value with regard to developing countries. Pierre Gourou initiated a literature concerning the tropics and agriculture in 1947 with *Les Pays tropicaux*, followed by P. Monbeig (1952), G. Sautter (1966), and P. Pélissier (1966).

An important motif in the regional undertaking was historical geography that explained the origins of agrarian landscapes (Chevalier, 1956; Juillard, 1953; Meynier, 1958). Studies from French historians, perhaps inspired by Lucien Febvre, relieved geographers of some of their work. Yet historical geography came to have less meaning for French geographers: "for most French geographers what still matters is the explanation of the present; a study bearing less and less relation to contemporary concerns will only attract a minority of colleagues" (Claval, 1984:24). This led to a reexamination of regional thinking, which perhaps began with André Cholley in 1942 and was continued by Le Lannou (1949), who attempted to follow la Blache and to develop geography around the notion of human occupance. Throughout the late 1940s and the 1950s the demographic point of view was much in evidence in geographers' work.

Mention must be made of Fernand Braudel, who wrote most of his dissertation on the history of the western Mediterranean in the sixteenth century while in a prisoner of war camp. The definitive study, published shortly after war's end, propelled Braudel to a position as one of the finest French historians—but one advocating very close proximity to the study of geography (preferably Vidalian geography)—and assuring him leadership of the Annales school earlier assumed by Lucien Febvre. Claval has written (2003:38–39)

> Braudel's role in shaping geohistoire was important for both geographers and historians: historical geography had been an application of the procedures of contemporary geography to the past; geohistory was based on a different hypothesis —that history could not be understood independently from its geographic setting.

Although other historians shared similar thought, including G. Duby, P. Goubert, and LeRoy Ladurie, Braudel was the only one to systematize this dimension of French historical method. More spectacular change came in the 1950s when geographers turned to the economic sciences. At this time Marxists may have been more visible in geography than in any other discipline in France. Yet theirs was an ideology and not a methodology suited to the needs of the geographic discipline. For example, it shed little light on the origin of spatial patterns. When applied geography began to emerge in the 1960s (Labasse, 1966; Phlipponeau, 1960), some geographers with Marxist inclinations denied its legitimacy. The Marxist group, distancing itself from geographers in the United States, retarded the progress of the new theoretical movement initiated in the late 1950s. However, the Marxists had attempted to introduce science into French geography, which previously had been high art. Nevertheless, French geographers became involved in comprehensive planning processes, which are a marked feature of the contemporary French scene. After much debate in the 1960s and early 1970s over the question of applied geography (*la géographie appliqueé*) as a contradiction of traditional values in the profession (McDonald, 1964, 1975), the younger generation has embraced it as both a theoretical stimulant and a significant source of employment.

Meanwhile, fewer French spoke German, and fewer north European, British, and U.S. geographers spoke French. This lessening of communication further insulated French geography from developments abroad. Then came the student uprising of 1968, at which time the meaning of French institutions was challenged. From 1968 until 1974 the number of students studying geography dropped substantially. The decade was marked by a continued diminution of the primacy of Paris, which, although it continued to attract French academics, lost much of its dominance: The appeal of life in the metropolis had waned, and the vitality (and employment opportunities) of numerous regional centers had increased. Geographic departments at Strasbourg, Nancy, Grenoble, Lyon, Aix-Marseille, Rouen, Rennes, Lille, and other regional centers all experienced an impressive expansion in terms of both education and involvement in practical regional affairs. The numerous regional geographical journals in which much of the best original work appeared were ample evidence of this trend (McDonald, 1965).

With the decline of Marxism, the increase in the use of the English language, and the greater transatlantic exchange, especially with Quebec, new ways of thinking geographically were imported from the Anglo-American realm. Model building and concern for the development of theory increased but met with resistance from the strong traditionalists within French geography. The positivism implicit in much of the Anglo-American contribution of the 1960s did not appeal, as elsewhere, to a profession whose roots were in unique traditions of the human-land complex through historic time. And so some Marxist works, some search for theory, and a quest for new directions characterize recent geographical inquiry in France. Paul Claval (1984) suggests that French geography has been rethinking itself around the theme of spatial organization (Auriac, 1982; Dauphiné, 1979; Noin, 1976), and systematic studies are flourishing, especially in population, medical, and transportation geography. Claval (1991) also suggests that after 1968 geographers began to explore economics, anthropology, and history for new ideas and sought foreign experiences. Simultaneously, traditional German influences declined, while the influence of Anglo-American geography and geographers increased. Some traditional geographical ways, including induction and description, have been retained.

The radical approach was less popular in France than in the English-speaking world because of the disillusionment of many French geographers with the forms of Marxism that prevailed in France after World War II and because Lévi-Strauss, Barthes, Derrida, and others appeared less attractive to French scholars than to their American and English colleagues. There was, resultantly, a parallel with the phenomenological approach as developed in the English-speaking world. Its roots were more literary than philosophical, as exemplified by Frémont's interest in *l'espace vécu*, the lived experience of space (Frémont, 1976; Berque, 1982). Important groups worked from the late 1970s on the problems of perception (Bailly, 1977; Debarbieux, 1998), the role of the human body and human senses in the experience of the world (Pitte, 1983; 1991), and territoriality (Bonnemaison, 1981). Systematic studies were devoted to the idea of landscape (Berque, 1990; 2001) and the cultural approach in geography (Claval, 1995a; Bonnemaison, 2001). From the late 1970s a strong revival of political geography and geopolitics was experienced (Sanguin, 1975; Lacoste, 1976; Claval, 1979).

Postcolonial studies did not have the same impact in France as in Britain or the United States since French scholars as early as the 1950s and 1960s had integrated the views of *négritude* expressed by Léopold Senghor and Aimé Cesaire and Franz Fanon's critique of colonialism. Gender studies are also lacking in French geography —they are also lacking in French-Canadian geography and in French history and sociology. It is not because French geographers (as well as French-Canadian ones and French sociologists and anthropologists) neglected this domain, but because French scholars have not developed the same attitudes in the face of modernization and postmodernization of societies.

French geography evolves along lines parallel to those of Anglo-America, but its path remains different in some respects. The contemporary renewal of French geography has produced a social science with an emphasis on economy in the 1960s, social and political problems in the 1970s and 1980s, and culture in the 1990s. During the last 15 years the emphasis in physical geography shifted from geomorphology and climatology to ecology. As elsewhere, geographers are trying to find ways to facilitate sustainable growth.

The movement of ideas in the profession has encouraged introspection and an interest in geographical thought. Significant works have come from Paul Claval (1964, 1972, 1975, 1978, 1981, 1991, 1998), André Meynier (1952, 1969), the Comité National de Géographie (1972, 1980), Philippe Pinchemel (1968, 1980, 1981), Vincent Berdoulay (1981, 1988, 2001), André-Louis Sanguin (1993), Marie-Claire Robic (1993, 2000) and Baudelle (2002). From beyond France Anne Buttimer (1971), H. F. Andrews (1984, 1986), G. S. Dunbar (1978), and Anne Godlewska (1998) are specially noteworthy.

REFERENCES: CHAPTER 9

Andrews, H. F. 1984. "L'oeuvre de Paul Vidal de la Blache: notes bibliographiques." *Canadian Geographer*: 1–18.
———. 1986. "The Early Life of Paul Vidal de la Blache and the Makings of Modern Geography." *Institute of British Geographers. Transactions [London]* New series. Vol. 11. No. 2:174–182.
Auriac, F. 1982. *Système économique et espace: un exemple langue-docien*. Paris: Economica.
Bailly, A. 1977. *La Perception de l'espace urbain*. Paris: Centre de Recherche et d'Urbanisme.
Baker, S. 1988. Vidal de la Blache, Paul Marie Joseph, 1845–1918. *Geographers: Biobibliographical Studies* 12:189–201.

Baudelle, G., et al. 2002. *Géographies en pratiques (1870–1945). Le terrain, le livre, la cité.* Rennes: Presses Universitaires de Bretagne.

Beaujeu-Garnier, J. 1951. *Le Morvan et sa bordure.* Paris: Armand Colin.

———. 1956–1958. 2 vols. *Géographie de la population.* Paris: Librairie de Medici.

———. 1976. *Methods and Perspectives in Geography.* London, New York: Longmans (translated by J. Bray).

———. December 1990. "French Geography since 1950." *Scottish Geographical Magazine [Edinburgh]* 106, no. 3:130–134.

Berdoulay, V. 1977. "Louis-Auguste Himly, 1823–1906." *Geographers: Biobibliographical Studies* 1:43–47.

———. 1981. *La formation de l'école géographique française (1870–1914).*

———. 1988. *Des Mots et des lieux. La dynamique du discours géographique.* Paris: CTHS.

———. 2001. "Geography in France: Context, Practice, and Text." In G. Dunbar, ed. *Geography: Discipline, Profession and Subject since 1870.* Netherlands: Kluwer Academic Publishers; pp. 46–78.

Berque, A. 1982. *Vivre l'espace au Japan.* Paris: PUF.

———. 1990. *Médiance.* Montpellier: Reclus.

———. 1996. *Les Raisons du paysage. De la Chine antique aux environnements de synthèse.* Paris: Hazan.

———. 2000. *Ecoumène.* Paris: Belin.

Blanchard, R. 1906. *La Flandre: étude géographique de la Plaine Flamande in France, Belgique, et Pays-Bas.* Paris: Armand Colin.

Bonnemaison, J. 1981. "Voyage autour du territoire." *L'Espace géographique,* 10, 4: 249–262.

———. 2001. *La Géographie culturelle.* Paris: CTHS.

Brunhes, J. 1910. *La Géographie Humaine.* Paris: Armand Colin.

———. 1920. *Human Geography.* Translated by I. C. LeCompte. Edited by I. Bowman, R. E. Dodge and J. Brunhes. New York: Rand McNally and Co.

Buttimer, A. 1971. *Society and Milieu in the French Geographic Tradition.* Chicago: Rand McNally.

Capot-Rey, R. 1946. *Géographie de la circulation sur les continents.* Paris: Gallimard.

Chevalier, M. 1956. *Les Pyrénées ariégeoises.* Paris: Marie-Thérèse Genin.

Claval, P. 1964. *Essai sur l'évolution de la géographie humaine.* Cahiers de géographie de Besançon, No. 12. Paris: Les Belles Lettres.

———. 1972. *La pensée géographique. Introduction à son histoire.* Paris: Sedes.

———. 1975. "Contemporary Human Geography in France." *Progress in Geography* 7:253–292.

———. 1976. "Contemporary Human Geography in France." In C. Board et al. (eds.). *Progress in Geography.* Vol. 7, pp. 235–292. London: Edward Arnold.

———. 1978. *Espace et pouvoir.* Paris: PUF.

———. 1981. "Les Géographes et les réalités culturelles." *L'Espace géographique* 10:242–248.

———. 1984. "France." In R. J. Johnston and P. Claval, eds., *Geography since the Second World War.* London: Croom Helm, pp. 15–41.

———. 1991. "France." In G. S. Dunbar, ed., *Modern Geography: An Encyclopedic Survey.* New York and London: Garland Publishing, Inc., pp. 56–58.

———. 1995. *La Géographie culturelle.* Paris: Nathan.

———. 1998. *Histoire de la géographie française de 1870 à nos jours.* Paris: Nathan.

———. 2003. "Fernand Braudel, 1902–1985." *Geographers: Biobibliographical Studies* 22:28–42.

Comité National de Géographie. 1972. *Recherches géographiques en France.* Montreal and Paris.

Dauphiné, A. 1979. *Espace, région et système.* Paris: Economica.

Debarbieux, B. 1998. "Les problématiques de l'image et de la représentation en géographie." In Bailly, Antoine, ed., *Les Concepts de la géographie.* Paris: A. Colin, pp. 199–211.

Demangeon, A. 1905. *La Picardie et les régions voisines, Artois, Cambresis, Beauvaises.* Paris: Armand Colin.

de Martonne, E. 1902. *La Valachie, essai de monographie géographique.* Paris: Armand Colin.

———. 1909. *Traité de géographie physique.* Revised and enlarged, 1913, 1920; 3 vols. 1925–1927. Paris: Armand Colin.

———. 1917. "The Carpathians: Physiographic Features Controlling Human Geography." *Geographical Review* 3:417–437.

———. 1927. "Regions of Interior-Basin Drainage." *Geographical Review* 17:397–414.

Dickinson, R. E. 1969. *The Makers of Modern Geography.* London: Routledge & Kegan Paul.

Dunbar, G. S. 1978. *Elisée Reclus: Historian of Nature.* Hamden, Conn.: Shoe String Press.

Fischer, E., Campbell, R. D., and Miller, E. S. 1967. *A Question of Place, the Development of Geographic Thought.* Arlington, Va.: Beatty.

Freeman, T. Walter. 1967. *The Geographer's Craft.* Manchester University Press; New York: Barnes & Noble.

Gallais, J. 1967. *Le delta intérieur du Niger. Étude de géographie régionale.* Dakar: IFAN.

Gallois, L. 1908. *Régions naturelles et noms de pays: étude sur la région parisienne.* Paris: Armand Colin.

Giraud-Soulavie (Abbé). 1780–1784. *Histoire naturelle de la France méridionale,* Paris: 7 vol.

Godlewska, A. 1999. *Geography Unbound. French Geographic Science from Cassini to Humboldt.* Chicago: Chicago University Press.

Gottmann, J. 1946. "French Geography in Wartime." *Geographical Review* 36:80–91.

Gumuchian, H. 1991. *Représentations et aménagement du territoire.* Paris: Anthropos-Economica.

Harrison-Church, R. J. 1951. "The French School of Geography." In G. Taylor, ed., *Geography in the Twentieth Century.* New York: Philosophical Library, pp. 70–90.

Joerg, W. L. G. 1922. "Recent Geographical Work in Europe." *Geographical Review* 12:431–484.

Johnston, R. J. 1996. "Jean Gottmann: French Regional and Political Geographer Extraordinaire." *Progress in Human Geography* 20.2:183–193.

Juillard, E. 1953. *La Vie rurale dans la plaine de Basse-Alsace.* Paris: Les Belles Lettres.

Labasse, J. 1966. *L'Organisation de l'espace.* Paris: Hermann.

Lacoste, Y. 1976. *La Géographie, ça sert, d'abord, à faire la guerre.* Paris: La Découverte.

Le Lannou, M. 1949. *La Géographie humaine.* Paris: Flammarion.

Levainville, J. 1909. *Le Morvan.* Paris: Armand Colin.

L'Information géographique. 1957. *La Géographie française au millieu du XXe siècle.* Paris: Bailliere & Fils.

Martin, G. J. 1964. "The Region in French Geographic Thought, c. 1900–1930." In *Papers of the Michigan Academy of Science, Arts & Letters* 49:325–332.

McDonald, J. R. 1964. "Current Controversy in French Geography." *Professional Geographer* 16:20–23.

———. 1965. "Publication Trends in a Major French Geographical Journal." *Annals AAG* 55:125–139.

———. 1975. "Current Trends in French Geography." *Professional Geographer* 17:15–18.

McKay, D. V. 1943. "Colonialism in the French Geographical Movement, 1871–1881." *Geographical Review* 33:214–232.

Meynier, A. 1952. "Cinquante ans de géographie française." In *Volume jubilaire du laboratoire de géographie de Rennes*, pp. 47–52.

———. 1958. *Les Paysages agraires*. Paris: A. Colin.

———. 1969. *Histoire de la pensée géographique en France*. Paris: Presses universitaires de France.

———. 1972. *La Pensée géographique française contemporaine*. Presses universitaires de Bretagne.

Monbeig, P. 1952. *Pioneers et planteurs de São Paulo*. Paris: Armand Colin.

Noin, D. 1976. *L'Espace français*. Paris: A. Colin.

Pélissier, P. 1966. *Les Paysans du Sénégal*. Saint-Yrieix: Fabrègue.

Phlipponeau, M. 1960. *Géographie et action. Introduction à la géographie appliquée*. Paris: A. Colin.

Pinchemel, P. 1968. "Géographie—L'histoire de la géographie, évolution chronologique, les tendances de la pensée géographique." In *Encyclopaedia Universalis*. Vol. 7, pp. 621–625.

———. April–June 1980. "L'histoire de la géographie japonaise." *L'espace géographique* 9, no. 2:165–171.

———. ed. 1981. "Histoire et épistémologie de la géographie." France, Ministère des universités, Comité des travaux historiques et scientifiques, *Bulletin de la section de géographie*, no. 84.

Pitte, J.-R. 1983. *Histoire du paysage français*. Paris: Tallandier, 2 vol.

———. 1991. *Gastronomie française*, Paris: Fayard.

Robic, Marie-Claire, ed. 1993. *Jean Brunhes. Autour du monde. Regards d'un géographe, Regards de la géographie*. Boulogne: Musée Albert Kahn.

———. ed. 2000. *Le "Tableau de la géographie de la France" de Paul Vidal de la Blache*. Paris: CTHS.

Roger, A. 1997. *Court Traité du paysage*. Paris: Gallimard.

Sanguin, A.-L. 1977. *La Géographie politique*. Paris: PUF.

———. 1993. *Vidal de la Blache. Un géant de la géographie*. Paris: Belin.

Sautter, G. 1966. *De l'Atlantique au fleuve Congo: une géographie du souspeuplement*. 2 vols. Paris: Imprimerie Nationale.

Sion, J. 1908. *Les paysans de la Normandie orientale*. Paris: Armand Colin.

Sorre, M. 1913. *Les Pyrénées méditerranéennes*. Paris: Armand Colin.

———. 1948. *Les fondements de la géographie humaine*. Paris: Armand Colin.

Vallaux, C. 1906. *La basse Bretagne*. Paris: Armand Colin.

———. 1925. *Les sciences géographiques*, 2nd ed. 1929. Paris: Armand Colin.

Vidal de la Blache, P. 1899. "Leçon d'ouverture du cours de géographie." *Annales de géographie* 8:97–109.

———. 1903. *Tableau de la géographie de la France*. Vol. 1, E. Lavisse, ed. *Histoire de France*. Paris: Hachette. Published separately as *La France: tableau géographique*. 1908. Paris: Hachette.

———. 1910. "Les régions françaises." *Revue de Paris* 6:821–840.

———. 1913. "Des caractères distinctifs de la géographie." *Annales de géographie* 22:289–299.

———. 1917. *La France de l'Est: Lorraine-Alsace*. Paris: Armand Colin.

———. 1922. *Principes de géographie humaine*. Ed. E. de Martonne. Paris: Armand Colin. Trans. M. T. Bingham, *Principles of Human Geography*. 1926. New York: Henry Holt.

Vidal de la Blache, P., and Gallois, L., eds. 1927–1948. *Géographie universelle*. 15 tomes & 23 vols. Paris: Armand Colin.

10

THE NEW GEOGRAPHY IN
GREAT BRITAIN

> *A new scientific truth does not triumph by convincing its opponents and
> making them see the light, but rather because its opponents eventually die, and
> a new generation grows up that is familiar with it.*
>
> *Max Planck,* Scientific Autobiography and Other Papers

For the British the Age of Exploration did not come to an end with the voyages
of Captain Cook. The extension of knowledge of the earth's surface through
exploration and research has continued to be a major concern of British geographers and of the British geographical societies (Crone, 1964). Among these, the
most significant were the Royal Geographical Society, 1830; the Royal Scottish
Geographical Society, 1884; the Geographical Association, 1893; and the Institute
of British Geographers, 1933. A large percentage of the papers published in
nineteenth century British geographical periodicals were reports of exploration of
the still relatively unknown parts of the earth. Geographers and others were active
in founding a British Empire throughout the world. This involved exploration,
measurement, and mapmaking. Pride in empire, direct or indirect, undoubtedly
assisted geography in establishing for itself a place in the curricula of British educational institutions.

Geography started its growth in the universities of Britain after the appointment
of Halford J. Mackinder at Oxford in 1887. The period of major expansion came
after 1900 (Freeman, 1974, 1980; Withers, 2001; Johnston and Williams, 2003).

THE NINETEENTH CENTURY

In the nineteenth century, British geography as taught in the schools was generally
considered to be a dull and laborious subject. Uninspired, untrained teachers presented pupils with lists of places and products to be memorized. In the universities,
geography was offered by geologists, and lectures on geography as a background
for understanding the course of history were given by historians.

Nineteenth-century Britain, however, had the remarkable scholar Mary
Somerville (Baker, 1963:51–71; Neeley, 2001). A self-made geographer who read
widely and was in close touch with the leading scholars of her time, she was far in
advance of her contemporaries in her understanding of the nature of geography
as a field of study. After two earlier books on celestial mechanics and the physical
sciences, she started on her work *Physical Geography* in 1839. When it was ready

for publication, the first volume of Humboldt's *Kosmos* appeared, and Somerville had to be persuaded by her friends to publish her own work. The first edition of *Physical Geography* was published in 1848[1] prior to Darwin's *Origin of Species*. In it she described the surface features of the land, the oceans, the atmosphere, plant and animal geography, and man as an agent of change of the physical features of the earth. She worked on many revisions during her long life (she died at age 92 in 1872), adding new materials as they became known to her, including materials contained in the *Physical Atlas* by Keith Johnston based on the *Berghaus Atlas* (Freeman, 1961:189). In faraway Vermont, George Parkins Marsh found her observations about humankind's destructive use of the earth very stimulating, and he made frequent references to her work.

In 1877 Thomas Henry Huxley's *Physiography* was published. It had an important influence on the teaching of geography at all levels and strengthened the teaching of physical geography. The book exemplified post-Darwinian causal reasoning and linked learning to pupil experience. The book begins with the study of the River Thames at London Bridge and proceeds outward, terminating with the earth as a planet. The book's emphasis on local field work was largely responsible for including field work in the curriculum. *Physiography* established physical geography in Britain (Stoddart, 1975).

Another British scholar who made important contributions to geography was Francis Galton, who was a cousin of Darwin and better known for his studies of heredity than geography. For Galton, geography became a hobby to which he devoted a considerable amount of time and thought. After traveling in South Africa, he served on the council of the Royal Geographical Society from 1854 to 1893. His interest in the study of British weather led him to make the first British weather map in 1861, based on reports from 80 stations. He was the first to point out the weather patterns that could be revealed by plotting lines of equal air pressure on a map (isobars) and was also the first to recognize the nature of air circulation around a center of high pressure. The first weather map to be published in a newspaper was one that he prepared for *The Times* (April 1, 1875).

Galton's interest in plotting things on maps was bounded by no restrictive definition of the scope and nature of geography as a field of learning. He prepared a map of the world showing lines of equal travel time from London (an isochronous map) in 1881. He also made a map of "female beauty" in Great Britain based on his own observations. He identified three classes—good, medium, and bad. On the resulting map the high point in female beauty was London and the low point was Aberdeen, Scotland (Freeman, 1967:41). Studying the hereditary genius among British scholars, he found that 92 percent of the scientists who became famous had been born in places located within only half of the country (Gilham, 2001).

In 1885 Galton edited a pamphlet entitled *Cambridge Essays*, to which numerous members of the faculty at Cambridge made contributions. He wrote a brief piece

---

[1]The first three editions, published in 1848, 1849, and 1851, were in two volumes; later editions were in one volume—1858, 1862, 1870, and 1877. The last two editions were edited by Henry W. Bates (Baker, 1963:53).

entitled *Notes on Modern Geography*, in which he described geography as "a peculiarly liberalizing pursuit, which links the scattered sciences together and gives to each of them a meaning and significance of which they are barren when they stand alone" (Galton, 1855:81).

At the time when Somerville, Huxley, and Galton were making their contributions, there was no professional body of scholars to carry their ideas on because there were no clusters of geographers in the universities. Furthermore, in Great Britain the geologists included physical geography as a part of their own field and studied the influence of the physical features of a country on the people. Archibald Geikie wrote that the influence of physical features can be seen "(1) in the distribution of migration of races; (2) in the historical development of a people; (3) in industrial and commercial progress; and (4) in national temperament and literature" (Geikie, 1865).

The introduction of geography into the British universities resulted chiefly from the efforts of the Royal Geographical Society (Baker, 1963:64). In 1884 John Scott Keltie, then secretary of the Society, was asked to make a survey of the status of geography in Great Britain and to compare it with the position of geography in other countries. He reported that in the other countries of Europe and in America there were professors of geography in many of the universities and that Britain compared unfavorably in this field. In 1886 the president of the Society wrote to the authorities at Oxford and Cambridge, pointing to the findings of the survey and urged that something be done about it. The result was the appointment of a geographer at Oxford in 1887, at Cambridge in 1888, and thereafter at almost all other British universities (Keltie, 1886).

HALFORD J. MACKINDER

Halford J. Mackinder (Fig. 30) led the way in the expansion of geography offerings in the universities. He was appointed reader in geography at Oxford in 1887.[2] Mackinder's training was in natural science and history. He reached the conclusion that history without geography was mere narrative and that since every event occurred in a particular time at a particular place, history and geography, which deal respectively with time and place, should never be separated. In a lecture delivered before the Royal Geographical Society (Mackinder, 1887) he identified geography as the field that traces the interactions of humans and their physical environment. He said, "We hold that no rational political geography can exist which is not built upon and subsequent to physical geography" (Fischer, Campbell, and Miller, 1967:258–261; Mackinder, 1902).

---

[2]Mackinder was not the first to receive an appointment to teach geography in a British university. Baker observes that Baldwin Norton was a lecturer in geography at Oxford in 1540 and 1541. Richard Hakluyt was a lecturer in geography at Oxford after 1574. During the three centuries before Mackinder's appointment, lectures in geography were given by geologists and historians, but there was no place where future geographers could receive advanced training until the establishment of the School of Geography at Oxford in 1899, which Mackinder had recommended four years earlier (Baker, 1963:58–60, 119–129).

Figure 30 H. J. Mackinder

Mackinder was less interested in the details of human–land relations than in developing a world view.[3] His first major work, *Britain and the British Seas* (Mackinder, 1902), was an example of a regional study in a global context. Two years later he gave the now-famous lecture at the Royal Geographical Society on "The Geographical Pivot of History" (Mackinder, 1904), in which he announced the heartland theory as a concept of global strategy. His warning regarding the challenge to sea power by land power fell on deaf ears in a Britain securely in control of the world's oceans. In 1919 he elaborated the same theme in *Democratic Ideals and Reality* (Mackinder, 1919/1942), but his message went unheeded until the outbreak of World War II.

Mackinder's heartland theory was nothing less than a model to place the broad sweep of world history on the stage provided by global geography. He identified a world island consisting of the continents of Eurasia and Africa (Fig. 31). The least accessible part of the world island he called the heartland, throughout which the rivers flow either into inland seas, such as the Caspian, or into the frozen Arctic Ocean. And extending like a peninsula from one end of the heartland are the deserts

---

[3]Mackinder was both a scholar and a practical man of affairs. While he was at Oxford, he also held the position of principal of University College, Reading (1892–1903). In 1905 he was named director of the London School of Economics and Political Science. Between 1910 and 1922 he was a member of Parliament. From 1920 to 1945 he was chairman of the Imperial Shipping Committee and from 1925 to 1930 chairman of the Imperial Economic Committee. Incidentally, he was the first to climb to the summit of Mt. Kenya (17,040 ft.), which he did in 1899 (Baker, 1963; Blouet, 1987; Parker, 1982).

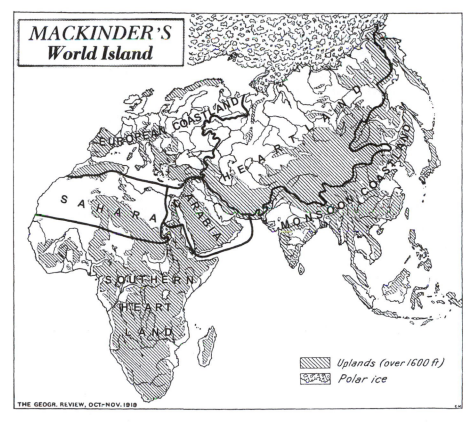

Figure 31 Mackinder's world island

of Arabia and the Sahara. In contrast to this curving area of generally thin population and difficult accessibility from the oceans are the coastlands on either side: the European coastland and the so-called monsoon coastland, both easily accessible from the sea. In these coastlands is found most of the world's population. Mackinder identifies Africa south of the Sahara as a southern heartland, of slight strategic importance, but, like the interior of Eurasia, inaccessible from the sea (Teggart, 1919). As C. R. Dryer pointed out, Mackinder did not include the Americas except to name them as satellites out on the margins of things, along with Australia (Dryer, 1920). Mackinder's main theme relates to the repeated invasions of the coastlands by conquerors coming from the heartland. He went back to the prehistoric migrations of humankind, spreading from the heartland in three directions: southeastward into the monsoon coastland and on to Australia; northeastward through Siberia and Alaska into the Americas; and westward into the European coastland and the southern heartland of Africa. Repeatedly throughout the course of history, the earlier migrants along these routes were invaded and conquered by migrants who came later. The coastlands, he insisted, had always proved vulnerable to attack from the heartland, and the heartland remained invulnerable because sea power could be denied access to it.

In 1919, after World War I, Mackinder argued for the formation of a buffer zone of small states to keep Germany and Russia apart. He summarized his view of global strategy with the famous dictum:

> Who rules East Europe commands the Heartland;
> Who rules the Heartland commands the World Island;
> Who rules the World Island commands the World (Mackinder, 1919/1942:150).

If Germany and Russia could form an alliance or if Germany could conquer Russia, the stage would be set for world conquest. In the 1930s the German geopolitician Karl Haushofer favored an alliance with Russia, an issue over which he had a difference of opinion with Hitler.

We can understand now that like all theoretical models that generalize geographic observations, Mackinder's heartland concept helps people to understand complex sequences of events by oversimplifying them. The model is based on selecting a few facts of location and a few sequences of events and ignores complicating details. It cannot provide a precise blueprint of things to come, yet it cannot be entirely ignored. In terms of its premises, its deductions are startlingly clear.

THE DEVELOPMENT OF BRITISH GEOGRAPHY TO WORLD WAR I

The idea that Mackinder set before the British geographers called for basic work in physical geography to provide a scientifically accurate description of the stage setting on which the human drama was to be played out. In this point of view he was close to the concepts of Richthofen and Hettner, but not Schlüter, and close to Vidal de la Blache and de Martonne, but not Brunhes. Mackinder himself was concerned with the global view, but most of his contemporaries and his followers turned their attention to the analysis of human–land relations in small areas.

Two of Mackinder's better-known contemporaries who also contributed to the development of British geography were George G. Chisholm and Hugh Robert Mill (Freeman, 1977). Chisholm was a pioneer in the field of commercial geography, and his *Handbook of Commercial Geography* (Chisholm, 1889) became a classic that appeared in many revised editions. In it he brought together a vast amount of information about world trade, and he also formulated the basic theory describing such trade. The 11th edition (edited by L. Dudley Stamp) starts as follows:

> The great geographical fact on which commerce depends is that different parts of the world yield different products, or furnish the same products under unequally favourable conditions. Hence there are two great results of commerce: the first, to increase the variety of commodities at any particular place; the second, to equalise more or less, according to the facilities for transport, the advantages for obtaining any particular commodity in different places between which commerce is carried on.

Hugh Robert Mill outlined the contrast between Mackinder and Chisholm in the following words:

Mackinder was brilliant, brushing aside all irrelevancies, sketching the broad outlines of the science with a masterly hand, and by his gifts of generalization and exposition often suggesting new lines of research to the more pedestrian votaries of Geography. . . . Chisholm was profoundly learned, laborious, accurate, and meticulous in definition and safe-guarding of every detail (Mill, 1951:85).

Mill was like neither Mackinder nor Chisholm. Plagued by ill health for much of his long life, Mill became an avid reader and a master of many of the varied branches of geography. He had a fine feeling for words and a sense of the poetry that infuses all aspects of physical and human geography. He sums up the major focus of his own research interests as follows:

The study of the part played by water in the economy of the world through the action of solar heat and terrestrial gravitation raising vapour from the sea and carrying the condensed moisture back over the land, sustaining all forms of life and furnishing hydroelectric power, the only inexhaustible supply of energy (Wrigley, 1950:660).

Although the study of water was the major focus of his interest, he describes his life as made up of nine interwoven strands. He never was able to take part in a polar exploring expedition, but polar exploration became one of his specialties. His accounts of polar exploration, including the biographies of famous explorers, were not only accurate recordings of events, but also literary accomplishments—among them *The Siege of the South Pole* (1905) and *The Life of Ernest Shackleton* (1923). The study of water as part of the "realm of nature" had fascinated him as early as 1891 (Mill, 1891), but it became a major interest when he was named joint director of the British Rainfall Organization in 1900 and director the following year. Under his direction rainfall maps of Great Britain were prepared on the basis of 50-year averages. Only approaching blindness made it necessary for him to retire from his post in 1919.

Mill recognized the need for detailed studies of Britain while he was working at the Royal Geographical Society. In 1896 he drew up a plan for using the sheets of the Ordnance Survey (1 inch to the mile) as bases on which to plot categories of land quality and land use for all of the British Isles (Mill, 1896). In 1900 he provided sample studies of two sheets to demonstrate the utility of such detailed field mapping (Mill, 1900). Later he said that the neglect of his proposal was one of the greatest disappointments of his life. The idea was realized by L. Dudley Stamp during the 1930s. Special mention must be made of *The International Geography* (1899). This was the compendium of 55 chapters written by 71 authors, of whom Mill was one, additional to his editing of the approximately 1100 pages. In the US the book was used as a text in a number of colleges and universities. Elsewhere it was regarded as a work of reference.

The geographers who followed Mackinder, Chisholm, and Mill gradually spread and strengthened British work in this professional field. Andrew J. Herbertson, who succeeded Mackinder as director of the Oxford School of Geography in 1905, remained there until his death at the age of 49 in 1915 (Jay, 1979). He was interested primarily in improving the teaching of geography, which for too long had been characterized by the study of encyclopedic collections of data organized by political units. Herbertson proposed a framework of natural regions for the study of world geography. On a global scale, he suggested, the great natural regions should be identified in terms of associations of surface features, climate, and vegetation. Here he was thinking along lines similar to Penck and Passarge. For his categories of surface features he went back to the work of Eduard Suess and especially to the translation and amplification of Suess's work rendered by Emmanuel de Margerie. For his major divisions of climate he relied on Alexander Supan. His 15 major natural regions (Fig. 32) revealed the regularities of climate because the same regions appeared in similar positions on each of the continents. With his wife he wrote two very successful textbooks and then with O. J. R. Howarth edited the *Oxford Survey of the British Empire* (Crone, 1964:201; Dickinson and Howarth, 1933:238–239; Herbertson, 1905; and Jay, 1979).

British Geography during World War I and after

The Admiralty War Staff Intelligence Division compiled a considerable number of regional handbooks. Emphases varied, and most were heavily illustrated. The volumes constituted reference material both during the war years and at the time of peace negotiations (Freeman, 1980; Darby, 1983; Clout, 2003). The work, supervised by H. Dickson, was accomplished by some 70 writers and more than a dozen draughtsmen. (This was a project parallel to The Inquiry undertaken in the United States. See Chapter 18). This much work in regional geography was to result in the years that followed in books, articles, and newly created departments of geography.[4]

Already some departments of geography existed, including those at the London School of Economics, 1895; University College, London, 1903; Birckbeck College, London, 1909; Liverpool, 1909; Aberystwyth, Wales, 1917. Others were to follow. Then came a flow of regional geography textbooks: P. W. Bryan on North America, E. Shanahan on South America, L. D. Stamp on Asia, W. Fitzgerald on Africa, M. Newbigin on southern Europe, and many others. Stamp published a number of regional books for different levels of schooling and then was joined by S. H. Beaver to provide yet more on the genre. There was no shortage of texts in regional geography, a genre that also embraced physical geography. And there was no shortage of textbooks and scholarly books in physical geography. In those years frequently works of scholarship were adopted as texts in the upper levels of some of the schools. Typical examples include P. Lake, *Physical Geography* (1915); W. G. Kendrew, *Climates of the Continents* (1922); S. W. Wooldridge and R. S. Morgan, *The Physical Basis of Geography: An Outline of Geomorphology* (1937).

The British geographers were influenced by the French and Germans and also by the American geographer William Morris Davis. Much discussion involved the relationship of physical geography and human geography. At first geomorphology, which was described as the last chapter of geological history, was accepted almost

---

[4]In 1921, 10 universities offered honors programs in geography (Joerg, 1922; Keltie, 1921), but by 1964 there were programs of advanced study in 32 universities (including 7 separate colleges of the University of London). *Orbis Geographicus* for 1964–1966 lists 464 professional geographers in Great Britain. By 2000 there were 109 institutions of higher education in Britain offering geography (Withers, 2001).

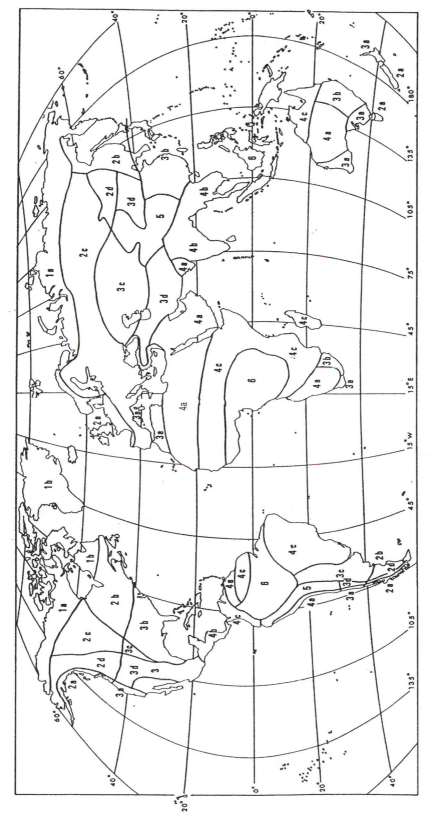

Figure 32 Herbertson's major natural regions (revealing regularity of climatic distribution). Courtesy of Robert P. Beckinsale

everywhere as a branch of geography and was a required part of every program of training. As a result, in the period since World War I a large number of studies in geomorphology by British authors have been published (Clayton, 1964; Gregory, 2003). Only since World War II has a movement appeared to reduce the emphasis on geomorphology, but this has been resisted by those who received their training in earlier decades (Stamp and Wooldridge, 1951; Wooldridge, 1956; Wooldridge and East, 1951; Gregory, 2003).

During the period of expansion following World War I, British geography developed five distinctive characteristics: (1) a continuing concern with exploration; (2) an emphasis on various kinds of regional studies; (3) the inclusion of field observation and map interpretation as essential parts of training programs; (4) an emphasis on studies in historical geography and a related concern with the history of geography; and (5) the study of geography because of its relevance to economic, social, and political policy problems.

Exploration. The professional periodicals of Great Britain devote a large proportion of their pages to accounts of exploration. The earlier expeditions had the attainment of a destination as major objectives. Ernest Shackleton and Robert F. Scott in their expeditions to the Antarctic did a large amount of pioneering work in geology, meteorology, and biology, yet their essential purpose was to reach the South Pole.[5] Similarly, when H. J. Mackinder ascended Mt. Kenya in 1899 and when Sir Edmund Hillary and Norgay Tenzing attained the summit of Mt. Everest in 1953, it was the climb to the top that had been their objective. Of course, behind the quest lay the epic theme of human versus nature and a desire to know the face of another part of the earth. The great advance of technology has robbed epic achievement of some of its meaning. Now to an ever increasing extent exploring expeditions are organized and financed for specific scientific purposes. Today people go exploring to study geology, botany, glaciology, geography, zoology, archaeology, or other subjects. Between 1960 and 1964 the Royal Geographical Society financed 184 expeditions. Of these 62 went to the Arctic or its fringe, 37 to Africa, 21 to South America, 19 to southwest Asia (which the British call the Middle East), and the remainder to a variety of previously unknown spots (Kirwan, 1964:223). In recent years the Society has sponsored exploration and exploratory research in Australia, Saudi Arabia, Venezuela, the Indian Ocean, and a large number of other places. It has additionally been involved in advisory seminars and workshops. There can be little doubt that the image of geography in Great Britain includes a continued concern with exploration.[6]

Regional Studies. There is fully as much confusion over the meaning of the words *regional studies* as there is over the German word *Landschaft* (Dickinson, 1976). At least three different meanings are attached to regional studies as a characteristic of British geography: (1) regional studies whose purpose is to divide the

---

[5]In January 1912 Scott reached the South Pole only to find that the Norwegian explorer Roald Amundsen had reached it little more than a month earlier. The Scott party perished during the return to their base.

[6]The traditional concern with exploration at Cambridge is reflected in the foundation of the Scott Polar Research Institute at that university (Crone, 1964:206).

surface of the earth into homogeneous areas or regions of varying size; (2) regional studies that are descriptions of segments of the earth surface; and (3) regional studies produced by an individual geographer, who devotes a large part of a professional career to the continued study of different aspects of one part of the earth. And none of these three kinds of regional study is necessarily involved with the regional concept. Numerous attempts have been made to clear up the semantic complexity that surrounds these words and the ideas for which they stand. But it is important to appreciate that to a certain extent the obscurity is a result of British experience with regions.

Experiments with the classification of very general regions for use at the global scale were started when A. J. Herbertson proposed his scheme of major natural regions (Fig. 32) in 1905. These natural regions were defined as associations of surface features, climate, and vegetation and were intended to provide an empirical generalization regarding the arrangement of these features for teaching purposes.

Following Herbertson, other British geographers experimented with a variety of different ways of classifying regions. In 1912 Marion I. Newbigin, editor of the *Scottish Geographical Magazine*, suggested an approach to the definition of regions that was a clear reflection of the ideas of Vidal de la Blache, Lucien Gallois, and Jean Brunhes. She started with a question: "Why is it easier for men to make their living at some places than at others?". The answer, she believed, could be found by examining the relationship between the *genre de vie* and the productivity of the land. The classification of regions, therefore, should be based on the kinds of relationships observed between human communities and their natural surroundings. Herbert J. Fleure of Aberystwyth carried this idea forward by defining seven kinds of global scale regions.[7] He postulated that all human activities are primarily directed toward accomplishing three functions: nutrition, reproduction, and the increase of well-being. In seeking criteria for delimiting human regions, he eliminated the first two because without them "a race would perish." He then classified his regions according to the measure of the earth's response to human efforts in the pursuit of well-being (Fleure, 1917, 1919). The seven types of regions Fleure identified were:

Regions of hunger
Regions of debilitation
Regions of increment
Regions of effort
Regions of difficulty
Regions of wandering
Industrialized regions

---

[7]Fleure was trained as a zoologist and was elected to the Royal Society as an anthropologist. He insisted on the need for close cooperation between geography, history, and anthropology. He was greatly influenced by the ideas of the French sociologist Frédéric Le Play and by the Scottish regional planner Patrick Geddes. With H. J. Peake he was the author of the series of books entitled *Gorridors of Time*, which demonstrated the method of treating the factors of time, type, and place together in one work. See T. W. Freeman, 1987, "Herbert John Fleure, 1877–1969." *Geographers: Biobibliographical Studies* 11:35–51.

He recognized that regions of increment, while requiring human effort, rewarded effort so liberally as to leave a surplus of both food and leisure. But when certain advanced societies enlarged their expectations for the "good life," some former regions of increment dropped to lower categories. Fleure recognized that the new technology and finance of industrialized regions had so modified the relation of human beings to their environment that such regions could replace any of the others. Newbigin and Fleure's experiments were imaginative efforts to formulate illuminating empirical generalizations; they were offered at a time when neither the statistical data nor the electronic equipment with which to store and process such data were available.

Another experimental regional scheme was proposed by John F. Unstead in 1916 (Unstead, 1916). In his classification of geographic regions, physical and human factors were to be given equal weight. Furthermore, Unstead recognized that regions could be defined at different degrees of generalization: Starting with the immediately observable units of area, which he called *stows*, he moved on to somewhat larger regions called *tracts* (which were roughly the equivalent of the French *pays*). Then these were combined into *subregions*, *minor regions*, and *major regions*.

In time, two difficulties with Unstead's scheme became apparent. The first difficulty was the concept of a uniform unit of area, a stow, that could be used as the basic building block in erecting a structure of world regional divisions. The idea of a unit area so homogeneous that it cannot be further subdivided is found in many languages. Yet it is clear that the stow is indivisible only because it is so conceived by the observer. Actually, it is necessary to start with the observable fact that no two microscopic points on the face of the earth are identical and that any area enclosed by a line and described as homogeneous is homogeneous only with respect to selected features. The face of the earth is an intricate system of interconnected features that forms a continuum of varying aspect. Regions are defined and drawn to illuminate some aspect of the problems geographers are concerned about. In fact, it would be impossible to reach any comprehension of the nature of the earth's surface without recognizing areas of partial homogeneity. The colors of the spectrum form a continuum, yet we find no intellectual difficulty with arbitrarily defining a certain segment of the spectrum and calling it red. The difficulty arises when the region is identified as a unit area, an indivisible segment of space.[8] If the geographer were reduced to the size of an ant, he would soon start subdividing the indivisible stow.

The second difficulty with Unstead's scheme was his announced intention to recognize homogeneous associations of physical and human factors (Unstead, 1916:241). This idea is logically derived from the theory that the way people live is a reflection of their natural surroundings. According to this theory, an area that is homogeneous with respect to its natural features will also be homogeneous in the ways people adjust to these features. The French geographers recognized these difficulties and, following Vidal, defined regions in terms of the way people live. Human use, said Vidal, creates homogeneity even where the natural features are not homogeneous. But the German geographers had difficulty in thinking of the different

---

[8]See D. L. Linton, "The Delimitation of Morphological Regions," in Stamp and Wooldridge, 1951:199–217; ref. 209.

Robert P. Beckinsale

Richard J. Chorley

Henry C. Darby

Herbert J. Fleure

Thomas W. Freeman

Peter Haggett

Ron Johnston

Hugh Robert Mill

Alan G. Ogilvie

L. Dudley Stamp

Michael J. Wise

Sidney W. Wooldridge

features associated in an area—in part because of the nature of the German language. As early as 1913 Lionel W. Lyde warned against attempts to make regions of human settlement fit exactly into the areas defined as natural regions (Lyde, 1913).

After Unstead, the British geographers began to specify their criteria for defining regions more precisely and to avoid regional schemes that involved the associations of too many diverse factors. In 1919 Charles B. Fawcett prepared a map of the service areas of the major cities of England (Fawcett, 1919). This was the first identification of functional regions (Fig. 33). In 1932 he mapped the continuously built-up urban areas of Britain (conurbations, as Patrick Geddes had called them). In 1937 a committee with Unstead as chairman (including John L. Myres,

Figure 33 Fawcett's provinces of England

Percy M. Roxby, and L. Dudley Stamp) reviewed the numerous schemes for dividing the world into regions, not only in Britain but also in other countries (Unstead et al., 1937).

Meanwhile, a very different kind of regional study was being produced. These were book-length treatments of specific parts of the world also stimulated by the similar studies by German and French geographers. For the Twelfth International Geographical Congress, which met in Cambridge, England, in 1928, the British geographers prepared a volume of regional essays, each dealing with a region of Great Britain. A total of 24 regions were marked off on the basis of surface features and underlying geological formations, but no map of regional boundaries was included in the hope that futile discussions about such boundaries could be avoided. "The purpose of regional geography," the committee decided, "is to describe the regions of the country as they are and to discover the causes that have made them what they are" (Ogilvie, 1928:1). Today we can understand that what a region is depends on what concepts are in the mind of the observer, but in 1928 geographers still thought they were dealing with objective reality. Unfortunately, perhaps, Albert Demangeon had published *Les Isles Britanniques* the previous year. It was the first in the series of *La Géographie Universelle* and was quite excellent. The fact that it was not translated for several years deferred the otherwise inescapable comparison. Yet in some ways the two books complemented each other, and at a time when knowledge of Great Britain and the British Isles was not so readily available, both were doubtless welcome additions to the existing literature. The Ogilvie volume put British geography on display. For many years these essays stood as a model for this kind of writing. It is interesting to compare the essays edited by J. Wreford Watson and J. B. Sissons on the occasion of the Twentieth International Geographical Congress in 1964. At this time the geographical study of Great Britain was organized topically rather than regionally (Watson and Sissons, 1964).

Many regional monographs, some published since 1960, are included in the bibliography in Minshull's book *Regional Geography* (Minshull, 1967). The traditional regional study follows a more or less standard outline of topics. It begins with the bedrock geology, the surface features, the climate, and the vegetation and soils. The treatment of the stage setting is then followed by a history of the course of settlement, starting with the earliest human inhabitants. More recently, however, the authors first announce a general theme that characterizes the aspect of the region to be investigated (as indeed Fleure did in his chapter on Wales in the 1928 volume). The regional study is then tightly organized around materials relevant to the theme.[9]

The third kind of regional study is, perhaps, a form of applied geography. When geographers devote a large part of their professional lives to studying different aspects of one part of the world, they become known as regional specialists, and their publications, each dealing topically with some aspect of the area, are called regional studies. There are many examples of such specialists among the British geographers,

---

[9]Compare, for example, Freeman (1950) or Monkhouse (1959) with Cole (1960), Longrigg (1963), Harrison-Church et al. (1964), or Prothero (1969).

but perhaps we can illustrate this kind of work by two examples. One was David G. Hogarth, who devoted his life to the study of the people and problems of what the British called the Near East.[10] He began traveling in Turkey and Arabia in 1887, first as an archaeologist and later as an observer of the people and their problems (Hogarth, 1902). The informed advice he gave to his government regarding the treatment of the Turks after World War I was, unfortunately, not acted on promptly.

Another regional specialist was Percy M. Roxby of Liverpool, who spent his life in the study of China[11] (Freeman, 1967:156–168; Freeman, 1981). In 1912 he was awarded a fellowship that enabled him to travel to China among other places. He developed a lifelong fascination with the Chinese and published numerous papers dealing with that country (Roxby, 1916, 1925, 1938). During World War II he was employed by the British Naval Intelligence to prepare the *Handbook of China*, which was completed in three volumes in 1944 and 1945.[12] He was also one of Britain's important historical geographers. His treatment of East Anglia in the volume of regional essays in 1928 is an outstanding example of this method (Ogilvie, 1928:143–166). His published papers also include influential methodological studies (Roxby, 1926, 1930).

British geographers continued to contribute to the literature of regional studies, at least of the second and third types long after 1945. The recognition of the difficulty of defining homogeneous areas led to more sophisticated methods of identifying and analyzing regions (Haggett, 1966:241–263). Papers and books in the literary tradition, seeking to present the "personality" of a region continued for many years after Kimble's essay "The Inadequacy of the Regional Concept" (1951), though in decreasing numbers.

Field Studies and Map Interpretation. The third distinctive characteristic of British geography prior to World War II was the continued requirement for training in field observation and map interpretation as essential parts of training programs at all levels. Even children in the elementary grades are expected to practice the observation of things out of doors, very much as recommended by Pestalozzi. There are exercises in the reading of topographic maps and the following of maps

---

[10]Hogarth was first an explorer and archaeologist, but he gradually turned his attention to geography. He was at one time director of the British School of Archaeology at Athens. From 1908 to 1927 he was the director of the Ashmolean Museum at Oxford. One of his best known students was Lawrence of Arabia, Thomas E. Lawrence.

[11]Roxby was a student of history at Oxford. In 1912–1913 he received a fellowship to permit him to travel in the United States, China, and India. As a result, he devoted his career to the study of China (and also of Chinese workers who were living in ghetto conditions around Liverpool). He was appointed to the staff at the University of Liverpool in 1904 and taught geography there until after World War II. Unfortunately, he died in China in 1947 before his intended definitive book on Chinese geography was written.

[12]The *Handbook of China* included: Vol. 1, *Physical Geography, History, Peoples* (1944); Vol. 2, *Modern History and Administration* (1945); Vol. 3, *Economic Geography, Ports and Communications* (1945).

in the field, and then further training is given in the interpretation of detailed maps of unfamiliar areas. Perhaps this early familiarity with the use of maps and the resulting attention to the features of the surrounding landscape have led to the widespread British habit of taking long walks in the country or cycling to more distant places as a form of recreation.

This attention to field observation and map interpretation that is a part of the school experience of most British people, even those who live in cities, is carried on at a more sophisticated level in the training of geographers in the universities. It has long been customary to require theses dealing with the geography of small areas as part of undergraduate programs, and the more elaborate studies of regions have been the tradition for dissertations. Examinations for honors candidates regularly include map interpretation. Even in the contemporary period of using electronic devices for gathering and analyzing information, the British geographers continue to make use of direct field observation as a basic method.

Historical Geography. The fourth distinctive characteristic of geography as it developed in Great Britain after World War I has been a continued emphasis on the historical method (Baker, 1972; Williams, 2003). This emphasis seems to have been derived quite directly from the geographic writings of some of the nineteenth-century historians. In 1838 Thomas Arnold published his famous *History of Rome* in which he included chapters on the natural surroundings.[13] He also included a map and discussion of the area of occurrence of malaria in Roman times (Baker, 1963:33–50). Some of Arnold's pupils continued to insist on the importance of geography as the basis for understanding history. As a result, there were lectures on geography in British universities long before there were geographers on the faculties; some of these lectures were given by geologists and some by historians. Even now it is not uncommon to find books on historical geography written by anthropologists or economic historians (Beresford, 1954; Fox, 1932).

The development of historical geography by geographers came after Mackinder's work on the British Empire (Mackinder, 1902). In his training as an historian Mackinder learned to appreciate the need for looking at the story of man's settlement of the land from the perspective of time. He insisted that the geographer should attempt to re-create past geographies and show how sequences of change have led to the presently observable features (East, 1951:80). Otherwise geography would become the mere description of contemporary features. Adding the time dimension permits the study of processes of change and reveals that the present geography is only the latest stage in a sequence of stages. This is the way the observed features of

---

[13]Thomas Arnold was appointed professor of modern history at Oxford in 1841 but died the next year. His pupils included A. P. Stanley and E. A. Freeman, who published on historical geography. Chapter 3 of the first volume of Thomas Macaulay's *History of England* (London, 1848) included a description of England in 1685 that Baker describes "as a model of what historical geography should be" (Baker, 1963:36). H. T. Buckle, however, went far beyond Montesquieu in relating the characteristics of people to climate. Bibliographies of the historical writings of this period are contained in Baker (1963:33–50) and Clark (1954:79–80).

geography in an area could be explained.[14] After Mackinder most of the British geographers included accounts of historical geography in their regional studies. Among the many important contributions to historical geography, special mention should be made of Marion I. Newbigin (1926), E. G. R. Taylor (1930, 1934), E. W. Gilbert (1933), W. Gordon East (1935, 1951), and H. C. Darby (1936, 1940a, 1940b, 1951, 1952).

Applications of Geography.   Geography in Britain has also included some notable contributions toward the solution of practical problems. Dickinson has traced the influence of the French sociologist Frédéric Le Play on the Scottish regional planner Patrick Geddes (Dickinson, 1969:197–207). Geddes developed the concept of the regional survey of potential land quality and land use as the basis on which to draw up a plan for economic development. If one is inclined to wonder why someone did not set forth such an idea many decades earlier, it is well to remember the historical context. Before there could be regional surveys there had to be standard sets of detailed (topographic-scale) maps on which to plot the data. In 1896 Hugh Robert Mill suggested using the sheets of the Ordnance Survey (1 inch to the mile), and four years later he provided an example of how such mapping of relevant information could be done and how it could be used (Mill, 1896, 1900). His suggestion was discussed, but no one did anything about it.

The British geographer who put these earlier ideas into practice was L. Dudley Stamp[15] (Wise, 1988). Returning in 1926 to the London School of Economics from some three years in Burma, he turned his attention to the study of Great Britain. Convinced that Mill and Geddes were right about the need for a survey of Britain as a basis for planning, he began searching for ways to carry out such a survey by plotting categories of land quality and land use on the Ordnance maps at 6 inches to the mile. The sheets of this map include the field boundaries, which greatly facilitated the work of plotting the data. He found that some surveys of this kind had already been attempted. For example, some maps of a parish in Leicestershire were done by schoolchildren under supervision of their teachers. Stamp recognized that quite aside from the practical importance a survey would have as a basis for planning, the work of making it would be an excellent educational experience for the children.

---

[14]For a long time there was confusion between historical geography and the history of geography; in fact, bibliographies used to list both of these categories under one heading. Historical geography can now be viewed as the re-creation of past landscapes and the tracing of geographic changes through time (Clark, 1954:72–73). Geographical history, then, is a study of the effect of geography on the course of history. The history of geography has to do with the development of geographic concepts and the progress of geographical studies.

[15]L. Dudley Stamp completed two degree programs at King's College, University of London, in geology and geography. He then went to Burma for a petroleum company and from 1923 to 1926 was professor of geography and geology at the University of Rangoon. From 1926 until his retirement in 1958 he was at the London School of Economics and Political Science.

With great energy and patience, Stamp undertook to organize and direct what became known as the British Land Utilisation Survey (Stamp, 1947). With professional geographers as advisers, his first job was to define the categories to be plotted on the maps. Then it was necessary to "sell" the project to the directors of education in the counties and then teach the field workers how to do the work. Some 22,000 volunteer schoolchildren were asked to give three or four days' time, after which the maps had to be checked. It took eight weeks for a skilled cartographer to redraw the maps on a scale of 1 inch to the mile and make them ready for publication. Geography students and staffs at various British universities cooperated in providing the professional personnel. Stamp reports that the hardest part of the project was securing funds to print the maps and to publish county reports explaining what the maps showed. The work was started in the summer of 1931, and by the end of 1935 the mapping was essentially complete.

When World War II began in 1939, the vital importance of the maps was quickly appreciated. Britain had to undertake a rapid program of agricultural expansion because there were not enough ships to bring in the usual supplies of food. The Ministry of Agriculture recommended that funds be appropriated to publish the maps as rapidly as possible, and this was done between 1939 and 1945. It would scarcely have been possible to increase the production of wheat so rapidly had it not been for the existence of these field studies. Many years later, in 1965, Stamp was knighted in recognition of his contribution to the survival of his country.[16]

The survey maps were used for a variety of purposes in addition to the planning of emergency crop expansion during the war. After the war they were the basis for the reconstruction of Britain. In the universities several studies were made of the historical geography of agriculture by reconstructing the crop patterns of former times and comparing them with the survey maps (Stamp, 1947).

WORLD WAR II AND THE COMING OF CHANGE

Once again, in time of war geographers found themselves in demand. The government needed and wanted a large set of regional geographies. They were to be produced as swiftly as possible. A total of 58 volumes were printed under 31 titles, each covering countries specified by the authorities. Some of the work was accomplished at Oxford and some was accomplished at Cambridge. At both centers the universities had excellent library holdings. All the volumes would contain maps and photographs (Clout, 2003), and all were produced at a high level of quality. Typically, 2000 to 4000 copies of each of the handbooks was published which, 10 years after cessation of hostilities, were sold to universities and libraries.

The pattern was rather similar to that followed in World War I. Regional geographies were found to be of great value especially with regard to military and economic matters. Just as was the case in World War I, the United States established a center for such studies in World War II. (This was known as the Office of

---

[16]On April 15–16 and May 10, 1941, air raids destroyed nearly 50,000 of the already printed sheets as well as the set of plates from which they had been printed. Maps stored on a farm and at publishing houses survived.

Strategic Services. By September 1943 there were 75 geographers working in this section of the government.)

Many of the studies accomplished in England assisted faculty members in their lectures for a number of years after the war. They also facilitated the development of college-level textbooks for a number of years. Robert and Monica Beckinsale published *Southern Europe* in 1975, which may have been the last of this group of books deriving from war years activity. Other geographers found themselves serving abroad during the war in difficult circumstances, though certainly seeing other parts of the world that otherwise they would have not. Such "travel," of course, might be extremely hazardous (Balchin, 1987). In some cases it led to the production of excellent works such as C. A. Fisher's *South-East Asia* (1964), O. H. K. Spate's *India and Pakistan* (1954), and others.

Further regional works were provided by geographers with special reference to the British Isles. O'Dell wrote on Scandinavia (as would Meade), J. M. Houston on the western Mediterranean, Wooldridge on the Weald, and R. E. Dickinson three books on Germany. There were many other texts that could be termed regional. In part, this was due to the war years following a depression that led to a pent-up demand. And, in part, it reflected paper and apparatus shortages immediately following the war. Yet regional geography persisted and, approximating the arrival of the 20th International Geographical Congress in London, three significant books in the regional genre were published: *The British Isles*, by J. W. Watson and J. B. Sissons, *Field Studies in the British Isles* by J. A. Steers, and *Great Britain: Regional Essays* by J. Mitchell (ed.). Then regional geography began to fade as it was replaced by a new geography. In part, reduction of the genre was brought about by the essays typified by Kimble ("boundaries that don't exist around areas that don't matter") and Buchanan on the "Poohscape," both direct assaults on the regional concept. And, in part, the emergence of and growing preference for systematic studies exerted an erosive effect on regional geography in the classroom. Then came the numerate and model-building new geography by a younger generation in quest of science. The concept of region did not die, but it was much reduced from its previous standing.

It should not be forgotten, however, that regional geography dominated geography in Britain for some 70 years, occupied the best minds in the profession, was taught to large numbers of school children and university students throughout Great Britain and the empire, and was a geography consonant with the public notion of what was geographic. It was very much more than the lists of subheads that have occasionally caricatured this genre in more recent years. It permitted and even required the integration of physical geography with human geography encouraging a unity of discipline not missed until it was dismissed. All fields of study are, of course, subject to revision, and geography is no exception. Yet it was a stage in the evolution of the discipline whose significance may have been underestimated. Some geographers continued to write regional geography long after the developments of the 1960s.

Another "New" Geography

The waves of innovation in geographic study that distinguish what might be called the new geography and that will be discussed in Chapters 17 to 20 reached Britain in the 1960s (Haggett and Chorley, 1967). Not that the five distinctive traits of British geography were swept away. Rather, new traits have been added. Regional geography was still dominant in the curriculum of both school and university students, and both new and revised texts of this genre were published.

The 1960s ushered in the theoretical and quantitative revolutions in geography. A number of British geographers had visited the United States either as faculty members or as graduate students. Some stayed; those who returned to Britain took with them something of the ferment that at the time characterized American geography. This revolutionary group of American geographers hailed especially (though by no means exclusively) from the geography departments at Iowa, Evanston, Seattle, and Madison. This departure from the regional concept was taught to British students and spread to colleagues, especially by way of the Study Group in Quantitative Methods of the Institute of British Geographers. The use of statistics was not new in British geography, but the surge in its use in the 1960s far surpassed that in previous eras. S. Gregory wrote *Statistical Methods and the Geographer* (1962), and J. P. Cole and C. A. M. King wrote *Quantitative Geography* (1968): Both were read by undergraduates studying the ways and means of the new geography. Much attention was given to quantification, to statistical description of patterns, and to statistical manipulation and testing of hypotheses.

The center of innovation was the University of Cambridge, where the leaders were R. J. Chorley and P. Haggett, both of whom had visited the United States.[17] Chorley studied under Strahler, held a university post at Brown, traveled widely in the United States, and began to gather archival data concerning W. M. Davis. Haggett was also to travel in the United States and participate with geographers. These two geographers edited two books that would become instrumental in introducing the new geography to both the academic and pedagogic communities. The first of these books, *Frontiers in Geographical Teaching* (Chorley and Haggett, 1965), was the product of the Madingley Lectures, which were given in 1963. A variety of authors were represented in this book, which demonstrated the application of quantitative

---

[17]Three British geographers, all of them graduates or faculty of Cambridge, were of chief importance as leaders of this new movement. Richard J. Chorley, a geomorphologist, was one of the first geographers to make use of General System Theory in the study of landforms and also one of the first to point to the utility of such theory in human geography. David W. Harvey completed his graduate study at Cambridge in 1962 after spending the year 1960–1961 studying with Torsten Hägerstrand at Lund. Harvey was appointed assistant lecturer at Bristol University in 1961 and was promoted to lecturer in 1964. In 1965–1966 he was at Pennsylvania State University, and in 1969 he joined the staff of Johns Hopkins University. In 1987 he accepted an appointment as the second holder of the Halford Mackinder chair, School of Geography, Oxford. Peter Haggett completed graduate study at Cambridge in 1960. He taught at the University of London from 1951 to 1957 and was appointed to the staff at Cambridge in 1957. In 1966 he was appointed to a second chair of geography at Bristol (where Ronald Peel had been professor of geography since 1957). One of the lecturers at Cambridge who played a role in stimulating the younger generation to experiment with innovations of method was Alfred Caesar, who from 1944 to 1946 worked in the Ministry of Town and Country Planning. He was appointed lecturer at Cambridge in 1949.

methods to the understanding of geographical problems. It had a positivist stance and helped spread this posture, probably on a larger scale than previously in Britain. Editors Chorley and Haggett concluded by urging "the importance of the construction of theoretical models." Their next book, *Models in Geography* (1967), dealt with the utility and application of models to a variety of research areas. This very large book, republished as a series of paperback volumes, was designed for undergraduate students. It demonstrated the scientific methods available to geographers, and at the same time it presented a variety of review articles that were also to be read for their content. Chorley and Haggett included explicit (and implicit) support for the view that the discipline would progress more rapidly if the nomothetic, rather than the earlier idiographic, approach was used. It would be hard to overestimate the impact of this book, which inspired work of similar genre for some years.

A third very important book published in Britain during the 1960s was Peter Haggett's *Locational Analysis in Human Geography* (1965). This, too, was an important example of the new geography, emphasizing hypothesis testing by way of inferential statistics. Haggett divided spatial structure into what he considered its component parts: nodes, hierarchies, networks, flows, and surfaces. (A sixth component, diffusion, was added in the 1977 edition.) The author sought to generalize about spatial order within society; in this undertaking, the function of distance was of the first importance.

Finally, mention must be made of David Harvey's *Explanation in Geography* (1969), published while Harvey was a resident of Johns Hopkins University, but in considerable measure a product of his experience at Cambridge, Bristol, and Lund. The book had considerable impact on those British geographers who leaned toward positivism and on others in search of the scientific method. Harvey attempted to render and explain the component parts of science as viewed by the geographer. It was the first book of its kind written by a geographer, and it appeared at a most appropriate time. Much of what had been written in the 1960s was the heady product of an innovative geography whose method sometimes dominated its product (probably more so in the United States than in Britain). As a result, Harvey's book emerged as a primer in scientific method and procedure when it was needed.

As R. J. Johnston has suggested, the revolution comprised several interrelated components: a concern for scientific rigor, an argument that quantitative methods formed a necessary component of this more rigorous approach, a claim that human geographers should focus on searching for spatial order in the patterning of human activities, and a desire that human geographers' work should be applied to a wide range of "real-world" problems (Johnston, 2003).

These changes were both methodological and epistemological. There had been precursors such as Fawcett (1919 and mentioned earlier in this chapter), Dickinson, who had studied hierarchies of functional regions (Johnston, 2001)[18] (similar to work

---

[18]See "Robert Eric Dickinson 1905–1981" by L. R. Pederson. *Geographers: Biobibliographical Studies*. 8(1984):17–25. The R. E. Dickinson papers are on deposit with the department of geography, University of California, Berkeley.

done by the American geographer R. S. Platt [1928]), and the work of R. O. Buchanan and S. Beaver at the London School of Economics, B. Law at University College, London, and W. Smith at the University of Liverpool. But, of course, it was not until the 1960s that this thinking was shared by larger numbers of geographers. Haggett's *Geography: A Modern Synthesis* (1972) was published as an undergraduate text and was followed by other texts offering the new geography. They worked their way through the British educational system, transforming the geography offered. Then, by the 1970s a generation of regional enthusiasts retired and were not replaced by their kind. The tradition of regional geography continued with the occasional book, but most of this sort of work found its way into studies by historical geographers (Clout, 2003).

The geography of the 1960s is referred to as spatial science. It espoused a point of view that could be shared by both physical and human geographers. The adoption of quantitative data helped to reveal and explain patterns. Both physical and human geographers shared the same literature, which by 1969 included Haggett and Chorley's *Network Models in Geography*. By the early 1970s, however, physical and human geographers had begun to drift apart. The physical geographers' focus had shifted from spatial form toward analysis of the processes that created those forms. They adopted the systems approach in order to facilitate the comprehension of processes at work. Nevertheless, the physical geographers also became more specialized, and studies of landforms, biogeography, and climatology each won their own adherents. Physical geography remains strongly entrenched in British geography and thus retains a long-standing tradition.

Human geographers became disenchanted with the geography of the 1960s. Exploitation of the models derived from the work of J. H. von Thünen, Alfred Weber, Walter Christaller, and August Lösch did not produce universal satisfaction. The models developed could not replicate a satisfactory environmental platform, neither could they realistically characterize decisions made by humankind. The positivists who had generalized so much about humans behavior were now to be nudged by the introduction of a more humane human geography.

Regional, historical, and cultural geography stood largely apart from these changes in methodology. One reason may be that the traditional British geographies were still prominent and were taught and written of by established academics reluctant to change their mode. Another reason may be that the subjective and empiricist spirit is not given to the positivist mode of science. Regional geographers remain empirical, continuing to produce useful works, and have been championed by William R. Mead, J. H. Peterson, and Michael J. Wise. Both H. Clifford Darby and W. Gordon East have argued that the subject of historical geography is more nearly one of art, and not new methods of science. Cultural geography entered Britain in part from Germany, in part from Berkeley, and in part it was indigenous. It borrowed little from the recently introduced scientific method.

An alternative response to positivist human geography in Britain came with D. M. Smith's work on welfare geography. David Harvey had addressed this same topic in *Social Justice and the City* (1973) by way of Marxist theory. Marxist geographers began an active search for the social good and have been written of by D. Gregory in *Ideology, Science, and Human Geography* (1978a).

British geography was fragmented, and an eclectic pluralism now characterized the discipline. With the removal of the region as mainstream, the core had disappeared. In its place was left a variety of contending points of view. Whether this diversity should be regarded as a sign of vitality or a discipline in disarray is in the eye of the beholder. Methodology alone seemed to provide common ground in the 1960s and 1970s. The search for some sort of core began. Some geographers urged the bridging of physical and human geography, which had now divided. (Physical geographers had grouped themselves with the physical and life sciences, and human geographers had grouped themselves with the social sciences.) Rapprochement of the two groups has been sought in the environmental issue, in preservation of the landscape, in applied geography, and elsewhere in the disciplinary gamut.

It should be noted that in Britain, applied geography has been particularly strong. Its tradition extends back to the assemblage of the British Empire, publication of Keltie's *Applied Geography* (1890), the years of the Paris Peace Conference (1918–19), L. D. Stamp's Land Utilisation Survey of the 1930s, Coleman's Land Use Survey in the 1960s, the Admiralty Handbooks of World War II, Stamp's *Applied Geography* (1960), and work in town and country planning and in land-use-transport matters, all of which led to the founding of the periodical *Applied Geography* (1981). The applied sector will probably expand given a strong following in the school and university systems, in which the production of geographers outstrips the demand in educational institutions.

Historical Geography

Historical geography had a rich history prior to World War II. After that interlude of hostilities, historical geography continued as one of the strengths of British geography. H. C. Darby was the dominant figure, producing *A New Historical Geography of England* (1973), the Domesday geography of all the main parts of England (1962–71), *Domesday England* (1977), "Historical Geography in Britain, 1920–80: Continuity and Change" (1983), *The Changing Fenland* (1983), and, posthumously, "The Relations of History and Geography: Studies in England, France and the United States" (2002). His work was the product of integrating in a very meaningful way the fields of both geography and history. He suggested (1953) four ways in which the two disciplines could be combined; the geography behind history, past geographies, the history behind geography, and the historical element in geography. Significantly, Darby worked in an area of limited size and one which had retained records for the distant past. He received a knighthood in 1988 for his remarkable accomplishment, which brought attention to geography. Others worked at this task of explaining the creation of humanized landscapes. The historian W. G. Hoskins had already provided *Making of the English Landscape* (1955) and arranged the publication of 20 county volumes, largely by historians of the making of landscapes. Hoskins himself wrote 28 books centering on landscape and local history, then produced a television film and a series of shorter films that bought the public close to his subject. Among numerous other authors who contributed to this genre, mention must be made of A. R. H. Baker, R. A. Donkin, H. C. Prince, and M. Williams. Some British historical geographers worked in Australia. Others, such as M. P. Conzen and D. Cosgrove, worked in the United States, where C. O. Sauer,

A. H. Clark, D. W. Meinig, and D. Lowenthal made their own excellent contributions that married with the British undertaking. This branch of geography proved so fruitful that the *Journal of Historical Geography* was founded in 1975, and the CUKANZUS meetings were initiated (The term "CUKANZUS" derives from Canada, United Kingdom, Australia, New Zealand and US. This is now known as the International Congress of Historical Geographers.).

Another function of historical geography during the post-1960 period was to aid and abet a struggling regional geography. The latter was thought to be additive, its components to be studied in sequence but without realizing synthesis. At any rate, the new geography reduced the regional undertaking. Yet regional writing persisted in the elaboration of many works in historical geography. Since historical geography was little affected by the new mode, regional geography had, in part, secured a niche.

Related to this continued use of the historical method is the interest of British geographers in the history of geography as a field of learning. The bibliographies of the early chapters of this book, which deal with ancient geography and with the progress of exploration, contain many references to British writings.[19] Interest in writing on different aspects of the history of geography continues in British scholarly work. A leading historian of geography in Britain was J. N. L. Baker.[20] The first three parts of a definitive history of the study of landforms have been issued and will eventually consist of four volumes. Volume 1 deals with geomorphology before Davis (R. J. Chorley, A. J. Dunn, and R. P. Beckinsale, 1964); Volume 2 deals with W. M. Davis (R. J. Chorley, R. P. Beckinsale, and A. J. Dunn, 1973); and Volume 3 reveals post-Davisian geomorphological thought in *Historical and Regional Geomorphology, 1890–1950* (R. P. Beckinsale and R. J. Chorley, 1991). A fourth volume, currently in progress, will complete a monumental inquiry into the history of the study of landforms.

Other British writers who have published on the history of geography include Dickinson and Howarth (1933), Mill (1951), Freeman (1961, 1967, 1980, 1982), Crone (1964), Kirwan (1964), and Dickinson (1969). More recently, R. W. Steel wrote *The Institute of British Geographers: The First Fifty Years* (1984). Steel also edited a series of essays concerning British geography at a variety of institutions and in a variety of settings: *British Geography* (1987). D. R. Stoddart edited a series of essays entitled *Geography, Ideology and Social Concern* (1981) and in 1986 gathered a number of his essays published elsewhere and contributed *On Geography and Its History*. Stoddart states enthusiastically that his geography "springs from Forster, Darwin, Huxley: and it works." In seeking to put the geography back in the bio-, Stoddart offers a distinctive viewpoint and reveals much

---

[19]For example, see books by C. R. Beazley, E. H. Bunbury, Frank Debenham, Richard Hakluyt, G. H. T. Kimble, John Needham, Percy Sykes, E. G. R. Taylor, J. Oliver Thomson, and H. F. Tozer.

[20]J. N. L. Baker went to Oxford as assistant to the reader H. O. Beckit in 1923 after serving in the army in France in 1915–16 and in the army in India in 1918–19. He remained at Oxford until his retirement in 1962. His collected writings were published in a single volume in 1963 (Baker, 1963).

about the history of geographical thought. The work of the very prolific R. J. Johnston is of importance. His post-1945 studies in the philosophy and history of geography, with special reference to American and British human geography, are a vital source for anyone studying the post-World War II Anglo-American geographical scene. Special mention must be made of his *Geography and Geographers: Anglo-American Human Geography since 1945* (fifth edition, 1997). C. J. Withers has published on an earlier period with special reference to Scotland. R. J. Mayhew has published on British geography typically from the seventeenth to the nineteenth centuries. M. Heffernan has written on early twentieth-century British geography with reference to World War I. Others have also contributed to this skein of thought, which has seen much growth since the 1970s. Some four dozen studies of British geographers have been contributed to *Geographers: Biobibliographical Studies*.

Contemporary Geography in Great Britain

Geography is currently very strong in Great Britain. The debate concerning the new and the old geographies has long-since quietened, and numeracy and model building have been applied where relevant with benefit. Currently, research advances on many fronts: institutional, environmental, historical, regional, place, space, applications, disease distribution, urban dwellers, feminism, ethics and social concern, and many other branches of the discipline.

These strengths (varying throughout the decades) have been shared especially with American geographers in the post-1960 period. Geographers crossed the Atlantic in ever increasing numbers to attend meetings, to hold temporary or permanent academic appointments, to give lectures, and the like. Whereas early in the century the United States imported ideas from Britain, over the last three decades the United States has been able to export ideas to the United Kingdom. Of course, ideas travel back and forth, being revised, improved, and revised once more. Use of a common language undoubtedly facilitated the diffusion of thought and literature to many other parts of the world. It also meant a very large increase in English-language publications. Many of these were devoted to specialist branches of geography, and others were devoted to interdisciplinary matters, the latter a product of centrifugal forces within the discipline. Of particular significance was the inauguration in 1969 of *Progress in Geography*. In 1977 this was converted into two "Progress" journals, *Progress in Human Geography* and *Progress in Physical Geography*. These are only two of a list of 3445 geographical serials, past or present, provided by Harris and Fellman (1980). Nevertheless, they represent and typify new growth in British geography (and also in the Anglo-American realm).

British geography is very strong at the preuniversity level (a level that represents one of the largest problems for geography in the United States), and typically students are tested by national examinations at the ages of 16 and 18 years. The tradition of empire has also helped establish a good image for geography in Great Britain. With this firm foundation, students who proceed to the university are able to accomplish a great deal; it is not surprising that British academic geography has been so successful. Despite the economic recession of the late 1970s and early 1980s, as well as the cutbacks in higher education, geography was left relatively unscathed. By the 1990s the number of universities had increased, departments were larger (e.g., the University of Southampton had 24 staff in 2001, Durham had 41, University College, London had 35), and more funds were available, which led to more research activity. Two problems continued to hamper further development. The first of these was fragmentation. Professional geography had become "a community of subcommittees," the product of centrifugal forces (a problem also confronting U.S. geography). The second problem confronting British geography was the image of the field projected from the past and from vernacular comprehension. Partly as an offset to this, and more significantly as progress reports, since 1956 quadrennial reports have been published in *The Geographical Journal* revealing both progress and impediments. It is a self-searching exercise that has become a significant tool in the analysis of vicissitudes in the history of British geography during the last half century. More recently, central authority has imposed evaluation schemes for faculty in both teaching and research functions (Withers, 2001). The implementation of these assessment techniques consumes time, adds to concern, and with regard to published research, may in some cases encourage haste at the expense of more desirable properties. Even so, British geography is both rigorous and vibrant, the spirit of which is captured in Johnston and Williams's *A Century of British Geography* (2003).[21]

REFERENCES: CHAPTER 10

Baker, A. R. H. 1972. "Historical Geography in Britain." In A. R. H. Baker, ed., *Progress in Historical Geography*. Pp. 90–110. Devon: David & Charles.

Baker, J. N. L. 1963. *The History of Geography*. New York: Barnes & Noble.

Balchin, W. G. V. 1987. "United Kingdom Geographers in the Second World War. *Geographical Journal* 153:159–180.

———. 1993. *The Geographical Association: The First Hundred Years*. Sheffield, U.K.: Geographical Association.

Beckinsale, R. P., and Chorley, R. J. 1991. *The History of the Study of Landforms or the Development of Geomorphology*. Vol. 3: *Historical and Regional Geomorphology 1890–1950*. London and New York: Routledge.

Beresford, M. W. 1954. *The Lost Villages of England*. New York: Philosophical Library.

Blouet, B. 1987. *Halford Mackinder, A Biography*. College Station: Texas A&M University Press.

Buchanan, K. M. 1968. A preliminary contribution to the geographical analysis of a Poohscape. *IBG Newsletter* 6:54–63. Reprinted in *Progress in Human Geography* 23 (1999):253–266.

Chisholm, G. G. 1889. *Handbook of Commercial Geography*. 18th ed. 1966. L. D. Stamp and S. C. Gilmour, eds. London: Longmans, Green.

Chisholm, M. 1962. *Rural Settlement and Land Use*. London: Hutchinson University Library.

———. 1975. *Human Geography: Evolution or Revolution?* Harmondsworth: Penguin Books.

Chorley, R. J., ed. 1973a. *Directions in Geography*. London: Methuen.

---

[21] In this context, special mention must be made of the first two chapters of this book by D. N. Livingstone and R. Johnston which jointly provide an excellent historical summary of British geography from c. 1500 to the present.

———. 1973b. "Geography as Human Ecology." In R. J. Chorley ed., *Directions in Geography*. Pp. 155–170. London: Methuen.

Chorley, R. J., and Haggett, P., eds. 1965. *Frontiers in Geographical Teaching*. London: Methuen.

———. 1967. *Models in Geography*. London: Methuen.

Chorley, R. J., Beckinsale, R. P., and Dunn, A. J. 1973. *The History of The Study of Landforms, or the Development of Geomorphology*. Vol. 2: *The Life and Work of William Morris Davis*. London: Methuen.

Chorley, R. J., Dunn, A. J., and Beckinsale, R. P. 1964. *The History of the Study of Landforms, or the Development of Geomorphology*. Vol. 1: *Geomorphology Before Davis*. London: Methuen.

Clark, A. H. 1954. "Historical Geography." In P. E. James and C. F. Jones, eds., *American Geography: Inventory and Prospect*. Pp. 70–105. Syracuse, N.Y.: Syracuse University Press.

Clayton, K. M., ed. 1964. *A Bibliography of British Geomorphology*. London: George Philip & Son.

Clout, H. 2003. "Place Description, Regional Geography and Area Studies: The Chorographic Inheritance." In Johnston and Williams. 2003. *A Century of British Geography*. Oxford: Oxford University Press, pp. 247–274.

Clout, H., and Gosme, C. 2003. "The Naval Intelligence Handbooks: A Monument in Geographical Writing." *Progress in Human Geography* 27:153–173.

Cole, M. M. 1960. *South Africa*. London: Methuen.

Coleman, A. 1961. "The Second Land Use Survey: Progress and Prospects." *Geographical Journal* 127:168–186.

Crone, G. R. 1964. "British Geography in the Twentieth Century." *Geographical Journal* 130:197–220.

Darby, H. C., ed. 1936. *A Historical Geography of England Before A.D. 1800*. Cambridge: At the University Press.

———. 1940a. *The Draining of the Fens*. Cambridge: At the University Press.

———. 1940b. *The Medieval Fenland*. Cambridge: At the University Press.

———. 1951. "The Changing English Landscape." *Geographical Journal* 117:377–398.

———. 1952. *The Domesday Geography of Eastern England*. Cambridge: At the University Press.

———. 1953. "On the Relations of Geography and History." *Transactions and Papers, Institute of British Geographers* 19:1–11.

———, ed. 1973. *A New Historical Geography of England*. London: Cambridge University Press.

———. 1977. *Domesday England*. London: Cambridge University Press.

———. 1983. "Academic Geography in Britain: 1918–1946." *Transactions, Institute of British Geographers*. N. S. 8:14–26.

Davies, W. K. D. 1972. "Geography and the Methods of Modern Science." In W. K. D. Davies, ed., *The Conceptual Revolution in Geography*. Pp. 131–139. London: University of London Press.

Dickinson, R. E. 1969. *The Makers of Modern Geography*. London: Routledge & Kegan Paul.

———. 1976. *Regional Concept: The Anglo-American Leaders*. London: Routledge & Kegan Paul.

Dickinson, R. E., and Howarth, O. J. R. 1933. *The Making of Geography*. Oxford: Clarendon Press.

Dryer, C. R. 1920. "Mackinder's 'World Island' and its American 'Satellite.'" *Geographical Review* 9:205–207.

East, W. G. 1935. *An Historical Geography of Europe*. London: Methuen.

———. 1951. "Historical Geography." In S. W. Wooldridge and W. G. East, eds., *The Spirit and Purpose of Geography*. Pp. 80–102. London: Hutchinson University Library.

Fawcett, C. B. 1919. *The Provinces of England*, rev. ed. 1960. London: Hutchinson University Library.

———.1932. "Distribution of Population in Great Britain." *Geographical Journal* 79:100–116.

Fischer, E., Campbell, R. D., and Miller, E. S. 1967. *A Question of Place, the Development of Geographic Thought*. Arlington, Va.: Beatty.

Fleure, H. J. 1917. "Régions humaines." *Annales de géographie* 26:161–174.

———. 1919. "Human Regions." *Scottish Geographical Magazine* 35:94–105.

Fox, C. 1932. *The Personality of Britain: Its Influence on Inhabitant and Invader in Prehistoric and Early Historic Times*. Cardiff, Wales: National Museum.

Freeman, T. W. 1950. *Ireland: A General and Regional Geography*. London: Methuen.

———. 1961. *A Hundred Years of Geography*. Chicago: Aldine.

———. 1967. *The Geographer's Craft*. New York: Barnes & Noble.

———. 1974. *The British School of Geography*. Unpublished manuscript. 12 pages.

———. 1977. "Hugh Robert Mill 1861–1950." In *Geographers: Biobibliographical Studies*. Vol. 1, pp. 73–78. London: Mansell.

———. 1980a. *A History of Modern British Geography*. New York: Longmans.

———. 1980b. "On RGS History." In *Geography, Yesterday and Tomorrow*, E. H. Brown, ed. London: Oxford University Press.

———. 1981. "Percy Maude Roxby, 1880–1947." *Geographers: Biobibliographical Studies* 5:109–116.

———. 1982. "Charles Bungay Fawcett, 1883–1952." In *Geographers: Biobibliographical Studies*. Pp. 39–46. London: Mansell.

Galton, F. 1855. "Notes on Modern Geography." In *Cambridge Essays*. Pp. 79–109. London: Parker.

Garnett, A. 1983. "I.B.G.: The Formative Years—Some Reflections." *Transactions of the Institute of British Geographers*, n.s. 8, no. 1:27–35.

Geikie, A. 1865. *The Scenery of Scotland Viewed in Connection with Its Physical Geology*, 2nd ed. 1887. London: Macmillan.

Giddens, A. 1981. *A Critique of Contemporary Historical Materialism*. London: Macmillan.

———. 1984. *The Constitution of Society*. Oxford: Polity Press.

Gilbert, E. W. 1933. *The Exploration of Western America, 1800–1850, An Historical Geography*. Cambridge: At the University Press.

———. 1960. "The Idea of the Region." *Geography* 45:157–175.

———. 1972. *British Pioneers in Geography*. Newton Abbot: David and Charles.

Gillham, N. W. 2001. *A Life of Sir Francis Galton: From African Exploration to the Birth of Eugenics*. Oxford: Oxford University Press.

Gold, J. R. 1980. *An Introduction to Behavioural Geography*. Oxford: Oxford University Press.

Gregory, D. 1976. "Rethinking Historical Geography." *Area* 8:295–299.

———. 1978a. *Ideology, Science and Human Geography*. London: Hutchinson.

———. 1978b. "The Discourse of the Past: Phenomenology, Structuralism, and Historical Geography. *Journal of Historical Geography* 4:161–173.

———. 1980. "The Ideology of Control: Systems Theory and Geography." *Tijdschrift voor Economische en Sociale Geografie* 71:327–342.

———. 1981. "Human Agency and Human Geography." *Transactions, Institute of British Geographers* NS6: 1–18.

————. 1985. "People, Places and Practices: The Future of Human Geography." In R. King, ed., *Geographical Futures*. Pp. 56–76. Sheffield: The Geographical Association.

————. 1989. "Areal Differentiation and Post-modern Human Geography." In D. Gregory and R. Walford, eds., *Horizons in Human Geography*. Pp. 67–96. London: Macmillan.

Gregory, K. 2003. "Physical Geography and Geography as an Environmental Science." In Johnston and Williams. 2003. *A Century of British Geography*. Pp. 93–136. Oxford: Oxford University Press.

Gregory, S. 1963. *Statistical Methods and the Geographer*. London: Longman.

Grigg, D. B. 1977. "Ernst Georg Ravenstein: 1834–1913." In *Geographers: Biobibliographical Studies*. Vol. 1, pp. 79–82. London: Mansell.

Haggett, P. 1966. *Locational Analysis in Human Geography*. New York: St. Martin's Press.

————. 1990. *The Geographer's Art*. Oxford: Basil Blackwell.

Haggett, P., and Chorley, R. J. 1967. "Models, Paradigms, and the New Geography." In *Models in Geography*. Pp. 19–42. London: Methuen.

————. 1969. *Network Models in Geography*. London: Edward Arnold.

Harris, C. D., and Fellman, J. D. 1980. *International List of Geographical Serials*. Third edition. University of Chicago Department of Geography Research Paper No. 193.

Harrison-Church, R. J., et al. 1964. *Africa and the Islands*. New York: John Wiley & Sons.

Harvey, D. 1969. *Explanation in Geography*. London: Edward Arnold.

————. 1973. *Social Justice and the City*. London: Edward Arnold.

Herbertson, A. J. 1905. "The Major Natural Regions: An Essay in Systematic Geography." *Geographical Journal* 25:300–312.

Hogarth, D. G. 1902. *The Nearer East*. New York: D. Appleton.

Hoskins, W. G. 1955. *The Making of the English Landscape*. London: Hodder & Stoughton.

Huxley, Thomas H. 1877. *Physiography: An Introduction to the Study of Nature*. Macmillan: London.

Jay, L. J. 1979. "Andrew John Herbertson: 1865–1915." In *Geographers: Biobibliographical Studies*. Vol. 3, pp. 85–92. London: Mansell.

————. 1986. "John Scott Keltie: 1840–1927." In *Geographers: Biobibliographical Studies*. Pp. 93–98. London: Mansell.

Joerg, W. L. G. 1922. "Recent Geographical Work in Europe." *Geographical Review* 12:431–484.

Johnston, R. J. 1976. "Anarchy, Conspiracy and Apathy: The Three 'Conditions' of Geography." *Area* 8:1–3.

————. 1978. "Paradigms and Revolutions or Evolution: Observations in Human Geography Since the Second World War." *Progress in Human Geography* 2:189–206.

————. 1979. *Geography and Geographers: Anglo-American Human Geography Since 1945*. New York: John Wiley & Sons.

————. 1981a. "Paradigms, Revolutions, Schools of Thought and Anarchy: Reflections on the Recent History of Anglo-American Human Geography. In B. W. Blouet, ed., *The Origins of Academic Geography in the United States*. Pp. 303–318. Hamden, Conn.: Archon Books.

————. 1984b. "The Region in Twentieth Century British Geography." *History of Geography Newsletter* 4:26–35.

————. 1986a. *On Human Geography*. Oxford: Basil Blackwell.

————. 1986b. "Four Fixations and the Quest for Unity in Geography." *Transactions, Institute of British Geographers* NS11:449–453.

————. 2001. "City Regions and a Federal Europe: Robert Dickinson and Post-war Reconstruction." *Geopolitics* 6:153–176.

————. 2002. "Robert E. Dickinson and the Growth of Urban Geography: An Evaluation." *Urban Geography* 23:135–144.

———. 2003. "The Institutionalisation of Geography as an Academic Discipline." In Johnston and Williams. 2003.

———. 2003. "Order in Space: Geography as a Discipline in Distance." In Johnston and Williams. 2003. *A Century of British Geography*. Oxford: Oxford University Press, pp. 303–345.

Johnston, R. J., and Williams, M. 2003. *A Century of British Geography*. Oxford, New York: Oxford University Press.

Keltie, J. S. 1886. *Report of the Proceedings of the Royal Geographical Society in Reference to the Improvement of Geographical Education*. London: John Murray.

———. 1921. *The Position of Geography in British Universities*. New York: American Geographical Society.

Kimble, G. H. T. 1951. "The Inadequacy of the Regional Concept." In L. D. Stamp and S. W. Wooldridge, eds., *London Essays in Geography*. Pp. 151–174. Cambridge, Mass.: Harvard University Press.

Kirwan, L. P. 1964. "The R. G. S. and British Exploration, a Review of Recent Trends." *Geographical Journal* 130:221–225.

Kuhn, T. S. 1962. *The Structure of Scientific Revolutions*. Chicago: University of Chicago Press.

———. 1977. "Second Thoughts on Paradigms." In F. Suppe ed., *The Structure of Scientific Theories*. Pp. 459–482, plus discussion 500–517. Urbana: University of Illinois Press.

Livingstone, D. N. 1984. "Natural Theory and Neo-Lamarckism: The Changing Context of Nineteenth Century Geography in the United States and Great Britain. *Annals of the Association of American Geographers* 74:9–28.

———. 2003. "British Geography 1500–1900: An Imprecise Review." In Johnston and Williams. 2003. *A Century of British Geography*. Oxford: Oxford University Press, pp. 11–44.

Longrigg, S. H. 1963. *The Middle East, a Social Geography*. Chicago: Aldine.

Lyde, L. W. 1913. *The Continent of Europe*. London: Macmillan.

Mackinder, H. J. 1887. "On the Scope and Methods of Geography." *Proceedings of the Royal Geographical Society* 9:141–174.

———. 1902. *Britain and the British Seas*. New York: D. Appleton.

———. 1904. "The Geographical Pivot of History." *Geographical Journal* 23:421–437.

———. 1919. *Democratic Ideas and Reality*. New York: Henry Holt (republished 1942).

Mair, A. 1986. "Thomas Kuhn and Understanding Geography." *Progress in Human Geography* 10:345–370.

Mayhew, R. J. 1998. "The Character of English Geography c. 1660–1800: A Textual Approach. *Journal of Historical Geography* 24:385–412.

Meller, Helen. 1990. *Patrick Geddes: Social Evolutionist and City Planner*. Vancouver: University of British Columbia.

Middleton, D. 1977. "George Adam Smith: 1856–1942." In *Geographers Biobibliographical Studies*. Vol. 1, pp. 105–106. London: Mansell.

Mill, H. R. 1891. *The Realm of Nature*. London: John Murray.

———. 1896. "A Proposed Geographical Description of the British Isles." *Geographical Journal* 7:345–365.

———. 1900. "A Fragment of the Geography of England—Southwest Sussex." *Geographical Journal* 15:205–227, 353–378.

———. 1951. *An Autobiography*. London: Longmans, Green.

Minshull, R. 1967. *Regional Geography, Theory and Practice*. London: Hutchinson University Library.

Monkhouse, F. J. 1959. *A Regional Geography of Western Europe*. London: Longmans, Green.

Neeley, K. A. 2001. *Mary Somerville: Science, Illumination and the Female Mind*. Cambridge: Cambridge University Press.

Newbigin, M. I. 1926. *Canada: The Great River, the Lands, and the Men.* New York: Harcourt, Brace.

Ogilvie, A. G., ed. 1928. *Great Britain: Essays in Regional Geography.* Cambridge: At the University Press.

Oughton, M. 1978. "Mary Somerville: 1780–1872." In *Geographers: Biobibliographical Studies.* Vol. 2, pp. 109–111. London: Mansell.

Parker, W. H. 1982. *Mackinder: Geography as an Aid to Statecraft.* Oxford: Clarendon Press.

Priestley, R., Adie, R. J., and Robin, G. deQ. 1964. *Antarctic Research, a Review of British Scientific Achievement in Antarctica.* London: Butterworth.

Prothero, R. M., ed. 1969. *A Geography of Africa: Regional Essays on Fundamental Characteristics, Issues and Problems.* New York: Praeger.

Robson, B. T., and Cooke, R. U. 1976. "Geography in the United Kingdom, 1972–1976." *Geographical Journal* 142:3–72.

Roxby, P. M. 1916. "Wu Han, the Heart of China." *Scottish Geographical Magazine* 32:266–278.

———. 1925. "The Distribution of Population in China." *Geographical Review* 15:1–24.

———. 1926. "The Theory of Natural Regions." *Geographical Teacher* 13:376–382.

———. 1930. "The Scope and Aims of Human Geography." *Scottish Geographical Magazine* 46:276–299.

———. 1938. "The Terrain of Early Chinese Civilization." *Geography* 23:225–236.

Smith, D. M. 1977. *Human Geography: A Welfare Approach.* London: Edward Arnold.

———. 1988. "Towards an Intepretative Human Geography." In J. Eyles and D. M. Smith, eds., *Qualitative Methods in Human Geography.* Pp. 255–267. Cambridge: Polity Press.

Stamp, L. D. 1947. *The Land of Britain, Its Use and Misuse,* 2nd ed., 1950. London: Longmans, Green.

———. 1960. *Applied Geography.* Harmondsworth, Middlesex: Penguin.

Stamp, L. D., and Wooldridge, S. W., eds. 1951. *London Essays in Geography, Rodwell Jones Memorial Volume.* Cambridge, Mass.: Harvard University Press.

Steel, R. W. 1974. "The Third World: Geography in Practice." *Geography* 59:189–207.

———. 1982. "Regional Geography in Practice." *Geography* 67:2–8.

Steel, R. W., ed. 1987. *British Geography, 1918–1945.* Cambridge: Cambridge University Press.

Stoddart, D. R. 1965. "Geography and the Ecological Approach: The Eco-system as a Geographic Principle and Method." *Geography* 50:242–251.

———. 1967. "Growth and Structure of Geography." *Transactions, Institute of British Geographers* 41:1–19.

———. 1975. "'That Victorian Science': Huxley's *Physiography* and Its Impact on Geography." *Transactions of the Institute of British Geographers,* no. 66:17–40.

———. 1981. "Ideas and Interpretation in the History of Geography." In D. R. Stoddart, ed., *Geography, Ideology and Social Concern.* Pp. 1–7. Oxford: Basil Blackwell.

———. ed. 1981. *Geography, Ideology and Social Concerns.* London: Basil Blackwell; Totowa, N.J.: Barnes & Noble.

———. 1986. *On Geography: and Its History.* Oxford: Basil Blackwell.

Sykes, P. 1934. *A History of Exploration from the Earliest Times to the Present Day.* 2nd ed., 1935; 3rd ed., 1950. London: Routledge & Kegan Paul (reprinted, New York: Harper Bros., 1961).

Taylor, E. G. R. 1930. *Tudor Geography, 1485–1593.* London: Methuen.

———. 1934. *Late Tudor and Early Stuart Geography.* London: Methuen.

———. 1946–1947. "Geography in War and Peace." *The Advancement of Science* 4:187–194.

Teggart, F. J. 1919. "Geography as an Aid to Statecraft, an Appreciation of Mackinder's 'Democratic Ideals and Reality'." *Geographical Review* 8:227–242.

Thompson, J. M. 1929. *Historical Geography of Europe, 800–1789*. London: Clarendon Press.

Unstead, J. F., 1916. "A Synthetic Method of Determining Geographical Regions." *Geographical Journal* 48:230–249.

———. et al. 1937. "Classifications of Regions of the World." *Geography* 22:253–282.

Watson, J. W., and Sissons, J. B., eds. 1964. *The British Isles, a Systematic Geography*. London: Thomas Nelson.

Whitehand, J. W. R. 1970. "Innovation Diffusion in an Academic Discipline: The Case of the 'New' Geography." *Area* 2, no. 3:19–30.

———. 1984. "The Impact of Geographical Journals: A Look at ISI Data." *Area* 16:185–187.

———. 1985. "Contributors to the Recent Development and Influence of Human Geography: What Citation Analysis Suggests: *Transactions, Institute of British Geographers* NS10:222–223.

Williams, M. 2003. "The Creation of Humanized Landscapes." In Johnston and Williams, *A Century of British Geography*. Oxford: Oxford University Press, pp. 167–212.

Wilson, L. S. 1946. "Some Observations on Wartime Geography in England." *Geographical Review* 36:597–612.

Wise, M. J. 1975. "A University Teacher of Geography." *Transactions, Institute of British Geographers* 66:1–16.

———. 1977. "On Progress and Geography." *Progress in Human Geography* 1:1–11.

———. 1988. "Laurence Dudley Stamp, 1898–1966." *Geographers: Biobibliographical Studies* 12:175–187.

———. 1997. "Sir Dudley Stamp and the World Land Use Survey." In A. L. Singh, ed., *Land Resource Management*. New Delhi: DK Publishers, 1–8.

Withers, C. W. J. 2001. "A Partial Biography in the Formalization and Institutionalization of Geography in Britain since 1887." 79–119. In *Geography: Discipline, Profession and Subject since 1870: an International Survey*. Ed. G. Dunbar.

Wooldridge, S. W. 1950. "Reflections on Regional Geography in Teaching and in Research." *Transactions, Institute of British Geographers* 16:1–11.

Wooldridge, S. W. 1956. *The Geographer as Scientist, Essays on the Scope and Nature of Geography*. London: Thomas Nelson.

Wooldridge, S. W., and East, W. G. 1951. *The Spirit and Purpose of Geography*. London: Hutchinson University Library.

Wrigley, E. A. 1965. "Changes in the Philosophy of Geography." In R. J. Chorley and P. Haggett, eds., *Frontiers in Geographical Teaching*. Pp. 3–20. London: Methuen.

Wrigley, G. M. 1950. "Hugh Robert Mill: An Appreciation." *Geographical Review* 40:657–660.

11

The New Geography in Russia, the Soviet Union, and the Commonwealth of Independent States

> *Geographic science in the Soviet Union has traveled a prolonged and many-sided path of development. Arising after the Great October Socialist Revolution, Soviet geography received as a great and valuable heritage from prerevolutionary Russian geography not only an immense store of geographic facts, but also a whole system of fruitful progressive scientific traditions, schools, and concepts, many of which have become classic. During its existence, Soviet geography has repeatedly enlarged its scientific heritage. It has gathered new factual materials, continued and enriched the progressive classic scientific trends, and created new theoretical concepts which have developed on the basis of scientific Marxism-Leninism and in close connection with the practice of socialist construction.*
> —*I. P. Gerasimov, "Geography in the Soviet Union"*

The impact of German geographical ideas during the last quarter of the nineteenth century had quite different results in Russia than it had in France and Britain. By this time Russia had a long history of geographical work, including the production of maps and atlases and the writing of regional monographs. Although many explorers and scholars from Germany and other countries of western Europe lived and worked in Russia and were influential in promoting geographical studies, most of those who made the maps, gathered the statistical data, and wrote reports describing different parts of the national territory were Russians. Because of the language barrier, which is also an alphabet barrier, an understanding of the importance of the work of Soviet geographers and of their predecessors in pre-1917 Russia was delayed in reaching the geographical scholars of western Europe and the United States for many decades.[1]

The exploration and mapping of the vast extent of land area that was eventually included in the Russian Empire were carried out mostly by Russians, but with considerable assistance from skilled mapmakers from the West. Peter the Great, who ruled Russia from 1682 and 1725, recognized the vital importance of having accurate geographical information to guide the eastward expansion of the empire.

---

[1]The transliteration of Cyrillic into Roman letters requires some arbitrary decisions regarding the spelling of names. This book follows the system recommended by Theodore Shabad, one-time editor and translator of *Soviet Geography*.

248

He gave his support to exploring expeditions and to the publication of the results of their discoveries. The southern part of European Russia was surveyed in the late seventeenth century, and the resulting maps were published in Amsterdam (Bagrow and Skelton, 1964:170–176). In 1719 all official Russian mapmaking activities were placed under the direction of Ivan Kirilov, Secretary of the Governing Senate. It was from this highest administrative office of the Russian state that surveys and mapping of the Russian territory were ordered by Kirilov. He knew the French cartographer Joseph Nikolas Delisle but there is no documentary evidence that he helped Kirilov in his atlas of 1734. Publication of further sheets of this atlas ceased with Kirilov's death in 1737. A more important atlas project was "Atlas Rossiyskiy . . ." (Russian atlas) published in 1745 in both Russian and Latin by the St. Petersburg Academy of Sciences. This very large atlas was completed under the supervision of Leonhard Euler (1708–83) and provided a much larger view of the country than the work of Kirilov. There were a number of omissions. Mikhail Vassilyevich Lomonosov remarked of it "looking at the then Geographic Archives and this atlas it is easy to see how much more accurate and fuller it could have been" (Postnikov, 1996). Nevertheless, it was the most thorough atlas of Russia available at that time.

Meanwhile, the Moscow Mathematical and Navigational School was founded in 1701. Students from this school were employed as navigators in the navy or as cartographers, land and sea surveyors, engineers, and artillery experts. When the Naval Academy was opened in 1715 at St. Petersburg, the role of the Moscow school was abbreviated. By 1727 there were 229 students attending the Naval Academy. Two English instructors hired to teach in this school left behind in Russian cartography a tradition of theodolite surveying.

Vassily N. Tatishchev (1686–1750, Alexandrovskaya, 1982) assumed the role of Kirilov and began to focus on local surveying and providing more details on maps. In 1737 Tatishchev was appointed "to expand and correct" maps and compile the map of Russia. He at once ordered six groups of geodesists to survey and correct work already accomplished and find the right longitudes and latitudes in the more remote parts of the country. Instrumental surveys replaced word-of-mouth data. Tatishchev also insisted that maps and written descriptions should supplement one another. By the 1740s nearly 85 percent of the land surface of Russia had been mapped, but it had not been accomplished in a logical manner. To integrate this large collection of maps via a new geodetic foundation, J. N. Delisle (1688–1768), French astronomer and scientist, was invited to St. Petersburg. Delisle sought to put in place a triangulation network similar to that developed by Cassini in France, but the plan was attacked and not brought to fruition.

Initially, the expeditions sought the location of rivers, mountains, coastlines, minerals, and other natural resources. But the great Russian encyclopedist Mikhail V. Lomonosov urged that the exploring parties be charged with the systematic collection of information about the physical character of the land, the population, and the condition of the economy.[2] In 1758 he became head of the world's first

---

[2]Lomonosov was a universal scholar in the classical sense. From 1736 to 1741 he studied at the University of Marburg, Germany. In 1755 he was one of a group of scholars who

officially named Department of Geography, which was in the Russian Academy of Sciences (Gerasimov, 1968b). But it was not until 1768, three years after the death of Lomonosov, that the Academy of Sciences sent out the first expedition for the specific purpose of gathering and reporting on the physical and economic geography of a part of the national territory (Nikitin, 1966:7).

Gerasimov has claimed that Lomonosov's contribution to the advancement of the science of geography was inestimable. Lomonosov had a deep appreciation for the importance of science. A particularly influential work by Lomonosov, *On the Strata of the Earth* (1763), demonstrates that his thought was far ahead of his time. He wrote on the mutability of nature, the importance of studying causation, and the respect that scientists and humankind should bestow on earth's resources. His work constitutes the basis for Russian geographical work: Numerous of his followers, including J. I. Lepekhin, P. S. Pallas, S. P. Krasheninnikov, and I. P. Falk, would make notable contributions to the discipline (Alexandrovskaya, 1982).

In the 1720s the Russian Empire launched a campaign to expand the empire's boundaries. It began with the First (1728) and Second Kamchatka expeditions and the voyages to the islands in the Bering Strait and the northern shore of Alaska by M. S. Grozdev and I. Fedorov (1732). With these voyages the Russians studied the northern shores of Siberia, established the separation of Asia and North America, and generated thought that led to Russian colonies in Alaska and on the Aleutian Islands.

In 1819 M. M. Speransky was appointed governor-general of Siberia. He established a new administrative system that led to the development of local studies in natural history and considerable mapping of this little-known land. Quite soon it became obvious that local resources would not adequately fund these studies over such a vast and unsettled territory. Speransky moved to St. Petersburg in 1821 and established a special Siberian commission whose task would include the development of a program for organizing the administration and study of natural resources within Siberia. Gavriil Stepanovich Batenkov, manager of the Siberian commission office, prepared a "Draft of a Plan for Exploration of the Siberian Territories," which presented a program of research for the geographic, cartographic, and geodetic study of Siberia (Postnikov, 2002, 251).

I. F. Kruzenstern and Y. F. Lisyansky organized expeditions around the world, and later V. M. Golovnin and O. E. Kotzebu traveled largely in the Pacific Ocean. Fabian Bellingshausen and M. P. Lazarev journeyed to the Antarctic (1819–21) (Debenham, 1945) and F. P. Lütke to coastal Asia (1821–25). Russian names were now sprinkled liberally onto the world map. In addition, expeditions were conducted throughout the Russian Empire, especially the White, Baltic, Black, Caspian, and Aral Seas. G. I. Nevelskoi (1849–55) contributed substantially to knowledge of the Sea of Okhotsk and the Sea of Japan (Bassin, 1994). These findings were

---

founded the Moscow State University. At that time Pushkin described Lomonosov as "a university in himself." Before he was named head of the new Department of Geography in the Russian Academy of Sciences in 1758, he had been head of the Department of Chemistry. He was also famous as a linguistic scholar and as a poet.

assembled and supplemented by Russian expeditions in the second half of the eighteenth century.

The Academy of Sciences provided an institutional focus for the great variety of scholarly works, many of which were geographical in purpose. The Germans were very influential in this effort. Anton Friedrich Büsching was the pastor of a German Lutheran church in Saint Petersburg from 1761 to 1765, and in 1766 the part of his *Neue Erdbeschreibung* dealing with Russia was translated into Russian. It was he who first described the division of European Russia into latitudinal zones of differing natural conditions—north, middle, and south. The Russian geographers quickly adopted his suggestion that the national territory should be divided into natural regions for the practical purposes of administration.[3] At first most of the scholars named as academicians were foreigners who, like Büsching, were living in Russia. In the course of time the proportion of Russian academicians increased. Before 1800 numerous regional descriptions were already in existence. Early in the nineteenth century two persistent characteristics of Russian geography had become established. One was the emphasis on regions as the basis of organization for geographical work and as real entities that could be objectively defined. Between 1800 and 1861 there were 15 different regional divisions of European Russia that were proposed and defended. The second persistent characteristic was that the study of these regions was undertaken for practical purposes. The Russian intellectuals of that period were disturbed by the poverty of the serfs and sought to undermine the system of bondage that tied most of the rural population to the aristocracy of landowners. For example, K. I. Arsenyev wrote a *Short Universal Geography* that was organized by regions and had a strong economic emphasis, reflecting the author's concern with improving the peasants' quality of life. The book was first published in 1818 and went through 20 editions by 1848. In 1832 the same writer published a monograph on the historical geography of Russian cities in which he offered a classification of cities in terms of their economic functions.

Another distinctive characteristic of Russian geography has been the continued use of that name to cover a wide variety of specialties. While classical geography was undergoing analysis in Germany as each academic discipline sought to establish its separate existence, in Russia the tendency was for scholars with diverse interests to come together as geographers. In the 1840s this convergence of specialties on geography created a need for some kind of institution to provide a forum for the presentation and discussion of different kinds of studies dealing with the physical earth and its human inhabitants. In 1845 Fedor Petrovich Lütke was largely instrumental in founding the Imperial Russian Geographical Society. In addition to promoting geographical studies as such, the society also assumed responsibility for studies in geology, meteorology, hydrography, anthropology, and archaeology. During the period from 1845 to 1917 the society published some 400 volumes of papers and monographs. The diverse specialties represented in the society were known collectively as "the geographical sciences" (Hooson, 1968).

---

[3] Catherine II decreed in 1784 that officers and noblemen should wear uniforms of distinctive color in each of the three zones (Nikitin, 1966:6).

GEOGRAPHY IN RUSSIA BEFORE 1917

In Germany the deaths of Humboldt and Ritter marked a break in the continuity of geographical study until the new geographers, such as Richthofen, introduced the new geography. In Russia there was no such break. For this reason it is difficult to select any one scholar as the "grand old man" of Russian geography. It would be more realistic to select four grand old men: one grandfather, Semenov Tyan-Shanski, and three fathers, Voeikov, Dokuchaiev, and Anuchin. These formed the nucleus of Russian geography before the October Revolution of 1917.

Petr Petrovich Semenov-Tyan-Shanski

Petr Petrovich Semenov Tyan-Shanski provided the continuity between the scholars of the classical period, such as Lomonosov, Büsching, and Arsenyev, and the scholars of the modern period (after 1870), who could no longer claim competence in all branches of geographical study. In 1853–54 Semenov attended Ritter's lectures at Berlin and worked with Richthofen to prepare for exploratory work in Central Asia. In 1858 he explored the Dzungarian Basin and the bordering Altai Mountains to the north and the Tien Shan Mountains to the south. He was the first European to cross the latter range, and for this exploit the czar granted him and his family the right to add Tyan-Shanski as a suffix to his name in 1906. In 1888 he explored the desert of Turkestan east of the Caspian Sea.

In 1873 Peter Semenov was appointed the society's vice president replacing Admiral F. P. Lütke, who had held the position since 1845. As vice president Lütke and Semenov were supervisors of the society's activities (Postnikov, 2003). At the time when Peter Semenov was appointed vice president one of the emperor's brothers—the grand duke Konstantin Romanov—was the society's official president. Peter Semenov held the position in St. Petersburg for more than 40 years. Semenov was also a member of the Committee for the Emancipation of the Peasants from the Bonds of Serfdom. (The serfs were emancipated in 1861.)

Semenov was much more than an explorer. He had been attracted to the study of geography by his contacts with Ritter, but, like Reclus, he did not embrace Ritter's teleological philosophy. He was much more concerned about using geography as a means of decreasing the poverty of the rural people. That is, he wanted to emphasize the practical importance of geographical study, or what we would today call its "social relevance." He himself wrote a number of regional monographs, including a five-volume work on Russia[4] that has been described as a "perceptive blend of natural, historical, and economic phenomena" (Hooson, 1968:259). In 1871 he published a general work on the historical geography of Russian settlement. He was also a member of the committee that planned and directed the first Russian census of population in 1897. A man of diverse interests and competence, he was ideally suited to guide the fortunes of such a composite institution as the Geographical Society, and he was able to preserve its unity when in other hands it might well have fallen apart into separate specialties. By the time he died in 1914, he had set a distinctive

---

[4]P. P. Semenov Tyan-Shanski, *Geographical-Statistical Dictionary of the Russian Empire*, 5 vols. (St. Petersburg, 1863–1883).

stamp on Russian geography, giving it unity in spite of the variety of its parts and pointing it toward practical and remedial objectives (Thomas, 1988).

The Followers of Semenov

The new geography came into Russia between 1880 and 1914. The ideas of Richthofen, Ratzel, and Hettner were familiar to Russian geographers because many of them studied in Germany. The revolutionary ideas of Charles Darwin were perhaps less intoxicating than they were in Britain because of the earlier studies of evolution by the Russian biologist K. F. Rul'ye. In any case the Russians rejected the more extreme forms of environmental determinism stemming from Herbert Spencer and also the use of biological analogy to describe sequences of landforms as proposed by the American geographer William Morris Davis. To be sure, some of the historians did support the ideas of climatic influence on national character or of the critical importance of the large Asian rivers in providing the setting for the development of early civilizations (Matley, 1966). But the geographers as a whole avoided these difficulties.

In the period between 1880 and 1914 three outstanding followers of Semenov helped to give modern Russian geography its distinctive stamp. Two of these, A. I. Voeikov and V. V. Dokuchaiev, were primarily research workers whose innovative studies of climates and soils gave them international reputations. One, D. N. Anuchin, was primarily an educator who established geography as a major university subject and who drew up the curricula for the primary and secondary schools (Esakov, 1978).[5]

Aleksandr Ivanovitch Voeikov[6] (1842–1916) was a scholar who followed Semenov in the wide range of his interests (Fedosseyev, 1978). His doctoral dissertation, "Direct Insolation in Various Parts of the Globe," was accepted at Göttingen University in 1865. His studies of the earth's heat and water balances continued for the rest of his life. His studies in climatology were directed to the improvement of agriculture. Hence, when he experimented with different ways to measure the depth of snow, he was partly concerned with the effect of snow cover on temperature and to a very large extent with the forecast of crop yields for the following summer. He was the originator of what came to be called snow science. In the 1870s he traveled in the United States and Asia and thereafter was a regular correspondent with the Smithsonian Institution in Washington, D.C. In 1886 he published a discussion of James H. Coffin's *The Winds of the Globe* (Smithsonian

---

[5]A geographer of that same generation who never held a university post was Prince Peter Kropotkin. A member of the Russian aristocracy, he was nevertheless a devoted anarchist and a close friend of the French geographer-anarchist Elisée Reclus. His lifelong interest was in the study of geography, and in 1876 and from 1886 to 1917, while he was in exile from Russia, he lived in Great Britain. Most of what Reclus had to say about Russia came from Kropotkin. "I cannot conceive of a physiography," he said, "from which man has been excluded."

[6]Sometimes spelled Woeikoff or Voyeikov. He studied at Berlin, Göttingen, and Heidelberg. In 1884 he went to the University at St. Petersburg as docent and was promoted to professor in 1887.

Contributions to Knowledge, vol. 20 [Washington, D.C., 1875]). His concern with the improvement of Russian agriculture led him to compare the farm practices in areas with climates similar to those of European Russia. These were the first systematic studies of climatic analogs. Based on his advice, tea was successfully planted in Georgia (east of the Black Sea), cotton in Turkestan, and wheat in Ukraine. His book, *The Climates of the World*, was published in Russian in 1884, but it became known to climatologists of other parts of the world when it was translated into German and published in Germany in 1887. Similarly, a monograph, *Distribution of Population on the Earth in Relation to Natural Conditions and Human Activities* (published in Russian in 1906), appeared in German in brief form that same year. He also became known to the rest of the world through papers and books published in France.[7]

One of Voeikov's most important contributions to international geography, however, was his insistence on the importance of studying the influence of man on the environment. He was one of the first Europeans to recognize and report on the destructive effects of man's use of the land. In fact, he criticized Richthofen for not calling attention to the man-caused gullying in the loess lands of China. In his opinion a variety of changes follow the removal of the cover of natural vegetation —in some places the changes are disastrous. He pointed to the overgrazing of some of the Russian steppes, with a consequent acceleration of gully erosion. The clearing of the forests in the north could produce a change in the climate toward increasing drought. He was always enthusiastic about what irrigation could do to improve the productivity of arid or semiarid lands (Voeikov, 1901).[8]

V. V. Dokuchaiev, who became the first professor of geography at St. Petersburg in 1885 (the same year in which Voeikov was appointed docent), was less known beyond Russia than Voeikov because his works appeared only in Russian. But Dokuchaiev deserves a major place among the world's geographers because of his innovative studies of soil. Looking more closely at the natural zones that Büsching had outlined, he was the first scientist to realize that soil is not just disintegrated and decomposed rock. The geographers of Germany, France, and Great Britain conceived of the soil as a faithful reflection of the underlying geological formations. So they spoke of pre-Cambrian soil, or Devonian soil, or glacial soil (derived from glacial deposits). But Dokuchaiev, working on the plains of European Russia, saw that the parent material only provided the substance in which a soil would form. He noted that different kinds of soil could be identified by looking closely at the layers or horizons, which differed because of differences in the soil-forming

---

[7]These translations into German and French include: *Die Klimate der Erde* (Jena: H. Costenoble, 1887); "De l'influence de l'homme sur la terre," *Annales de géographie* 10(1901):97–114, 193–215; "Le groupement de la population rurale en Russie," *Annales de géographie* 18(1909):13–23; *Le Turkestan Russe* (Paris: Armand Colin, 1914). The brief summary of his study of world population is in *Petermanns Geographische Mitteilungen* 52(1906):241–251, 265–270.

[8]It is interesting that he made no mention of George Perkins Marsh in his writings. Yet Marsh's *Man and Nature* (1864) was published in Russian translation in 1866. Perhaps he came into contact with Marsh's ideas through Elisée Reclus.

processes. Soil is formed by water percolating through the loose material at the surface and carrying away soluble minerals; it is also formed by the mixture of organic matter from plants and animals with the upper horizon. The soil, said Dokuchaiev, reflects the extraordinarily complex interaction of climate, slope, plants, and animals, with the parent material derived from the underlying geological formations (Gerasimov et al., 1962:111). A soil that had been exposed to all these conditions for a long time would more closely reflect the complex of climate and vegetation than it would the parent material.

Dokuchaiev's ideas made slight impact on geographers in western Europe not only because he wrote in Russian, but also because there was little opportunity to observe such soils in the West. Especially in France and Britain it seemed quite clear that soils of different kinds were produced on rocks of different types. There were no large expanses of plain on which the arrangement of climate zones could produce observable soil differences. Dokuchaiev's doctoral dissertation, on the other hand, was a careful and detailed study of the Russian chernozem (or black earth) published in 1883. In 1889 he published his concept of soil-forming processes and of the natural zonation of soils according to climate. He recognized, as Voeikov did, that man was a major agent of change on the surface of the earth and that the transformation of natural zones into agricultural regions involved such factors as the attitudes of the people and their technical skills. His concept of natural zones transformed by man comes very close to Schlüter's concept of the landscape type. Dokuchaiev, in fact, described geography as "landscape science" (*landschaftovedenie*).

Dokuchaiev trained a number of students at St. Petersburg who continued to develop the master's ideas. L. I. Prasolov became the editor in chief of a soil map of Russia, compiled from a variety of detailed studies on a scale of 1/1,000,000—a project that was later carried forward by I. P. Gerasimov (Gerasimov et al., 1962:113). N. M. Sibirtsev, who died in 1900 at the age of 40, contributed the concept of *zonal* soils as distinct from *azonal* soils—the zonal reflecting climatic patterns and the azonal more closely related to underlying parent materials or such local conditions as poor drainage. The scholar who did the most to make Dokuchaiev's work known outside of Russia was Konstantin D. Glinka. In 1908 he published a major work on the zonal soils of the world that extended the concepts of Dokuchaiev to types of zonal soils not found in Russia. This work was translated from Russian into German and published in German in 1914; it was from the German edition of Glinka that Curtis F. Marbut, an American soil geographer, published an edition in English in 1927.[9] Dokuchaiev's ideas about soils were enthusiastically received in the United States, where, unlike western Europe, there are large expanses of plain on which zonal soils can be observed.[10] Concern for the issue of soils, food, and population

---

[9]K. D. Glinka, *The Great Soil Groups of the World and Their Development*, translation by C. F. Marbut from *Die Typen der Bodenbildung* (Ann Arbor, Mich.: Edwards Bros., 1927).

[10]Before the unwary reader uses this as an example of environmental determinism, let us observe that neither the steppe nomads of Russia nor the plains Indians in America had the necessary technical skills and theoretical knowledge to recognize the existence of zonal soils.

D. N. Anuchin

V. A. Anuchin

Nikolai N. Baranskiy

Lev S. Berg

Vasily V. Dokuchaiev

Innokenti P. Gerasimov

P. P. Tyan-Shanski

Alexander I. Voeikov

were present. Although a country vast in area, there was still population pressure on the means of subsistence. This led men such as Nikolay Vavilov to become specialists in plant genetics. It was felt that such studies could increase crop production (Esakov, 1980; Harris, 1991).

Meanwhile, geographical courses were being offered at Moscow State University. In 1887 Dmitry N. Anuchin was named head of the new Department of Geography and Ethnography. Anuchin was trained at Heidelberg in anthropology and anthropogeography, and his point of view in both writing and teaching was strongly human oriented. His textbooks provided a new kind of geography to teach in the schools; after 1912, when the authorities recognized geography as a field in which students could major, Anuchin's students at Moscow State University were the ones who were appointed to most of the new positions throughout the country. Lev S. Berg, for example, went to St. Petersburg, where he continued the development of Dokuchaiev's ideas concerning landscape science and where he remained active after 1917. There was also the geologically trained V. P. Semenov Tyan-Shanski (the son of Petr), who outlined a system of geomorphic regions of the Russian plains with special reference to human-land relations. His 19-volume semipopular regional geography of Russia was published in St. Petersburg between 1899 and 1914. He was a regional geographer who also advanced economical and statistical surveys. Gerasimov has claimed that Semenov Tyan-Shanski's work provides the basis for economic geography in the Soviet period.

SOVIET GEOGRAPHY SINCE 1917

The October Revolution of 1917 affected all phases of Russian life, including the pursuit of geographical knowledge. Only those scholars could survive who were able to bring their ideas into line with the concepts of Marx, Engels, and Lenin. Marx was ambiguous concerning his geographical principles, and perhaps it is fortunate that passages from his writings can be found to support a great variety of points of view. Lenin, on the other hand, was much more specific. As an economic determinist, he was strongly opposed to any suggestion that the natural environment could in any way control human destiny. For him geography was the necessary foundation on which the design of a new kind of economy had to be based. The most important product of geographical study was the identification of rational regions within which the segments of the new national economy could be constructed. Lenin considered geography an essentially practical subject, but it was so important that in 1921 he decreed that the subject should be taught in all the schools. Unfortunately, his efforts to provide for instruction in geography were not very successful for the usual reasons: the schoolteachers were not trained in the concepts of geography, and there were no teacher-training institutions where they could study geography.

Yet in spite of the new directions of geography after 1917, most of the distinctive characteristics of Soviet geography can be traced back to the prerevolutionary period. The tendency to look at natural landscapes as systems of interrelated parts is typically Russian and derives from geographers such as Semenov, Voeikov, and Dokuchaiev. The continued interest in such physical processes as heat and water balances goes back to Voeikov. The preoccupation with the drawing of regional boundaries began in the eighteenth century, as did the concern with practical problems of economic development. And to find that geography is a focus of diverse specialties rather than a remnant left over from the separation of the disciplines is in line with longstanding Russian tradition.

The relations between the ruling Communist party and the members of a profession such as geography are not always clear to people unfamiliar with the Soviet system. The kind of academic freedom that was developed in Germany in the nineteenth century and lost for a time in the 1930s never existed in the Soviet Union. In the Stalin era scholars who ventured to express disagreement with party policies could be arrested and punished. But after the death of Stalin in 1953 there was more freedom to express critical views on policy questions. Disagreements within the profession concerning professional matters were given wide publicity, even in the public press. Discussion went on until some kind of a consensus could be discerned, then the leaders of the Communist party announced the decision. Disagreements on methodological questions involved the usual semantic traps when scholars sometimes mistook word symbols for reality and when certain words and phrases were repeated without clearly defined referents. Words in both Russian and English carry rich loads of connotations; when words are translated, many shades of meaning are lost. As a result, the international discussion of theoretical or philosophical questions is not always fruitful. The best language for such discussion is mathematics.

Geography as a professional field was in a strong position in the Soviet Union in 1990. We will attempt to trace the steps by which it reached this status.[11]

The Early Soviet Years

In the years immediately following the October Revolution the continuity of geographical teaching and research was greatly aided by Lenin himself. In 1918 he presented his "Draft of a Plan of Scientific and Technical Work." In this document he called on the Academy of Sciences to work on the development of a plan for "a rational location of industry in Russia from the point of view of proximity to raw materials and a minimal expenditure of labor from the processing of the raw materials through all subsequent stages of the processing of semifinished goods to the finished product" (Saushkin, 1966:5).

---

[11]Such an attempt for those who do not read Russian becomes possible because of the English translations of Russian studies since 1959. The periodical *Soviet Geography* was translated and edited for many years by Theodore Shabad and published monthly (except July and August) by the American Geographical Society. There is also the volume translated by Lawrence Ecker and edited by Chauncy D. Harris, *Soviet Geography, Accomplishments and Tasks*, a symposium of 50 essays by 56 leading Soviet geographers edited by Gerasimov and others (Gerasimov et al., 1962). The issue of *Soviet Geography* for September 1967 is a directory of Soviet geographers. In 1992 the title *Soviet Geography: Review and Translation* was changed to *Post Soviet Geography*; in 1996 the title was again changed to *Post Soviet Geography and Economics*; in 2002 it was retitled *Eurasian Geography and Economics*.

The geographers in the Academy of Sciences were ready to make practical use of the large amount of information that had been gathered during the preceding half century under the guidance of Semenov and Anuchin. In 1918 the Commission for the Study of Natural Productive Forces was established in the Academy of Sciences, and one of its subdivisions was the Department of Industrial Geography. The first assigned task was to make an inventory of Russia's natural resources. At the same time, in 1918, the first graduate school in geography in the Soviet Union, the Institute of Geography at Leningrad University, was started by L. S. Berg, A. A. Grigoriev, and others.[12]

A geographer who appeared on the scene at this time in a key position was Nikolai N. Baranskiy.[13] As a student at Tomsk at the turn of the century, Baranskiy had been involved in underground activities against the government, and by 1917 he had established himself as one of the leading Bolsheviks. As Saushkin puts it:

> . . . geography (and economic geography in particular) was for Baranskiy a logi-cal continuation of his revolutionary party work. He saw in it one of the powerful tools for remaking the world, for building communism, for indoctrinating and educating the people. Baranskiy was clearly aware of the role that geography played in the apprehension of the universe, and of the strength and party orientation of its scientific ideas (Saushkin, 1966:20).

Furthermore, Baranskiy was a close personal friend of Lenin. For the development of geography in the Soviet Union, here was the right man in the right place at the right time. He, as much as any one person, was responsible for the development of Soviet economic geography.

Lenin was a strong supporter of the kind of economic regionalization proposed by Baranskiy. In 1920, in setting up the studies on which the construction of a net-work of electric power stations and transmission lines was to be based, Lenin specified that he wanted "a geography of each of the sections of the electrification plan" and that he expected to see a map of the main power stations. He insisted that "the map should clearly show the regions to be served by the central power stations, the type of industry to be included, and everything associated with these regional stations" (Saushkin, 1966:9).

The State Planning Commission (GOSPLAN) was established in 1921. After some difficulty in gaining general acceptance for any regional division of the Soviet Union, a special commission on regionalization was appointed and given authority to recommend a rational plan for the division of the natural territory into functional

---

[12]The Institute of Geography was reorganized as the Faculty of Geography at Leningrad in 1925. The Commission for the Study of Natural Productive Forces of the Academy of Sciences in 1930 became the Institute of Geomorphology under Grigoriev; in 1934 it became the Institute of Physical Geography; and in 1936 it became the Institute of Geography.

[13]N. N. Baranskiy was born in Tomsk in 1881 and attended the Law School at the University of Tomsk. Like many young people in those days, Baranskiy was committed to the cause of social revolution. In 1901 he was expelled for Marxist activities, and before 1917 he had been arrested three times by the czarist police. At the university he had become interested in economic geography because of its relevance to social problems.

units. The commission formulated the concept of the economic region in 1922 as follows:

> A region should be a distinctive territory that would be economically as integrated as possible, and, thanks to a combination of natural characteristics, cultural accumulations of the past, and a population trained for productive activity, would represent one of the links in the entire chain of the national economy. This principle of economic integration makes it possible to construct further, on the basis of a properly selected complex of local resources, capital assets brought in from outside, new technology, and the national plan of economic development, a regional development plan that would make optimal use of all possibilities at minimal cost. This will also help achieve other important results: the regions will specialize to a certain extent in those activities that can be developed most fully in them, and exchange between regions will be limited to a strictly essential amount of purposefully directed goods. Regionalization will thus help establish a close link between natural resources, working skills of the population, and assets accumulated by previous cultures and new technology, and yield an optimal productive combination by insuring a division of labor among regions and, at the same time, organizing each region as a major economic system, thus evidently insuring optimal results (Saushkin, 1966:12).

GOSPLAN divided the Soviet Union into 21 such regions and then proceeded to a detailed study of each of them. The work was done by a large number of young men and women with a variety of professional backgrounds, including engineering, economics, and geography. Some of those who worked on the initial planning went on to become leaders in economic geography. It was in 1920 that one of these newly created geographers, L. L. Nikitin, made his first investigation of the writings of the prerevolutionary geographers. The older data were combined with the latest information regarding natural resources, physical conditions, population, and types of economy in each of the regions. "Never before in Russian geography had such a vast, diversified body of material been generalized on a regional basis" (Saushkin, 1966:15).

Baranskiy applied the ideas of economic regionalization to the theory of economic geography, to the teaching of it in the schools, to the establishment of research tasks in the universities, and to forecasting the territorial organization of productive forces (Saushkin, Kosmachev, and Bykov, 1980). Baranskiy was the first Soviet geographer to demonstrate that the science of economic geography consisted of a system of economic regions related each to the other by a territorial division of labor. Baranskiy substituted the notion of studying complexes of productive forces in regional settings; that was a new conception in Soviet geography, and it met with some criticism, especially in the 1930s. Baranskiy and his opponents, especially V. A. Den, enjoined debate. In the end Baranskiy triumphed, referring to Den's "political deviation" and "leftist deviation," both synonyms for failure. Baranskiy then brought to the attention of his fellow geographers the works of non-Soviet geographers including Alfred Weber, Alfred Hettner, Henri Baulig, Pierre Gourou, Eugeniusz Romer, and Preston James.

Meanwhile, other geographers not directly employed by GOSPLAN were also working on problems of industrial location and resource development. Of special interest were the plans for the organization of interregional combines. N. N. Kolosovskiy was the author of the plans for the great Ural-Kuznetsk Industrial Combine. A theoretical model for an integrated industrial region combining basic resources with steel production and with industries using the steel had been drawn up in 1927 for the Dnieper Basin, but Kolosovskiy's plan called for moving raw materials and finished goods between regions. The plan was immediately attacked by other geographers on the basis of central-place theory, but in 1931 such controversy was dangerous because the loser could be arrested for anti-Soviet activities.

As might be expected, the geographers who worked on these practical problems of economic planning advanced in their profession more rapidly than did those who remained in the universities in teaching positions. The methodological and philosophical discussions among the university geographers seemed irrelevant to those who were using their knowledge of geography to find answers to "real" problems. In the universities some even reverted to the traditional way of teaching economic geography by topics instead of by regions. Baranskiy, who headed the new section of economic geography in the Research Institute at Moscow, fought persistently for the regional approach because this seemed to him the only way that geography could make a useful contribution to practical questions of economic development.[14]

Some of the university geographers even continued to teach various forms of environmental determinism in spite of authoritative decisions outlawing such ideas. Lenin himself had taken vigorous exception to the notion that the steppes north of the Black Sea could never be used for agriculture because of climatic conditions. This point of view, he said, was based on the existing level of technology and ignored the inevitability of technological change (Nikitin, 1966:36).[15] Yet in 1923 A. A. Kruber, then head of the Department of Geography at Moscow, wrote in a textbook: "Like all living things on earth, man is subject to the same forces of nature,

---

[14]At Moscow State University, where the Department of Geography and Ethnography had been established by D. N. Anuchin in 1887, a Department of Geography was established in 1919 headed by A. A. Kruber. Anuchin became the head of a Department of Anthropology and Ethnography. These were still undergraduate departments concerned primarily with teaching. In 1923 a government decree (dated December 1922) established the Scientific Research Institute of Geography, which was headed at different times by A. A. Kruber, A. A. Borzov, N. N. Kolosovskiy, and B. F. Kosov. In 1929 this institute was divided into two sections; one was to work on problems in physical geography, the other, under Baranskiy, was to work on economic geography. This was the first time that postgraduate training in economic geography as such became possible (Ryabchikov, 1968:348). The Department of Geography became the Faculty of Geography and Soils in 1933 and the Faculty of Geography in 1938, under A. A. Borzov. Borzov, a physical geographer, and Baranskiy, an economic geographer, led numerous field expeditions to various parts of the country; their close and friendly cooperation was a major factor in assuring the unified approach of geography during the years before World War II (Ryabchikov, 1968:349–350).

[15]This statement by Lenin seems to give authoritative support for the principle that the significance of the physical and biotic features of humans' natural surroundings is a function of the attitudes, objectives, and technical skills of people themselves—which is a widely accepted bourgeois concept.

which, with irresistible inevitability, determine both the conditions of settlement and way of life of man" (Saushkin, 1966:16).

There were also repeated efforts to separate physical geography from economic geography. The argument is the familiar one that the "laws" governing the physical world are not at all the same as those governing economic behavior, and, therefore, these two fields of study cannot logically, or even practically, be included in the same discipline. It was all very well for Borzov and Baranskiy to work together in harmony (much as Vidal de la Blache and de Martonne worked together at the Sorbonne), but their students could find little common ground between physical and economic studies. O. A. Konstantinov (who went on to become one of the leading economic geographers of the Soviet Union) when he was a young man in 1926 wrote one of those statements that young people the world over often write—to their later embarrassment:

> We reject the possibility for economic geography to be simultaneously part of two quite different systems of science (the geographic and economic sciences). We hold that ours is not only an economic discipline, but a purely economic discipline. In other words we are for a complete break with geography, in the sense of complete outlawing of geographic approaches (Saushkin, 1966:23).

Baranskiy led the opposition to these challenges to the unity of geography. In the 1930s Stalin's policy of strong centralized control of the economy made regional study seem less relevant and seemed to negate the whole idea of an economic region as a "major territorial production complex with a specialization of national significance."

The Decisions of 1934

Baranskiy was victorious over his challengers. Little by little in the late 1920s and early 1930s, the number of geographers who supported his regional approach and his belief in the unity of geography increased. In 1933 at Moscow State University, a Faculty of Soils and Geography was established.[16] Included in this faculty were departments covering both physical and economic geography (a good way of saying both yes and no to questions about the unity of geography).

Then in 1934 two decrees gave official recognition of the professional support that had been given to Baranskiy. The Council of People's Commissars and the Central Committee of the All-Union Communist Party issued a decree on May 16, 1934, concerning the teaching of geography in the elementary and secondary

---

[16]In a Soviet university a department was one unit organized to teach and do research in one specialty. A faculty included several departments, and a school was still larger. The Faculty of Soils and Geography in 1933 included the following departments: soil science, physical geography (Borzov), economic geography (Baranskiy), physical regional geography, and cartography. In 1934 the Department of Economic Geography was split into a Department of the Economic Geography of the USSR (Baranskiy) and a Department of the Economic Geography of the Capitalist Countries (Vitver). In 1938 the faculty was split into the Geography Faculty (Borzov, dean) and the Geology and Soils Faculty (Ryabchikov, 1968:349).

schools. The decree restored the teaching of physical geography, with emphasis on the use of maps. It also specified the teaching of economic geography. On July 14, 1934, the presidium of the Committee on Higher Technical Education of the Central Executive Committee of the USSR specified the kind of geography to be taught in colleges and universities. Here is some of what was specified:

> ... economic-geography teaching should concentrate on a large body of concrete economic geography material, systematically represented on maps. The specifics of economic geography, the location of productive forces, and economic region-alization should be of central concern to teaching personnel. Most of the content of economic geography should be taught on a regional basis. In the course on the economic geography of the U.S.S.R., in particular, at least seventy percent of the time should be devoted to economic regions (quoted by Saushkin, 1966:30–31).

The result of the publication of these decisions was an accelerated growth of the fields included in geography. Even the newspaper *Pravda* published an editorial on September 10, 1937, entitled "Know Your Geography." In both Moscow and Leningrad new departments and institutes were established, as they were also in other universities throughout the Soviet Union. In 1938 the Faculty of Geography at Moscow included departments covering general physical geography, physical geography of the USSR, physical geography of foreign countries, economic geography of the USSR, economic geography of the capitalist countries, geodesy, and carto-graphy. But during and after World War II (by 1948), the Faculty of Geography included the following additional departments: geomorphology, meteorology and climatology, hydrology, geography of the polar lands, geography of soils, biogeography, paleogeography, and oceanography (Ryabchikov, 1968). Furthermore, studies of the geography of population had been included under economic geography. Baranskiy himself offered a seminar on the economic geography of the United States in which several specialists on this subject were trained (Saushkin, 1966:37).

Progress in Physical Geography

In 1963 Innokenty P. Gerasimov, the director of the Institute of Geography at the Academy of Sciences, himself an outstanding physical geographer specializing in the study of soils, presented the following summary of the important progress being made in the various specialties of physical geography and biogeography:

> The particular physical-geographic disciplines made extraordinarily rapid pro-gress throughout the Soviet period, both in the development of theory and in the formation of new approaches and research methods. In climatology, for example, Soviet scholars developed the theoretical principles of forecasting and a typology of climatic phenomena based on dynamic meteorology; they developed the concept of an integrated climatography and, in recent years, gave rise to the promising study of the radiation budget and the moisture cycle and their role in the formation of climates. In the field of hydrology, Soviet scholars worked out the theory of the water budget, the relationships between components (surface water, soil water, and groundwater) and methods of transforming one into the other. Glaciology saw the development of the physical theory of glaciation processes, based on the study of heat and mass exchange in various types of glaciers. Geomorphologists established the dynamic character of many exogenous processes (erosion, deflation, abrasion, etc.) and, on the basis of a general theory of external

and internal forces and the study of recent crustal movements, developed the morphotectonic or morphostructural approach to geomorphology. Soil scientists identified many new soil types characteristic of taiga, desert, and mountain areas, and worked out new approaches (for example, physical and biogeochemical) to the study of the dynamics of soil-forming processes and the circulation of substances in the natural environment. Biogeographers gave emphasis to the ecological and biocoenotic approaches in the study of plant groupings and of animal populations; in recent years these methods have been enriched by an analysis of nutritional relationships and quantitative patterns of formation of biomass in various environments.

. . . Although the achievements of the particular disciplines provided the main elements of modern theory of physical geography, the formation of such general theory was also complicated by the growing differentiation of the physical-geographic disciplines, in which scholars concentrated their attention increasingly on specific components of the natural geographic environment (Gerasimov, 1968b:242; see also Kalesnik, 1958).

Philosophical Discussion

In 1960 many of the differences of opinion regarding the scope and nature of geography that had been smoldering for a long time in spite of official decisions were brought out into the open (Matley, 1966). The lively discussions of the 1960s were started with the publication of a book by V. A. Anuchin entitled *Theoretical Problems of Geography* in 1960. Hooson suggests that this book "for the first time in Soviet history, set out to investigate the theoretical basis of geography as a whole through historical and philosophical analysis" (Hooson, 1962:469). V. A. Anuchin (who is a distant relative of D. N. Anuchin), a vigorous supporter of the idea of a unified geography, attacks both an "inhuman" physical geography and an "unnatural" economic geography. He rejects the idea of geographical determinism, which he equates with bourgeois geography, and he also attacks the other extreme of indeterminism, which he associates with the name of the American geographer Robert S. Platt. The geographical method is best illustrated, he says, in studies of territorial complexes (regions) in which the physical features, the history of settlement, the population, and the economy are in balance.[17]

V. A. Anuchin's book produced an immediate reaction among the Soviet geographers. It was warmly praised by Baranskiy and Saushkin and attacked by Gerasimov, Kalesnik, Konstantinov, and others. Konstantinov characterized it as "unscientific and anti-Marxist," which is a curious echo of his earlier effort to remove economic geography entirely from geography. The methodological controversy continued for about a decade, much of it available for study in *Soviet Geography*.[18]

---

[17]In 1956 V. A. Anuchin published a regional study of Transcarpathia that is an excellent example of the balanced approach to a territorial complex (Hooson, 1959:79).

[18]V. A. Anuchin submitted the book for the higher degree, Doctor of Science (the first degree, independent of the second, is the Candidate of Science degree). The work was rejected in Moscow in 1962. A large number of spectators were present. A further two-day-long defense by Anuchin at Moscow State University in 1964 resulted in his acquisition of the Doctor of Science degree.

Some geographers reacted with impatience. For example, V. V. Volskiy protested that the Anuchin discussions took valuable time away from the important tasks of economic construction (Volskiy, 1963).

Constructive Geography

Constructive geography is geography applied to the practical purposes of building the socialist economy. Geographical concepts and methods have meaning in terms of what they contribute to economic development planning. As a result, the mathematical procedures as developed in regional science in the United States were eagerly adopted in the Soviet Union. The term may be *landscape science* instead of *regional science*. For example, A. G. Isachenko makes use of new terms with old meanings to focus attention on applied geography as landscape science, which he relates to the Dokuchaiev "law of zonality" (Isachenko, 1968). The landscape, he points out, is a dynamic system in which matter and energy are circulating and in which there are rhythmic (seasonal) changes of heat and water balance and biological productivity. A classification of landscapes with a hierarchy of scales or degrees of generalization was worked out for the whole Soviet Union to be mapped on a scale of 1/4,000,000.[19] These landscape definitions will provide an objective basis for defining physical regions. The chief purpose of doing this is not just to provide maps for teaching purposes, but to help identify regions useful for planning purposes.

Landscape science, says Gerasimov, goes back to the ideas of Humboldt and Dokuchaiev (but not Schlüter) that were brought together in the writings of L. S. Berg (Gerasimov, 1968a). A landscape is a combination of interrelated environmental components (local climate, landforms, soils, plants, and animals) occupying a discrete territory. It exists objectively in the natural environment. But studying landscapes just for the purpose of describing them is not enough: Constructive geography must use this knowledge for the effective transformation of nature. Gerasimov offers four examples of constructive geography (Zimina, 1988):

1. The study of the geophysics of natural and cultural landscapes of the wooded steppe in the central chernozem zone. The purpose was to investigate the heat and water budget on the earth's surface in virgin land and in cultivated land and then to experiment with various technical devices for controlling natural processes and so increase agricultural productivity.

2. The study of the irrigated lands of Central Asia. The purpose was to find ways to control salt accumulation, to use water more efficiently, and to increase the irrigated area. Questions investigated include the fate of the Aral Sea and what the drying up of this body of water would mean for the total economy of the region.

---

[19]He suggests that a landscape is made up of discrete parts. For example, the slopes of a valley side or the flat valley bottom are units known as *facies*; the whole valley, combining two or more *facies*, makes up a *urochishche*. This seems to repeat Linton's notion of indivisible unit areas.

3. The study of means of reclaiming the swamps of the Ob Valley by using properly placed dams and diversion canals. The study also includes the hydroelectric potential.

4. Studies of the water in Lake Baikal for the specific purposes of reducing pollution, regulating the flow through the Angara River, and finding new ways of putting this natural resource to better use.

Urban Studies

Until after World War II Soviet geographers had paid little attention to studies of cities. Although as early as 1910 V. P. Semenov Tyan-Shanski had pointed to the need to classify cities in terms of their economic functions, the research studies of the 1920s and 1930s were mostly devoted to problems in physical or regional geography. In 1946 N. N. Baranskiy again suggested the need to develop a method of classifying cities and made reference to urban studies in the United States as examples of what might be accomplished. This time the suggestion was taken up, and a large number of urban studies resulted. In 1962 Y. G. Saushkin wrote that: "The geography of cities is the most rapidly developing branch of Soviet population geography. The literature on methodology, methods for studies of cities, systems of cities of the country as a whole and of its regions, as well as individual cities of the U.S.S.R. is abundant" (Saushkin, 1962:34).

Most of these studies were undertaken for the practical purpose of providing the background for planning projects (Fuchs, 1964). Chauncy D. Harris in a monograph on Soviet cities also reviews the Soviet literature in this field. He reports that by 1970 some 400 Soviet geographers had published studies in urban geography (Harris, 1970:28). He credits O. A. Konstantinov of Leningrad with "playing a leading role in defining the methodology and philosophy of Soviet urban geography" (Harris, 1970:402). Harris summarizes his monograph as follows:

On the basis of data assembled from such diverse sources as the census publications, administrative handbooks, encyclopedias, gazetteers, atlases, maps, and scholarly publications on Soviet urban geography, 30 variables (characteristics) were recorded for 1,247 cities. These data were subjected to statistical analysis [earlier]. Three principal components were found to be highly significant in that they measured a high proportion of the variation for the thirty characteristics analyzed. The first factor to be extracted in the principal components analysis showed highest association with the logarithm of the population in 1959; it is called the *size factor*. The second exhibited the highest association with the logarithm of the urban population potential of each city within its major economic region; it is called the *density factor*. The third revealed highest association with the percentage increase in population 1926–1959; it is called the *growth factor* (Harris, 1970:403).

What Holds Geography Together?

In spite of the integrating effect of focusing geographical studies on practical problems, a tendency to fly apart into separate disciplines appeared from time to time among academic people. Y. G. Saushkin and T. V. Zvonkova identify recurring cycles with an amplitude of some 25 to 30 years during which centrifugal and centripetal tendencies can be observed. It seems that geographers are brought together because of their concern with the geographical landscape (variously defined) and with specific segments of earth space. The processes going on in a landscape are inter-connected spatial systems, and the study of them separately is not the same as the study of them as systems. In studying the economy of an area, full consideration must be given to physical factors. There is also need to study the changes in the landscape resulting from human action, whether planned or unplanned. The basis of physical geography is landscape science. The general and the particular com-plement and enrich each other. Especially in field study does the essential unity of the geographical approach become apparent (Ryabchikov, 1968:354–355).

The training of young people for advanced degrees in universities in what was the Soviet Union as elsewhere in the world perpetuated the recognition of geography as a field of study. Within the system of higher education in this country, geo-graphy was represented in 36 of the roughly 70 universities that trained both research geographers and geography teachers as well as in 74 of the 185 teachers colleges. Economic geography was also taught in a few specialized schools of eco-nomics, such as the Plekhanov Institute of National Economy in Moscow and the Finance-Economics Institute in Leningrad.

Twenty of the 36 universities had separate faculties (schools) of geography consisting of two or more specialized geography departments; in other universities, geography was combined with geology, botany, chemistry, geophysics, or other natural sciences. Russian geographers say that the world's largest educational and research institution in geography is the geography faculty of Moscow University. In 1953 the imposing new buildings of the university were completed on Lenin Hills overlooking the city of Moscow. The faculty occupies six floors, and above them, in the central tower, is the Earth Science Museum. In the mid-1970s the faculty was broken down into 14 departments and 30 teaching and research laboratories. It had a student body of 1700 and a teaching and research staff of 1780, of whom 40 percent were employed on outside contracts.[20] Since then, further growth has been accomplished.

SOVIET GEOGRAPHY

In the early 1980s Soviet geography remained primarily a "natural" rather than a "social" discipline, strongly oriented to serving the practical needs of the planned economy, with a high predominance of applied over theoretical studies. Interesting general observations about the state of Soviet geography in the middle 1980s were made by Roger Brunet (1990) on the basis of the *Directory of Soviet Geographers*, which was edited by Chauncy D. Harris on the basis of materials collected by Theodore Shabad (1988). The directory covers the period 1946–87 and lists more than 3700 Soviet geographers, with information on their location and their specializations.

---

[20]The institutional structure of geography in the Soviet Union is described in *Soviet Geography* 18(1977):540–556, following a useful directory of Soviet geographers, pp. 433–538.

According to the directory, in 1987 3394 persons in the Soviet Union were living geographers.[21] The majority of them—2303 persons (68 percent)—were specialists in physical geography, against only 880 (26 percent) in human geography. The remainder, 211 persons (6 percent), specialized in cartography, databases, history of geography, and teaching. The distribution of specialists within physical geography is also interesting: 41 percent of all physical geographers were found in hydroclimatic studies (climatology, hydrology, limnology, and oceanography).[22] Biogeography and soil studies with 29 percent was the second-largest area of interest in Soviet physical geography; geomorphology came third, claiming 24 percent of all Soviet physical geographers. Soviet physical geographers specialized in studies of the territory of the USSR itself; among regional specialities, polar studies, both Arctic and Antarctic, figured prominently. The profile of Soviet human geography was also very different from its Western counterpart. Among human geographers, 365 were listed as experts in narrow economic specialties, 356 in the general "economic geography,"[23] and another 165 in urban and regional studies, population geography, geography of tourism, and so on. Political, social, and general theoretical issues were little mentioned. The majority of Soviet human geographers were engaged in the study of the USSR, with 120 persons (14 percent) interested in the study of foreign countries. Remote sensing occupied a relatively important place in technical methods, with approximately 100 geographers listing this interest. Computer cartography developed late and was mentioned as a specialty in 1987 by only five geographers.

The regional distribution of geographers followed the pattern typical for Soviet science: It revealed a heavy concentration of researchers in the capitals, Moscow and Leningrad (now Saint Petersburg again). In 1987 a total of 1419 geographers, or 42 percent of those listed in the directory, worked in Moscow, and another 374 persons, or 11 percent, worked in Leningrad. In Moscow geographers were occupied in such main centers of Soviet geographic science as the Faculty of Geography at the Moscow State University (more than 500 researchers), the Institute of Geography of the Academy of Sciences (about 300 researchers), and other research institutes and centers, ministries, and planning agencies. In Leningrad the main centers of geographers' concentration were the Faculty of geography at the Leningrad State University (more than 100 researchers), the Institute of the Arctic and Antarctic Studies, the Institute of Limnology, and others. Leningrad also houses the headquarters of the Soviet (now Russian) Geographical Society. In the list of

---

[21] This directory represents a very broad approach in its selection criteria: Some persons cited here, especially in human geography, were neither trained nor worked as geographers and can hardly be counted as professional geographers.

[22] Traditionally, limnology and oceanography have been considered as geographic subdisciplines in Russian geography.

[23] Since the beginning of the 1930s and until the *perestroika* reforms, the term *economic geography* was often related to the whole domain of human geography. Research carried on under the title of economic geography might be devoted to issues in regional, urban, or even cultural geography.

the main Soviet geographic centers, Moscow and Leningrad were followed by the capitals of other former Soviet republics including Kiev, Tashkent, and Tbilisi. The city of Irkutsk was the main geographic center for Siberia and the Russian Far East, followed by Vladivostok and Novosibirsk. Other important centers of geographic research include Kazan, Voronezh, Obninsk, Perm, Kaliningrad, Rostov, and Yakutsk (the Russian Federation); Odessa, Lvov, and Simferopol (Ukraine); and Tartu (Estonia). Most geographers working in smaller centers teach in pedagogical colleges or work at field stations.

In the first half of the 1980s Soviet geography methodologically followed the trends of the previous decade. In spite of calls from its conservatives "to carry out fundamental and applied research in accordance with the methodology of the macroeconomic plan" (Alayev, 1984), Soviet geography strove to enlarge its scope. In 1982 one of the eminent geographers at the Leningrad State University, Sergey Lavrov (1982), identified two main trends in the development of contemporary Soviet geography: ecologization and socialization. These two terms mean, respectively, the growing interest in issues in nature-society interaction and in the study of social and cultural aspects of human spatial activity. Nevertheless, much space was devoted in methodological debate to the problems of overcoming fragmentation of the discipline and creating a "unified" geography (Gerasimov et al., 1981). The increase in ecological concerns corresponded favorably with the quest to unify elements in geographic research, the historical focus on bioclimatic processes, the longstanding tradition of landscape studies (*landshaftovedenie*), and the composite study of natural complexes, with the stress on links among diverse elements (Hooson, 1968). Expansion of ecological research stimulated, in turn, the development of such fields as remote sensing, mathematical modeling, and the use of system theory (Gvozdetskiy, 1982).

In the 1970s and 1980s Soviet geographers, both physical and human, still devoted a great deal of their work to the problems of regionalization. In fact, the problems of regionalization (*rayonirovaniye*) can be considered one of the most distinctive methodological features of Soviet geography. From the works of Baranskiy, Grigoriev (Zabelin, 1981), and Kolosovskiy came the concept of objectively existing regions, both natural and human. Early in the 1980s a group of geographers headed by Yulian Saushkin from Moscow State University suggested a new natural-economic region grid for the territory of the Soviet Union (Saushkin, 1982). Traditions of the landscape approach were revealed in the creation of the integral landscape map for the whole of the USSR (Isachenko, 1982). The problems of global environmental change were studied by several geographers, most notably a meteorologist, M. Budyko.

Human geography slowly broadened its scope from narrow economic studies to other social and cultural issues such as education, well-being, and leisure time. Among the pioneers in social geography should be mentioned the Georgian geographer V. Gachechiladgze, who conducted research on such politically sensitive topics as the geography of crime and comparative analysis of well-being in the union republics (Gachechiladze, 1985 and 1988) in the 1970s and early 1980s, prior to the *perestroika* and *glasnost* reforms.[24] One traditionally popular field in Soviet

---

[24]perestroika = restructuring; glasnost = openness.

geography, urban geography, still draws considerable attention in the Commonwealth of Independent States. A number of studies have analyzed the system of urban and rural settlement in the former USSR, specifically its economic, social, and ecological aspects. The complex analysis of the Moscow metropolitan region is the most notable example.[25]

The general development of Soviet geography, especially with regard to methodological issues, remained under the control of the governing state ideology until 1985, when Gorbachev's reforms began. These reforms, shaking the very foundations of Soviet society, led to a drastic reassessment of the goals, values, and methods of Soviet science, especially its social and humanitarian branches, which had suffered most from ideological intrusions. Geography was not an exception. In 1986 a group of leading geographers of the Institute of Geography in Moscow's Academy of Sciences confirmed that

> Over the past 30 years, the attention of [Soviet] geography has been focused on relatively narrow issues, most frequently of a practical nature. Their solution was based on conceptions of geographical science which took shape in the thirties and which already do not fully conform to the current level of the science and practical demands. Emphasis in current research must be placed on elaboration of fundamental principles of science which could serve as a basis for the further resolution of problems of the interaction of society and nature (Kotlyakov et al., 1987).

Then the Scientific Council of the Institute outlined the five major research areas for the future development of the Institute's work:

> (1) development of the theoretical foundation of general, complex integrated physical geography as the physical basis for the solution of problems of interaction between society and nature; (2) complex research on the factors and mechanisms of stability and instability of geosystems as a basis for the rational utilization of nature and for environmental protection; (3) discovery of regularities governing the territorial organization of society; (4) the elaboration of theoretical principles for the territorial organization of society; and (5) development of theory and methodology for geoinformation, cartographic modelling, and remote sensing (Kotlyakov et al., 1987).

This plan established a distinctive program for the development of geography. It was clearly focused on the problems of people-environment interaction, use of a system approach, the shift of Soviet physical geography away from the primary task of inventorying the country's natural resources, and suggestions for the most effective use of resources in solving the most crucial ecological problems.

The *perestroika* reforms in the late 1980s continued to bring new issues to geographers' attention. Among the most critical new themes were regional disparities in levels of social and economic development, the question of territorial justice in future development programs, imbalanced development in resource frontier regions, problems of local administrative autonomy, analysis of the administrative-

---

[25]An issue of a Soviet geography magazine was devoted to studies of the Moscow metropolitan region: *Soviet Geography* 29, no. 1 (1988).

territorial division of the country, problems of interethnic conflicts, and, certainly, environmental issues (Lavrov, 1990). One important issue that has drawn particular attention is the Aral Sea problem in former Soviet Central Asia, where poor land use techniques and careless use of water for irrigation purposes caused an ecological calamity affecting the whole region.

The democratization of Soviet society brought some profound changes in the methodology of Soviet geography. These changes included: (1) reappearance of a genuine plurality of viewpoints and unrestrained discussions; (2) restoration of the critical functions of geography as a discipline; (3) free exchange of ideas with Western geography; and (4) rediscovery of the methodological and practical potential of the prerevolutionary Russian geography. Human geography, which suffered the most from ideological imposition, was especially affected. The field of political geography, virtually nonexistent in the USSR, quickly appeared.[26] A monograph edited by L. V. Smirnyagin, N. V. Petrov, and V. A. Kolosov containing the analysis of the first democratic elections to the Soviet Parliament may be considered the first Soviet work in electoral geography (Berezkin et al., 1989). An interest in the new fields of the geography of perception should also be mentioned (Polyan, 1985). The monograph *Regions of the USA* (1989) by Leonid Smirnyagin is a practical as well as a methodological account of the longstanding tradition of Russian and Soviet regional geography.

The disintegration of the Soviet Union at the end of 1991 has created a totally different situation for geographers. Instead of one integrated network of Soviet geography, today there are 15 divisions. An enormous diversity of economic, cultural, and political conditions now prevails. The intensive search for national identities that has begun in the newborn sovereign states will encourage the process of building new national schools of geography in these former Soviet republics (Hooson, 1994). However, as George Demko and Matthew Sadgers (1992) indicate, the common intellectual heritage of Soviet and Russian geography will persist as the major strand of continuity in post-Soviet geography in the former Soviet realm.

THE COMMONWEALTH OF INDEPENDENT STATES[27]

It is thought that the previous geography of the USSR being better funded, and channeled without interference in the interest of a totalitarian state, could accomplish more than geographers scattered in a Russian federation of 15 divisions. Data collected during the existence of the USSR could be shared throughout the country, whereas some of the studies rendered in the post-1991 period tend to be fragmentary.

---

[26]See a special issue on political geography in the USSR: *Soviet Geography* 30, no. 8 (1989).

[27]This summary of post-Soviet geographical developments has been provided by Alexey Postnikov, professor and doctor of science, and deputy director of the Institute of the History and Science of Technology, Russian Academy of Science. Much of the literature cited is in the Russian language but has here been rendered in English.

Against the fragmentation, however, much of the work of the Soviet 1980s retains currency. This owed in considerable measure to the liberating effect of the Gorbachev regime. Yet the Chernobyl tragedy, pollution surrounding Baku, and the reduction of the size of the Aral Sea were only some of the problems contributing to the instability of a multiethnic confederation. Hence, the geography of the post-1991 period benefited from the more liberal inclinations of the 1980s. This includes some of the work accomplished in physical geography, still the leading branch of Russian geographical science. Methods and theory of natural resource study were vital themes in these works, which include Yulig F. Kuizhnikov's *Main Foundations of the Distance Sounding (air-space) Methods for Geographical Studies* (1980) and Yu. A. Isakov, N. S. Kazanskaya and D. A. Panfilov's "Classification, Geography and Anthropogenic Transformation of Geographical Systems (1980)." F. N. Milkov produced the three-volume *Physical Geography* (1987). Also notable were Yu. G. Puzachenko's *Ecoinformation Science* (1987) V. S. Skulkin's *Forest Zones of the USSR Vegetation's Structure: A System Analysis* (1980), M. A. Glazovskay, *A General Pedology and Soils' Geography* (1981). A collective work, *Geography and Natural Resources*, was established by the Siberian branch of the Academy of Sciences of the USSR in 1980. In 1981 a new scientific research vessel was sent on its first expedition. Many years of active Soviet and foreign oceanographic studies were analyzed in O. L. Leontev's monograph *The Physical Geography of the Ocean* (1982). In 1987 two Soviet deep-sea human-manipulated diving vehicles, *Mir-1* and *Mir-2*, achieved a depth of 6000 meters. Later, in 1994, these vehicles were used by the joint Russian-British expedition aboard the scientific research vessel *Academic Mstislav Keldysh* in the middle Atlantic, where a new hydrothermal zone was found, unknown before 1994. During seven years' study aboard the same ship ecologically significant monitoring of the sunken nuclear submarine *Komsomolets* was accomplished. The work was completed in 1995.

During the 1990s a decline in the country's scientific activity relating to resource studies, transportation development, and other practical tasks of the state, meant that opportunities for advancement were lost (Postnikov, 2003). Yet social, human, and economic geography received a stimulus even in the late 1980s due to liberalization of government ideology. Suggestive of this revised viewpoint are Lev N. Gumilyev's *Old Rus' and the Great Steppe* (1987), *Ethno-genesis and the Earth's Biosphere* (1989), and *Ethnos' Geography in the Historical Period* (1990). Further distance from the earlier totalitarian viewpoint came in 1993 with the founding of a new Russian Academy of Sciences and Ministry of Culture Institute of the Cultural and Native Heritage. Increased interest in human geography was demonstrated by publications relating to the Russian history of geography and cartography. These included A. Postnikov's *Russian Maps: A History of Geographical Study and Cartography of Our Country* (1996) and the first full-color facsimile edition of the outstanding example of traditional Russian cartography, *The Drawing Book of Siberia* (2003), compiled in 1701 by the Siberian Semen Ul'yanovich Remezov.

Matters concerning geographical ecology began to assume more significance, especially in regions with problems resulting from environmental damage caused by human misuse. N. F. Glazovsky studied the origins of environmental troubles in *Aralian Crises* (1990). Other geographers, including K. N. D'yakonov, T. V. Zvonkova, and V. G. Linnik, published their results on ecological matters. To help protect the natural environment of the steppe a research institute was founded in Orenburg by the Ural branch of the Russian Academy of Sciences. Further concern for the environment was demonstrated by a long-term glacial project (1974–97) that was summarized in *Atlas of Snow and Ice Resources of the World* and completed under the supervision of V. M. Kotlyakov. Subsequently Kotlyakov published six books in serial form: *Antarctic Glaciology* (2000), *Snow Cover and Glaciers of the Earth* (2000), *Geography in the Changing World* (2001), *Ice and Hypotheses* (2001), *In the World of Snow and Ice* (2002), and *Science is Life* (2003). The latter is not only a history of the life of one of Russia's outstanding geographers but could also be considered a history of Soviet and Russian geography in the twentieth century. In 2002 the scientists who were involved in this project were honored with the State Prize of the Russian Federation.

Technology is also part of Russian geography. Adoption and use of geographic information systems has intensified since the 1980s. Significant works of this genre include Yu. G. Puzachenko's *Ecoinformation Science* (1987), A. M. Berlyants *Image of the Area: Map and Information* (1986), and *Cartography and GIS Science* (1991). The theory of geographical systems as a joint development of organic and non-organic matter in nature was developed by A. Yu. Reteyum in *Winter Worlds* (1988).

With the new era in Russian society, the opportunity for young geographers is brighter than perhaps for many decades (Hooson, 2001). This presupposes the development of strong educational systems, an improving economy to fund serious research and travel for exchange with others, and the opportunity to undertake studies in political geography, which were for so long discouraged, if not prevented.[28]

REFERENCES: CHAPTER 11

Agafonov, N. T., Anuchin, V. A. and Lavrov, S. B. June 1983. "The Present Tasks of Soviet Geography. *Soviet Geography*:411–422.

Alayev, Enrid. 1984. *Soviet Geography Today: Social and Economic Geography*. Moscow: Progress Publishers.

Alexandrovskaya, O. A. 1982. "Mikhail Vasilyevich Lomonosov 1711–1765." In *Geographers: Biobibliographical Studies*. Vol. 6, pp. 65–70. London: Mansell.

———. 1982. "Vasili Nikitich Tatishchev, 1686–1750." *Geographers: Biobibliographical Studies* 6:129–132.

———. 1983. "Pyotr Alexeivich Kropotkin, 1842–1921." In *Geographers: Biobibliographical Studies*. Vol. 7, pp. 57–62. London: Mansell.

———. 1991. "N. I. Vavilov, 1887–1943." In *Geographers: Biobibliographical Studies*. Vol. 13, pp. 109–116. London: Mansell.

Annenkov, V. V., and Demko, G. J. 1984. "Development of Relations between Geographers of the United States and the USSR from the 1950s to the 1980s. *Soviet Geography* 25:749–757.

---

[28]For an assessment of contemporary matters in the C.I.S. see Bater, J. H. 1996. *Russia and the Post-Soviet Scene: A Geographical Perspective*. New York: Wiley.

Anuchin, V. A. 1977. *Theoretical Problems of Geography* (English edition edited by R. Fuchs and G. Demko). Columbus: Ohio University Press.

Bagrow, L., and Skelton, R. A. 1964. *History of Cartography*. Cambridge, Mass.: Harvard University Press (see reference in Chapter 3).

Bassin, M. 1983. "The Russian Geographical Society, the 'Amur Epoch', and the Great Siberian Expedition, 1855–63." *Annals of the Association of American Geographers* 73:240–256.

———. 1991. "Russia Between Europe and Asia: The Ideological Construction of Geographical Space." *Slavic Review* 50, no. 1:1–17.

———. 1992. "Geographical Determinism in *fin-de-siècle* Marxism: Georgii Plekhanov and the Environmental Basis of Russian History." *Annals of the Association of American Geographers* 82, no. 1:3–22.

———. 1994. "Russian Geographers and the 'National Mission' in the Far East." In Hooson, ed. *Geography and National Identity*. Oxford UK: Blackwell, pp. 112–133.

Berezkin, A. V., et al. 1989. "The Geography of the 1989 Elections of People's Deputies of the USSR." *Soviet Geography* 30:607–634.

Berg, L. S. 1950. *Natural Regions of the U.S.S.R.* (English edition edited by J. A. Morrison and C. C. Nikiforoff). New York.

Brunet, R. 1990. "Soviet Geographers." *Soviet Geography* 31:257–262.

Budyko, M. I., Vinnikov, K. Ya., Drozdov, O. A. and Yefimova, N. A. September 1979. "Impending Climatic Change." *Soviet Geography*:395–411.

Chappell, J. E. 1965. "Marxism and Geography." *Problems of Communism* 14:12–22.

———. 1975. "The Ecological Dimension: Russian and American Views." *Annals of the Association of American Geographers* 65:144–162.

Debenham, F., ed. 1945. *The Voyage of Captain Bellingshausen to the Antarctic Seas, 1819–1821*. 2 vols. London: printed for the Hakluyt Society.

Demko, G. 1965. "Trends in Soviet Geography." *Survey* 55:163–170.

Demko, G., and Sagers, Matthew. 1992. "Post-Soviet Geography." *Post-Soviet Geography* 33:1–3.

Esakov, V. A. 1978. "Dmitry Nikolaevich Anuchin, 1843–1923." In *Geographers: Biobibliographical Studies*. Vol. 2 pp. 1–5. London: Mansell.

Esakov, V. A. 1980. "Vasily Vasilyevich Dokuchaev, 1846–1903." *Geographers: Biobibliographical Studies* 4:33–42.

Fedosseyev, I. A. 1978. "Alexander Ivanovitch Voyeikov, 1842–1916." In *Geographers: Biobibliographical Studies*. Vol. 2, pp. 135–141. London: Mansell.

French, R. A. 1968. "Historical Geography in the USSR." *Soviet Geography: Review and Translation* 9:551–553.

———. 1969. "Lifting the Iron Curtain." In R. U. Cooke and J. H. Johnson, eds., *Trends in Geography: An Introductory Survey*. Pp. 268–274. Oxford: Pergamon Press.

Fuchs, R. J. 1964. "Soviet Urban Geography—An Appraisal of Postwar Research." *Annals AAG* 54:276–289.

Fuchs, R. J., and Demko, G. J., eds. 1977. *Theoretical Problems of Geography* by V. A. Anuchin (introduction by D. J. M. Hooson). Columbus: Ohio State University Press.

Gachechiladze, R. G. 1988. "Analysis of Regional Difference in Material Welfare of the Population: The Georgian SSSR." *Soviet Geography* 29:413–419.

———. 1985. "Geography and the Social Consumption Funds." *Soviet Geography* 26:19–25.

Gerasimov, I. P. September 1966. "The Past and the Future of Geography." *Soviet Geography*:3–14.

———. 1968a. "Constructive Geography: Aims, Methods and Results." *Soviet Geography* 9:739–755.

————. 1968b. "Fifty Years of Development of Soviet Geographic Thought." *Soviet Geography* 9:238–252.

Gerasimov, I. P., et al., eds. 1962. *Soviet Geography, Accomplishments and Tasks.* (A symposium of 50 chapters by 56 leading Soviet geographers, edited by a committee of the Geographical Society, I. P. Gerasimov, chairman. English translation by L. Ecker, edited by C. D. Harris.) New York: American Geographical Society.

————. 1981. *Soviet Geography Today: Aspects of Theory.* Moscow: Progress Publishers.

Gerenchuk, K. I. 1980. "On Theoretical Geography." *Soviet Geography* 21:42–47.

Glinka, K. D. 1914. *Die Typen der Bodenbildung, ihre Klassifikation und geographische Verbreitung.* Berlin: Trans. C. F. Marbut, 1927. *The Great Soil Groups of the World and Their Development.* Ann Arbor, Mich.: Edwards Bros.

Gvozdetskiy, N. A. 1982. *Soviet Geography Today: Physical Geography.* Moscow: Mysl.

Harris, C. D. 1970. *Cities of the Soviet Union, Studies in Their Functions, Size, Density, and Growth.* Chicago: Rand McNally.

————. 1975. *Guide to Geographical Bibliographies and Reference Works in Russian or on the Soviet Union.* University of Chicago Department of Geography Research Paper No. 165. Chicago: University of Chicago Press.

————. 1980. *Annotated World List of Selected Current Geographical Serials.* 4th ed. Chicago: University of Chicago Press.

————. 1991. "N. I. Vavilov, 1887–1943." In *Geographers: Biobibliographical Studies.* Vol. 13, pp. 117–132. London: Mansell.

Hooson, D. J. M. 1959. "Some Recent Developments in the Content and Theory of Soviet Geography." *Annals AAG* 49:73–82.

————. 1962. *"Methodological Clashes in Moscow." Annals AAG* 52:469–475.

————. 1968. "The Development of Geography in Pre-Soviet Russia." *Annals AAG* 58:250–272.

————. 1984. "The Soviet Union." In R. J. Johnston and P. Claval, eds. Pp. 79–106. *Geography since the Second World War.*

————. 1994. "Ex-Soviet Identities and the Return of Geography." In *Geography and National Identity.* D. Hooson ed. Pp. 134–140. Oxford: Blackwell.

————. 1996. "The New Political Geography in Russia: Antecedents and Application." In V. Berdoulay and J. A. van Ginkel, eds. *Geography and Professional Practice.* (Nederlandse Geografische Studies 206) Utrecht.

————. 2001. "Geography in Russia: Glories and Disappointment." In G. Dunbar, ed. *Geography: Discipline, Profession and Subject since 1870: An International Survey.* Pp. 225–243. Dordrecht, Netherlands: Kluwer Academic Publishers.

Ilyina, T. D. 1977. "Vladimir Leontyevitch Komarov: 1869–1945." In *Geographers: Biobibliographical Studies.* Vol. 1, pp. 55–58. London: Mansell.

Isachenko, A. G. 1968. "Fifty Years of Soviet Landscape Science." *Soviet Geography* 9:402–407.

————. 1982. "Classification of Landscapes in the USSR." In *Soviet Physical Geography Today: Physical Geography,* compiled by N. A. Gvozdetskiy. Pp. 120–139. Moscow: Progress Publishers.

Kalesnik, S. V. 1958. "La géographie physique comme science et les lois géographiques générales de la terre." *Annales de géographie* 67:385–403.

————. 1968. "The Development of General Earth Science in the U.S.S.R. during the Soviet Period." *Soviet Geography* 9:393–402.

Konstantinov, O. A. 1968. "Economic Geography in the U.S.S.R. on the 50th Anniversary of Soviet Power." *Soviet Geography* 9:417–424.

Kostinskiy, G. 1990. "Regional Preferences of Soviet Youth." *Soviet Geography* 31:732–752.

Kotlyakov, V. M., Preobrazhenskiy, V. S., and Alayev, E. B. 1987. "Major Research Areas in the Institute of Geography in the Twelfth Five-Year Plan." *Soviet Geography* 26:535.

Lavrov, S. B., ed. 1981. *Soviet Geography Today: Aspects of Theory.* Moscow: Progress Publishers.

———. 1982. *Geografiya i Sovremennost* [Geography and the Modern Time]. Leningrad: Leningrad State University Press.

———. 1990. "Regional and Environmental Problems of the USSR: A Synopsis of Views from the Soviet Parliament. *Soviet Geography* 31:477–499.

Markov, K. K. 1968. "Methodological Principles of the Curriculum of a Geography Faculty." *Soviet Geography* 9:358–367.

———. January 1977. "Recollections about Lev Semyonovich Berg (on the centennial of his birth)." *Soviet Geography*:19–22.

———. March 1979. "Soviet Economic Geography: Six Decades of Development and Contemporary Problems." *Soviet Geography*:131–139.

Matley, I. M. 1966. "The Marxist Approach to the Geographical Environment." *Annals, Association of American Geographers* 56:97–111.

———. September 1982. "Nature and Society: The Continuing Soviet Debate." *Progress in Human Geography*:367–396.

Morris, A. S. June 1967. "Mikhail Lomonosov and the Study of Landforms." Institute of British Geographers, *Transactions*, no. 41:59–64.

Murzayev, E. M. 1981. "Lev Semenovich Berg, 1876–1950." In *Geographers: Biobibliographical Studies.* Vol. 5, pp. 1–7. London: Mansell.

Nikitin, N. P. 1966. "A History of Economic Geography in Prerevolutionary Russia." *Soviet Geography* 7:3–37.

Pokshishevskiy, V. V. 1966. "Relationships and Contacts between Prerevolutionary Russian and Soviet Geography and Foreign Geography." *Soviet Geography* 7:56–76.

Polyan, P. M. 1985. "Geography and the Inspirational Resources of Nature." *Soviet Geography* 26:229–238.

Polyan, P. M., and Thomas, C. 1991. "V. P. Semenov-Tyan-Shanskiy, 1870–1942." In *Geographers: Biobibliographical Studies.* Vol. 13, pp. 73–79. London: Mansell.

Postnikov, A. 1995. *The Mapping of Russian America: A History of Russian–American Contacts in Cartography.* Milwaukee: University of Wisconsin Press.

———. 1996. *Russia in Maps: A History of the Geographical Study and Cartography of the Country.* Moscow: Nash Dom—L'Age D'Homme.

Pryde, P. R. 1972. *Conservation in the Soviet Union.* London: Cambridge University Press.

———. 1983a. "The 'decade of the environment' in the U.S.S.R." *Science* 220:274–279.

———. 1983b. Visit by American geographers to the USSR. *Soviet Geography* 24:771–775.

Rakhilin, V. K. 1983. "Alexander Nikolayevich Formozov, 1899–1973." In *Geographers: Biobibliographical Studies.* Vol. 7, pp. 39–46. London: Mansell.

Ryabchikov, A. M. 1968. "Geography at Moscow University over the Last 50 Years (1917–1967)." *Soviet Geography* 9:343–357.

Saushkin, Y. G. 1962. "Economic Geography in the U.S.S.R." *Economic Geography* 38:28–37.

———. 1966. "A History of Soviet Economic Geography." *Soviet Geography* 7:3–104.

———. 1982. "Natural-economic Regions of the Soviet Union." *Soviet Geography* 23:137–148.

Semenov Tyan-Shanski, P. P. 1900. *La Russie Extra-Européene et Polaire*. Paris: P. Dupont.

Semenov Tyan-Shanski, V. P. 1928. "Russia: Territory and Population, a Perspective on the 1926 Census." *Geographical Review* 18:616–640.

Shabad, T. 1983. "The Soviet Potential in Natural Resources: An Overview. In R. G. Jensen, T. Shabad, and A. Wright, eds., *Soviet Natural Resources in the World Economy*. Pp. 251–274. Chicago: University of Chicago Press.

———. 1986. "N. N. Baranskiy, 1881–1963." In *Geographers: Biobibliographical Studies*. Vol. 10, pp. 1–16. London: Mansell.

Shabad, T., and Harris, C. D., eds. 1988. *Directory of Soviet Geographers 1946–1987*. Silver Spring, Md.: V. H. Winston.

Smirnyagin, L. V. 1989. "Rayony SSh'A" Moscow: Mysl.

Spate, O. H. K. 1963. "Theory and Practice in Soviet Geography." *Australian Geographical Studies* 1:18–30.

Szava-Kovats, E. 1966. "The Present State of Landscape Theory and Its Main Philosophical Problems." *Soviet Geography* 7:28–40.

Thomas, C. 1988. "Peter Petrovich Semenov-Tyan-Shansky, 1827–1914." *Geographers: Biobibliographical Studies* 12:149–158.

Voeikov, A. I. 1901. "De l'influence de l'homme sur la terre." *Annales de géographie* 10:97–114, 193–215.

Volskiy, V. V. 1963. "On Some Problems of Theory and Practice in Economic Geography." *Soviet Geography* 4:14–25.

Zabelin, I. M. 1981. "Andrew Alexandrovich Grigoryev, 1883–1968." *Geographers: Biobibliographical Studies* 5:55–61.

Zimina, R. P. and Ya. G. Mashbits. 1988. "Innokenti Petrovich Gerasimov, 1905–1985." *Geographers: Biobibliographical Studies* 12:83–93.

Zvonkova, T. V., and Saushkin, Y. G. 1968. "Problems of Long-Term Geographic Prediction." *Soviet Geography* 9:755–765.

Zvonkova, T. V., and Saushkin, Y. G. April 1977. "Interaction between Physical and Economic Geography." *Soviet Geography* 18:245–250.

12

THE NEW GEOGRAPHY IN CANADA

Journey over all the universe in a map without the expense and fatigue of traveling, without suffering the inconveniences of heat, cold, hunger, and thirst.
—*Cervantes*

In the late nineteenth century Canada was being subjected to exploration. From 1898 to 1902 Otto Sverdrup mapped part of the coast of Ellesmere Island and discovered yet other islands; Roald Amundsen completed the Northwest Passage between 1903 and 1906 and thereby accomplished a feat that had defeated other explorers for some 400 years; Robert E. Peary, arguably, reached the North Pole in 1909; W. H. Hobbs published *Characteristics of Existing Glaciers* in 1911, advancing his glacial anticyclone hypothesis; and Vilhjalmur Stefansson traveled 10,000 miles from 1913 to 1918 over sea and land, reaching a latitude of 81°N and returning to write *The Friendly Arctic* (1921). The people of Canada, situated in northerly latitudes, had a keen interest in these activities relating to the frozen frontier. Exploration of the interior had also been a rugged undertaking. From the 1670s European exploration of the Canadian West was an essential part of the fur trade. Indians, explorers, and voyageurs were active in learning of the physical domain. The accomplishments of Alexander Mackenzie, Philip Turnor, and Peter Pond are well known; accomplishments of the likes of David Thompson are less well known. The latter provided the first comprehensive map of the Canadian West (Jackson, 1998).

The Geological Survey of Canada was founded in 1841 (earlier than the founding of the U.S. Geological Survey) and at once began accumulating information about a little-known land. Confederation came in 1867, and people, knowledge, and literature began a growth process. By the 1890s population density averaged a little more than one person per square mile, which in part bespoke thin soils (the product of glaciation), severe climate, and a neighbor to the south accepting large numbers of immigrants into a less hostile physical environment. Against this backdrop the study of geography could prove valuable.

In the late nineteenth century geography courses were taught in English-speaking Canada in both elementary and secondary schools, and they were modeled largely after the British system. Geography books were written by both British and Canadian teachers. Developments in Quebec were somewhat different, being inspired by French geography and geographers and put into practice by French Canadians. The school systems gave only casual and sporadic attention to geography, and when the subject was taught in the universities, it was based on other than a demand welling up from school geography.

GEOGRAPHY IN FRENCH-SPEAKING CANADA

Geography in the French-speaking schools of Quebec has its roots in that subject as taught in France. It is reported that beginning in the 1660s, the Jesuits offered geography as part of mathematics and philosophy in the schools of New France. The explorer Samuel de Champlain has since been regarded as a practicing geographer (Trudel, 1958), while Sebastien Vauban exerted particular influence on French Canada. The first French-Canadian geography textbook was published in 1804. In 1835 the teaching of geography was recommended in the primary and secondary schools of Quebec (then known as Lower Canada). In 1836 the Montreal Normal School began instruction in geography. Geography was, therefore, not an unknown subject to students in Quebec schools.

In 1877 the first geographical society in Canada was founded in Quebec (Morissonneau, 1971). The society published a bulletin from 1880 to 1934. In 1881 Quebec sent geographers to represent Canada at the Third International Geographical Congress, held in Venice. L. E. Hamelin has suggested that in these years French-Canadian geography was influenced by the migration from Quebec to the United States, by the settlement of the Canadian prairies, and by exploration and colonization in the subarctic region of Quebec. Henry Laureys, a Belgian scholar, was appointed the first professor at the School of Higher Commercial Studies (affiliated with the University of Montreal) in 1910. In 1914 he published *Essai de géographie du Canada* in Brussels, and it was adopted as a textbook in Quebec. Although geography was not at that time given departmental status, staff were added. Benoit Brouillette, a Canadian who was formally educated in the French university system, joined the staff in 1931, and Raymond Tanghe was added to the faculty after he completed his graduate work in France.

After World War I geography was taught in other faculties of the University of Montreal. Courses were established in the Faculty of Social Sciences in 1921 by Emile Miller; his *Terres et peuples du Canada*, a pioneer text, facilitated the study of Canadian geography in the classroom. From 1925 to 1927 Jean Brunhes, one of the best known French geographers, offered geography at the University of Montreal, working with his well-known book, *Géographie humaine*.

Another renowned geographer from France, Raoul Blanchard, made annual visits in the years 1929–38 and 1945–49, during which he undertook field work and sometimes lectured. Peter Nash has observed that Blanchard traversed the Atlantic by ship 20 times, spending much of his time during his 15 visits to Canada in Quebec. He was the last surviving protégé of Paul Vidal de la Blache and the champion of Francophone Canadian geographers. Blanchard was an influential figure in Canadian geography through his students, courses, and literature. His publications numbered more than 400 and were largely written at Grenoble, Quebec, and Cambridge (at Harvard); his volumes on the French Alps and on the Province of Quebec are especially notable. His methods, philosophy, and student-disciples are testimony to his powerful presence. The work of Raoul Blanchard was recognized in a special issue of *Cahiers de Géographie de Québec* in 1986 on the 50th anniversary of the publication of his first book on French Canada.

Further links with France came with Pierre Deffontaines, who brought French human geography (rather than regional geography) to Quebec; he made extensive contributions over many years (Hamelin, 1986). Both Henri Baulig[1] and André Siegfried visited Canada, Siegfried coming on several occasions prior to 1940. Baulig wrote the prodigious *Amérique septentrionale* in 1935 and 1936 (two books constituting one tome), and Siegfried published *Le Canada, puissance internationale* (1937) translated as *Canada* (1937). Although both of these books were published in Paris, they were read widely in Quebec, and also by an international community. The two books by Baulig were considered to be among the very finest examples of regional geography as it was then practiced. Other French geographers had previously published works on Canada, most notably E. Reclus (Joerg, 1936).

Despite these contacts with academic geography in France, geography was not established at the graduate level in Quebec universities. It was offered as part of general education for businesspeople and other French-Canadians, where it had very considerable influence.

GEOGRAPHY IN ENGLISH-SPEAKING CANADA

The first appearance of geography in a university curriculum in English-speaking Canada was at the University of New Brunswick in 1800 (Williams, 1946). New Brunswick also appointed the first provincial geologist in 1838. In 1841 the Canadian Geological Survey was organized under Sir William Logan. This development is of interest to more recent geographers because the members of the survey provided data via exploration on which regional description early in the twentieth century would rely. Geography was offered in the secondary schools in the mid-nineteenth century and perhaps owed something to the American example (Mayo, 1965). Prior to 1850 Jedidiah Morse's books were widely used in Ontario (then known as Upper Canada). However, Morse's work and other American textbooks gave little attention to Canada and presented an antimonarchic viewpoint. As a result, textbooks from Ireland were introduced into Upper Canada. One of the first of these textbooks was *Geography Generalized* by Robert Sullivan (Croal, 1940). The *Irish National Readers* were introduced into Upper Canada and by 1860 may have been the chief source of geographic information for pupils. In 1857 one of the first Canadian geography texts was published: *The Geography and History of British America* by J. C. Hodgins. Much attention was given to physical geography, but by 1896 commercial and economic geography began to enter the Ontario school curriculum.

After 1900 the Committee of Ten exerted an influence on Canadian geography. Ralph S. Tarr's textbooks were adopted in British Columbia after 1900, but within a few years physical geography in the high schools of both Canada and the United States was in decline. In British Columbia the course in physical geography was absorbed into general science and was dropped from the high school curriculum by 1920–21. In Ontario, however, a textbook entitled *Ontario High School Physical Geography* was extended through several revisions and was studied by several

---

[1] Henri Baulig studied with and then associated with W. M. Davis at Harvard, 1904–10. At that time he traveled North America extensively.

generations of students. This course survived in Ontario until 1937, when it was finally dropped. In addition, in Ontario economic geography entered the curriculum in 1896, and by 1914 the first years of high school required Commercial and Map Geography, a course that emphasized the study of Canada and the British Empire. Later, George W. Cornish of the Ontario College of Education, University of Toronto, wrote a series of texts for the course that emphasized commodities, Canada, and empire. The social studies movement that emerged in the United States in the late teens had less impact in Canada. Although the subject lost ground in the secondary schools, it at least remained at the elementary level. Teachers of geography were needed, and, in the absence of university departments of geography, they were hard to produce. Nevertheless, Canadian students seem to have studied more geography than their American counterparts. In a 1923 study in which approximately 90,000 students in six provinces were involved, nearly 36,000 high school students were enrolled in one or more geography courses, and almost 14,000 more were taking a course in physiography. If physiography is included in geography, more than half of the total sampled were taking a course in geography (Dyde, 1929).

Of considerable significance were the books relating to the geography of Canada that were published in English. Especially noteworthy were *The International Geography* by H. R. Mill (1899), which included "The Continent of North America" by W. M. Davis and "Dominion of Canada" and "Newfoundland and Labrador" by J. B. Tyrell (formerly of the Geological Survey of Canada); *North America* (1901), an anthology assembled by F. D. and A. J. Herbertson; and the extended Canadian entry in *The Oxford Survey of the British Empire* (A. J. Herbertson and O. J. R. Howarth, eds., 1914).

GEOGRAPHY IN THE FEDERAL GOVERNMENT OF CANADA

In the federal government in Ottawa, geography referred to work with maps and atlases. In 1880 John Johnson was appointed the first federal government geographer. He was in charge of all government cartographic work, which later led to production of the first *Atlas of Canada* under the direction of James White, then chief geographer, in 1906. This was a superb cartographic accomplishment and was the second national atlas produced anywhere in the world. The atlas was revised and updated under the direction of J. E. Chalifour in 1915. The third edition of the *Atlas of Canada* was published in 1957. It was a more comprehensive work than previous editions owing to larger amounts of data being available. The fourth edition of this national atlas was published as a set of 127 separate maps between 1969 and 1973, also published as a bound volume in 1974. A fifth edition comprised 93 separate maps published between 1978 and 1995. Each of these editions made use of the most recently available data. The collective editions of this atlas have doubtless been of inestimable help in economic development since 1906.

Canada was represented at the series of international geographical congresses that had begun in Antwerp in 1871. Sir Sanford Fleming, chancellor of Queen's University, surveyor and cartographer, was one of the attendees of the Venice Congress of 1881. He was arguably the most strenuous supporter of the Greenwich Meridian as the prime meridian, and he is often credited with the conception of the modern system of standard time zones (Nicholson, 1959).

Canada also took part in international geographical meetings through federal government representatives. The country was represented at the International Geographical Congresses in Paris in 1889 and in London in 1895. In London Sir Charles Tupper, then high commissioner of the United Kingdom and one of the Canadian delegates, was elected one of the honorary vice presidents.

Canadian participation in the International Geographical Congresses and the Union (the latter formed in 1922) also involved the government. In 1904 six Canadian geographers participated in the congress held initially in Washington, D. C., and for the first time presented papers. Thereafter Canadians participated regularly. Canada officially joined the International Geographical Union in 1934. The Canadian Geographical Society, which had been formed in 1929, was Canada's official voice. Montreal would host the Twenty-second International Geographical Congress in 1972. Six regional volumes were published to commemorate the congress (Trotier, 1972). Two years later, in 1974, the Canadian Association of Geographers and the Canadian Committee for Geography prepared a special issue of *The Canadian Geographer* to mark the occasion of the Twenty-third International Geographical Congress, held in Moscow in the summer of 1976.

Geography in other government agencies became known. For example, the Natural Resources Intelligence Service, established in 1917, published many booklets relating to the economic and industrial geography of Canada and its regions. Although none of the geographers in the service had been professionally trained, these publications were intensely geographic and were read widely. It was not until 1943 that a branch of the federal government hired its first professional geographer. The lack of geographical expertise in government was noted by H. L. Keenleyside, deputy minister of Mines and Resources. In the summer of that year, Trevor Lloyd, a Canadian lecturing in geography at Dartmouth College, undertook work for the Canadian Wartime Information Board. A few months later the Northwest Territories Administration of the Department of Mines and Resources appointed J. Lewis Robinson "to compile, organize and analyze information about Northern Canada for wartime purposes and peacetime development." In 1945 an assistant was added to the staff and helped with the publication of articles relating to the geography of northern Canada. In 1947 a Geographical Bureau in the Department of Mines and Resources was established. Trevor Lloyd, on leave of absence from Dartmouth College, was the first director, and in 1949 J. Wreford Watson (previously head of the Geography Department at McMaster University) became chief of the Bureau. In 1950 the Bureau was raised to the category of one of five branches in the reorganized Department of Mines and Resources. The Joint Intelligence Bureau was established in 1947 in the Defence Research Board, Department of National Defence. The directors of the branches—Mines, Surveys and Mapping, Observatories, Geological Survey, and Geographical—reported directly to the deputy minister. This arrangement persisted until the late 1950s, when the post of director general of scientific services inserted a step between the deputy minister and the directors. The administrative hierarchy became more complex, although the potential for geographic contribution remained impressive. Already in the 1950s branch workers were expected to work toward a doctorate.

The influence of the Geographic Branch on the development of geography in Canada was wide ranging. By the early 1960s geography in the universities had begun to grow more important. This was a happy development, for exploratory government geography was beginning to diminish as funds were reduced. Regional studies were giving way to quantitative studies, the DEW (Distant Early Warning) Line had focused attention on the north, and geographers were past the innovative stage of air photo and mosaic methods. Meanwhile, the government was demanding more practical results from the programs it supported. The branch was heavily involved in academic glaciology and glacial geology in the Arctic but gave little attention to southern Canada, where most Canadians lived. In 1967 the Geographical Branch was terminated, and personnel and functions were shifted elsewhere in other government departments. Many geographers went to a new department, which was called Environment Canada, in 1970. Initially, it discharged traditional services such as the Meteorological Service, Forestry, and Fisheries. However, after Kenneth Hare assumed the position of science adviser, geographers were viewed as useful scientists and were employed in the Lands Directorate, in glaciology, climatology, and parks. What seemed to pose a disaster for the discipline when the Geographic Branch was terminated emerged as a blessing because geographers were soon dispersed throughout some 30 government departments and did much useful work. The departments included Statistics, Regional Economic Expansion, Urban Affairs, Indian Affairs, and Northern Development (Fraser, 1983). Worthy of special mention is the *Glacier Atlas of Canada* (1969–72). This was initiated as part of Canada's contribution to the International Hydrological Decade (1965–74) program for a world inventory of glaciers. Data concerning Canada's ca. 100,000 glaciers was established and the atlas was developed, though not completed. A few bound limited edition copies have been deposited in selected reference libraries. The work was begun in the Inland Waters Branch of government and later transferred to the Department of the Environment.

Geographers were also employed by the provincial governments, and ever more so in the area of planning and development, land utilization, and conservation. By the early 1970s geographer Pierre Camu was president of the St. Lawrence Seaway Authority and administrator of the Canadian Marine Transportation Administration. Numerous other geographers held governmental posts of responsibility.

Meanwhile, Isaiah Bowman, who was born in Canada and who was director of the American Geographical Society, had been developing a research program concerning Pioneer Belts and Pioneer Fringe throughout the world. Bowman had felt that geographers should head this project and that the frontier was a priceless laboratory. By September 1928 the Pioneer Belts Committee of the Social Science Research Council had agreed to help sponsor a Canadian-American Pioneer Belts Project, the funding and control of which resided with the American Geographical Society. In fact, the Council of the Society made funds available conditional on a pledge from Canadian authorities that they would complete the Land Classification Survey in the northern portion of the prairie provinces. William A. Mackintosh of Queens University, Kingston, Ontario, was appointed director of the Canadian

Pioneer Problems Committee, a distinguished group that included Dawson (McGill), Martin (Manitoba), Lower (Queens), and Innis (Toronto). In this committee geographers worked with economists, historians, and sociologists. An advisory committee included Frederick Merk (chairman), Oliver E. Baker, Kimball Young, and W. L. G. Joerg. The work progressed in admirable fashion, with 13 authors participating in the publication of 3000 pages (in eight volumes) concerning frontier settlement and the exploitation of unsettled areas (Martin, 1980). The series of books entitled "Canadian Frontiers of Settlement" proved valuable to the Canadian government and contributed to the growth and development of the country in a variety of ways. It helped the cause of geography in Canada at a critical time in its history. Clearly, administrations were beginning to grapple with the place of geography in the university.

THE DEVELOPMENT OF GEOGRAPHY IN CANADIAN UNIVERSITIES

The University of British Columbia first offered geography courses in 1915 in the Department of Geology and Mineralogy. The geography offered was largely physical geography, and geologists such as Stuart J. Schofield, Edwin T. Hodge, and Reginald W. Brock taught it. In 1922 the title of the department was changed to Geology and Geography, and more courses were added. Brock died in 1935, and Gordon Davis was hired to teach the three geography courses in the curriculum.

As early as 1895 George M. Wrong, a historian at the University of Toronto, had offered a geography course, one part of which was entitled Commercial Geography, with the second part relating to transportation (Joerg, 1936). Several other courses relevant to geography were offered in the Department of Geology, and later the Department of Political Science offered several courses of a geographic nature including Economic Geography. Geography textbooks (such as *Physical and Commercial Geography* by H. E. Gregory, A. G. Keller, and A. L. Bishop) were adopted. But ultimately the interests and work of Harold A. Innis served as the catalyst that led to a Department of Geography. Innis perhaps remains better known as an economic historian, but he was intensely interested in geography. He attended the International Geographical Congress, Cambridge (England), in 1928 and visited geography departments in England and Germany. That autumn his academic title was changed to assistant professor of economic geography. Innis had completed a Ph.D. at the University of Chicago (1920) and retained connections with that university. After learning of Griffith Taylor's arrival in Chicago in 1928, Innis began to negotiate for Taylor's leadership of geography at Toronto.

Taylor had become chairman of the first university Department of Geography in the Southern Hemisphere in 1920 at the University of Sydney, Australia. Since that time he had written several works on a variety of subjects. In particular, he had perceived that the "boomer" position of the Australian government was ill founded. (The Australian government encouraged immigration to the island-continent without understanding the limitations imposed by the environment.) His opposition on this matter, coupled with a sabbatical leave and the opportunity to associate with the department of Geography at the University of Chicago, led him to Chicago in 1928, where he stayed for the next seven years. His creativeness and his prolificness were remarkable. Taylor was invited to give lectures at Toronto in 1935, and in the following year the formation of the Department of Geography was announced. It was the first such department created in a Canadian university (pursuant to the binomial Department of Geology and Geography at the University of British Columbia in 1922). His first two assistants were Andrew H. Clark and Ann Marshall, both of whom later became professional geographers. By the late 1930s Taylor had added Donald Putnam and George Tatham to the geography faculty, and they jointly lectured to more than 600 students a year. By 1940 an honors course in geography had been established, and graduate students were being admitted. The first doctorate in geography was awarded to Chun-Fen Lee (1944), and the following year the second doctorate went to J. Wreford Watson, then at McMaster University.

Other Ontario universities also started geography programs in the late 1930s. In 1938 Edward G. Pleva, from the University of Minnesota, began giving geography lectures in the Geology Department at the University of Western Ontario, in London. In 1939 at McMaster University, J. Wreford Watson from the United Kingdom offered geography courses. As enrollments increased, a Department of Geography was announced at McMaster University in 1942.

During World War II, the lack of geographers in Canada became apparent. This fact, coupled with the knowledge that world understanding was a prerequisite to lasting peace, encouraged the establishment of new departments after the war and the expansion of those that had already been established. In 1945 McGill University founded a department led by George H. T. Kimble from the United Kingdom, and in the next year F. Kenneth Hare and J. Ross Mackay were added to the faculty. In 1947 the McGill Summer School of Geography was opened in Stanstead, Quebec, bringing in leading U.S. and British geographers as summer lecturers. Later, the headquarters of the Arctic Institute of North America was established on the McGill campus. In 1950 Kimble left to become director of the American Geographical Society, and Hare became chairman of the department.

By 1950 there were eight geography (or joint) departments across Canada, and graduate work was offered in six of them.[2] During the 1950s geography departments were established in all the major universities, and in the 1960s geography was added in many of the smaller universities. By 1961 there were 19 geography departments in Canada, and by 1966, 35 universities included geography departments or geographers. In 1966 student enrollments were substantial. Led by the Universities of

---

[2]The sequence for the foundation of geography departments in Canada is: Toronto, 1935, Western Ontario, 1938, McMaster, 1942, McGill, 1945, Laval, 1946, British Columbia, 1946, Western Ontario, 1946, Montreal, 1947, Victoria, 1949, Ottawa, 1951, Manitoba, 1954, Edmonton, 1955, Carleton, 1957, Waterloo-Lutheran (Wilfred Laurier University), 1959, Queens, 1960, Saskatchewan, 1960, Waterloo, 1962, York, 1962, Victoria, 1962, Calgary and Sherbrooke, 1963, Brock, Winnipeg, and Simon Fraser, 1964, Windsor, 1965, Regina, 1966, and Trent, 1968. (The University of British Columbia offered geography in 1922, but there was not a department of geography.)

British Columbia, Toronto, and Alberta, undergraduate enrollment reached a total of 19,148 (Robinson, 1967).

Graduate enrollments also increased. Beginning in the 1960s, Canadians began to remain in Canada for their graduate education instead of studying in the United States or the United Kingdom. In 1966 there 304 graduate students in residence in English-speaking Canadian universities; of these, 243 students were studying for an MA degree and 64 were in Ph.D. programs. By 1967 some 20 geography departments throughout Canada were offering graduate-level work. McGill, Western Ontario, British Columbia, and Toronto had the largest number of graduate students. Correspondingly less graduate work was undertaken in the United States. A considerable number of these students came from Britain; they were initially attracted by scholarships for graduate study and then by opportunity in the new land.

Canadian departments expanded in the 1970s. The number and variety of graduate specializations increased in Canada, and strengths were developed in most branches of the discipline. The larger departments increased from 15 to 20 faculty members each to 30 and more by 1990, a figure larger than in most U.S. universities. This large size ensured production of larger numbers of both undergraduate and graduate students, and ever increasing specialization. By the year 2000 there were 48 university geography departments in Canada, with many departments including 15 to 25 members, the University of Toronto being the largest with 34 members, and the University of British Columbia with 33 (though, of course, these numbers are subject to change). As late as 1985–86 more professional geographers in the universities were socialized outside Canada than in it, and a majority also achieved their highest degree beyond Canadian boundaries (Porteous and Dyck, 1987). The process of Canadianization had begun to take place. This process had been speeded up in the 1970s when the Canadian government decreed a hiring policy of "Canada first." This policy reduced the number of faculty imported from the United Kingdom and the United States. Later, this policy was reviewed and revised.

GEOGRAPHICAL ASSOCIATIONS, SOCIETIES, AND PUBLICATIONS IN CANADA

Geographical societies and associations were established in Canada and published journals that would provide an outlet for the work of Canadian geographers and others. For example, the Royal Canadian Geographical Society was founded in 1930 and published the *Canadian Geographical Journal*,[3] a journal oriented to popular tastes. The Canadian Association of Geographers was established in 1951, and its publication was *The Canadian Geographer*. The first president was Donald F. Putnam from the University of Toronto; Raoul Blanchard and T. Griffith Taylor were designated honorary presidents in 1952. Two presidents of the Canadian association, F. Kenneth Hare and J. R. Mackay, were also made presidents of the Association of American Geographers. This association held its annual meetings in Montreal

---

[3]This publication, now entitled *Canadian Geographic*, had a circulation of 195,000 (1990). It is estimated that about 1 million people read *Canadian Geographic*.

and Toronto in 1956 and 1966, respectively, and in 1961 Canadian-born Andrew H. Clark was made honorary president of the association.[4] In the first 25 years of the Canadian organization, 7 of the 25 presidents were French-speaking geographers from Quebec universities.[5] The association's original policy was to rotate the presidency between English-speaking and French-speaking academic geographers. This bilingual policy broke down in the 1970s, when the English-speaking membership far outstripped the French-speaking. In 1972 the Canadian Association of Geographers granted its first awards for scholarly achievement and service to the profession.[6] The association initially published an annual *Newsletter*, which was later replaced by a *Directory*. In 1983 the association started a second publication, *The Operational Geographer*, which was aimed primarily at geographers in nonacademic positions. The Federal Geographical Branch published *The Geographical Bulletin*. French-Canadian geographers published *Cahiers de géographie du Quebec* and *La Revue canadienne de géographie*, both of which were written mainly in French, although some articles were in English.

The association also was responsible for a number of excellent publications. In 1976 while the Société de Géographie du Quebec was celebrating its centennial, the Canadian association was celebrating its 25th anniversary. Honoring that occasion, Hamelin and Beauregard provided *Retrospective, 1951–1976* (1979). Recognition of Canada's centennial inspired the very successful *Canada: A Geographical Interpretation* (John Warkentin, ed., 1967). Then, planning for a series of works in the 1980s led to a fine four-volume Canadian Association of Geographers Series in Canadian Geography, for whom the editor was R. Cole Harris. These were published between 1993 and 1997 and include *Canada's Cold Environments* (eds. French and Slaymaker, 1993); *The Changing Social Geography of Canadian Cities* (eds. Bourne and Ley, 1993); *Canada and the Global Economy: The Geography of Structural and Technological Change* (ed. Britton, 1996); and *The Surface Climates of Canada* (eds. Bailey et al., 1997).

Several universities also published discussion papers, and monographs were established at different institutions, for example, *The Ontario Geographer* (University of Western Ontario), *The Alberta Geographer* (University of Alberta, Edmonton), and *The Western Geographical Series* (University of Victoria).

THE GROWTH OF CANADIAN GEOGRAPHY SINCE 1935: AN OVERVIEW

Canadian geography seems to have passed through three distinct phases. The first was led by the Department of Geography at Toronto. Here T. Griffith Taylor

---

[4]William F. Ganong was the first Canadian to be invited to join the Association of American Geographers as a founding member; he declined.

[5]P. Dagenais (1952); B. Brouillette (1954); P. Camu (1956); F. Grenier (1964); L. Beauregard (1968); L. E. Hamelin (1971); L. Trotier (1974).

[6]In 1990 the association had more than 1100 full members and several hundred student members. By the year 2003 that figure was approximately 800, with fewer student members.

initiated a series of papers on Canadian geography much as he did for Australian geography earlier in the century. In 1936 Taylor published "Fundamental Factors in Canadian Geography," an excursus in geological control that relates topography, climate, and communications. His physiographic division of Canada, supportive of the regional position, remains useful to this day. As was his style, the paper was speculative and assertive on the issue of population distribution, with Taylor suggesting that ultimately Canada might support a population of 100 million people. This article proved to be the genesis of a sustained geographic literature on Canada. One year later, in 1937, Taylor wrote "The Structural Basis of Canadian Geography," which demonstrated Canada's place in the "World Plan." Two further articles were "Topographic Control in the Toronto Region" and "Climate and Crop Isopleths for Southern Ontario." In 1937 he published *Environment, Race and Migration: Fundamentals of Human Distribution, with Special Sections on Racial Classification and Settlement in Canada and Australia*. This was a revision of *Environment and Race: A Study of Evolution, Migration, Settlement of Status of the Races of Man (1927)*. The revised book, speculative and energetic, attracted both attention and criticism in the review literature. Taylor was a pioneer, with many hypotheses he was anxious to advocate and propound but less anxious to confirm. Such a geography was bound to attract attention; it was unlikely to produce a school, but it could and did generate student and other admirers. The spectacle of this scholarship brought attention to geography in Canada. He then wrote "Structure and Settlement in Canada" (1940), "The Climates of Canada" (1941), *Canada's Role in Geopolitics, a Study in Situation and Status* (1942), "Towns and Townships in Southern Ontario" (1945a), "A Mackenzie Domesday, 1944" (1945b), "A Yukon Domesday, 1944" (1945c), and a host of other articles. Meanwhile, he had produced *The Geographical Laboratory, A Practical Handbook for a Three Years' Course in North American Universities* (1938), *Canada, Study of Cool Continental Environments and Their Effects on British and French Settlement* (1947), and *Canada and Her Neighbors* (with D. J. Seivewright and T. Lloyd, 1947). The *Practical Handbook* was widely used by students, and his *Canada* was widely read in the universities and beyond. Significant, too, was his interest in the Arctic, especially in the years of World War II, which encouraged others to share and further this work. It is interesting to observe that each of the established national geographies has developed around the work of one (or more) outstanding individuals; for Canada, that individual was Taylor.

Taylor brought Donald F. Putnam (1938) and George Tatham (1939) into the department at Toronto. Putnam, a native of Nova Scotia, had a splendid grasp of agriculture and physiography. He was a first-rate field worker, and he also contributed substantially to an understanding of Canada's regional geography. His *Physiography of Southern Ontario*, published in 1951 and written with Lyman Chapman (of the Ontario Research Foundation), is a book of exceptional merit, still of value today. Tatham, educated at Liverpool University, where he was much influenced by P. M. Roxby, taught in the British university system until he moved to Toronto in 1939. There he developed courses on human geography, Europe, the history of geographic thought, political geography, and topographic map interpretation. His two essays in Taylor's *Geography in the Twentieth Century* (1951)—"Geography in the Nineteenth Century" and "Environmentalism and Possibilism"—live on, a tribute to their

Raoul Blanchard

Louis-Edmond Hamelin

F. Kenneth Hare

R. Colebrook Harris

Peter H. Nash

Donald F. Putnam

George Tatham

T. Griffith Taylor

excellence. The significance of this department in the rise of geography as a university subject in Canada cannot be overestimated. Taylor noted that development in an unpublished document, "The First Fourteen Years." His insistence on extensive and continuous field and laboratory study rendered Canadian geography somewhat different from American geography. Putnam's life and accomplishments have been well recorded (Putnam and Sanderson, 2000) and a biobibliography published (Sanderson, 2001).

Taylor's leadership of geography at Toronto was an essential part of Canadian geography, although it was not the whole of it. Physical geography and regional geography were vigorously developing, and British influence, learning, and personnel continued. However, there was room for models that were already being adopted and philosophy that was being developed. It was a beginning. But as J. W. Watson has stated, "Canadian geography prided itself on *real* and not *theoretical* geography."

The second phase of Canadian geography began in the 1950s. At this time regional geography began a modest decline. Systematic geography became fragmented, and cartography, historical geography, and applied geography surged in popularity. Most of the geography retained a close proximity to the real world. There was little activity in the theory of methodology. The declining concern with the regional enterprise led J. Lewis Robinson to urge Canadian geographers to return to the regional concept (Robinson, 1956, 7, 49). In his presidential address before the Canadian Association of Geographers, "Geography and Regional Planning" (Robinson, 1956, 8, 1–8), he demonstrated that despite a respect for regional geography, little regional work was being accomplished. More and more, specialization had become the order of the day. A considerable literature emerged concerning the justification of geography. Hamilton (1950, 1, 7–10) suggested that the schools emphasize the regions of Canada and that the universities develop systematic geography. The Canadian Association of Geographers established an education committee to investigate goals recognizing that the success of university-level geography depended on the success of geography in the high school. T. Hills (1957, 9, 55–59) accepted geography as "the study of the world to show how physical and human phenomena are related in space and how variations in these relationships give rise to distinctive regions of landscapes, which are the core of geographic study." School and university geography were thoughtfully related to each other and were not separated as they were in the United States.

One paper that commanded attention was that by Pierre Dagenais (1953, no. 3, 1–15): "The Status and Tendencies of Geography in Canada." He stated that geographers should emphasize systematic rather than regional geography and seek "scientific precision." He cited Andrew Clark's survey of university theses in Canadian geography, which revealed that 167 of 181 theses were regional. By then, however, trends were changing. In the early to middle 1950s, approximately 10 percent of the articles published in *The Canadian Geographer* were regional, but nearly two-thirds were systematic. In this same period, there was little emphasis on theory, methodology, or qualitative analysis. A new point of view arrived in 1958 with "The Interactance Hypothesis" (1958, 11, 1–8) by Ross Mackay. This essay demonstrated that interactions could be studied by way of a theoretical model expressing relationships between variables. This work marked the beginning of quantitative geography in Canada. The quest for precision was underway, although it lagged behind the movement in the United States. Geographers began to study quantitative methodology and thereby became more involved in the study of mathematics than previously. Several impressive contributions were published in *The Canadian Geographer*, namely, Lukermann on probability theory (1965), Werner on the law of refraction in transportation (1968, vol. 12) and two models from Horton's Law (1972), Curry on spatial regression (1972) and Wayne Davies on factorial ecology (1973). Ian Burton (1963) had already published "The Quantitative Revolution and Theoretical Geography."

In addition, geography departments became heavily involved in preparation of provincial atlases and in city, regional, and provincial planning. Physical geography remained a traditional area of strength. Less activity was evident in foreign area studies, economic geography, and cultural geography.

After the 1970s studies concerning geography as a discipline became more numerous, and geographers involved themselves in the process of explanation. F. E. Lukerman offered "Geography as a Formal Intellectual Discipline and the Way It Contributes to Human Knowledge." (Though not Canadian, Lukermann was well received in Canada.) Edward C. Relph published "Relation Between Phenomenology and Geography" (1970:193–201). L. Guelke offered "Problems of Scientific Explanation in Geography (1971:38–53), and D. J. Walmsley wrote "Positivism and Phenomenology in Human Geography" (1974:95–107). Geographers sought to offer explanation and the contribution of this to philosophical discourse. Much less attention was accorded the history of geographical thought. In this important period specialization began to characterize both teaching and research; this development led to divisions of knowledge among coworkers. Larry S. Bourne, D. Michael Ray, James W. Simmons, Paul Villeneuve, and Maurice Yeates borrowed from quantitative analysis and wrote about the structuration of Canada's future urban society. Urban geography became very significant in Canadian geography, and many excellent contributions were made.

In recent years some interesting urban research has been undertaken on the differences between Canadian and U.S. cities. Perhaps Canadian geographers have too long used the model of the American city to help them understand the Canadian city. The legitimacy of this approach was strongly challenged by M. A. Goldberg and J. Mercer in *The Myth of the North American City: Continentalism Challenged*, a theme Bunting and Louise Filion elaborated on in *The Monograph* (1988). Of much value was the *Canadian Geographer*'s special volume for 1967 commemorating the centenary of the founding of Canada. Numerous articles were written about Canadian geography (W. A. Dean), population changes (T. R. Weir), geography in universities (J. Lewis Robinson), geography in secondary schools (R. G. Putnam), historical geography (R. Colebrook Harris), man and landscape (J. G. Nelson), quantitative geography (Leslie Curry), geomorphology (J. T. Parry), biogeography (John M. Crowley), resource development policy (Derrick Sewell and Ian Burton), urban geography (James W. Simmons), geography and planning (John N. Jackson), and economic geography (Richard S. Thoman). This publication constitutes a form of "Inventory and Prospect"; certainly, it provides excellent coverage for the years 1950–67.

Later, analyses of trends in *Canadian Geography* were offered by J. Lewis Robinson (1986:9, 15–18) and J. W. Watson (1981:393–398). Canadian geographers have focused heavily on Canadian geography and physical geography (largely geomorphology and climatology) and have shown rigorous attention to urban studies and articles employing mathematical symbols. P. J. Smith (1975) elaborated the role of geographer and urban planning in the urban planning process and encouraged geographers to contribute to the task. Ralph R. Krueger urged integration of the best in the old and the new geography and unity within diversity among geographers (Krueger, 1980). Once the core of regional geography had been removed, there was little to hold the discipline together. Krueger urged the search for a new framework, so that the diversity of the field would be its greatest strength. This point has since been reiterated by Brenton Barr, who has urged a return to regional studies and regional courses. Leslie J. King shares the same concern (King, 1988).

Canadian geography has embraced high technology (Krueger, 1989). Computers, tied to satellite and geographic information systems technology, have enabled geographers to undertake resource analysis, landscape interpretation, and environmental monitoring as never before. Canada had a significant role in developing France's satellite earth observation system; this technology was also sold to India and Thailand. Within the Department of Geography at the University of Waterloo, the Digital Remote Sensing Group, identified as the Earth-Observations Laboratory, analyzes the need for remote sensing some years from the present. Remote sensing has become an established part of the university's geography curriculum. This has meant excellent employment prospects for students with these skills, which, in turn, encourages the expansion of these programs.

The Geography Department at the University of Waterloo merged with the university's Faculty of Environmental Studies in 1969. Geography became the largest component of the new group; other components included Environment and Resource Studies, a School of Architecture, and a School of Urban and Regional Planning. Many joint appointments have been made. The enterprise has grown rapidly, with the unit doubling its original size with more than 80 faculty and approximately 1500 undergraduate students. All the deans and associate deans have been geographers.[7] In addition, the Geography Department of the University of Waterloo and the Geography Department at Wilfred Laurier in Waterloo have combined their graduate offerings and now probably offer the largest graduate geography program in North America. The applied nature of the offerings renders the graduate students suitable for technical appointments all over the world.

Historical geography has remained a strong concern in Canadian geography. D. Meinig (1986) revisited the early exploration and settlement of the Atlantic Canadian coastal area: the colonization of eastern Canada and reorganization of British North

---

[7]Special mention might be made of Peter Nash, Founding Dean of the Faculty of Environmental Studies, 1970. A Festschrift, *Abstract Thoughts: Concrete Solutions, Essays in Honour of Peter Nash*, (eds. L. Guelke and R. Preston) was published in 1987 as an expression of appreciation for his large geographic knowledge and his leadership at Waterloo.

America after the American Revolution is well revealed. R. Cole Harris (1987) edited the *Historical Atlas of Canada: From the Beginning to 1800*. This was at once recognized as an authoritative source concerning the physical environment and early settlement in Canada. J. Gibson has developed a substantial part of the historical geography of the Pacific Northwest (1991). Numerous other studies were made concerning early mining, communications, transportation, and other matters.

The 1980s and 1990s were marked by a renewed interest in global thinking: Global interaction and change have been brought to the intellectual level. (Geography always had a global mission in the classroom but not in research.) The World Conservation Strategy of 1980 and the Bruntland Report, Our Common Future, 1987, came out at a time when the Royal Society of Canada had initiated the International Geosphere-Biosphere Program. The Canadian Association of Geographers has established a Global Change Committee, and universities are now hiring faculty with global perspectives. Global warming, desertification, destruction of genetic and forest resources, demographic change, and the spread of disease are just some of the concerns that have encouraged geographers to return to the macro-perspective they once held. Yet, foreign area specialists are limited in number. The larger focus is on Canada.

The history of geography in Canada seems to hold little interest, perhaps because the university departments of geography are still very young, or because much of the work of the geographical community has been done by imported scholars. In the earlier decades T. G. Taylor and George Tatham occasionally offered essays in the history of geography. Taylor's *Geography in the Twentieth Century* (1951) is an example of such a contribution. The preface contains the statement that Taylor had been approached "with a view to producing a study of the growth, fields, techniques, aims and trends of geography. Characteristics of its evolution during the period since 1900 were specially to be described." In "Griffith Taylor and Canadian Geography" (1967), George Tomkins presents an interesting dissertation concerning biography, thought, and pedagogy. Unfortunately, this work never reached a wide audience. Marie Sanderson has also written about the life and thought of T. G. Taylor using archival documentation and has supplemented Taylor's autobiography, *Journeyman Taylor*. She has also written about the life & thought of Donald J. Putnam. John Warkentin wrote of the life & thought of George Tatham. J. Lewis Robinson has written many shorter pieces relevant to the history of Canadian geography. His works relate to geography at the University of British Columbia, numbers of geographers in Canada, the production of doctorates, geography in Canada from the 1950s to the 1970s, trends in Canadian geographical thinking, and so on. Vincent Berdoulay, who until 1980 was at Ottawa, has written on La Blache and Harold Innis. Joseph May covered the history of geography with special reference to Immanuel Kant. Both Leonard Guelke and Anne Buttimer have dealt with the philosophy of geography. For Canadian francophone geography, L. E. Hamelin and L. Beauregard have made meaningful contributions, as has Peter Nash with his series of contributions concerning Raoul Blanchard.

B. M. Barr warns that lack of proficiency in foreign languages and foreign area studies are two weaknesses in the national geography. (Since 1978 Canada has shipped more goods across the Pacific than the Atlantic.) Barr also points out that

40 percent of all immigrants to Canada in the early 1980s were from Asia. It would seem that Canada's new global viewpoint would help comprehend these developments. Meanwhile, geographers are dealing with dimensions of the people-environment theme in Canada. Canadian university geography departments are large, vigorous, and excellent and are in the process of developing an autochthonously rich Canadian geography.

REFERENCES: CHAPTER 12

Balchin, W. G. V., and Coleman, A. M. 1966. "Geography in Canadian Universities." *Nature* 209, no. 5027:960–963.

Bailey, W. G., Timothy R. Oke, and Wayne R. Rouse, eds. 1997. *Surface Climates of Canada.* Montreal: McGill-Queen's University Press.

Barr, B. M. 1986. "Canadian Geography in a Multilingual World: The Implosion of Relevance." *The Canadian Geographer* 30:290–301.

Baulig, H. 1935 and 1936. *Amerique Septentrionale.* 2 vols. Paris: A. Colin.

Beauregard, L. 1979. "Epistémologie de la Géographie au Canada français." In L.-E. Hamelin and L. Beauregard, eds., *Retrospective 1961–1976.*

Berdoulay, V. 1987. "Le possibilisme de Harold Innis." *The Canadian Geographer* 31:2–11.

Bird, J. B. 1967. *The Physiography of Arctic Canada.* Baltimore, Md.: Johns Hopkins University Press.

Bladen, V. W., Easterbrook, W. T., and Willits, J. H. March l953. "Harold Adams Innis, 1894–1952." *American Economic Review* 43:1–25.

Blanchard, R. 1935. *L' Est du Canada français.* 2 vols. Montreal: Beauchemin.

———. 1948. *Le Centre du Canada français.* Montreal: Beauchemin.

———. 1949. *Le Québec par image.* Montreal: Beauchemin.

———. 1953–1954. *L' Ouest du Canada français.* 2 vols. Montreal: Beauchemin.

———. 1960. *Le Canada français, étude géographique.* Montreal: Fayard.

Bourne, L. S., and D. F. Ley, eds. 1993. *The Changing Social Geography of Canadian Cities.* Montreal: McGill-Queen's University Press.

Britton, John N. H., ed. 1996. *Canada and the Global Economy: The Geography of Structural and Technological Change.* Montreal: McGill-Queen's University Press.

Bunting and Filion. 1988. *The Monograph* 39, no. 2 and 39, no. 4.

Burton, I. 1963. "The Quantitative Revolution and Theoretical Geography." *Canadian Geographer* 7, no. 4:151–162.

Cahiers de Géographie de Québec. 1986. Numéro spécial: La géographie du Québec cinquante ans après Raoul Blanchard.

Camu, P., Week, E. P., and Samets, Z. W. 1964. *Economic Geography of Canada.* Toronto: Macmillan.

Chapman, A. D. 1921. "Position of Geography in Canada." *The Geographical Teacher* 11:52–54.

Chapman, J. 1966. "The Status of Geography." *Canadian Geographer* 10, no. 3:133–44.

Clark, A. H. April 1950. "Contributions to Geographical Knowledge of Canada since 1945." *Geographical Review* 40:285–312.

Creighton, D. G. 1957. *Harold Adams Innis: Portrait of a Scholar.* Toronto: University of Toronto Press.

Davis, W. M. 1899. "North America." In *The International Geography.* Ed. H. R. Mill. London and New York: D. Appleton & Co. Pp. 664–678.

Dean, W. G., et al. 1967. "Canadian Geography, 1967." *Canadian Geographer* 11, no. 4:195–371.

Dyde, W. F. 1929. *Public Secondary Education in Canada*. New York: Columbia University Press.

Evenden, L. J. 1976. "The Western Division of the Canadian Association of Geographers." *Canadian Geographer* 20, no. 3:329–333.

Floch-Serra, M. 1989. "Geography and Post-modernism: Linking Humanism and Development Studies." *Canadian Geographer* 33, no. 1:66–75.

Fraser, J. Keith. 1983. "The Road Less Travelled: Reflections on a Career in Geography." *Canadian Geographer* 27, no. 4:305–312.

French, Hugh M., and Olav Slaymaker, eds. 1993. *Canada's Cold Environments*. Montreal: McGill-Queen's University Press.

Grenier, G. 1965. "Raoul Blanchard, 1877–1965." *Canadian Geographer* 9, no. 2:101–103.

Guelke, L. 1971. "Problems of Scientific Explanation in Geography." *Canadian Geographer* 15:38–53.

Guelke, L., and Preston, R. E. eds. 1987. *Abstract Thoughts: Concrete Solutions—Essays in Honour of Peter Nash*. Waterloo, Ont.: University of Waterloo, Department of Geography, Publication Series 29.

Hamelin, L. E. 1961. "La géographie de Raoul Blanchard." *Canadian Geographer* 5, no. 1:1–9.

———. 1963. "Petite histoire de la géographie dans le Québec." *Cahiers de géographie de Québec* 13:137–153.

———. 1964. "Géomorphologie–géographie globale–géographie totale associations internationales." *Cahiers de géographie de Québec* 16:199–218.

———. 1986. "Les carrieres Canadiennes de Raoul Blanchard et Pierre Deffontaines." *Cahiers de Géographie du Quebec* 30:137–150.

———, and Beauregard, Ludger, eds. 1979. *Retrospective 1951–1976*. Montreal: Canadian Association of Geographers.

Harris, R. Colebrook 1997. *The Resettlement of British Columbia: Essays on Colonialism and Geographical Change*. Vancouver: UBC Press.

Harris, R. Colebrook and Geoffrey J. Matthews. 1987. *Historical Atlas of Canada*. Toronto: University of Toronto Press.

Hodgins, J. George. 1857. *The Geography and History of British America, and of the Other Colonies of the Empire; to Which Is Added a Sketch of the Various Indian Tribes of Canada, and Brief Biographical Notices of Eminent Persons Connected with the History of Canada*. Toronto: Maclear and Co.

Innis, H. A. 1930. *The Fur Trade in Canada: An Introduction to Canadian Economic History*. New Haven, Conn.: Yale University Press.

———. 1935. "Canadian Frontiers of Settlement: A Review." *Geographical Review* 25:92–106.

———. 1951. *The Bias of Communication*. Toronto: University of Toronto Press.

Innis, H. A., and Broek, J. O. M. 1945. "Geography and Nationalism: a Discussion." *Geographical Review* 35:301–311.

Jackson, C. Ian. 1998. "David Thompson, 1770–1857." *Geographers: Biobibliographical Studies* 18:94–112.

Joerg, W. L. G. 1936. "The Geography of North America: A History of Its Regional Exposition." *Geographical Review*: 26:640–663.

Kerr, D. P., 1960. "The Tasks of Economic Geography." *Canadian Geographer*: 15:1–9.

Kimble, G. H. T. 1945. "The Craft of the Geographer." *Canadian Geographical Journal* 3:257–263.

―――. 1946. "Geography in Canadian Universities." *Geographical Journal* 108:114–115.

King, L. J. 1988. "Geography in a Changing University Environment." *Canadian Geographer* 32, no. 4:290–295.

Krueger, R. R. 1972. "Notes on Applied Geography in Canada in General and in Ontario in Particular." In R. E. Preston, ed. Pp. 374–379. *Applied Geography and the Human Environment*. Proceedings of the Fifth International Meeting Commission of Applied Geography International Geographical Union. University of Waterloo Department of Geography.

―――. 1980. "Unity out of Diversity: The Ruminations of a Traditional Geographer." *Canadian Geographer* 24, no. 4:335–348.

―――. 1989. "Recent Developments in Canadian Geography." *The Monograph* 40, no. 2:20–24.

Lukermann, F. E. 1965. "The 'calcul des probabilités' and the école française de géographie." *Canadian Geographer* 9, no. 3:128–137.

Marshall, A. 1972. "Griffith Taylor's Correlative Science." *Australian Geographical Studies* 12:184–192.

Martin, G. J. 1980. *The Life and Thought of Isaiah Bowman*. Pp. 112–114. Hamden, Conn.: Archon Books.

Mayo, W. L. 1965. *The Development and Status of Secondary School Geography in the United States and Canada*. Ann Arbor: University Publishers.

Meinig, D. W. 1986. *The Shaping of America: A Geographical Perspective on 500 Years of History*. New Haven, Conn.: Yale University Press.

Morissonneau, C. 1971. *La Société de Géographie de Québec 1877–1970*. Québec: Les presses de L'Université Laval.

Nash, P. H. 1986a. "Mark I Role Model and Inspiring Catalyst: Some Characteristics from a Perspective of Four Decades." *Cahiers de Géographie de Québec* 30:151–61.

―――. 1986b. "The Making of a Humanist Geographer: A Circuitous Journey." *Geography Publication Series* No. 25, Waterloo, Ont.: University of Waterloo, Department of Geography: 10–33.

―――. 1990. "A Still Revered 'Tomodachi' After Half A Century: Reflections on Raoul Blanchard and Epistemological Changes in Canada." In R. Preston and B. Mitchell, eds., *Reflections and Visions, 25 Years of Geography at Waterloo*. Pp. 61–75.

Nicholson, N. L. 1959. "Canada and the International Geographical Union." *Canadian Geographer* 14:37–41.

Parker, I. 1988. "Harold Innis as a Canadian Geographer." *Canadian Geographer* 32:63–69.

Porteous, J. D., and Dyck, H. 1987. "How Canadian Are Canadian Geographers?" *Canadian Geographer* 31:2177–2179.

Putnam, D. G., ed. 1952. *Canadian Regions: A Geography of Canada*. Toronto: Dent.

―――. 1964. "Obituary, Griffith Taylor." *Canadian Geographer* 7:197–200.

Putnam, R. G., and Sanderson, M. 2000. *Down to Earth: A Biography of Geographer Donald Fulton Putnam*. University of Toronto, Department of Geography.

Reeds, Lloyd G. 1964. "Agricultural Geography: Process and Prospects." *Canadian Geographer* 8, no. 2:51–63.

Robinson, J. Lewis. 1951a. "The Development and Status of Geography in the Universities and Governments of Canada." *Yearbook of the Association of Pacific Coast Geographers* 13:3–13.

―――. October 1951b. "Geography in the Universities of Canada." *Canadian Geographical Journal* 2:89–183.

―――. March 1966. "Geography in the Canadian Universities." *Professional Geographer* 18, no. 2:69–74.

————. 1967. "Growth and Trends in Geography in Canadian Universities." *Canadian Geographer* 11, no. 4:216–229.

————. 1986. "Geography in Canada." *The Professional Geographer* 38:411–417.

————. 1989. *Concepts and Themes in the Regional Geography of Canada.* 2nd ed. Vancouver: Talonbooks.

————. 1991. "The Beginning of Geography at the University of British Columbia." *The Operational Geographer* 9, no. 2:9–10.

Romey, W. D. 1989. "Study of Canadian Geography." In Gary L. Gaile and Cort J. Willmot, eds., *Geography in America.* Pp. 563–580. Merrill Publishing Co.

Sanderson, M. 1988. *Griffith Taylor: Antarctic Scientist and Pioneer Geographer.* Pp. xi–218. Ottawa: Carleton University Press.

————. 2001. "Donald Fulton Putnam, 1903–1977." *Geographers: Biobibliographical Studies* 21:72–84.

Siegfried, A. 1937. *Le Canada, puissance internationale.* Paris: A. Colin: *Canada.* Translated by H. H. and Doris Hemming. New York: Harcourt, Brace & Co.

Smith, P. J. 1975. "Geography and Urban Planning: Links and Departures." *Canadian Geographer* 19, no. 4:267–278.

Stefansson, V. 1921. *The Friendly Arctic: The Story of Five years in Polar Regions.* New York: Macmillan, 1921.

Sullivan, R. 1845. *Geography Generalized: or An Introduction to the Study of Geography on the Principles of Classification and Comparison.* 4th ed. Dublin: William Curry, Jun and Co.

Taylor, T. Griffith. March 1936. "Fundamental Factors in Canadian Geography." *Canadian Geographical Journal* 12:161–171.

————. May 1937. "The Structural Basis of Canadian Geography." *Canadian Geographical Journal* 14:297–303.

————. 1938. "Geography at the University of Toronto." *Canadian Geographical Journal* 23.

————. October 1940. "Structure and Settlement in Canada." *The Canadian Banker* 48:42–64.

————. October 1941. "The Climates of Canada." *The Canadian Banker* 49:34–59.

————. March 1942. "Environment, Village and City: A Genetic Approach to Urban Geography, with Some Reference to Possibilism." *Annals, Association American Geographers* 23:1–67.

————. April 1945a. "Towns and Townships in Southern Ontario." *Economic Geography* 21:88–96.

————. May 1945b. "A Mackenzie Domesday, 1944." *Canadian Journal of Economics and Political Science* 11:189–233.

————. August 1945c. "A Yukon Domesday, 1944." *Canadian Journal of Economics and Political Science* 11:432–449.

————. 1958. *Journeyman Taylor: The Education of a Scientist.* Edited by Alasdair Alpin MacGregor. London: Robert Hale Ltd.

Tomkins, G. S. 1967. "Griffith Taylor and Canadian Geography." Unpublished Ph.D. thesis, University of Washington, Seattle.

Trotier, L., gen. ed./rédacteur en chef. 1972. *Studies in Canadian Geography/Etudes sur la géographie du Canada,* 6 vols. Toronto: University of Toronto Press.

Trudel, M. June 1958. "La Géographie Champlain, fondateur de Québec." *La Revue Canadienne de Géographie* 12:3–6.

"University Dissertations, Theses and Essays on Canadian Geography." July–October 1950. *La Revue Canadienne de Géographie* 4:1–8.

Vanderhill, B. G. 1982. "The Passing of the Pioneer Fringe in Western Canada." *Geographical Review* 72:200–217.

Walmsley, D. J. 1974. "Positivism and Phenomenology in Human Geography." *The Canadian Geographer* 18, no. 2:95–107.

Warkentin, J., ed. 1968. *Canada: A Geographical Interpretation*. (Canadian Association of Geographers). Toronto: Methuen.

Watson, J. Wreford. December 1950. "Geography in Canada." *Scottish Geographical Magazine* 66:170–172.

———. 1981. "The Development of Canadian Geography: The First Twenty-five Volumes of the Canadian Geographer." *Canadian Geographer* 25, no. 4:391–398.

White, J. and Department of the Interior. 1906. *Atlas of Canada*. Toronto: Toronto Lithographing Co.

Williams, C. T. 1946. *Philosophy of Educational Method*. Seattle, WA: University of Washington, College of Education, Bureau of Educational Research and School Service.

Wolforth, J. 1986. "School Geography—Alive and Well in Canada?" *Annals of the Association of American Geographers* 76:17–24.

13

THE NEW GEOGRAPHY IN SWEDEN

The empirical treatment of quantities by Statistics, the rational treatment of quantities by Mathematics, the treatment of logical thinking by Philosophy, the treatment of position or distribution in space by Geography, and the treatment of development and distribution in time by History—all these would seem to be applicable to all kinds of objects, and would seem to some extent to be operations that most men of science need to carry out.
—Sten De Geer, "On the definition, method and classification of geography"

The five countries of Norden have made a large contribution to geography that is wholly disproportionate to the aggregate size of their population. The cultural similarities of these peoples have produced a convergence of opinion that supposedly derives from the notion that geography is larger than the sum of its parts, that the concept of region remains widely supported, and that a strong feeling for the countryside persists (Mead, 1987).

GEOGRAPHICAL INSTITUTIONS

The Royal Danish Geographical Society was founded in 1876, the Swedish Society for Anthropology and Geography in 1877, the Geographical Association of Finland and the Geographical Society of Finland in 1888, and the Norwegian Geographical Society in 1889. Perhaps the earliest of the local societies were those at Uppsala (1895) and Göteborg (1908), while the first of the university departments of geography was established in Copenhagen in 1883. The largest contributions to geography from Scandinavia have apparently come from Sweden.

Several antecedents to modern geography in Sweden may be cited. First are the archives and current output of the 350-year-old Land Survey Board, with headquarters in Stockholm and its technical plant in Gävle. (Official mapping by the Swedish Topographic Survey began in 1628.) The second source is the population statistics maintained through the centuries by the parish clergy and currently made available through the Demographic Data Base at Umeå University. The Royal Swedish Academy of Sciences was founded in 1739. This institution provided a forum for the works of natural scientists such as Linnaeus and Celsius and for the work of generations of topographers and the hydrographic activity of Gustav Klint. In 1749 Sweden became the first country in the world to start a regular national census.

The author would like to acknowledge the kindness of Tom Mels (University of Kalmar, Sweden) in offering suggestions for improvement of this chapter.

(At that time the population of Sweden was 1,763,700.) And in 1810 the field survey for the topographic mapping of Sweden was initiated. The first sheet was engraved in 1826 and the last in 1924.

In 1885 popular demand for travel literature led to the organization of the Swedish Touring Club, with an initial membership of 67 persons. It not only supplied the demand for information about foreign countries, but it also undertook to develop facilities, such as trails and rest houses, for the many Swedes who wanted to journey into the most remote parts of their own country. By 1900 the club had a membership of more than 25,000, which increased to 98,000 by 1922 (Anrick, 1923).[1] Since 1886 it has published a yearbook on the different provinces of Sweden. In 1881 *Ymer*, the longest running Swedish journal was established, and in 1919 *Geografiska Annaler* was founded (divided into a physical and human series in 1965). From 1922 to 1978 another periodical, *Globen*, was published. In 1925 the South Swedish Geographical Society began publication of *The Swedish Geographical Yearbook*. (Papers are normally published in Swedish, although short summaries are offered in English.) Research series were published by a number of university geography departments: Göteborg (1929), Stockholm (1929), Uppsala (1936–1968), and Lund. (Predominantly dissertation publication began in 1929, and Lund Studies in Geography began in 1949.) *Geografiska Notiser* was founded in 1943. These publications were the result of the Swedish people's expressed interest in geography.

Prior to 1900, the periodical *Ymer* kept geography before the public. This periodical of the Swedish Society for Anthropology and Geography helped establish geography in the educational institutions of Sweden. The articles contained in this publication tended to interest a wide audience and were published in Swedish. Frequent contributors included Karl Ahlenius, Gunnar Andersson, E. W. Dahlgren, Gerard De Geer, Sven Hedin, Gustaf Retzius, Hugold von Schwerin, and Hjalmar Stolpe. The work of foreign geographers was frequently noticed and included that by Friedrich Ratzel and Ferdinand von Richthofen, Edward Brückner, Henry G. Bryant, James Fairgrieve, Lucien Gallois, Charles Rabot, H. W. Fairbanks, A. E. Frye, and Wallace W. Atwood.

Geography was introduced at Lund on a gradual basis. Examination began in 1892 and was administered by the professor of history. According to Olof Wärneryd, the first student to be examined was Emil Sommarin, who would later become professor of economics at Lund. A chair was formally created in 1895, and Hans Hugold von Schwerin was appointed to it from 1897 to 1912. Karl Ahlenius was awarded a professorial chair at Uppsala University (1902–06). Gothenburg followed in 1905 and Stockholm in 1929; each had a Department of Geography that included a professorship. At last geography had established itself as a profession.

EXPLORATION

There was a keen interest in exploration during the years when geography was becoming established in Sweden. For the Swedes Scandinavia and the polar world held a

---

[1]The club issued circulars, maps, guide books, and a "Province Series" of yearbooks. By 1923 the club had published 550 maps.

special interest, though not to the exclusion of warmer lands. The first important step toward Arctic exploration was taken by Baron Adolf Erik Nordenskiöld, who in 1858 was appointed chief of the division of mineralogy of Sweden's National Museum of Natural History. In 1868, with F. W. von Otter (who later became Sweden's minister of the navy and prime minister), he reached 81°42′N, which represented an Arctic record. After an unsuccessful attempt to reach the North Pole in 1872–73, Nordenskiöld concentrated on the possibility of navigating the Arctic Ocean from Europe to the Pacific. This was to become known as the Northeast Passage. Nordenskiöld eventually accomplished this goal in 1878–79 on the *Vega*. His accounts contain a wealth of information concerning the people along the coast of the Arctic Ocean and their ways of life. His report on the voyage extended to five volumes. *Scientific Observations of the Vega Expedition* was translated into several languages and was packed with new knowledge. Nordenskiöld was also a very capable geologist and mineralogist who contributed two monumental books on the history of cartography: *Facsimile-Atlas to the Early History of Cartography* and *Periplus: An Essay on the Early History of Charts and Sailing Directions*. Both works were published simultaneously in Swedish and English in 1889 and 1897, respectively.

From 1898 to 1902 the Norwegian Otto N. Sverdrup traveled the Canadian Arctic archipelago, while fellow countryman Roald Amundsen achieved the Northwest Passage between 1903 and 1906. Peary achieved the North Pole in 1909 and Amundsen the South Pole in 1911. These events fascinated the people of the northlands and provided a focus to the thinking of geographers: exploration, northern latitudes, areas of deprivation (just as the tropics and colonialism had provided impetus, if not focus, for the geographers of some west European countries).

Not all the exploring was in the northlands. Sven Hedin graduated from high school in 1885 and at once took a post as tutor to the son of a Swedish engineer in the Baku oilfields on the Caspian shore. At the close of this work, he journeyed to Tehran, crossed Persia to Baghdad, and returned to Sweden, where he would study geography and geology with W. C. Brögger of Stockholm's School of Commerce and A. G. Högbom of Uppsala University. He continued his studies in Berlin under Ferdinand von Richthofen, the German authority and explorer of Asia. Hedin presented his doctoral thesis to the University of Halle in 1892.[2] Having completed his studies, he embarked on the first of his scientific expeditions to Central Asia in 1893, sponsored in good measure by King Oscar of Sweden. He traveled overland to the Tarim Basin and its terminal lake, Lop Nor, and thence to Beijing. His second expedition began in 1899. He traveled by boat down the Tarim River, mapping its course, and made further studies in the area of Lop Nor. He then mapped part of the mountain system that forms the northern boundary of the Tibetan Plateau.

Hedin's third expedition (1905) took him to western and central Tibet. His fine sketches, maps, and notes were to prove valuable in developing reliable maps for that part of the world. Later he published his findings on these travels: *Scientific Results of a Journey in Central Asia* (1904–07) and *Transhimalaya* (1909–13). His fourth expedition, begun in 1926, was initiated by Lufthansa, the German airline,

---

[2] Sven Hedin, "Der demarend nach eigener Beobachtung."

which wanted a survey made of an air route from Berlin to Beijing. Hedin took a large party, which included meteorologists who were to establish ground weather stations. After this work had been completed, Hedin organized in the field a Sino-Swedish expedition to the northwestern provinces of China. He was in charge of a distinguished group of specialists that studied Central Asia and surveyed a possible motor road from east China to the province of Sinkiang. The group (called a mobile university) remained in the field until 1935. Upon his return to Sweden, Hedin and then others wrote numerous reports published in an extended series of volumes; they have now been translated into several languages and have greatly extended our knowledge of that part of the world. Although Hedin had not been associated with a university post, this kind of activity excited and inspired the national conscience and encouraged the development of institutionalized geography.[3]

SOME ORIGINS OF SWEDISH GEOGRAPHY

Swedish geography retained an interest in the natural environment. From the 1880s to the early years of the 20th century, Baron Gerard De Geer and Ragnar Lidén attempted to establish a date for the last retreat of the ice age by counting the varves (bands of coarse and fine material) in clays deposited in glacial lakes. De Geer undertook extended field trips in geochronology and inspired much work in glaciology. He pieced together the records of some 1500 outcrops lying between the southern tip of the Scandinavian peninsula and the modern proglacial lakes in central Sweden. From this evidence he inferred that the retreating ice exposed southern Sweden about 13,500 years ago. Based partly on De Geer's work, it has been calculated that maximum expansion of the last glaciation in Europe occurred approximately 35,000 years ago. De Geer's contribution was original in geological science and joined the work of those otherwise interested in dating climatic change.

De Geer first visited the United States in 1891, when he explained his geochronological investigations to university audiences and undertook considerable field work. At this time, however, he was still attempting to elaborate the time scale by following the ice recession from the margin to the core of the former ice sheet in Sweden. He was busy at this process from 1878 to 1915 (De Geer, 1926:262). As a consequence of this work, he brought the Swedish expedition of 1920 to the United States, seeking to test the international applicability of the Swedish time scale. He brought his two most experienced assistants with him, Ragnar Lidén and Ernst Antevs.[4] His wife, Ebba Hult De Geer, a capable geochronologist in her own right, followed shortly thereafter. When the party returned to Sweden, Antevs remained in New England, where he began to work out a chronology of North America in late glacial times. For that purpose he studied the recession of the last ice sheet in

---

[3]For a complete bibliography of Hedin's works until 1935, see Sven Rinman, a publication honoring Sven Hedin on his 70th anniversary, February 19, 1935, "in series with *Geografiska Annaler*," Stockholm, 1935.

[4]Three of G. De Geer's best known students were Hans W: son Ahlmann, Carl Cizon Caldenius, and Ragnar Sandegren. They published "The Quaternary History of the Ragunda Region Jämtland," *Geol. Fören. Förhandl* 34(1912):353–364.

New England and New York state. Antevs worked in the Connecticut River Valley from Hartford, Connecticut, to St. Johnsbury, Vermont.

Antevs worked out a chronology of the ice retreat, and the whole was published in *The Recession of the Last Ice Sheet in New England* (Antevs, 1922). It was the first such work accomplished in North America. Antevs studied more than varves in his attempt to secure registration of climatic change in two continents. He also examined tree rings, migrations of plants and animals, the nature and cause of crustal warping during the withdrawal of the ice, and the emergence and submergence of the coast. He had sought to determine whether the glaciers in various parts of the world were synchronous. His book was one of a series published by the American Geographical Society, which placed the problems of the Quaternary before academics and the reading public. This also meant that Gerard De Geer's varve-measuring techniques in Sweden became known in North America. Glaciological study was indigenous to Scandinavian geography. While in the United States, Ernst Antevs shared an office with Ellsworth Huntington at Yale University; Huntington was then attempting to date climatic changes of the past and had already created a climatic curve from measurements of the Big-Trees.

The method of studying the landscape was largely derived from Germany. The earliest occupants of chairs in geography in Sweden were usually educated by German geographers. (The work of German geographers received constant attention in *Ymer* and *Geografiska Annaler* until perhaps the 1940s). The anthropogeographic line of thought then in vogue was largely inspired by the work of Ratzel. It was a way of thinking and of seeking human consequences deriving from the physical environment. (In the United States such studies were called "determinist.") An alternate viewpoint was provided by Otto Schlüter and Alfred Hettner, who offered a chorology. Anthropogeography sought causes and consequences, whereas chorology provided the bases for study of the culture scape. Both geographies were large enough to embrace an emerging regional geography. Anthropogeography began to recede with the international tide after World War I, but the chorologic perspective remains popular to this day.

Hugold von Schwerin, who became Sweden's first professor of geography in 1897 (he had been university lecturer in geography and political science since 1884) at the University of Lund, learned much about human dependence on nature from the works of Oscar Peschel in the 1860s and 1870s; von Schwerin thought Ratzel too doctrinaire (Bergsten, 1984). In 1892 von Schwerin published "Mohammedanism in Africa," one of the earlier demonstrations of the Swedish interest in global matters. In the late 1800s little geography was available in the Swedish language, and so geographers were obliged to read in the German literature. Karl Ahlenius, the first professor of geography in Uppsala and an enthusiast of Ratzel, first introduced to the Swedish public Ratzel's thought as expressed in his *Anthropogeographie* (1882). Ahlenius (1897) also classified human races on the basis of skin color and other physiological characteristics. In 1902 S. Lönborg published "The Finnish-settled Regions in the Middle of Scandinavia," a work that investigated Finnish settlement in central Sweden and adjacent parts of Norway that was the product of immigration from Finland during the seventeenth and eighteenth centuries. Place names, education, religion, language, and the physique of the people were studied. However, the amount of social geography accomplished in this period was limited inasmuch as the academicians were qualified in natural science, giving their work a physical rather than a social orientation.

THE EMERGENCE OF REGIONAL GEOGRAPHY

In 1913 Helge Nelson studied the history of settlement in the mining district of central Sweden; this was one of the earliest studies in historical geography executed in the spirit of German regional studies, *Die Landschaftskunde* (Godlund, 1986). Ahlenius had initially influenced the thinking of Helge Nelson. Nelson became professor at Lund in 1916. Application had been made for the chair in 1913, and competition ensued between Nelson and Sten De Geer (son of Gerard De Geer). Nelson argued for a historically oriented regional undertaking, whereas De Geer advocated a science of spatial distribution of contemporary time.[5] Nelson remained at Lund until 1947, continually affirming the value of regional investigation. In 1918 he published an excellent paper on the human regions of Sweden in which human agglomerations were the basis of subdivision, since they are outgrowths of the region in which they reside.[6] He traveled in the United States and Canada in 1921, 1925, 1926 and 1933; in 1926 he was a visiting member of the faculty of the University of Chicago. He wrote considerably on North America and its Swedish settlement and contributed more than 500 items to the literature.

In 1926 Nelson published a two-volume work, *Nordamerika: Natur, bygd och svenskbygd* (North America and the Swedes), which is a fine account of the physiography of North America, with the human presence noticed in each region. The book was dominated by the theme of Swedes in North America, and it was recognized as the first geographically detailed account on that subject (although Kendric C. Babcock had previously published *The Scandinavian Element in the United States*, 1914). Nelson had already published *Canada nybyggarlandet* (*Canada, Settlers' Country*) and had an unusually thorough grasp of this northland. *The Swedes and the Swedish Settlements in North America* (1944) was the culmination of more than 20 years' work. The book included an atlas of 73 maps and became the standard work on the subject. Nelson sought to learn the social status of the emigrants, why they emigrated, and from what home districts they had come. He calculated that between 1861 and 1940, 1,119,300 Swedes emigrated to the United States and 18,800 to Canada. Of these 202,000 returned to Sweden from the States and 5600 from Canada. The book was indeed a compendium of data based on much field work and many interviews, as well as his encyclopaedic knowledge of this subject.

While Nelson was at Lund, Sten De Geer, son of Gerard De Geer, held a post at Stockholm's School of Commerce (from 1911 to 1928), and at the University of Göteborg (from 1928 until his premature death on June 2, 1933). He pioneered in urban geography and was especially interested in towns and ports on the Baltic.

---

[5]T. Hägerstrand, "Proclamations about Geography from the Pioneering Years in Sweden," *Geografiska Annaler* 64B(1982):119–125.

[6]This essay, with maps, was reviewed in *Geographical Review* 11(1921):144–145.

Well known to English readers is his "Greater Stockholm: A Geographical Interpretation" (1923b). In 1927 he made a second study of the Baltic harbors, comparing their uses in 1912–13 with those of 1923–24. Dot maps were used in both articles. In 1919 he published a "Map of the Distribution of Population in Sweden," a publication of 12 plates with an accompanying text of nearly 300 pages.[7] In this work De Geer demonstrated technical advances with the use of dots and sized globes for numbers of people. This study was widely recognized and applauded. Since that time his method of population mapping has been developed in several countries. In 1923 the Swedish geographer Alfred Söderlund published a similar distribution map of the population in Norway. De Geer's interest in greater Stockholm became the forerunner to other work relating to Stockholm.

After the Treaty of Versailles of 1919, De Geer published an extended article on the peace settlement in *Ymer*, which was expanded and published as a book, *Det Nya Europa*, in 1922. Geographers in many countries wrote on this same subject. In that year De Geer taught in the summer school at the University of Chicago. In August on his return from Chicago, De Geer stayed with Mark Jefferson in Ypsilanti, Michigan. They had already exchanged correspondence on glaciation, river widths, meander processes, and urban geography. The pair motored through Ohio and Pennsylvania studying the distribution of population. While on board the vessel returning him to Sweden, De Geer wrote a version (later much revised) of "The American Manufacturing Belt," which Jefferson reviewed for *The Geographical Review*.[8] De Geer was interested in the notion that America had developed a single manufacturing belt as had Europe. He considered the location of many manufacturing groups—their production, employment, initial site advantage, and other characteristics—and compared them to the European industrial complex. This was the first study of its sort, and it was of special value to geographers in North America.

Meanwhile, De Geer had been thinking about the nature of geography. He had begun his career with studies in physical geography but had become particularly interested in a more human geography. In 1922 before his arrival at the University of Chicago, where he could discuss the subject with a variety of geographers, he had reviewed N. Fenneman's "The Circumference of Geography."[9] In the following year De Geer offered "On the Definition, Method and Classification of Geography" (1923a). He stated that his definition of the field had served him in a practical manner; briefly, "Geography is the science of the present day distribution of phenomena on the surface of the earth" (1923:2). He suggested that the past needed to be considered only in order to understand the present. The essential purpose of his mapping distributions was to construct a synthetic regional view of the world. He wrote: "geographical provinces and regions form a synthesis of characteristic complexes of important distribution phenomena within limited parts of the earth's

---

[7]See Sten De Geer, "A Map of the Distribution of Population in Sweden: Method of Preparation and General Results," *Geographical Review* 12(1922):72–83.

[8]See Mark Jefferson, "'The American Manufacturing Belt' by Sten De Geer," *Geographical Review* 18, No. 4 (1928):690.

[9]*Geografiska Annaler* 4(1922):218–220.

surface" (1923a:10). De Geer sought a world regional scheme and divided the land area into 27 regions and the sea into 17. (He gave more attention to the sea than would be common later in the twentieth century.) Other geographers who had developed world schemes included Alfred Hettner (1908), Ewald Banse (1914), and Robert Seiger (1921).

De Geer attempted regional synthesis especially in the period 1926 to 1928 when he offered "The Kernel Area of the Nordic Race within Northern Europe."[10] J. Leighly felt this work deserved close study as an example of method. In 1928 De Geer published "The Subtropical Belt of Old Empires," which involved political and climatological thought. It was part of De Geer's global phase and perhaps represented his curiosity about human physiology and its different capacity. Certainly, this sort of investigation was in vogue and attracted the attention of workers such as Ellsworth Huntington, J. Russell Smith, and S. Colum Gilfillan. Both the "Nordic Race" and the "Subtropical Empire" studies had the effect of linking Swedish and American geographers more closely, just as his father's (and Antevs's) work on glaciation had done. American geographers such as Wallace W. Atwood, B. J. S. Cahill, Mark Jefferson, and Derwent Whittlesey published in *Geografiska Annaler*. Jefferson had already learned Swedish and served as a reviewer of Swedish literature for the *Geographical Review*.[11] Apparently, Carl Sauer had been impressed with the writings of De Geer and suggested that Leighly study with him. Leighly went to Sweden in 1925–26 and "with much help from De Geer" completed his dissertation in 1927 (Leighly, 1983). Entitled "The Towns of Mälardalen in Sweden: A Study in Urban Morphology," the dissertation was published in 1928 as Volume 3 of the newly established *University of California Publications in Geography*. In 1929–30 he returned to the Baltic and wrote "The Towns of Medieval Livonia" (Latvia and Estonia), which was not published until 1939. Later, in 1950, Leighly lectured at the Universities of Uppsala and Stockholm by invitation and in Swedish.

In 1928 Otto Nordenskiöld (nephew of A. E. Nordenskiöld), professor of geography at the University of Gothenburg, died at the age of 58. His contributions to polar geography were considerable and included *Polarnaturen* (Stockholm, 1918), which was published in English as *The Geography of the Polar Regions*.[12] Nordenskiöld had invested much time in South Polar research and produced *Wissenschaftliche Ergebnisse der Schwedischen Südpolar Expedition 1901–1903*.[13] Apart from his keen interest in the polar world, he had developed some curiosity about foreign colonization in South America. He traveled through Chile and Peru

---

[10]See pp. 162–171 in H. Lundborg and F. J. Linders, *The Racial Characters of the Swedish Nation* (Uppsala, 1926).

[11]Each published essays in *Geografiska Annaler* between 1929 and 1934. W. W. Atwood, "Geography and International Good-Will," 11(1929):101–104; D. Whittlesey, "A Locality on a Stubborn Frontier at the Close of a Cycle of Occupance," 12(1930):175–192: B. J. S. Cahill, "A World Map to End World Maps" 16(1934):97–108; and M. Jefferson, "The Problem of the Ecumene" 16(1934):146–158.

[12]American Geographical Society Special Publication, no. 8, 1928, New York.

[13]Published in seven volumes, Stockholm, 1908–1921.

during 1920–21 and published *Südamerika: Ein Zukunftsland der Menscheit* in 1927.[14] After Nordenskiöld's death, De Geer was given the post at Gothenburg. There he built a strong center for geography, although his emphasis was different from Nordenskiöld's. He encouraged rejuvenation of the local geographic society and in 1932 the foundation of the journal *Gothia*. In the first issue of this journal, De Geer explained the work of the geographical institute and invited foreign and particularly English-speaking participation in the institute's work. He died, five years to the day after Otto Nordenskiöld, at the age of 47. Meanwhile, De Geer had written on racial types, regions of Sweden, and political geography. In the last-named investigations, De Geer was perhaps somewhat influenced by Rudolf Kjellén.

Kjellén (1864–1922) was a political scientist, geographer, and politician at the University of Uppsala who became a member of the Swedish Parliament. He borrowed from the work of Ratzel, although his own work contained much original political theory. Kjellén's book *The Great Powers of Today* appeared in German translation in 1914. Of the American, the Russian, and other dangers that threatened Europe, he wrote, "In such a situation Germany appears as the most natural leader, geographically as well as culturally. Such would mean for Germany that it, as administrator of the right of primogeniture, should accept the position of world ruler" (Van Valkenburg, 1951). Kjellén's book went through 19 editions during the war and was followed by *The Great Powers and the World Crisis* in 1921. For Kjellén the state was an organism that possessed more strength than the sum of the individual constituent forces that make it what it is. The state was therefore expansive by nature.

Kjellén's belief in the strength and power of the successful state and the subordination of the weak state left him without support in Sweden. He titled what was perhaps his most significant theoretical book *Staten som Lifsform* (*The State as an Organism*), first published in 1916 and translated into German as *Der Staat als Lebensform* in 1917 (4th edition, Berlin, 1924). Written during World War I, this work presented his theory that the state constitutes five organs: Kratopolitik (government structure), Demopolitik (population structure), Sociopolitik (social structure), Oekopolitik (economic structure), and Geopolitik (physical structure). It was through this book that the term *geopolitik* entered into German thinking and German geography. Defeated German geographers and political scientists took hold of this instrument of thought and used a form of Darwinian selection to revivify an older German political philosophy. Geopolitik became a tool for rebuilding Germany, and then both Japan and Italy. The link between Swedish and German geography had been close for many years. Although Kjellen's work had an unfortunate impact, it nevertheless represented a certain nomothetic stream of thought in the social geography present at the time.

In the 1920s human geography in Sweden began to emerge from an anthropo-geographical posture. Two philosophical approaches became apparent (Mead, 1988). One concentrated on the totality of the landscape (the regional approach) and the other on distributions and correlations (the systematic approach). The study

---

[14]Published in Stuttgart, 1927.

of 200 years of landscape changes in Scania by Anna Kristoffersson (1924) may have been the first Swedish regional monograph. Gerd Enequist's[15] work on the Lule Valley and K. E. Bergsten's two volumes on Östergötland represented a large number of other such studies, revealing Sweden to geographers as never before. John Frödin in "The Fäbod-District around Lake Siljan" (1925) studied transhumance and the saeter system in an alpine grazing region in central Sweden near the Norwegian border. He anticipated the reduction of this very old system from one of much agricultural significance to a factor of minor significance. The saeter did decline; sometimes the forest overran the area, and occasionally urban dwellers exploited these areas as summer residences.

Economic geography was not ignored: H. W. Ahlmann wrote "The Economical Geography of Swedish Norrland" (1921),[16] and Olof Jonasson, a Swedish doctoral student at Clark University, wrote the first in a series of articles entitled "Agricultural Regions of the World" for the newly founded periodical *Economic Geography*. Jonasson wrote "Agricultural Regions of Europe" in two parts.[17] It was the first such detailed statement on the subject to be made available in American geography, and it included a larger colored "fold-out" map. In this précis of work, Jonasson became one of the first geographers in North America to take note of the von Thünen model. Other Swedish geographers turned their research attention abroad.

Special mention should also be made of the work of Gunnar Andersson, who was professor of economic geography at the Stockholm School of Commerce in Stockholm and coeditor and a vigorous secretary of the Swedish Anthropological and Geographical Society and editor of *Ymer*. He was an authority on plant geography, especially on the evolution of Sweden's postglacial vegetation. He then turned his attention to economic geography and wrote on the possibilities of white settlement in Australia, natural resources in Sweden, and the geography of alimentation.

Physical geography continued to be a vital part of Swedish geography, with study of regional physiographies, glaciers, rainfall, and meteorology most prominent. Notices and book reviews frequently treated geomorphologic and meteorologic matters, the first frequently being the work of eminent German authors and the second inspired by the Bergen school and the Bjerknes'. Hans W:son Ahlmann and Axel Wallén were dominant in writing these reviews and notices.

In the 1930s and 1940s Swedish geography became more preoccupied with territory, structure, and function (Godlund, 1986). Once again De Geer's work was preeminent, although he died prematurely in 1933, the same year in which the third volume of *Géographie universelle* was published. Entitled *États Scandinaves: Régions polaires boréales* by Maurice Zimmermann, this fascinating work brought geographical attention to Sweden. De Geer had become deeply interested in a combination of structural and functional geography in built-up areas, He had become

---

[15]Gerd Enequist, the Uppsala geographer, was the first Swedish woman to become a professor.

[16]This was abstracted in the *Geographical Review* 12(1922):132–133.

[17]This derived from Jonasson's dissertation, "Agricultural Regions of Europe," completed at Clark University in 1926.

Gerard De Geer

Sten De Geer

Torsten Hägerstrand

William William-Olsson

very knowledgeable of urban morphology and city problems and the distribution of population, trade, and industry. The mapping of umlands (urban hinterlands) by way of a variety of indicators was accomplished for Stockholm, Jönköping, and Gothenburg. Within Stockholm geographers undertook research in functions and regions. In 1928 De Geer moved to Gothenburg; Hans Ahlmann was appointed in Stockholm.

Ahlmann, a glaciologist who was interested in geomorphology, was made responsible for human geography. Recognizing the need for studies in urban geography, he became very enthusiastic over his discovery of the nine-volume work *The Regional Survey of New York and Its Environs 1924–1928* by R. M. Haig et al. This work aroused interest in the growth of Stockholm. William William-Olsson joined three students, who with Ahlmann published the first volume of the Stockholm investigation, *The Internal Differentiation of Stockholm* (in Swedish, 1934). The study included large amounts of data about each house and household, shop, employed person, and numerous occupations in Sweden. The work focused on people and not dwellings. Most details were mapped, and generalizations were made only in the final stages of the work. The geographers consulted with architects and town planners as well as urban authorities from other countries. In 1937 William-Olsson published his doctoral thesis, "The Geographical Development of Stockholm, 1850–1930." In this work he wrote about the inner differentiation and changes in the city. As a result, the officials responsible for the long-term planning of Stockholm asked William-Olsson to assist them. He did so, establishing a program that demonstrated the interplay between Stockholm and the rest of Sweden. In 1941 he published *The Future Development of Stockholm*, a work that presented a theory concerning the relationship between town and countryside based on basic and nonbasic activities. With this work William-Olsson realized that generalizations can be unsafe. His forecasts for Norrland and a number of small towns eventually proved quite accurate, however.

During the 1940s the number of geography chairs in Sweden doubled, and chairs in both physical and human geography were created in some institutes (Uppsala in 1948 and Lund in 1949). In 1948 the Swedish Council for Social Science Research was created in order to fund geography centered on human-social themes and methodology. By 1950 there were nine professors of geography in Sweden, and geography had attained a sound position in the school curricula. Societies and publications were performing soundly. However, the regional geography of earlier years and other traditional icons began to pale. Gerd Enequist published "The Distribution of Main Occupational Groups in the Communities of Sweden, 1930" in 1943, and three years later produced "The Distribution of Main Occupational Groups in the Communities of Sweden, 1940." These two studies of the economically active population were innovative and were of considerable value to regional planning.

THE BEGINNINGS OF CONTEMPORARY SWEDISH GEOGRAPHY

Another innovative occasion in the context of traditional Swedish geography was the arrival of Edgar Kant from Estonia in 1945. Kant introduced Walter Christaller's central-place theory and the work of von Thünen to Swedish geography. He also introduced his own work concerning the Estonian town of Tartu (1926). In 1945–46 Kant's research assistant was Torsten Hägerstrand, who was also interested in migration processes. In addition, Kant brought the Swedish geographers into contact with the French geographers, perhaps especially through Georges Chabot. In 1948 Kant wrote "About Sociological Regionalism, Social Time and Social Space," an article that had considerable impact on Swedish geography at a time when change was imminent. As Anne Buttimer (1987) has pointed out, Kant had a large network of geographer friends and correspondents, innate ability, capacity in several languages, and the desire to enjoy scholarship. He brought an outsider's view to Swedish geography, the impact of which is still unclear, and he also brought knowledge of particular significance to the historian of geographical thought. He traced the origins of the idea of central-place theory to an eighteenth-century study on the size and spacing of towns and the approximately hexagonal structure of hinterlands by a Swedish economic historian, Carl Brunckman, in 1756. Kant produced evidence that several works—for example, Frère de Montizon's "Philosophical Map of French Population" (1830)—had previously adopted the "dot" map method of Sten De Geer.

Already in the 1930s a numerate geography had become visible. By the end of the 1940s migration studies of an individual nature were being produced. In 1947 Torsten Hägerstrand published a work of exceptional methodological significance concerning a rural parish in Östergötland and the contact of the individuals concerned with the outside world. Hägerstrand followed this work with another study of Simrishamn, in Scania, in 1949. At a symposium in Lund in 1954 Hägerstrand and other scientists talked about their studies, and the product of this symposium was published as a report in the series "Lund Studies in Geography," 1957. Included in this report were a study by Reino Ajo on patterns of migration in Finland, another by Esse Lövgren (Uppsala) entitled "Mutual Relations between Migration Fields: A Circulation Analysis," and Hägerstrand's contribution, "Migration and Area. Survey of a Sample of Swedish Migration Fields and Hypothetical Considerations on Their Genesis."

Meanwhile, in 1953 Hägerstrand examined innovations and diffusions and wrote a dissertation on the subject, later translated by Allan Pred (1967) as *Innovation Diffusion as a Spatial Process*. Using simulated Monte Carlo models, Hägerstrand compared the facts of geographic contact with theoretical constructs. He focused on the process of innovation and adopted mathematical and statistical methods that were new to this time in this combination. He examined the diffusion of innovations among the population in a part of central Sweden. Examples ranged from car ownership to the improvement of pasture. His empirical findings are of less significance, however, than his analysis of the diffusion process. Using random samples (Monte Carlo simulation), Hägerstrand was able to build a stochastic model of the process of diffusion. The Hägerstrand model permitted simulation and testing against empirical data. It could be seen that circumstances at one stage in the process would influence distribution forms at subsequent stages. The model therefore would be of utility to planners. Hägerstrand represented a systematic geography that used regions as analytical units for studying social change (Hägerstrand, 1953). This moved Swedish geography further from traditional and descriptive geography and established geography as a social science of special value to regional planners. This process had been initiated in 1950, when physical and human geography were made two separate areas of study, and human geography was placed in the social science faculty.

Kant's work had a special impact on Hägerstrand. It was Kant who "brought Europe to Lund: he had been a student in Hungary, Austria, Switzerland, Germany, Holland and France" (Hägerstrand, 1983:246), and who introduced Hägerstrand to von Thünen, J. G. Kohl, Tord Palander, Walter Christaller, and Pitirim Sorokin. Kant's lectures in social geography were particularly significant to Hägerstrand, for they "paved the way in my mind for my later work on diffusion of innovation." Hägerstrand was made a full professor in 1957, after which he and Sven Godlund began to train planners and to investigate regional planning problems. He "felt it as our obligation to try to open a new labour market for our advanced students" (Hägerstrand, 1983:252). The disadvantage of this program was that virtually a generation of graduate students entered planning and were lost to the geographical profession. Yet these made-over geographers could and did demonstrate their usefulness to society. In this context Lund has played a major role, and Hägerstrand (and later Gunnar Törnquist) made significant contributions to geography, to time-geography and location theory, and to regional planning. The government created an agency, the Expert Group for Regional Studies, that helped foster the impression that geography was a planners' profession.

Meanwhile, in 1960 Stockholm was the host city to the Nineteenth International Geographical Congress. This was the first time the congress had met in Scandinavia, and it helped to bring attention to the geography and geographers of Scandinavia. *The Geography of Norden*, edited by Axel Sømme, included contributions by numerous geographers. One of them, K. E. Bergsten, wrote the chapter on Sweden. The whole work was cast in an interesting traditional regional mode and was widely read. Symposia were held in different centers in Sweden, which helped bring the early phases of the new geography of the United States into contact with Swedish geography. Another institutional development that strengthened Swedish geography was the establishment of a foundation by the Central Bank of Sweden on its 300th anniversary. This foundation was created to sponsor scientific investigation dealing with the human condition and future. The universities of Lund and Gothenburg were funded for a project entitled "The Process of Urbanization."

During the 1960s Hägerstrand portrayed individual behavior in time and space, adopting three-dimensional models. This research marked the initiation of what has come to be called time-geography. According to this geographic genre, time and space are resources that constrain activity—time restricting all individuals and space being more or less restrictive, depending on economic status and other individual circumstances. Time-geography brought even more prominence to Lund as a center for geographical innovation. Many studies were made in time-space geography and have been collected by T. Carlstein et al. in *Timing Space and Spacing Time* (1978). Further work was conducted by Tommy Carlstein in *Time Resources, Society and Ecology: On the Capacity for Human Interaction in Space and Time in Preindustrial Societies* (1982); Bo Lenntorp, *Paths in Space-Time Environments* (1976); Solveig Mårtensson, *On the Formation of Biographies in Space-Time*

Environments (1979); and Allan Pred (ed.), *Space and Time in Geography: Essays Dedicated to Torsten Hägerstrand* (1981).

What has been so frequently called the quantitative revolution that began in the United States in the late 1950s made itself known in Sweden. Visits were made by U.S. and British geographers (including W. Bunge, P. Gould, W. Isard, R. Morrill, J. Nystuen, A. Pred, and E. Ullman) and by W. Christaller, C. van Paassen, and A. Kuklinski. Hägerstrand traveled in Britain, continental Europe, and the United States. Swedish geography could be more adequately described as an applied social science rather than as a quantitative spatial science. Paradigmatic pluralism was maintained. Although spatial analysis remained dominant, other geographies were being exercised. Other geographers were heavily involved in model development with an increasing mathematical component and theory building. Studies on the location of industry and hospitals as well as revision of local government boundaries were undertaken. The division of institutional geography into physical and human branches led to a debate between W. William-Olsson and Carl D. Hannerberg concerning human geography. W. William-Olsson argued for a traditional holistic view, while Hannerberg explained the field as a complex of special disciplines. At this time Å. Sundborg at Uppsala was undertaking experimental work on fluvial processes. Physical geographers were employed in the Swedish environmental protection administration. Physical geography as a genre began to affiliate with earth science and became a little more preoccupied with people and less with theory. New trends developed in agrarian and rural research with reference to human territoriality, typified by Hannerberg (1976) and Malmberg (1980).

During the 1960s and 1970s time-geography, based on the life of the individual, began to have a gradual impact on the character of geography as people, and not money (as in a commodity value), began to be more the object of study. Wahlstrom urged that research and planning concerning the environment needed a "united historical-ecological-geographic perspective." This is typified by research in recreation. Opportunities did begin to emerge for those individuals who had specialized in the evolution of the rural landscape, for interest in land use and landscape conservation had increased. Simultaneously, there developed a keen interest in local history and historical geography, although specialists were not available to take advantage of this interest. Rapid technological developments in computer cartography and remote sensing were shared with the international community. Radical geography showed itself in pockets of thought but did not seem to be part of the mainstream.

By the 1980s Swedish geography had been moving away from its standing as an applied social science to a more theoretical social science. This thrust toward concern with methodology and theory was in keeping with the trends of many other national geographies. T. Lundén has pointed out that "the surge of applied geography in the 1950s–1970s produced important knowledge . . . but it never produced a theoretical discussion of the discipline" (1986:185). Professors have perhaps been encouraged to emphasize the application of research rather than the generation of thought concerning the evolving nature of the geographer's quest. Owing to the Swedish university structure, faculty are very dependent on external funding, which leads them to practical problem solving while teaching heavy loads. This has the inevitable effect of reducing critical long-range geographical research. In addition, the division of the discipline into physical and human geography has not been accepted by the older generation, for whom integration of these two fields was indispensable. This division has also stirred discussion concerning the unity of the discipline, and efforts have been made in the direction of methodology and philosophy at the expense of substantive research.

This thrust was assisted by the arrival of Anne Buttimer in Lund in 1977.[18] She contributed two interesting essays, both of which were in a philosophic vein: "Reason, Rationality, and Human Creativity" and "Musing on Helicon: Root Metaphors in Geography" (Buttimer, 1979 and 1982). Encouraged by Hägerstrand, she undertook a study in autobiography that sought to achieve a more human understanding of thought and accomplishment. Illustrative essays based on this project were published in 1983, and the book functions as something of a cross-section of the history of geographical thought for the previous 50 years. Of particular significance for an understanding of Swedish geography is Hägerstrand's autobiographical statement entitled "In Search for the Sources of Concepts" (Buttimer, 1983:238–256). In 1978 Buttimer and Hägerstrand initiated the Dialogue Project. "Invitation to Dialogue" was offered in 1980 and launched a series of discussions, many of which were filmed. These films serve as an archival source and as a repository of valuable data concerning the history of geography since World War II (Buttimer, 1986, 1993). Of particular value for an understanding of Scandinavian (and Swedish) geography is *Geographers of Norden: Reflections on Career Experiences* by Hägerstrand and Buttimer (1988). Interviews with K. E. Bergsten, G. Enequist, G. Hoppe, and K. G. Izikowitz represent the Swedish component of the book. This work has been welcomed for the perspective it offers on disciplinary evolution.

CONTEMPORARY GEOGRAPHY AND SOME SPECIALIZATIONS IN THE SWEDISH UNIVERSITIES

Feminism and gender issues arrived in the Sweden of the 1980s and 1990s. Gender dimensions were explored in the context of physical planning, regional policy, and labor markets. These studies revealed a glaring absence of consideration for male and female realities and the predominance of masculine discourse in Swedish planning practices and social relations. While thus still moving in familiar geographical realms, these gender studies transcended the confines of planning discourse and the dualism of work and everyday experience. Revealed was the gendered structuring of bureaucratic practices, wage-labor, and family life. Time-geography was creatively reworked in a feminist direction, which made it possible to visualize the daily paths and projects of work and everyday life in a way that would reveal patriarchal relationships or gender myopia in conventional physical planning. Other

---

[18]Little had been published previously in the history of geographical thought. Notable exceptions are Leo Bagrow, "The Origins of Ptolemy's Geographia," in *Geografiska Annaler* 27(1945):318–387 and Otto Nordenskiold's history of the development of scientific geography during the nineteenth century (1921). See *Geographical Review* 12(1922):324.

scholars turned their energies to consideration of the economic restructuring of Swedish society and the consequences of cutbacks and plant closures on the male and female labor force at different locations and in different regions (Forsberg, 1994; Friberg, 1993).

In most cases feminist researchers (overwhelmingly women) found disciplinary and personal inspiration and connection in the realms of sociology, history, and economic history rather than in the male-dominated corridors of human geography. It was only in the late 1980s under the influence of Nordic and Anglo-American debates on regions and patriarchal structures that the spatial dimension in feminist theory was treated seriously. Increasingly, female scholars gained better positions in the academic hierarchy; women-specific conditions, gender differences, and unpaid reproductive and household work are now accepted as significant research themes.

A second major theme in contemporary Swedish geography is the ongoing interest in environmental issues, revived during the 1980s and 1990s. Environmental concern expressed itself through a new consideration for and questioning of twentieth-century modes of life and resultant pollution. Methodological emphasis was initially placed on modified forms of systems analysis and time-geography and more recently on reflexive approaches (Hallin, 1994; Hultman, 1998). Some were concerned that, paralleling developments of a so-called ecological modernization ideology abroad, Swedish practices of production, politics, and consumption had become much greener in text, speech, and image—loudly reproducing the organic lexicon of recycling, sustainability, and ecology—but had accomplished little in reality. Even the (global) environmental movement of the late twentieth century seemed to be more concerned with coordinating interests with other transnational actors and states than with critical scrutiny and design of serious alternatives. Others offered more empirical illustrations of the difficulties and strategies of planning for nature preservation and environment in various Swedish settings.

Gender and environment were also major themes in development geographies of Africa and Asia, especially in the 1990s (Dahl, Drakakis-Smith, and Narman, 1993; Lundqvist, 1998; Malmberg, 1988). Otherwise, the geographies of development have been diverse, ranging from the Green Revolution to water supply, urban expansion, education, and earthquakes. Development geography has made useful contributions through case studies and critical examination of (Swedish) development programs but has been less interested in fashionable theoretical debates on, for example, postcolonialism.

A relatively novel theme in Swedish human geography has been the issue of immigration, multiculturalism, racism, ethnic segregation, and integration. Some of this literature was downright critical of the Swedish planning system and global social as well as economic mechanisms that excluded or disadvantaged "foreigners" in an increasingly multicultural society (Andersson, 1993; Tesfahuny, 1998). This kind of research came together well with developments in society and frequently embraced analysis of the powerful practices, mechanisms, and postcolonial discourses of the "Other."

In recent decades an informal division of labor—frequently overlapping—has developed in university departments of geography. At Stockholm University Staffan Helmfrid (during the 1980s vice chancellor of the university) is leading a team that is producing a new *National Atlas of Sweden*, which now comprises 20 volumes in Swedish and English (also computerized). This atlas is being produced by all geography departments in Sweden as well as a large number of government boards.

In the area of research, two older traditions are being further developed at Stockholm University. First is the historical geography of Sweden (its preindustrial landscape), a study that now includes not only old maps and other written sources but also work in the field in cooperation with archaeologists. A hypermodern Geo Processing Unit is now used for landscape analysis. Mats Widgren (2000) and colleagues have initiated comparative historical landscape research in the arid lands of Africa. This is an important contribution to environmental history. A second thrust is toward polar research, with an emphasis on the connection between glaciology and climatology. A new activity is remote sensing, which has developed into a strong section serving both physical and human geography. It is clearly an important tool for the production of Sweden's atlas.

At Uppsala University physical geography is well developed, with special reference to the study of rivers, particularly erosion and sedimentation. The Geography Department has a well-equipped laboratory and earlier included Åke Sundborg, an international authority in the field. In human geography population and development are presently of great importance. Professor Sture Öberg worked with the staff of Mauritz Strong for two years in preparing the environmental conference of 1992 in Rio de Janeiro. System analysis and simulation have been developed as instruments at Uppsala. At Umeå University, situated in the north of Sweden, Professor Erik Bylund (a student of Gerd Enequist) created a tradition of regional policy studies, with an emphasis on the problems of remote and sparsely populated areas. At Göteborg University the dominant specialty is the geography of transportation. In physical geography a significant theme has been urban climatology.

Lund University has developed two themes in physical geography: microclimatology and soil conservation in developing countries, with particular reference to Africa. In economic geography primary attention has been given to Sweden's relation to the Common Market and the possible regional consequences of a closer integration. In human geography an emphasis has been placed on development problems in Africa (participation in the organization of censuses). In addition, various themes relating to the concept of sustainable development as seen in the Swedish perspective have been adopted. An example is how a new energy budget would influence lifestyles in the future. (Time-geography is of relevance here.) Hagerstrand led a Danish-Norwegian-Finnish-Swedish team that studied urban metabolism—that is, the flow of matter in industrial society and the environmental consequences of these flows.

Hagerstrand (1992) has written that

> our old tradition to take up empirical studies with a rather strong applied inclination is still dominating. Social theory is a side issue. The big theoretical question today is how to integrate the physical and human branches of geography in ways which do not repeat the simplicities of the 20s and 30s. I would say that the central theme is "the human use of the Earth." Clearly an understanding requires some sort of "social theory" but not one which is defining away the material world.

What we need—as I see it—in order to be able to make sense of the physical-human question is a theory of technology.[19]

In the last decade both physical and human geography have been returned to the schools and teacher training curricula. Many smaller universities have begun courses and programs in tourist studies. GIS research and education is now established at most universities. Nordplan has been reorganized into an applied research organization called Nordic Center for Spatial Development. The linkages between human geographers and pragmatic issues of planning and regional development remain. Economic geography and geographies of development have also surged. And the separation of physical and human geography in the 1950s has found rapprochement since the 1970s.

Geography in Sweden is now ensconced in a larger profession than previously, shares viewpoints with the initial community, simultaneously undertakes research on many fronts always with its concern for sensible occupation of the Earth by its life forms.

REFERENCES: CHAPTER 13

Ahlmann, H. W. 1921. "The Economical Geography of Swedish Norrland." *Geografiska Annaler* 3:97–164.

Ahlmann, H. W., and Friberg, N. 1938. "Neue Strömungen in der nordischen geographischen Forschung." *Geographische Zeitschrift* 44, nos. 7–8:307–315.

Andersson, R. 1993. Immigration Policy, and the Geography of Ethnic Integration in Sweden. *Nordisk Samhällsgeografisk Tidskrift* 16:14–29.

Anrick, C. J. 1923. "A Popular Geographic Club of Sweden, The Swedish Touring Club and Its Activities." *Geographical Review* 13:608–612.

Antevs, E. 1922. *The Recession of the Last Ice Sheet in New England.* American Geographical Society Research Series No. 11, W. L. G. Joerg, ed.

———. 1925. "Swedish Late-Quaternary Geochronologies." *Geographical Review* 15:280–284.

———. 1928. *The Last Glaciation: With Special Reference to the Ice Retreat in Northeastern North America.* American Geographical Society Research Series, No. 17.

Asheim, B. T. 1979. "Social Geography—Welfare State Ideology or Critical Social Science?" *Geoforum* 10:5–18.

———. 1987. "A Critical Evaluation of Postwar Developments in Human Geography in Scandinavia." *Progress in Human Geography:* 333–353.

Blüthgen, J. 1941. "Entwicklung, Stand und Aufgaben der Geographie in Schweden (Sammelreferat)." *Zeitschrift für Erdkunde* 9, no. 3–4:65–88.

Buttimer, A. 1979. "Reason, Rationality and Human Creativity." *Geografiska Annaler* 61B:43–49.

———. 1982. "Musing on Helicon: Root Metaphors in Geography." *Geografiska Annaler* 64, no. 2:89–96.

———. 1983. *The Practice of Geography.* London and New York: Longman Group Ltd.

---

[19]Letter from T. Hägerstrand to G. Martin, July 3, 1992.

————. 1986. *Life Experience as Catalyst for Cross Disciplinary Communication.* University of Lund, Department of Geography.

————. 1987. "Edgar Kant 1902–1978." *Geographers: Biobibliographical Studies.* 2:71–82. London: Mansell.

————. 1993. *Geography and the Human Spirit.* Baltimore, Md.: Johns Hopkins University Press.

Carlstein, T., et al. 1978. *Timing Space and Spacing Time.* 3 vols. London: Arnold.

Claval, P. 1976. "La brève histoire de la nouvelle géographie." *Rivista Geografica Italiana* 83, no. 4:395–424.

Dahl, J., Drakakis-Smith, David & Närman, Anders, eds. 1993. *Land, food and basic needs in developing countries.* Meddelanden från Göteborgs universitets geografiska institutioner series B 83.

De Geer, G. 1912a. "A Geochronology of the Last 12,000 Years." *Comptes Rendus Congrès Géol. Internatl. à Stockholm*: 241–253.

————. 1912b. "Geochronologie der letzten 12,000 Jahre." *Geol. Rundschau.* Vol. 3, pp. 457–471. Leipzig.

————. 1921. "Correlation of Late Glacial Clay Varves in North America with the Swedish Time Scale. *Geol. Fören. Förhandl.*, 43:70–73.

————. 1926. "On the Solar Curve as Dating the Ice Age, the New York Morain, and Niagara Falls through the Swedish Timescale." *Geografiska Annaler* 8:253–284.

De Geer, S. 1908. "Befolkningens fördelning på Gottland." *Ymer* 28:240–253.

————. 1919. *Karta över befolkningens fördelning i Sverige.* Stockholm: Wahlström and Widstrand.

————. 1922a. "A Map of the Distribution of Population in Sweden: Method of Preparation and General Results." *Geographical Review* 12:72–83.

————. 1922b. "Storstaden Stockholm ur geografisk synpunkt." *Svenska Turistföreningen Arsskrift*: 155–168.

————. 1923a. "On the Definition, Method, and Classification of Geography." *Geografiska Annaler* 5:1–37.

————. 1923b. "Greater Stockholm: A Geographical Interpretation." *Geographical Review* 13:497–506.

————. 1927. "The American Manufacturing Belt." *Geografiska Annaler* 9:233–359.

————. 1928a. "Das geologische Fennoskandia und das geographische Baltoskandia." *Geografiska Annaler* 10:119–139.

————. 1928b. "The Subtropical Belt of Old Empires." *Geografiska Annaler* 10:205–244.

Folke, S. 1985. "The Development of Radical Geography in Scandinavia." *Antipode* 17:13–18.

Forsberg, G. 1994. Occupational Sex Segregation in a "Women-Friendly" Society: The Case of Sweden. *Environment and Planning A* 26:1235–1256.

Freeman, T. W. 1967. *The Geographer's Craft.* New York: Barnes & Noble.

Friberg, T. 1993. *Everyday Life: Women's Adaptive Strategies in Time and Space.* Lund Studies in Geography, series B: Human geography 55 Lund: Lund University Press.

Godlund, S. 1986. "Swedish Social Geography." In John Eyles, ed., In *Social Geography in International Perspective.* Pp. 96–150. New York: Barnes & Noble.

Gren, M. 1996. Visible/Invisible Boundaries. *Nordisk Samhällsgeografisk Tidskrift* 22:89–98.

Hägerstrand, T. 1952. "The Propagation of Innovation Waves." *Lund Studies in Geography*, Series B, No. 4.

————. 1953. *Innovationsförloppet ur korologisk synpunkt.* Lund: C. W. K. Gleerup. Trans. A. Pred.

————. 1967. *Innovation Diffusion as a Spatial Process*. Chicago: University of Chicago Press.

————. 1973. "The Domain of Human Geography." In R. J. Chorley, ed., *Directions in Geography*. Pp. 65–88. London: Methuen.

————. 1982. "Proclamations about Geography from the Pioneering Years in Sweden." Pp. 119–126. *Geografiska Annaler*, Series B:64.

————. 1983. "In Search for the Sources of Concepts." Pp. 238–256. In *The Practice of Geography*.

Hägerstrand, T., and Buttimer, A., eds. 1988. *Geographers of Norden: Reflections on Career Experiences*. Lund Studies in Geography, Series B. Human Geography 52. Lund University Press.

Hallin, Per-Olof. 1994. Energy, Lifestyles and Adaptation. *Geografiska Annaler* 76B:173–186.

Helmfrid, S. 1972. "Historical Geography in Sweden." In A. Baker, ed., *Progress in Historical Geography*. Pp. 63–89. Newton Abbot: David & Charles.

————. 1976–1977. "Hundra år svensk geografi." *Ymer*: 360–372.

Hultman, Johan. 1998. *The Eco-ghost in the Machine: Reflexions on Space, Place and Time in Environmental Geography*. Meddelanden från Lunds universitets geografiska institutioner Avhandlingar 134. Lund: Lund University Press.

Kant, E. 1953. "Migrationernas klassifikation och problematik." In *Svensk Geografisk Arsbok*. Pp. 180–209. (Reprinted and translated as "Classification and Problems of Migrations." In P. L. Wagner and M. W. Mikesell, eds., 1962, *Readings in Cultural Geography*. Pp. 341–354. Chicago: University of Chicago Press.)

Leighly, J. B. 1933. (Obituary of Sten de Geer) in *Geographical Review*, 23:685–686.

————. 1983. "Memory as Mirror." Pp. 80–89. In *The Practice of Geography*.

Lundén, T. 1982. "Stockholm 1980—Stalemate in Planning." *Geografiska Annaler*, Series B, 64:127–133.

————. 1984. *The Role of Political Planning in Determining Land Use—The Case of Järfälla Municipality, Sweden 1958–1983*. Stockholm: University of Stockholm International Graduate School.

————. 1986. "Political Geography around the World. VI. Swedish Contributions to Political Geography." *Political Geography Quarterly* 5, no. 2:181–186.

Lundqvist, J. 1998. *Avert looming hydrocide* Stockholm: Sida publications on water resources 12.

Malmberg, A. 1996. "Industrial Geography: Agglomeration and Local Milieu." *Progress in Human Geography* 20:392–403.

Malmberg, G. 1988. *Metropolitan Growth and Migration in Peru*. Umeå: Geographical reports 9.

Mead, W. R. 1972. "Luminaries of the North: A Reappraisal of the Achievements and Influence of Six Scandinavian Geographers." *Transactions of the Institute of British Geographers*, no. 57:1–13.

————. 1983. "Autobiographical Reflections in a Geographical Context." Pp. 44–61. In *The Practice of Geography*.

Mels, T. 2002. "Nature, Home, and Scenery: The Official Spatialities of Swedish National Parks." *Environment and Planning D: Society and Space* 20:135–154.

Nelson, H. 1933. *Svensk Geografiska Årsbok*. Lund, pp. 185–192. (With an English summary.)

Nordenskjöld, O. 1920. *Geografisk Forskning og geografiske Opdagelser i det nittende Aarhundrede*. Copenhagen: Nordsik Forlag.

————. 1921. (Swedish edition). *Geografisk Forskning og geografiske Opdagelser: det nittende Aarhundrede*.

Olsson, G. 1991. *Lines of Power/Limits of Language*. Minneapolis: University of Minnesota Press.

———. 1993. "Chiasm of Thought-and-Action." *Environment and Planning D: Society and Space* 11: 279–294.

Pred, A. 1977. "Planning-related Swedish Geographic Research." *Economic Geography* 53, no. 2:207–221.

———. 1984. "From Here and Now to There and Then: Some Notes on Diffusions, Defusions and Disillusions." In M. Billinge, D. Gregory, and R. Martin, eds., *Recollections of a Revolution: Geography as Spatial Science*. Pp. 86–103. New York: St. Martin's Press.

Ström, E. T. "Svensk kulturgeografisk forskning me särskild hänsyn till tiden efter 1940."

Swedberg, S., and Tengstrand, K. G. 1934. "Life of S. de Geer." *Gothia*, Göteberg, 3, 1–21, 23–8 (in Swedish).

Tesfahuney, M. 1998. "Mobility, Racism and Geopolitics." *Political Geography* 17:499–515.

Van Valkenburg, S. 1951. "The German School of Geography." In T. G. Taylor, ed., *Geography in the Twentieth Century*. Pp. 91–115. New York. Philosophical Library, London: Methuen.

Widgren, M. 2000. "Islands of Intensive Agriculture in African Drylands: Towards an Explanatory Framework." In Barker, Graeme & Gilbertson, David, eds. *The Archaeology of Drylands*. Routledge, London, pp. 252–267.

William-Olsson, W. 1940. "Stockholm: Its Structure and Development." *Geographical Review* 30:420–438.

———. 1975. "A Prelude to Regional Geography." *Geografiska Annaler*, Series B, 57:1–19.

———. 1983. "My Responsibility and My Joy." Pp. 153–166. In *The Practice of Geography*.

Zimmermann, M. 1933. *Etats Scandinaves. Régions polaires boréales*. Paris: Libraire Armand Colin.

14

THE NEW GEOGRAPHY IN JAPAN

The more one studies any subject, but especially, perhaps, a subject that one at first thinks mistakenly has been relatively neglected, the more one realizes the depths of one's own ignorance.
—*Sir Hugh Cortazzi*, Isles of Gold: Antique Maps of Japan

Geography in Japan is many centuries old. In earlier times geography was indigenous and related to both a knowledge of the surface of the earth and celestial mechanics. With the Meiji Restoration in 1868, the government began to import science and technology from Western countries and from Chinese sources.[1] By 1873 a number of scholars had been invited to Japan to help those in the newly founded educational system. Some of these scholars were from Germany, and on March 22 (the German emperor's birthday) they founded the German East-Asiatic Society. This organization became a forum for lectures and publications. It became an intellectual link in the development of Japanese—German relations and helped bring new ideas to a society wishing to modernize.

In 1869 an Office of Geography was established in the Ministry of Civil Affairs. This office was combined with the Surveyor's Office within the Ministry of Industries, and in 1876 it became the Geography Section of the Ministry of Home Affairs. In the next 18 years an embryonic modern geography was established: the Hydrographical Department of the Ministry of the Navy in 1870; the Topographic Survey and Cartographic Office in 1871; the Tokyo Meteorological Observatory in 1875; the Tokyo Astronomical Observatory in 1888; and the Land Survey Department of the Army, also in 1888.

A modern educational system was established in 1872. In 1877 Tokyo Imperial University (the University of Tokyo) was formed by the amalgamation of several schools. Normal schools were built, and in 1886 higher normal schools were established. Typical of the higher normal schools was the Tokyo Women's Higher Normal School (1890). Each of these institutions offered courses in geography. Until approximately 1887 foreigners were employed in government and teaching. Earth scientists who helped lay the foundation for the development of geography were Benjamin Lyman (from the United States) and Edmund Naumann (from Germany).

---

[1]From 1637 until 1853 the only outside contacts that were permitted were made through a Dutch trading post on an island in the harbor of Nagasaki. In 1853 Commodore Matthew Perry of the U.S. Navy sailed into Yokohama harbor, and thereafter outside contacts became more and more frequent. With the Meiji Restoration of 1868, which returned the emperor to the throne, a policy of rapid modernization was instituted.

Geography soon became a respected field because it encouraged the Japanese to work harder and to modernize their country, which was shown to be less prosperous than the Western countries. As a result, a substantial demand arose for geography teachers and for geographic literature (Takeuchi, 1990). In 1869 a compilation of information on the countries of the world was published and gained immediate popularity. The writer was a Japanese scholar named Yukichi Fukuzawa, and his book on world geography was in the tradition of the universal geographies of the West. Its impact on the Japanese was enormous, for it introduced its readers to the world beyond Japan. In a decade interest in geography was sufficient to support the organization of the Tokyo Geographical Society (1879). This society was modeled on the Royal Geographical Society, and its main undertakings were research and exploration. The Tokyo Society published a journal entitled *The Journal of the Tokyo Geographical Society*, which strongly influenced geographical thinking in Japan.

Another book modeled on British geography, *A Handbook for Travellers in Central and Northern Japan* (Yokohama, 1881), was in the mode of *Murray's Handbook* (Suizu et al., 1980). This book was revised and enlarged and then published as *A Handbook for Travellers in Japan*. It was added to *Murray's Handbook*, for Europeans were beginning to travel to Japan.

By 1886 Japan's educational system had been "completed." Japanese returned from study abroad and began to staff the educational institutions. Japanese studies now began to replace works imported from abroad.

Geography as a professional field in the universities owes its first development to the geologists, historians, and agricultural specialists who became interested in the study of the interrelations between human settlement and the physical earth. The geologist who led the way was Bunjiro Koto, the founder of the Department of Geology at the Imperial University of Tokyo. Koto had been a student at Tokyo. After completing an undergraduate course in geology, he had been sent at government expense to study geology in Germany, first at Munich and then at Leipzig. After his return to Japan in 1890, he began to offer courses in geography in the new Department of Geology at Tokyo.

The first scholars to be appointed professors of geography in Japan were Koto's students, who had taken geology at Tokyo and then had studied geography in Germany. *The Journal of Geography* was established at Tokyo Imperial University, offering articles on Japanese problems and the Japanese environment. The literature began to grow. Yazu published *Physical Geography* (1890), and S. Shiga published *Lectures on Geography* 1889 (there were numerous later editions) and *Japanese Landscapes* (1894). By 1903 *Japanese Landscapes* had gone through 15 editions; it is available even today in paperback. In the same year, 1903, Uchimura wrote *On Geography*, which incorporated the work of Arnold H. Guyot, Mary Somerville, George P. Marsh, and other Westerners. The book was retitled *Man and Land* and outlined geography as conceived by Ritter. Makiguchi wrote *Life Geography* based on Ritter and Ratzel's work and created the framework for human geography in Japan. This work was to human geography what Yazu's was to physical geography. Then in 1911 Noguchi published *Outline of Geography: Human Studies*, and human geography began to replace physical geography as the dominant geography.

Two monumental works that announced the coming of age of Japanese geography were *Geographical Dictionary of Japan, 1900–1909*, compiled by Yoshida, and *Regional Geography of Japan* (10 vols.), 1903–1915 by Yamasaki and Sato. Of special interest to American readers are the travels and resultant publications of two female American geographers. Eliza R. Scidmore (corresponding secretary of the National Geographic Society) traveled to Japan in the 1890s and produced several articles and books including *Jinrikisha Days in Japan* (1891) and *East to the West: A Guide to the Principal Cities of the Straits Settlements, China and Japan* (1898). Ellen Semple traveled to Japan in 1911 and published two articles on Japanese agriculture and colonial methods. Meanwhile a Japanese translation of her *Influences of Geographical Environment* was published in 1917 and again in 1926. Ellsworth Huntington's *Asia: A Geography Reader* and *Civilization and Climate* were also translated into Japanese.

Two American geographers specialized in the study of Japanese geography, commencing in the 1920s: Robert B. Hall and Glenn T. Trewartha. (Other American geographers who wrote on the population of Japan included Mark Jefferson, Darrell H. Davis and John E. Orchard.) Special mention might be made of Trewartha's *A Reconnaissance Geography of Japan*, which was published in 1934. This work was revised and published in 1945 as *Japan, a Physical, Cultural and Regional Geography*. Twenty years later a further revised version was published as *Japan: a Geography*. This book seems to have been the main source of information concerning Japanese geography for American geographers especially in the 1930s and 1940s.

THE DEVELOPMENT OF GEOGRAPHY AS AN ACADEMIC SUBJECT, 1907–1923

Two wars that occurred late in the Meiji era—the Sino-Japanese War (1894–95) and the Russo-Japanese War of 1904–05—had an impact on the development of geography in Japan. In both of these Japanese victories, geography proved very useful, further stimulating the growth of the subject in the educational system.

The first professor of geography in Japan was appointed in 1907 in the Institute of History at the University of Kyoto. This was Takuji Ogawa,[2] whose geographical ideas were largely derived from the teaching of Eduard Suess at Vienna. Appointed with him as an assistant professor was a young historian named Goro Ishibashi. Ogawa and Ishibashi pioneered in the study and teaching of historical and regional geography. Ogawa's own writings included *Regional Geography of Formosa* (1896), *Study of Human Geography* (1928), *Studies of the Historical Geography of China* (1928), *Studies of Historical Geography* (1929), *Geography of War* (1934), *Life of a Geographer* (1941), and *Japanese Archipelago* (1944). He never lost his interest

---

[2]Takuji Ogawa graduated in geology from the University of Tokyo in 1896. He was then sent to Vienna for graduate study, where he worked with Eduard Suess. Ogawa was a very broadly trained scholar. Not only was he thoroughly qualified in geology and physical geography, but he also had a profound knowledge of the Chinese classics and was fluent in several Western languages.

in the physical earth, and, when an Institute of Geology was established at Kyoto, he was appointed to it, leaving Ishibashi to head the work in geography.

Fujita, another of Ogawa's students, wrote a *History of Japanese Houses* (1927) and *On Japanese Linear Measure* (1929). Another distinguished geographer of this period was Naomasa Yamasaki, who played a major role not only in providing trained geography teachers for the secondary schools but also in developing geography as a professional field requiring graduate training in the universities.[3] Yamasaki had derived his geographical ideas from Albrecht Penck in Vienna; the record of his studies and his teaching clearly reflects his teacher's wide range of interests. Yamasaki investigated the glacial landforms of the high mountains of Japan and the landforms produced along fault lines. Between 1903 and 1915 Yamasaki coauthored with Denzo Sato and others *The Regional Geography of Great Japan.* This work, published in 10 volumes and more than 10,000 pages long, was the first detailed regional study written about Japan. It is believed that as a result of this extraordinary work, Yamasaki gave lectures to the crown prince (later the emperor). At the request of the Ministry of Railways Yamasaki supervised *The Japan Guide Book* (published in the 1930s after his death). He also published extensively on volcanoes, glaciation, and seismology, and he is credited with developing geography from a subject matter to a discipline. He traveled abroad once a year for most of his life and thereby distinguished himself in Japanese geography as an internationalist. Yamasaki was the first government scholar sent to Europe to study geography. Akira Nakanome was the second such scholar. He also studied with Albrecht Penck and Eduard Brückner. Under the latter's influence he became enthusiastic about glacial climatology. Upon his return to Japan he joined with Takuji Ogawa to found the first chair of geography at Kyoto Imperial University. He continued in his post at the Higher Normal College of Hiroshima (Ishida, 2000).

In 1923 the Earth Science Association was established under the guidance of the Geology and Geography departments of Kyoto University. In the next year a periodical, *The Earth*, was published.[4] Then, in 1925, the Association of Japanese Geographers was formed in Tokyo. This association published the *Geographical Review of Japan*, which continues to this day.

---

[3] Naomasa Yamasaki graduated in geology from the University of Tokyo in 1895, one year earlier than Ogawa. After teaching in the undergraduate program at Sendai (Miyagi Prefecture) for three years, Yamasaki was sent to study in Germany and Austria. During the years 1898 to 1901, he studied with Rein at Bonn and with Albrecht Penck at Vienna. In 1902, when he returned to Japan, he was appointed to teach at the higher normal school in Tokyo (later Bunrika University, now Kyoiku University). Here he was a major factor in supplying well-trained teachers for the Japanese secondary schools. He continued to teach at the higher normal school even after he was appointed to teach economic geography in the Faculty of Law at the University of Tokyo in 1908. He became the first professor of geography at Tokyo in 1911, when he was appointed to this chair in the Institute of Geology. He was transferred to the new separate Department of Geography when it was created in 1919.

[4] In 1937 the Earth Science Association and its journal, *The Earth*, were terminated.

JAPANESE GEOGRAPHY AFTER 1923

The Great Kanto Earthquake of 1923 focused national attention on the geosciences. Cooperation between the disciplines was accomplished to facilitate planning for reconstruction. The first national population census of 1920 led to an interest in population geography.[5] In 1926 topographic maps for the whole of Japan at a scale of 1/50,000 were published; these were followed by maps at scales of 1/200,000, 1/25,000, and 1/10,000. These maps further supported interest in population geography. By the end of the 1920s came the economic depression and then the Manchurian affair in 1931. These incidents once again brought attention to geography.

Tokyo Bunrika University was founded in 1929, with a Geography Department that quickly became a major center for geographical research. This development facilitated the widespread acceptance of geography as an independent science (Noma, 1976). A number of systematic lectures and world outlines were published, which provided texts for various educational levels. When Yamasaki died in 1929, Tsujimura succeeded to his post at Tokyo Imperial University. Tsujimura also taught at the Tokyo Higher Normal School; in both institutions he educated many students. He also contributed substantially to the *Journal of Geography*. He specialized in geomorphology and published *Geomorphology* (1922), *Regional Geomorphology of Japan* (1929), *New Geomorphology* (1932–1935), and several studies on fault geomorphology.

Geomorphology introduced the concepts of W. M. Davis into Japan, and it contained illustrations taken from Japan, Europe, and the United States. When Tsujimura led the *Geographical Review*, geomorphology was its primary focus, although it did not neglect human geography. Tsujimura emphasized the significance of the morphology of the cultural landscape. Keiichi Takeuchi (1984) suggests that he was influenced by Berkeley geography and German *Landschaft*. Tanaka and Uchida directed the *Papers of the Otsuka Geographical Association*, published by Bunrika University. Tanaka studied the regions of Japan. Noma has suggested that his system resembled Hermann Lautensach's *Formenwandel*. Noma's practice produced followers who were known as the chorographic school. Uchida, with a historical approach, dealt with population, settlement, agriculture, and commerce and developed followers who would become known as members of the local studies school.

In 1932 Kyoto Imperial University sponsored *Monographs in Geography*, which emphasized human geography and was edited first by Goro Isibashi and later by Saneshige Komaki. Japanese geography had entered a very fruitful period in the 1930s. At Kyoto University Ogawa and Isibashi had also attained prestige in the area of historical geography; at Tokyo Imperial University under Yamasaki and then Tsujimura, geomorphology was emphasized; and in the Tokyo University of Science and Literature Tanaka and Uchida developed the regional method. Geomorphology "appears to have been strongly influenced by the doctrines of

---

[5]Interestingly, the American geographer Mark Jefferson had already published "The Distribution of People in Japan in 1913," *Geographical Review* 2, no. 5 (1916):368–372 and "Additional Notes on the Population Density of Japan," *The Geographical Review* 4, no. 4, (1917):320.

Davis and was qualitative, explanatory and deductive in character" (Kaizuka, 1976). Davisian thinking was helpful in the educational system, for it was both clearly defined and applicable. Other literature and geographical ideas came from European geography.

The most dominant European influence on Japanese geography in the first half of the century came from Germany, which exported the essence of geopolitik into Japan. Between the two world wars geographers felt that this kind of thinking, which had an impact on government policy, would help validate geography as a science (Takeuchi, 1980). Japanese geographers became familiar with the work of Rudolf Kjellén and then Karl Haushofer and Otto Maull. Geographers and journalists were caught up in the new geopolitik, adapted from the German model. Several of the German works, especially those by Haushofer, were translated into Japanese.[6] This philosophy proved disastrous for Japan, and after the war its proponents were removed from their academic appointments, with geography suffering some consequent loss of image.

Other notions imported from Germany included those from the work of J. von Thünen and Alfred Weber. The first reference to von Thünen's theory was made by an agricultural scientist, Tokiyoshi Yokoi, in 1892,[7] and the theory was presented to geographers by Tsunesaburo Makiguchi in 1903.[8] Curiously, it was not until the 1920s that von Thünen's work was described in English.[9] Japanese translations were made in 1916 and 1928 of both the 1826 and 1842 editions of von Thünen's *Der Isolierte Staat*. Not until 1966 was this book translated into English (edited by P. Hall). The early interest in von Thünen may have been due to the economists' interest and the role of agriculture in Japan. In Europe and the United States, urbanization was the dominant force, and agricultural goods could be obtained by opening new lands or importing from other countries. Another influence in Japan was the economist Alfred Marshall. Marshall may have had greater faith in the Ricardian theory of rent than in von Thünen's theory, which came from Germany, as did Alfred Weber's work concerning industrial location. During the 1920s in Japan, the influence of Wilhelm Windelband and H. Rickert was strong, and the work of Wilhelm Dilthey, Heidegger, and Husserl became known (Yamano, 1980). I. Watanuki introduced the work of Alfred Hettner, H. Kunimatsu introduced O. Graf, and T. Matsui proposed that economic geography was a branch of *Kulturwissenshaft*. Japanese geographers became concerned about philosophy and philosophers to the extent that they wished to comprehend the individual, the unique, and yet to submerge the individual in order to generalize. All this seems to have been inspired

---

[6]Between 1939 and 1948 two Japanese versions of Haushofer et al., *Bausteine zur Geopolitik* (1928), and three Japanese versions of Haushofer's *Geopolitik des Pazifischen Ozeans, Studien uber die Wechselbeziehungen zwischen Geographie und Geschichte* (1925) were published. Also, Haushofer's *Weltmeer and Weltmacht* (1937) and *Geopolitische Grundlage* (1939) were translated, respectively, in 1943 and in 1940.

[7]T. Yokoi, *Textbook of Agriculture* 1892:48–51.

[8]T. Makiguchi, *Geography for Human Life*, 1903.

[9]O. Jonasson, "Agricultural Regions of Europe," *Economic Geography* 1(1925):284–286.

by Germany geography. Also from Germany came the work of Leo Bagrow, an emigrant from St. Petersburg, who organized the first major exhibition on the cartography of Japan. This took place in 1932 at the Japan Institute in Berlin. He adopted the excellent work of Count Paul Teleki—*Atlas of the History of the Cartography of the Japanese Islands* (1909)—as a foundation for the exhibit.

In the 1930s the concept of the region also became significant in Japanese geography. M. Odauchi, in *A Study of Local Geography* (1930), introduced the concept of natural regions into Japan. Odauchi refers to the thought of Frédéric Le Play, Arthur Geddes, and Andrew J. Herbertson. Work by Unstead and Fleure, both of whom were advocates of a regional framework, was translated into Japanese (Hasegawa, 1980). In addition, the French school of geography, initially led by Vidal de la Blache, exerted a considerable influence on the concept of region. His *Principles of Human Geography* was translated into Japanese in 1933,[10] and other of his works began to appear in Japanese journals and literature. K. Iizuka studied in France and wrote a thesis concerning la Blache and his book. Iizuka also translated la Blache's *Principles of Human Geography* (1940) (Nozawa, 1980). Bibliographical notes were attached to the translation and conveyed much about French geography to Japan. Jean Brunhes, one of la Blache's disciples, who had written *Human Geography*, saw this book translated into Japanese.[11] François Ruellan, a French geographer who was teaching in Japan at the time, reinforced the concept of region through his publications in Japanese geographical periodicals.

THE WAR AND POSTWAR REESTABLISHMENT

Geopolitics flourished in the environment of the late 1930s. In 1941 the magazines *Geography* and the *Scientific Pen* both issued special numbers devoted exclusively to geopolitics. On November 10, 1941, the Association of Geopolitics of Japan was established and published a journal titled *Geopolitics*. Saneshige Komaki, at Kyoto University, wrote his "Declaration of Japanese Geopolitics" (1940), which attracted considerable attention. However, Komaki was reluctant to apply Haushofer's ideas to Japan without modification. During the war an Institute of National Land Planning was established, and it published a magazine, *National Land Planning*.

Many of Japan's professional geographers were attached to the General Staff Office, the Military Academy and military preparatory schools and research teams in the Japanese colonies and occupied territories. Even if this activity was not the wish of the individual, it was hardly possible to refuse the work. In this way few geographers perished in the war (Takeuchi, 2000, p. 148). Overall, the war was a time of regression for Japan and terminated in trauma that would change Japan forever.

After 1945 new work was commenced under the direction of American geographers in the General Headquarters, Supreme Commander for the Allied Powers.

---

[10]The translation was rendered by S. Yamaguchi.

[11]Translated by T. Matsuo, 1929.

Included were undertakings concerning land classification, land use, water conservation and development, and photo interpretation in geomorphology. Edward Ackerman, one of the geographers who held an advisory post among the American occupation personnel, employed a number of Japanese to study resources in Japan. The survey was published as *Japanese Natural Resources: A Comprehensive Survey* in 1949. In addition, the *Soil Survey of Japan* (1948–1951) was published. The organization of the Natural Resources Research Association in 1947 was one result of this work. Another American contribution was the establishment of the University of Michigan Institute of Japanese Studies under the direction of Robert B. Hall.[12]

EDUCATIONAL REFORM

The number of educational institutions in Japan increased substantially under American influence in the postwar years as part of the general educational reform of the nation. In 1951 Japan could claim 95 public universities and 142 private universities. Nearly all of these institutions had at least one chair of geography. Therefore, the demand for trained geographers accelerated. To facilitate the professional growth of geographers and to provide a forum in which papers could be read, exchanged, and discussed, the Association of Japanese Geographers opened its doors to anyone who wished to join.

Other associations and periodicals were soon founded. In 1948 the Human Geography Association of Japan was organized in Kyoto; its journal was entitled *Human Geography*. Initially, this association provided services for geographers in western Japan; later it became nationwide in scope. The Tohoku Geographic Association, centered at Tohoku University, provided for the needs of geographers in northeastern Japan. Other associations were formed, each of which publishes a journal: the Association of Photometry (1947), the Association of Historical Geographers in Japan (1958), and the Japan Cartographic Association (1962).

Another impact on the development of Japanese geography was the 1946 order of the Supreme Commander for the Allied Powers that stipulated that those geographers who had been involved in geopolitical support of the government were thenceforth removed from their posts. The Association of Democratic Scientists of Japan and the Science Liaison Committee were founded in 1946. Through these activities eventually Japanese geographers were admitted as members or corresponding members of the International Geographical Union commissions. Mention should also be made of expatriates returning from overseas assignments: A number of these people entered teaching as geographers and government agencies, including the Meteorological Agency, the Geographical Survey Institute, the Defense Agency, and the Ministry of Foreign Affairs.

---

[12]R. Hall, a member of the University of Michigan Geography Department, was a specialist on Japan. His study of the road connecting Yedo (Tokyo) and Kyoto and of the region that it ties together remains a masterpiece of historical geography. See "Tokaido: Road and Region," *Geographical Review* 27(1937):353–377.

POSTWAR GEOGRAPHY

International exchanges in the postwar period proved very important for Japan. From 1949 onward Japanese students studied in the United States through a program sponsored by the U.S. Defense Department and the Fulbright Program. By about 1955 Japanese students could also study in England, France, and West Germany. The economic recovery and the strength of the Japanese currency now made it more feasible for students to study and travel abroad. In turn, foreign tourists and geographers also traveled to Japan, the geographers usually as visiting scholars. Then in 1957 the International Geographical Union's regional conference was held in Japan. This event gave Japanese geographers the opportunity to create relationships with their peers abroad, thereby reducing Japan's cultural isolation.

Prior to and during the war, freedom to write and speak what one thought had been curtailed in Japan (Takeuchi, 1984). The more liberal, forward-looking philosophy ushered in during the 1950s now produced new opportunities in geography. Advanced analytical methods were introduced, specializations emerged, and interdisciplinary cooperation was practiced. Physical and human geography came to be considered separate disciplines. Within physical geography, geomorphology, hydrology, and climatology became discrete fields, as cultural, industrial, and agricultural geography became discrete fields within human geography.

In physical geography dynamic analysis became popular, and lowland geomorphology won wide acceptance. Quantitative approaches were adopted with regard to coastal landforms and fluvial processes. The work of river terraces in relation to mountain glaciation and changes in sea level and other matters relating to climatic change were more thoroughly investigated. Takeuchi (1984) notes growing strength in hydrology and climatology from the 1950s onward. M. M. Yoshino holds a premier position in climatological studies in Japan, T. Nishizawa has contributed much to the study of urban climate, and H. Suzuki to historical climatology. Approximately 40 percent of all geographical publications in Japan are devoted to physical geography.

POST-1950 TRENDS IN HUMAN GEOGRAPHY

After 1950 prewar thinking and philosophy were severely criticized, as was environmentalism, but ecological and Marxist viewpoints enjoyed some popularity. The emphasis that developed on socioeconomic and historical causalities encouraged empirical studies of small areas. The tradition of local field work continues in Japanese geography. This work contrasts rather starkly with the nationwide and international studies that are more common today than previously.

In historical geography Kyoto University ranks first in the nation. Its focus is on restoring the landscape of past times. After the war its interests centered on the prehistoric and ancient periods; more recently, Kyoto's emphasis has been on the medieval period. K. Uchida (1971) of Tokyo Bunrika University continued his work on the seventeenth and eighteenth centuries and has now been succeeded by his admirers, Y. Asaka, T. Kitamura, and T. Kikuchi (1958). Some historical geographers, for example, M. Senda (1980), have become interested in the perception of humans in past eras.

Takuji Ogama

Shigetaka Shiga

Keiichi Takeuchi

Naomasa Yamasaki

Economic geography has grown concomitantly with Japan's economic growth. Studies of industrial areas began to develop in the 1960s, and studies of nationwide industrial location became more popular in the 1970s. The analysis of spatial organization relating to manufacturing began in the 1970s. In total, these studies have revealed much about Japanese industry during the rapid growth period of the 1960s and the less spectacular growth since 1974. Agricultural geography, which was very active prior to the war, was largely taken over by the agricultural economists after 1945. Urbanization and industrialization bit deeply into the rural landscape, encouraging the immigration of rural workers to urban areas. Part-time farm households have been studied by M. Ishii (1980). A theme that has gained considerable attention is whether optimum production for the national economy is essential, or whether it is desirable to subsidize villages in marginal areas that have lost population (Fujita, 1981). Geographers do not seem to have made a firm decision on this question at the government level. Mining and mineral studies and marine studies have received geographers' attention in recent years. Population geography, which was already popular before 1941, became more sophisticated after the war. Most of this work was undertaken in an environmental framework. Both geographers and demographers seized hold of this practical study (Hama, 1978). Studies in distribution and circulation were facilitated by the numerical approach. For example, H. Morikawa (1974), adopting quantitative methods, pursued research concerning central-place patterns in Japan.

Urban studies also proliferated after the war. Initially, assessing the war damage in the cities and planning for reconstruction were the primary tasks. Later, industrialization and urbanization motivated the urban geographers (Kiuchi et al., eds.). New methods of urban research were introduced from the West, and studies in commuting (Arisu, 1968), internal structure of urban areas (Tanabe, 1971), satellite cities (Yamaga, 1964), the hierarchy in urban settlements (Watanabe, 1955), and shopping districts (Hattori and Sugimura, 1977) typified the new directions. Special urban geography sessions were frequently on the programs of the Association of Japanese Geographers as urban problems increased.

Takeuchi (2000, 193) observed that "modern Japanese geography has always been under the sway of achievements in geographical studies of western countries and a large number of Japanese geographers are interested in the methodology and history of western academic geography". Since the 1860s this has been true, but in the last 15 years investigations and translations of such works have been both more numerous and intense in disposition. NishiKawa (1988) wrote on the ecologic approach by Humboldt, Ritter, Richthofen and Troll. The significance for contemporary geography of Humboldt was sought by M. Tamura (1993, 1995), of Ritter by Takeuchi (1981), and of Humboldt and Ritter by A. Tezuka (1990, 1997). Tanaka (1996) presented Ratzel's writings on Japan. Nozawa has published (with commentary) extensively on French geography (1994) including a book on Vidal de la Blache (1998b). Tezuka translated some of the work of Humboldt, Ritter, Hettner, Schlüter, and de Martonne. Hisatake studied the Berkeley school of geography, and Nojiri studied the Chicago school of human ecology. Sugiura published on the social history of the quantitative revolution. Ishikawa published briefly on some AAG presidential addresses. Noguchi reported on the closure of the department of geography at the University of Michigan, and Nakayama considered the state of high school geography after 1984. And themes such as landscape, region, area study, methodology, and epistemology have received much attention. This is especially true with regard to the concept of region. Pursuant to the study of regional geography in Western countries, Japan now exhibits a newly reconstructed regional geography. Takeuchi ascribes this to the economic reduction of Japan in recent time (cf. the boom years of the 1980s and 1990s) which has driven responsibility and decision making from Tokyo back to the local regions, thus providing the latter with renewed authority and attention.

Japanese geography has clearly been modernized in the post–1945 period. Today geography has a prominent place in the schools and universities of Japan. (Large numbers of students graduate in geography and then undertake a career that does not relate to the subject at all.) Japanese geography is also making its way into the international literature. The 1957 International Geographical Union's regional conference was a help in this respect. The Twenty-fourth International Geographical Congress held in Japan in 1980 brought Japanese and non-Japanese geographers together as never before. With that meeting new attention was given to a geography once hidden by language, cultural, and geographical isolation.

Since that time indigenous Japanese geography has flourished based on the integration of work accomplished in the homeland and the ongoing importation of literature and ideas from Western countries and the Soviet Union/Commonwealth of Independent States.

REFERENCES: CHAPTER 14

Arisue, T. 1968. *Nihon no kotsu* [The Transportation of Japan]. Kokonshoin. Association of Japanese Geographers, ed. 1966. *Japanese Geography, Its Recent Trends*. Special Publication No. 1.

Fujita, F. 1981. *Nihon no sanson* [Mountain Villages in Japan]. Chijinshobo.

Hama, H. 1978. *Jinko mondai no jidai* [The Era of Population Problems]. NHK Books.

Hasegawa, K. 1980. "The Concept of 'Natural Regions' and Its Adoption in Japan." Geography Institute, Kyoto University, ed. *Geographical Languages in Different Times and Places*. Pp. 72–75. Japanese Commission on the History of Geographical Thought of the 24th IGU Congress in Japan.

Hattori, K., and Sugimura, N. 1977. *Shotengai to shogyochiiki* [Shopping Streets and Commercial Districts]. Taimeido.

Ikeda, Y. 1976. "Trends in Political Geography." In S. Kiuchi, ed., *Geography in Japan*. Pp. 211–214. Tokyo: University of Tokyo Press.

Inouye, Syuzi. July 1938. "Die Japanische Geographie der letzten zehn Jahre." *Geographische Zeitschrift* 44, nos. 7–8:284–289.

Ishida, H. 2000. "Akira Nakanome, 1874–1959." *Geographers: Biobibliographical Studies*. 20:68–76.

Ishii, M. 1980. "Regional Trends in the Changing Agrarian Structure of Postwar Japan." In the Association of Japanese Geographers, ed., *Geography of Japan*. Pp. 199–222. Teikokushoin.

Ishikawa, Y. 1980. "The Quantitative Studies in Japan in the Early Nineteen-Thirties." Geography Institute, Kyoto University, ed., *Geographical Languages in Different Times and Places*.

Pp. 82–84. Japanese Commission on the History of Geographical Thought of the 24th IGU Congress in Japan.

Ishimizu, T., and Ishihara, H. 1980. "The Distribution and Movement of the Population in Japan's Three Major Metropolitan Areas." In the Association of Japanese Geographers, ed., *Geography of Japan*. Pp. 347–378. Teikokushoin.

Kaizuka, S. 1976. "History of Geomorphological Research." In S. Kiuchi, ed., *Geography in Japan*. Pp. 35–45. Tokyo: University of Tokyo Press.

Kawabe, H. 1980. "Internal Migration and the Population Distribution in Japan." In the Association of Japanese Geographers ed., *Geography of Japan*. Pp. 379–389. Teikokushion.

Kikuchi, T. 1988. "Introduction to humanistic methodology in historical geography and human geography," *Historical Geography* 143:23–35.

Kiuchi, S. 1951. *Toshichirigaku kenkyu* [Studies of Urban Geography]. Kokonshoin.

——— ed. 1976. *Geography in Japan*. The Association of Japanese Geographers Special Publication No. 3, University of Tokyo Press.

——— Yamaga, S., Shimizu, K. and Inanaga, S., eds. 1964. *Nihon no toshika* [Urbanization of Japan]. Tokyo.

Kornhauser, D. H. 1979. *A Selected List of Writings on Japan Pertinent to Geography in Western Languages with Emphasis on the Work of Japan Specialists*. University of Hiroshima.

Masai, Y. 1976. "Researches On Anglo-America." In S. Kiuchi, ed., *Geography in Japan*. Pp. 277–280. Tokyo: University of Tokyo Press.

Matsuda, M. 1980. "The Adoption and Transformation of the Concept of Landscape in Japan." Geography Institute, Kyoto University, ed., *Geographical Languages in Different Times and Places*. Pp. 63–66. Japanese Commission on the History of Geographical Thought of the 24th IGU Congress in Japan.

Minamoto, S. 1984. "Shigetaka Shiga 1863–1927." In *Geographers: Biobibliographical Studies*. Vol. 8; pp. 95–105. London: Mansell.

Miyakawa, Y. 1977. *Kogyo haichiron* [The Location of Industry in Japan]. Taimeido.

Morikawa, H. 1974. *Chuchinchi kenkyu* [A Study on Central Places]. Taimeido.

Nakano, T. 1970. "Some Prevailing Trends of the Historical Development of Geosciences in the Far East." *Geoforum* 1, no. 3:63–80.

Nakayama, S. 1991. "Renaissance movement with regard to geographic education, in the United States of America." *Human Geography* 43:460–78.

Nishikawa, O. 1988. *Chikyu jidai no chiri shiso: Funboruto seishin no tenka (Geographical Thought in the Globalization Era: Development of the Spirit of Alexander von Humboldt)*. Tokyo: Kokon Shoin.

Noguchi, Y. 1985. "Background of the closure of the Department of Geography at the University of Michigan." 1,2, and 3, *Chiri* 30, 1:38–47; 30, 2:64–72; and 30, 3:72–79.

Noma, S. 1976. "A History of Japanese Geography." In S. Kiuchi, ed., *Geography in Japan*. Pp. 3–16. Tokyo: University of Tokyo Press.

Nozawa, H. 1980. "The French School of Geography in Japan, 1926–1940." Geography Institute, Kyoto University, ed., *Geographical Languages in Different Times and Places*. Pp. 67–71. Japanese Commission on the History of Geographical Thought of the 24th IGU Congress in Japan.

Nozawa, H. 1988. *Vidaru do ra Burashu kenkyu (Studies on Vidal de la Blache)*. Kyoto: Chijin Shobo.

———. 1994. "Camille Vallaux's geography." *Shien* 134:47–69.

Okuno, T. 1994. "The Trend of Human Geography in Japan: An Overview." *Tsukuba, Studies in Human Geography* 18:103–116.

————. 1996. "Geography in Japan." In *General Treatise of Geography*. Ed. O. Nishikawa. Tokyo: Asakura Shoten.

Pinchemel, P. April–June 1980. "L'histoire de la géographie japonaise." *L'espace géographique* 9, no. 2:165–171.

Science Council of Japan, ed. 1980. *Recent Trends of Geographical Studies in Japan*. Recent Progress of Natural Sciences in Japan. Vol. 5.

Senda, M. 1980. "Territorial possession in ancient Japan: The real and the perceived," in Association of Japanese Geographers (ed.) *Geography of Japan*. Tokyo: Teikoku Shoin.

Suizu, I., et al. 1980. *Geographical Languages in Different Times and Places: Japanese Contributions to the History of Geographical Thoughts* [sic]. Kyoto: Geography Institute, Kyoto University.

Takeuchi, K. 1974. "The Origins of Human Geography in Japan." *Hitotsubashi Journal of Arts and Sciences* 15:1–13.

————. 1976. "General Remarks on Human Geography during the Past Fifty Years." In S. Kiuchi, ed., *Geography in Japan*. Pp. 107–112. Tokyo: University of Tokyo Press.

————. 1978. "Some Remarks on the History of Regional Description and the Tradition of Regionalism in Modern Japan." *Hitotsubashi Journal of Social Studies* 10:36–44.

————. 1980. "Geopolitics and Geography in Japan Reexamined." *Hitotsubashi Journal of Social Studies* 12:14–24.

————. 1980b. "Geopolitics and Geography in Japan." Geography Institute, Kyoto University, ed., *Geographical Languages in Different Times and Places*. Pp. 85–88. Japanese Commission on the History of Geographical Thought of the 24th IGU Congress in Japan.

————. June 1980. "Some Remarks on the History of Regional Description and the Tradition of Regionalism in Japan." *Progress in Human Geography* 4, no. 2:238–248.

————. 1984. "Japan." In R. J. Johnston and P. Claval, eds., *Geography since the Second World War*. Pp. 235–263.

————. 2000. *Modern Japanese Geography: An Intellectual History*. Tokyo: Kokon Shoin Publishers.

Takeuchi, K., and Nozawa, H. 1988. "Recent Trends in Studies on the History of Geographical Thought in Japan—Mainly on the History of Japanese Geographical Thought." *Geographical Review of Japan* 61:59–73.

Tamura, A. 1993. "The idea of *Naturgemälde* in the first volume of *Kosmos* by Alexander von Humboldt." *Geographical Review of Japan* 66A:253–68.

————. 1995. "The art and science in Alexander von Humboldt's *Naturgemälde*." *Regional Studies* (The Rissho Geographical Association) 35, 2:1–17.

Tezuka, A. 1990. "Geographical ideas of Alexander von Humboldt." *Tsukuba Studies in Human Geography* 14:107–44.

————. 1997. *Zoku chirigaku no koten: Humboldt no sekai (Classics of Geography, Continued: World of Alexander von Humboldt)*. Tokyo: Kokon Shoin.

Toshiaki, O. 1980. "The Adoption of H. von Thünen's Location Theory in Japanese Geography." Geography Institute, Kyoto University, ed., *Geographical Languages in Different Times and Places*. Pp. 78–81. Japanese Commission on the History of Geographical Thought of the 24th IGU Congress in Japan.

Troll, C. March 1968. "Die deutsche geographische Japan-Forschung vor und nach der Meiji-Restauration. Zum Gedenken von Johannes Justus Rein (1835–1918), Philipp Franz von Siebold (1796–1866) und Engelbert Kaempfer (1651–1716)." *Erdkunde* 22, no. 1:7–13.

Tsujita, U. 1977. "Naomasa Yamasaki, 1870–1928." In *Geographers: Biobibliographical Studies*. Vol. 1, pp. 113–117. London: Mansell.

————. 1982. "Takuji Ogawa 1870–1941." In *Geographers: Biobibliographical Studies.* Vol. 6, pp. 71–76. London: Mansell.

Uchida, K. 1971. *Kinsei noson no jiukochiriteki kenkyu* [Population Geography of Rural Settlement in the Late Middle Age]. Chukokan.

Ukita, T., and Ashikaga, K. 1976. "Historical Geography." In S. Kiuchi, ed., *Geography in Japan.* Pp. 215–235. Tokyo: University of Tokyo Press.

Walter, L., ed. 1994. *Japan: A Cartographic Vision.* Munich: Prestel.

Watanabe, H. 1968. *Jorisei no kenkyu* [A Study of the Jori System]. Sogensha.

Watanabe, Y. 1955. *The Central Hierarchy in Fukushima Prefecture. A Study of Types of Rural Service Structure.* Science Report No. 4, Tohoku University Geography.

Yamaga, S. 1964. *Toshichirigaku* [Urban Geography]. Taimeido.

Yamano, M. 1980. "General Remarks on the Methodology of Human Geography in Japan (1930–1945)." Geography Institute, Kyoto University, ed., *Geographical Languages in Different Times and Places.* Pp. 75–78. Japanese Commission on the History of Geographical Thought of the 24th IGU Congress in Japan.

Yoshikawa, T. 1976. "General Comments on the Development of Physical Geography in Japan." In S. Kiuchi, ed., *Geography in Japan.* Pp. 29–33. Tokyo: University of Tokyo Press.

15

THE NEW GEOGRAPHY IN THE UNITED STATES BEFORE WORLD WAR I

*Reputations for originality are more often made by giving names to ideas already
coming into circulation than by inventing new ideas. In this way the fruits of many
men's thinking are appropriated by the more lucid and articulate among them.*
— *John P. Plamanatz*

When the "new geography" arrived in North America following its appearance
in Germany, an interest in geographical studies and in the teaching of geography in schools and colleges had already been demonstrated (Aay, 1981; Warntz,
1981). As in other parts of the world, geographical work was contributed by scholars
(e.g., Benjamin Franklin and Thomas Jefferson) for whom such studies constituted
only one of many intellectual interests. There were pioneers such as Jedidiah Morse,
George Perkins Marsh, and Matthew Fontaine Maury who brought new understanding to the study of the earth as the home of humankind. There were also the
endogenous ideas, which to some extent were the product of the wilderness conquest experience. In addition, there were ideas imported from Europe to the United
States by geographers including Carl Ritter, Alexander von Humboldt, Ernst
Kapp, Johann Georg Kohl, and Elisée Reclus. Correspondence between American
geologists, geomorphologists, and cryptogeographers and European geographers
developed. European books and periodicals were imported, and beginning in 1871
U.S. and European geographers met at the International Geographical Congresses.
European geographical ideas were also brought to America by such scholars as Louis
Agassiz at Harvard and Arnold Guyot at Princeton. Guyot's pupil William Libbey,
Jr., who succeeded Guyot as professor of physical geography at Princeton, carried
on the tradition of individual scholarship in his studies of oceanography, especially
of the relationship between the Gulf Stream and the Labrador Current. During the
nineteenth century, too, major advances in thematic mapping were introduced from
Europe through the efforts of such men as Lorin Blodget, Joseph C. G. Kennedy,
Daniel Coit Gilman, and Francis A. Walker.

In this brief survey of exogenous contributions, mention must also be made of
Franz Boas. In 1882 Boas unsuccessfully sought an appointment in geography at
Johns Hopkins University, where geographer Daniel Coit Gilman was president.
He then sought numerous posts in both American and German geography before
accepting an appointment as anthropologist at Clark University in 1889. He brought
with him something of the German geographic tradition and, more particularly
perhaps, that of Fischer and Ratzel (Koelsch, 2004; Cole, 1999).

338

An important part of the background for the development of the new geography in America was the tradition of the field survey and the resulting emphasis on induction from observations rather than deduction from theory. In the 1880s the Great Surveys of the West had just been combined in the U.S. Geological Survey. (For geographers at work in the federal government, see Friis [1981]; for Amerindian antecedents, see Lewis [1981].) The men who worked on these surveys had not received previous training in the concepts and methods of geography, and they had to find their own answers to the five questions listed at the beginning of Chapter 7 (what to observe, how to observe, how to generalize, how to explain, and how to communicate). Since these field men had not been indoctrinated with Lyell's ideas about marine planation or Werner's ideas about the origin of the earth, they were able to observe landforms and the processes that produced them without strong preconceptions. There was a practical motivation for their work. They distrusted theory and the findings of scholars that had been deduced from theory. Grove Karl Gilbert wrote: "In the testing of hypotheses lies the prime difference between the investigator and theorist. The one seeks diligently for the facts which may overthrow his tentative theory, the other closes his eyes to these and searches only for those what will sustain it" (Gilbert, 1886).

By 1880 the stage was set for the appearance of what we have called the new geography. As we have seen in Chapters 7 and 8, the formation of a professional field requires the existence of clusters of scholars working closely with graduate students in universities (Koelsch, 1981). The concept of the university as a community of scholars first appeared in America in 1876 when Daniel Coit Gilman became president of the newly founded Johns Hopkins University (Wright, 1961; Heyman, 2001); thereafter, the idea spread rapidly to other established universities. For the first time faculties qualified by advanced training and continued activity in research were selected to guide the training of younger generations. For the first time a professional group could lead and direct scholarly performance in each discipline, free from external interference.

Geographers began to participate in this new kind of university in Germany in 1874, and thereafter the innovation spread around the world. The pioneer who introduced the new geography in America was the geologist William Morris Davis, who had been appointed instructor of physical geography in the Geology Department at Harvard in 1878 (Beckinsale, 1981).[1] Davis set the course for geographical study: He helped found the discipline, and he also encouraged the development of a profession. Arguably these occurrences began around 1885–95. At first, geography was usually associated with geology; soon departments offering at least six

---

[1] W. A. Koelsch suggests that this was at a time when W. H. Tillinghast, a Harvard College Librarian knowledgeable in both history and the classics, wrote "The Geographical Knowledge of the Ancients," analytical of early Western geography. Tillinghast was part of an unsung Harvard "school" of humanistic geography based in Harvard learning and appreciation of the classics that encouraged the emergence of, for example, J. K. Wright (W. A. Koelsch, 2003. "William H. Tillinghast, John K. Wright, and Some Antecedents of American Humanistic Geography." *Journal of Historical Geography* 29, 4:618–630).

geography courses emerged at Columbia Teachers College (1899); Cornell University (1902); the University of California (1903); the University of Chicago (1903); the University of Nebraska (1905); Miami University, Ohio (1906); the University of Minnesota (1910); the University of Pittsburgh (1910); Nebraska Wesleyan University (1911); the University of Wisconsin (1911); Harvard University (1911); the University of Pennsylvania (1913); New York University (1913); Yale University (1914); and Denison University (1914).

Of these universities, Harvard, Yale, Pennsylvania, and Chicago were perhaps the major sources of the ideas involved in the scholarly competitive discussion from 1904 to 1914. At Harvard William Morris Davis was developing physical geography, although he was eventually to find a place for man in his construct. At Yale Herbert E. Gregory (one of Davis's own pupils) developed a fine departmental offering in human geography—notably owing to Ellsworth Huntington and Isaiah Bowman. At the University of Pennsylvania Emory R. Johnson and J. Russell Smith developed economic and commercial geography. At the University of Chicago in 1903 the first Geography Department in the United States offering advanced study to the doctorate was founded.[2] The Chicago program was strong in physical, human, and economic geography. (For sketches of other individual geographers of the period, see Dunbar, 1978; Koelsch, 1979a, 1979b; and Sherwood, 1977.)

In addition to the geography departments of the universities (and the still underestimated normal schools), numerous institutions had been established to support the geographic undertaking. Founding dates include the American Philosophical Society (1743); the American Association for the Advancement of Science (1848) (Section E); the Geographical Society of the Pacific (1881); the Geographical Society of California (1891); the Geographical Club of Philadelphia (1891) and renamed the Geographical Society of Philadelphia (1897); the Agassiz Association (1892); the American Alpine Club (1893); the Geographical Society of Chicago (1898); the Peary Arctic Club (1899); and the Geographical Society of Baltimore (1902). However, the two major American societies have been the American Geographical and Statistical Society (1851) and the National Geographic Society (1888). It is not possible in this brief space to indicate the very large contribution these two societies have made to geography. Prior to 1914 one is tempted to single out the *Bulletin of the American Geographical Society* (with some name changes) published by the American Geographical Society from 1852 to 1915. The National Geographic Society produced a magazine that reached a large audience—one-third of a million by 1914—and provided a knowledge of faraway lands and peoples to hitherto unknown numbers of people. Of vital significance to the development of academic geography in the United States were the International Geographical Congresses of 1892 (held on the occasion of the Columbian exposition and not as part of the series commenced at Antwerp in 1871) and 1904, both of which the National Geographic

---

[2]Of the Ph.D. degrees listed by Whittlesey in 1935 (Whittlesey, 1935), 19 were granted before 1916: five at the University of Pennsylvania (in economics); five at Chicago (starting in 1907); three at Johns Hopkins (in meteorology and geomorphology); two at Cornell; two at Yale; and two at Harvard.

Society sponsored. (The society also sponsored the International Geographical Union Congresses of 1952 and 1992.)

WILLIAM MORRIS DAVIS

William Morris Davis[3] spent his formative years as a scholar at Harvard with Nathaniel Southgate Shaler. He learned three distinctive habits of thought from Shaler. First, he developed the habit of careful field observation and made use of it in logical and impersonal argument. Second, he acquired the habit of seeing man and his works as part of the landscape, not separate from it. Third, he gained a clear appreciation of the importance of processes of change in explaining the varied features associated on the face of the earth. How were these habits of thought transmitted from Shaler to his young assistant?

The habit of careful observation and logical argument was derived, it seems, from working with Shaler in the field. Davis found out from personal experience that just going out to see "what an area is like" is less productive of useful results than going out to find the answers to questions. This was just the time when a major question among students of the earth had to do with the origin of the sand and gravel deposits that are so widespread in New England. Were they laid down during the Flood, as described in the Bible—which was the traditional explanation—or were they the results of the melting of the great ice sheets, as Agassiz insisted? Through his studies of the glacial history of Cape Cod, Shaler had given new support for the Agassiz hypothesis. Davis, working with Shaler, looked for evidence for or against the glacial hypothesis, and he learned to marshal his observations in terms of the questions he was asking. The results of field study had to be put together in logical form and communicated in scholarly—that is, impersonal—argument.

Shaler also transmitted to Davis a vision of the earth as the resource base on which the human inhabitants were dependent. Shaler was the first scholar since

---

[3] William Morris Davis was born of Quaker parents in Philadelphia in 1850. He graduated from Harvard in 1869 and a year later received the degree of master of engineering. From 1870 to 1873 he worked as an assistant at the Argentine Meteorological Observatory in Córdoba, Argentina. Returning to Harvard for further study in geology and physical geography, he was appointed assistant to N. S. Shaler in 1876 and was promoted to instructor in physical geography in 1878. In 1885 he was appointed assistant professor of physical geography and later promoted to professor. In 1899 he was named the Sturgis Hooper Professor of Geology at Harvard, a chair he continued to occupy until his retirement in 1912. In 1909 he was visiting professor at the University of Berlin and from 1911 to 1912 at the Sorbonne. After his retirement from Harvard, he held temporary appointments at the Universities of Oregon, California, Arizona, Stanford University, and the California Institute of Technology. He was one of the founders of the Association of American Geographers in 1904 and was its president three times—in 1904, 1905, and 1909. He was also president of the Geological Society of America and the Harvard Travelers Club. He never received the Ph.D. degree, but he was honorary doctor of many universities. He was a Chevalier de la Légion d'Honneur and received medals from many geographical societies of the world (Bryan, 1935).

George Perkins Marsh to point out how human activities changed the earth, especially through the depletion of unrenewable resources. Shaler has been described as a geologist by training but a geographer by instinct in that he was always conscious of studying the earth as the home of man.

The young Davis learned to think in terms of evolutionary change as a basis for scientific inquiry in the best possible way—by participating in the replacement of one hypothesis by a new one. Louis Agassiz was a persuasive and influential lecturer with a large following among the educated people of New England, who supported his work with substantial endowments. At the same time there was a rising tide of opposition to the notion that organisms, once created, remained without change. Asa Gray (1810–88), a botanist at Harvard, was amassing carefully evaluated evidence in support of the concept of organic evolution, which to the lay public was unpopular. Shaler, who had learned much from Agassiz about the observation of landforms, nevertheless joined those who supported the doctrine of evolution. Davis took part in the discussions in that exciting period when a widely accepted explanation was being rejected on the basis of careful scientific procedures.

Davis did not absorb and digest these ideas all at once. When he became Shaler's assistant in charge of field work in 1876, his teaching was not stimulating. His largely empirical approach to landforms left his students struggling with unordered detail (Davis and Daly, 1930:314–315). As a result, when the time came for his reappointment as an instructor in 1882, he received the following letter from President Charles W. Eliot, dated June 1:

> The Corporation offer you a reappointment as instructor in geology at a salary of $1200 a year.... The Corporation are quite aware that this position is not suitable for you as a permanency; but it is all that they are able to offer you now, with their present resources, and all that they expect to be able to afford for some time to come. In considering whether it is your interest to accept this offer temporarily, I hope that you will look in the face of the fact, that the chances of advancement for you are by no means good, although the Corporation have every reason to be satisfied with your work as a teacher.[4]

At this point another member of the Harvard faculty in geology, Raphael Pumpelly, who knew Davis as a student, offered him a position on the survey he was then conducting in Montana to identify the resources along the route of the Northern Pacific Railroad. Davis was assigned the task of describing the Montana coal measures. While this work was in progress, he began to visualize the outlines of a "cycle of erosion." He noted the existence of certain terraces above the Missouri River, which he interpreted as the result of the removal of an "unknown thickness of overlying strata," and the reduction of an earlier surface close to the baselevel of the drainage. The baselevel concept was derived from John Wesley Powell, and other insights into the process of river erosion came from Grove K. Gilbert and Clarence E. Dutton. But Davis, observing the landforms of Montana, began to formulate a theoretical model that would describe all such processes and

---

[4]Quoted from Chorley, Dunn, and Beckinsale, 1964:623. The letter from Eliot was copied by Davis and kept in his personal files. See also Davis and Daly, 1930:314–315.

surfaces (Chorley, Dunn, and Beckinsale, 1964:622). The concept of the cycle of erosion, first announced in 1884, was presented in revised form in 1899 (Davis, 1899a). But it is important to see what this observation did to Davis's teaching: the observed details of landforms in particular places, which had failed to interest his students a few years earlier, could now be related to a generalized model of landform development anywhere. In 1885, only three years after President Eliot's letter, Davis was appointed assistant professor of physical geography at Harvard.

There were two sides to Davis's work that can be examined separately, although, of course, he carried them on simultaneously. First, we may consider his contributions to geomorphology, especially his model of the cycle of erosion; second, we may look at his promotion of geography as a field of study in schools, colleges, and graduate schools.

Contributions to Geomorphology

Among the many contributions Davis made to geology and geomorphology, the one that was central to all the others was the concept of the cycle of erosion, which he called the geographical cycle (Davis, 1899a).[5] This was in the nature of a model, an ideal sequence of landforms that would take place during the erosion by running water of an upraised portion of the earth's crust. In his model Davis postulated that after the upheaval no further up-or-down movements would take place and that during the resulting cycle there would be no essential change of climate. When an initial surface is raised, rivers at once begin the work of erosion. The surface is cut by narrow V-shaped valleys that are extended headward as more and more of the initial surface is consumed. But rivers cannot cut down their valleys indefinitely. As Powell has pointed out, there is a baselevel below which rivers cannot cut—a level determined by the surface of the body of water into which a stream flows. Furthermore, before a valley is cut down all the way to baselevel, the river establishes a slope that is enough to permit the river to continue to flow. This is known as *grade*, an idea originally presented by Gilbert. Grade is an equilibrium between slope, volume of water, and the amount of load being carried. After the valley bottoms reach this graded condition, the rivers begin to widen their valleys, and the high country between the valleys is gradually reduced.

Davis provided an important terminology. When the initial surface is still undissected between the valleys, when the valleys are V-shaped, and when the rivers descend through them turbulently, this is a stage Davis described as *youth*. The greatest amount of relief occurs when the last remnant of the initial surface is dissected. Then the surface is gradually reduced, and the valleys begin to widen. This is the stage Davis described as *maturity*. When the rivers meander across wide valleys and the land between the valleys has been reduced to gently rounded slopes, this he called *old age*. The upheaved block of the earth's crust is worn down almost

---

[5]For a history of the Davisian concept of the cycle of erosion see "The Cycle of Erosion" in *The History of the Study of Landforms or the Development of Geomorphology. Volume 2. The Life and Work of William Morris Davis* (1973), pp. 160–197.

to a level plain, which Davis called a *peneplain*. The whole cycle, Davis pointed out, could start again with another uplift, resulting in *rejuvenation*.

Davis offered a formula for the description of landforms. These could be understood in terms of the interaction of three factors; *structure*, or the character and position of the underlying rock; *process*, or the combination of agents of erosion, such as running water, soil creep, underground solution, and ice; and *stage*, or the point in the sequence of landforms that had been reached at a particular time.

Davis insisted that this ideal sequence of landforms was not to be considered as rigid but rather that it could provide a theoretical framework in reference to which actually observed landforms could be described. This is what he called the explanatory description of landforms. He understood clearly that an infinite variety of disturbances could affect the ideal sequence because almost every region observed in specific detail would constitute a special case. This exemplifies the question of whether to describe the features of a particular place as unique or in terms of regularities or similarities among many places. Davis himself developed sequences for a variety of special conditions: for surfaces cutting across the upturned edges of tilted rock strata; for blocks bordered by fault scarps; and for places where the climate was arid. He pointed out that only rarely, if ever, would uplift take place quickly and be followed by no further movements, as postulated in his model. He showed how the theoretical sequence would be modified if further uplift and rejuvenation should take place at any stage in the cycle, and he provided innumerable examples of how the ideal cycle had been changed in specific circumstances. He also applied the evolutionary concept to the ideal sequence of forms in mountain regions sculpted by glaciers, in islands bordered by coral reefs (Davis, 1928), and in a limestone region with solution caverns (Davis, 1930a). Here is how he saw the utility of his scheme:

> In the scheme of the cycle of erosion . . . a mental counterpart for every landform is developed in terms of its understructure, of the erosional process that has acted upon it, and the stage reached by such action stated in terms of the whole sequence of stages from the initiation of a cycle of erosion by upheaval or other deformation of an area of the earth's crust, to its close when the work of erosion has been completed: and the observed landform is then described not in terms of its directly visible features, but in terms of its inferred mental counterpart. The essence of the scheme is simple, and easily understood; yet it is so elastic and so easily expanded or elaborated, that it can provide counterparts for landforms of the most complicated structure and the most involved history (Davis, 1899a).

Not only did Davis provide names for the stages of his cycle, but he also suggested technical terms for the various landforms, each term with an exact definition. He adopted Powell's three types of rivers: consequent, antecedent, and superimposed, and to these he added subsequent, obsequent, and resequent (Davis, 1909:483, 513). For the low mountains that stand above the general level of a peneplain, he gave the name *monadnock*, from Mt. Monadnock in New Hampshire, which stands above the New England peneplain (Davis, 1909:362, 591). He demonstrated again that when a name is given to a feature of any kind, students of the face of the earth at once begin to perceive it. This is an interesting example of the principle that percepts are closely related to concepts.

Davis demonstrated and defended the cycle of erosion and his terminology with such vigor that they were largely accepted throughout the world. He was invited to lecture in many countries, and his ideas were translated into several languages. In fact, the only complete statement of the cycle of erosion and its variations is contained in a German translation of his lectures at Berlin by Alfred Rühl (Davis, 1912). One of the most lucid presentations of Davis's ideas was written by the French physical geographer Emmanuel de Martonne (de Martonne, 1909; see also Baulig, 1950). Davis visited many countries where he made specific applications of his method of explanatory description to the landforms of particular places. He also experimented with methods of geographical writing (Davis, 1910), and in his famous paper on the Colorado Front Range he accompanied his description of the landforms with a commentary on the method he was using (Davis, 1911). In 1915 he published a lengthy paper on the principles of geographical writing that is important reading even these many years later (Davis, 1915).

Of course, the scheme was attacked, as all such hypothetical models should be and as Davis expected it to be. It is interesting that in Germany, where Davis published the only complete statement of the model, there was the largest resistance to his ideas. Alfred Hettner disagreed with Davis over his cycle of erosion on the basis that a theoretical model was too rigid and too specific to fit real world conditions as they exist in unique situations (particular places). Davis had offered a theoretical model, and Hettner pointed out that such models were misleading. The discussion that ensued, when examined closely by scholars familiar with both German and English, turns out to hinge on misunderstandings of word meanings. Siegfried Passarge, who wanted to describe landforms as a basic part of the study of landscape, was opposed to the Davis method of explanatory description. He called instead for a purely empirical treatment of the landform base. Davis took issue with Passarge's attempt to treat landforms empirically, pointing out the numerous inconsistencies that resulted (Davis, 1919). It may well be, however, that the Passarge-Davis controversy became heated because neither quite understood the basic objectives of the other. For Davis the natural history of landforms was the core of geographical work to which other elements of landscape could be related; for Passarge the goal was the treatment of landscape, including the many elements of diverse origin associated in an area of which the landforms were basic but not necessarily the most important. Davis himself had good reason to recall his own experience with the empirical approach to landforms and his own success with description based on an explanatory model.[6] Walther Penck also took issue with Davis and concluded

---

[6]Many attacks on the hypothesis of the cycle rather than the explanatory method were made in the period after World War I. Walther Penck, the son of Albrecht Penck, suggested that the initial steep slope on the margin of an upraised block would retreat parallel to its initial position, maintaining its steepness rather than flattening. Successive uplifts would be recorded by a series of scarps, each in process of retreat; and the rivers would pass over a series of "nick-points," each marking a change of baselevel and each in process of retreat upstream. More recently, L. C. King, the South African geomorphologist, has demonstrated the existence of a series of erosion levels bordered by scarps in Brazil and in South Africa. He describes the flattish erosion levels as *pediments* (Walther Penck, *Die Morphologische Analyse* [Berlin, 1922], translated as *Morphological Analysis of Landforms* [London, 1953]; L. C. King, "Cannons of Landscape Evolution," *Bulletin of the Geological Society of America* 64 [1953]:721–753; idem, *The Morphology*

his argumentation in *Die Morphologische Analyse* (1922) (Martin, 1974). Hettner, in correspondence, also found rigorous argument with Davis in the 1910s, and one has to wonder at the role of patriotism in science (Martin, 1985).

Contributions to Geographic Education

Davis also attempted to rescue the teaching of geography from too much attention to factual knowledge and not enough use of general concepts around which to organize the facts. This is precisely what Ritter attempted to change in 1817 and what Guyot pointed out to the Massachusetts Board of Education in 1848. In spite of these earlier efforts, Davis found the same emphasis on factual knowledge in the 1880s, and he proposed to do something specific about it. In 1932, looking back over a lifetime spent in the effort to improve the teaching of geography, he wrote:

> No geographer, therefore, need feel himself unfortunate because of the great diversity of facts that his composite subject requires him to study; for in the progress of his work he may discover relations and principles which bind his facts together in a thoroughly reasonable manner, and he may then concern himself, especially in his teaching, largely with those relations and principles, and introduce items of fact chiefly to illustrate the principles. Unhappily, geographers are often so impressed with the innummerable facts of their subject that much of their attention is given to individual occurrences in specified localities rather than to principles which the occurrences exemplify; and this is regrettable. Yet practically the same comment may be made on History, in which the mere sequence of events, or, still worse, the mere occurrence of events is often given great emphasis than their inherent importance warrants (Davis, 1932:214–215).

In a lecture delivered before the Scientific Association of Johns Hopkins University in 1889, Davis outlined a relatively simple model of landform development that a teacher could use to replace the bewildering description of detail then frequently attempted (Davis, 1909:193–209). This was essentially his concept of the cycle of erosion adapted to elementary and secondary schools. But in this address Davis was beginning to conceive of the study of geography as a means of introducing many

---

of the Earth [New York, 1962]; see also P. E. James, "The Geomorphology of Eastern Brazil as Interpreted by Lester C. King," *Geographical Review* 49 [1959]:240–246).

The part of the Davis scheme most subject to attack is the postulate of a single uplift. It is now understood that when a foot of rock material is removed from an area of the earth's surface, the result is an isostatic rise of the rock column beneath the surface by some 9 to 11 feet. This means that no surface could stand still long enough to permit the formation of a peneplain and explains why no peneplains are found undissected. As early as 1878 G. K. Gilbert wrote that the reduction of a drainage basin to a plain would demand a uniformity of conditions that nowhere exists (Gilbert, 1878). In modern times A. N. Strahler has expressed the opinion that "the cycle concept of Davis does not seem well adapted to express the dynamics of the erosion process. Instead, the concept of a steady state in an open system seems a logical replacement of the idea of 'maturity' while the stage of 'old age' may well be abandoned" (Strahler, 1950). (For other critical comments on the Davis system, see Chorley, 1965.)

It is also important to understand that Davis himself wrote critical reviews of his own scheme during the late years of his life and that he also contributed many additional original works (Davis, 1922, 1928, 1930a, 1930b).

kinds of physical science in a simple coherent framework. He was formulating the idea of a general earth science organized in terms of a dynamic model of earth-forming processes. About this same time Davis, who had learned from Shaler to see organic life, including man, as a part of the whole physical landscape, began to seek an even larger conceptual structure for geography. He began to seek cause and effect generalizations, "usually between some element of inorganic control and some element of organic response" (Davis, 1903:3–22). In 1906 he specified that

> any statement is of geographical quality if it contains a reasonable relation between some inorganic element of the earth on which we live, acting as a control, and some element of the existence or growth or behavior or distribution of the earth's organic inhabitants, serving as a response.
>
> . . . There is, indeed, in this idea of a causal or explanatory relationship the most definite, if not the only unifying principle that I can find in geography (Davis, 1906; quoted from Davis, 1909:8).

Davis juxtaposed what he called physiography and ontography and sought relationships of a causal nature. This search for environmental influences was transmitted to a number of his students, who in turn passed it on to their students or readers of their publications. The introduction of the causal notion may have had some unfortunate consequences; some geographers, and nongeographers as well, refer to the terms *influence* or *geographic factor* in describing a condition of the physical earth that influences human activities. Professional geographers would now rarely use these words, if at all, to refer to some factor of location. However, it is doubtful whether any national geography has ever reached a state of maturity without passing through a stage of inquiry characterized by the search for influences exerted by the physical environment on life forms.

Two famous Harvard scholars, a psychologist and a philosopher—Charles S. Peirce (1839–1914) and William James (1842–1910)—were laying the groundwork of pragmatism in the 1870s. They demonstrated that simple cause and effect relations were of questionable value as explanations of events because the existence of complex systems of functionally related parts made any simple explanation impossible. Pragmatism, as Peirce and James formulated it, was a method of determining the meaning of intellectual concepts; the meaning of any idea, they insisted, must be revealed in the practical consequences of the idea. The teaching of physical cause and human response was already outmoded before 1890. Furthermore, the pragmatic approach to knowledge was being introduced into elementary and secondary schools through the work of John Dewey.

By 1892 there was such a ferment of new educational ideas that the National Education Association appointed a Committee of Ten, headed by President Eliot of Harvard, to study the related problems of the content of precollege-level school programs and college entrance requirements. The committee organized nine different conferences, each to consider a specific field of study. One of the nine was directed to consider the content of courses in geography. The chairman of the Conference on Geography was T. C. Chamberlin, former president of the University of Wisconsin and in 1892 the chairman of the Department of Geology in the newly organized University of Chicago. The other members of the conference were geologists, meteorologists, and college or high school teachers of physical geography

Albert P. Brigham

William Morris Davis

Grove K. Gilbert

Mark S. W. Jefferson

Emory R. Johnson

John Wesley Powell

Rollin D. Salisbury

Ellen C. Semple

or natural history.[7] Davis had the chief hand in the resulting report. The recommendation that was passed and referred to the Committee of Ten concluded that:

> Physical geography should include elements of botany, zoology, astronomy, commerce, government, and ethnology, and . . . it should take on a more advanced form and should relate more specifically to the features of the earth's surface, the agencies that produce or destroy them, the environmental conditions under which these act, and the physical influences by which man and the creatures of the earth are so profoundly affected (Mayo, 1965:20–21).

The Committee of Ten expressed surprise when it received the report of the Conference on Geography, which presented many more far-reaching changes in the secondary school curricula than were presented by any of the other eight conferences. The committee wrote:

> Considering that geography has been a subject of recognized value in the elementary schools . . . , and that a considerable proportion of the whole school time of children has long been devoted to a study called by this name, it is somewhat startling to find that the report of the Conference on Geography . . . exhibits more dissatisfaction with prevailing methods . . . and makes the most revolutionary suggestions.[8]

Nevertheless, the major features of the report were adopted, and its radical proposals to raise geography from pure memory work to the status of a general science were recommended to the schools. Many schools introduced physical geography, or physiography as it came to be called, and many new textbooks were written to include Davis's ideas.[9] But there were also many secondary school teachers who were opposed to the new materials, and there were many more who were quite unprepared to teach them. Few indeed were the secondary school teachers who could identify a landform out of doors or who could make Davis's theoretical models come alive. Lacking proper training in the concepts of the new geography, they fell back again on recitations from memory. Within 10 years the new physiography was described as an uninteresting subject. Thereafter it was gradually pushed out by general science, social studies, and commercial geography.

---

[7] The Conference on Geography included: T. C. Chamberlin, University of Chicago, chairman; George L. Collie, Beloit College; William Morris Davis, Harvard University; Delwyn A. Hamlin, Rice Training School, Boston; Mark W. Harrington, Weather Bureau, Washington, D.C.; Edwin J. Houston, Philadelphia; Charles F. King, Dearborn School, Boston; Francis W. Parker, Cook County Normal School, Chicago; Israel C. Russell, University of Michigan.

[8] National Education Association, *Report of the Committee of Ten on Secondary School Studies* (New York: American Book Co., 1894), pp. 32–33.

[9] The "new geography" with its emphasis on the physical earth appeared in a series of school texts, such as A. E. Frye, *Elements of Geography* (1895); R. S. Tarr, *Elementary Physical Geography* (1895); J. W. Redway and R. Hinman, *Natural Advanced Geography* (1897); R. S. Tarr and F. M. McMurry, a series of geographies starting in 1900. See also Mark Jefferson's course in geography for some 1300 Cuban schoolteachers in the summer of 1900 at Harvard (reported in Martin, 1968). For a discussion of the teaching of geography in America up to 1924, see Dryer, 1924.

Figure 34 A landscape sketch by W. M. Davis: "Looking eastward down the normal early-mature valley of Fourmile Creek." (From Davis, 1911:81. Reprinted by permission of the Association of American Geographers.)

Davis himself was a remarkable teacher. He was a master of the lecture method and could capture the imagination of popular as well as scientific audiences. His pen and ink sketches of landscapes were superb (Fig. 34). In the field with a group of students, he could arouse a deep interest in deciphering the sequences of events that produced contemporary landscapes. But with his students he was sharply critical of inferior performance and discouraged many of his more sensitive students from further work in geography. Only the best students could "take it." Mark Jefferson, one of Davis's most effective disciples, had this to say about the master's teaching:

> Davis' teaching was the most interesting thing I ever met. Confronted with the world of out-of-doors his formulae proved up. I took all his courses at Harvard, a summer school with him and two other students in the Rocky Mountains. . . . The more you checked his teaching against the out-of-doors the sounder you found it. But he was not always easy to take. His was a school of intellectual hard knocks (Martin, 1968:41).

What Davis Started

We must give Davis credit for his tireless devotion to the advancement of what he thought of as geography. His students were among the outstanding geomorphologists and human geographers of the early twentieth century.[10] They were appointed

---

[10]There were six outstanding graduate students at Harvard with Shaler and Davis in 1891–1892: Albert P. Brigham, former minister in Utica who became a disciple of Davis and taught at Colgate University from 1892 to 1925; Richard E. Dodge, who taught at Teachers College, Columbia, from 1897 to 1916 and at Connecticut State College, Storrs, from 1920 to 1938, and founded the *Journal of School Geography* (later the *Journal of Geography*) in 1897; Curtis F. Marbut, who taught at Missouri from 1895 to 1910 and

to positions in many of the older eastern universities and colleges and to posts in the U.S. Geological Survey and the Soil Survey (see Krug-Genthe, 1903).

Those students who continued to work in physiography included Henri Baulig, John M. Boutwell, Reginald A. Daly, James W. Goldthwait, Frederick P. Gulliver, Thomas A. Jaggar, Douglas W. Johnson, George R. Mansfield, François E. Matthes, Walter C. Mendenhall, Philip S. Smith, and Lewis G. Westgate. Students who studied under Davis and continued their life work in ontography included Robert L. Barrett, Isaiah Bowman, Albert Perry Brigham, Alfred H. Brooks, Robert M. Brown, Collier Cobb, Sumner W. Cushing, Richard E. Dodge, Herbert E. Gregory, William M. Gregory, Ellsworth Huntington, Mark S. W. Jefferson, Curtis F. Marbut, Lawrence Martin, Vilhjalmur Stefansson, Ralph S. Tarr, Walter S. Tower, and Robert DeC. Ward. This formidable group of students was the primal generation and moving force in the establishment of American geography. Those individuals who took positions in universities and normal schools and with government agencies (two, Barrett and Stefansson, became traveler-explorers of accomplishment) also took with them Davisian learning. More than 90 percent of this group were accepted into membership in the Association of American Geographers in the period 1904–23. A Davis student (R. E. Dodge) edited the *Journal of Geography*, and another (I. Bowman) became director of the American Geographical Society. This diffusion of learning from the one hearth produced a direction for geography, albeit not a uniformity of thought. This circumstance facilitated advance of the discipline and cohesion of the profession.

Davis recognized that if geography were to become established as a professional field it would be necessary to organize a professional society in which members could present their ideas. When he was vice president of the American Association for the Advancement of Science (1904) for Section E—Geology and Geography, he used the occasion of the vice president's address to point out that although geology and geography had been granted equal status in the association for the previous 20 years, no vice president of Section E had ever spoken concerning geography. "He then proceeded to fling geography under the eyes or into the ears of the assembled geologists" (Brigham, 1924). He said that the study of geography could lead to no professional career outside of schoolteaching and that there were few opportunities in the universities. There was no organized body of mature

---

was in charge of soil surveys for the Department of Agriculture from 1910 to 1935; Ralph S. Tarr, who taught at Cornell from 1892 to 1912; Robert DeC. Ward, the climatologist, who taught at Harvard from 1890 to 1930; and Lewis G. Westgate, who carried out surveys for the U.S. Geological Survey chiefly in the West. Later distinguished students who worked with Davis at Harvard include: Alfred H. Brooks, who after 1903 was in charge of U.S. Geological Survey work in Alaska; Ellsworth Huntington, author of numerous books and research associate at Yale from 1919 to 1945; Mark Jefferson, who inspired many young scholars to become geographers when he taught at Michigan State Normal College at Ypsilanti from 1901 to 1939; Isaiah Bowman, director of the American Geographical Society from 1915 to 1935 and president of Johns Hopkins from 1935 to 1948; Douglas W. Johnson, geologist at Columbia; and James W. Goldthwait, geologist at Dartmouth.

scholars in geography and, therefore, no opportunity for the mutual encouragement that comes from professional fellowship. He proposed that geographers should organize a professional society similar to the Geological Society of America, "with criteria of expert training and ample publication as a basis for membership." He went on to point out where a nucleus of members could be found among teachers of geography, members of national or state weather services, and members of many government agencies dealing with geology, hydrography, biology, ethnography, and statistical studies. During the following months of 1904, A. P. Brigham called a group of interested people together, and from this meeting there emerged a plan for organizing such a society.

The first meeting and the act of incorporation took place in Philadelphia in December 1904.[11] Davis, as president, spoke about the objectives and opportunities for the new association (Davis, 1905). He was reelected president for 1905 (and for a third term in that office in 1909). It was at the 1905 meeting of the association that he gave his presidential address "An Inductive Study of the Content of Geography" (Davis, 1906), in which he identified geography as the study of the relationship between inorganic controls and organic responses.

Another of Davis's accomplishments on behalf of the profession was his arrangement of international field excursions. In 1911 Davis organized "the Liverpool-Rome Pilgrimage." The pilgrims numbered 32 and were drawn from 14 different countries. It constituted a form of traveling seminar: Much was learned and shared, and friendships were formed. In the following year Davis lectured at the Sorbonne and from there arranged the 1912 Transcontinental Excursion of the American Geographical Society (see Fig. 35). As a result of success especially in the previous year, Davis was able to secure financial help from Archer Huntington (one of the large unseen benefactors of American geography) and the cooperation of railroads, universities, chambers of commerce, university clubs, newspapers, scientific societies, government agencies, and business organizations all across the United States. The excursion included 43 European geographers from 13 different countries (see photograph of some of the participants). About 100 American geographers accompanied the Europeans for at least part of the excursion. The party left New York on August 22 on a train hired especially for the purpose and returned to New York on October 17 after covering 12,965 miles from coast to coast. Many professional papers were written on the basis of notes made during

---

[11] The charter members of the Association of American Geographers (from P. E. James and G. J. Martin, *The Association of American Geographers: The First Seventy-Five Years, 1904–1979* [Washington D.C., AAG, 1979], pp. 36–37.): C. Abbe, Jr.; Charles C. Adams; Cyrus C. Adams; O. P. Austin; R. L. Barrett; L. A. Bauer; A. P. Brigham; A. H. Brooks; H. G. Bryant; M. R. Campbell; F. E. Clements; H. C. Cowles; J. F. Crowell; R. A. Daly; N. H. Darton; W. M. Davis; R. E. Dodge; C. R. Dryer; N. M. Fenneman; H. Gannett; G. K. Gilbert; J. P. Goode; H. E. Gregory; F. P. Gulliver; C. W. Hall; R. A. Harris; A. Heilprin; R. T. Hill; E. Huntington; M. S. W. Jefferson; E. R. Johnson; M. Krug-Genthe; W. Libbey, Jr.; G. W. Littlehales; C. F. Marbut; F. E. Matthes; W. J. McGee; C. H. Merriam; R. W. Pumpelly; H. F. Reid; W. W. Rockhill; R. D. Salisbury; E. C. Semple; G. B. Shattuck; L. Stejneger; R. S. Tarr; R. DeC. Ward; B. Willis.

Figure 35 The transcontinental excursion, 1912

the excursion and published in many languages, but the greatest benefit was derived from the close personal friendships that developed among leaders of the geographic profession in the United States and Europe and from numerous unhurried professional discussions that were carried on. There has never been anything quite like this excursion.[12]

GEOGRAPHY IN THE EARLY TWENTIETH CENTURY

During the early years of the twentieth century, the movement to introduce a professionally acceptable kind of geography into schools, colleges, and universities gradually gathered strength. At the end of the nineteenth century, there were only five professors of geography in American universities: Davis at Harvard, Ralph S. Tarr at Cornell, William Libbey, Jr., the successor to Guyot, at Princeton, George Davidson at the University of California (Berkeley), and Richard E. Dodge at Teachers College, Columbia University. Many teachers' colleges around the country offered courses in geography, but only a few had courses that were taught by persons with some kind of geographical training.[13] Outstanding among the teachers' colleges was

---

[12]See the *Memorial Volume of the Trancontinental Excursion of 1912 of the American Geographical Society* (New York: American Geographical Society, 1915). See also Wright, 1952:158–166. The excursion was made possible by a substantial gift from Archer M. Huntington.

[13]C. E. Cooper, "The Status of Geography in the Normal Schools of the Far West," *Journal of Geography* 18(1919):300–305; idem, "The Status of Geography in the Normal Schools of the Middle States," ibid., 19(1920):211–222; idem, "The Status of Geography in the Normal Schools of the Eastern States," ibid., 20(1921):217–224. See also Dryer, 1924.

the Michigan State Normal College at Ypsilanti, where Charles T. McFarlane taught geography between 1892 and 1900 and was succeeded in 1901 by Mark Jefferson.

Geographical research studies, as reported to the meetings of the Association of American Geographers, increased in number and began to depart from the restricted paradigm of physical condition and human response set forth by Davis. The charter members of the association were geologists, climatologists, botanists, sociologists, teachers, but only a few of them had ever had any advanced training in geography (Brigham, 1924). Nevertheless, some, such as Douglas W. Johnson, who made important contributions to the evolution of shorelines, remained close enough to the newly emerging field of geography to aid in its development. Curtis F. Marbut, who became one of the world's leading authorities on soils, made use of Davis's ideas when he described soils as young or mature. His work in making the Russian soil studies known in the English-speaking world has been mentioned previously. Robert DeC. Ward, who taught climatology at Harvard for 40 years, was the president of the Association of American Geographers in 1917.[14] Many geographers who began their careers before World War I continued to work actively in the field in the period between the two world wars. Some of them require special attention both in this chapter and in Chapter 16. Next we will discuss at greater length the work of Mark Jefferson, Isaiah Bowman, Ellsworth Huntington, Ellen Churchill Semple, and Albert P. Brigham. We will also review the efforts made to define and outline physiographic regions for the United States. Then we will look at the beginnings of commercial and economic geography at the University of Pennsylvania.

Mark Jefferson

None of Davis's students did more to promote and improve the teaching of geography in the United States than Mark Jefferson, who was professor of geography at the Michigan State Normal College in Ypsilanti for 38 years, from 1901 to 1939. Jefferson deserves a special place in the history of geography not only because of the enthusiasm he kindled in his students, but also for his many contributions to the conceptual structure of geography.[15]

---

[14]R. DeC. Ward taught at Harvard from 1890 to 1930. In 1903 he published a translation of the *Handbuch der Klimatologie* by the Austrian scholar Julius Hann (Hann, 1903). In 1908 he published his book on the effect of different kinds of climate on human life (Ward, 1908). Thereafter most of his efforts were devoted to the organization of material on regional climatology (Ward, 1925; Ward and Brooks, 1936).

[15]Before Mark Jefferson graduated from Boston University in the class of 1884, he left to accept a position at the observatory in Córdoba, Argentina. He was in Argentina from 1884 to 1889, serving for two years as submanager of a surgarcane plantation near Tucumán. He then returned to Massachusetts, where he held various administrative posts in secondary schools and where he taught geography. In 1897–98 he studied geography at Harvard with Davis and was greatly stimulated by this experience. In the summer of 1900, when Harvard received a group of some 1300 Cuban schoolteachers seeking to find out about the new geography, Jefferson was appointed to lecture to them in Spanish. He gave 18 lectures and led 12 field trips. Davis was greatly impressed with Jefferson's ability as a teacher (Martin, 1968).

The Michigan State Normal College at Ypsilanti was already noted for its geography teaching before Jefferson went there in 1901. Lectures on geography had been offered since 1853 and after 1860 by John Goodison. When Goodison died in 1892, the vacancy was filled by Charles T. McFarlane, then 21 years old and newly graduated from the New York State Normal College. McFarlane turned out to be an excellent selection. Two years later his courses were described as scientific, partaking of "the close reasoning of physics and mathematics, and the rich insights into the ground of history and social life" (Martin, 1968:78). McFarlane's course of study in geography made use of the recommendations of the Committee of Ten.[16]

When McFarlane resigned from Ypsilanti in 1900, Davis recommended Jefferson as a replacement. Jefferson started teaching at Ypsilanti in June 1901 and only retired in 1939 at the age of 76 after the Michigan Board of Education had set an age limit for active teachers. During his 38 years at the Normal he had a hand in training a large number of teachers of geography, some of whom were to make important contributions to the profession.[17] Jefferson, who was a great admirer of Davis, nevertheless took issue with many of his teacher's ideas. He never accepted the concept of determinism. Furthermore, he disagreed strongly with the recommendations of the Committee of Ten regarding the content of school geography. Jefferson insisted that the focus of geography instruction should be "man on the earth," in that order—not "the earth and man." Nor was he willing to omit the teaching of the geographic conditions in certain particular countries, which we might describe as regional geography. He wanted to make these countries seem real to grade school pupils—real in landscape, life, and institutions (Martin, 1968:343–344). This kind of geography was not part of the recommended systematic approach that Davis favored. In 1904 Jefferson surveyed the geography instruction in Michigan schools, and among the 129 largest high schools he found less than a dozen following the Committee of Ten program.

Jefferson remained largely aloof from discussion of the question "What is geography?" He never began a course with a definition of the field but rather let the scope of geographical study emerge from the materials he covered. No one definition, he believed, could be more than partially inclusive. Here is what he wrote in 1931 in answer to a questionnaire:

---

[16]McFarlane went to Vienna on sabbatical leave in 1898 and studied with Albrecht Penck. In 1901 he became principal of the New York State Normal College at Brockport. In 1910, however, he went to the Teachers College at Columbia as comptroller, a post he held until his retirement in 1927. He also taught courses on the teaching of geography during that time. One of McFarlane's outstanding students at Ypsilanti was H. H. Barrows, who finished his work there in 1896. Barrows, teaching at Ferris Institute in Big Rapids, Michigan, had a student who later became a leader in the profession—Isaiah Bowman. Bowman had been inspired to seek a career in geography when he heard one of McFarlane's lectures.

[17]Among Jefferson's outstanding students were Isaiah Bowman (who came to Ypsilanti in 1901 to find that McFarlane had gone and who remained to find great stimulation from the teaching of Jefferson), Charles C. Colby, Darrel H. Davis, William M. Gregory, George J. Miller, Almon E. Parkins, and Raye R. Platt (see also Dryer, 1924).

Some one has said that anything that you can put on a map is subject matter of geography. That is what I would call locational or distributional geography. . . . But geographers are contemplative persons who cannot be satisfied with so meagre an account of the subject. . . . The nature of geography is the fact that there are discoverable causes of distributions and relations between distributions. We study geography when we seek to discover them. . . . But there is an art of geography—the delineation of the earth's features and inhabitants on maps—cartography, and a science of geography, which contemplates the fact delineated and seeks out causes of the form taken by each distribution and its relationship to others (Martin, 1968:319–321).

Jefferson amply demonstrated that vitality and effectiveness in teaching are related to research and to the communication of the results of research in scholarly publication. He taught 63 different courses during his lifetime (aided by an excellent collection of slides that were the product of his own field work), and his teaching load in many semesters was as much as six courses (18 semester hours). Nevertheless, he established an outstanding reputation as a productive scholar. In the period between 1909 and 1941 he had 31 papers published in the *Bulletin of the American Geographical Society* and its successor, the *Geographical Review* (Wright, 1952:294). This is by far the largest number of professional articles published in these prestigious periodicals during this time by any one scholar. A number of them were major contributions to the concepts of geography (Jefferson, 1909, 1915, 1928, 1939). He wrote on population distribution and urban structures, developed the concepts of ecumene, primate city, and anthropography, and contributed significantly to the advance of cartography and to regional geography (with special reference to North America, Europe, and Argentina).

Isaiah Bowman

One of Jefferson's oustanding students was Isaiah Bowman.[18] After completing his undergraduate work at Harvard with Davis in 1905, Bowman received an appointment in the Department of Geology at Yale, where the geologist Herbert E. Gregory (one of the charter members of the Association of American Geographers and its president in 1920) was gathering together a vigorous group of young

---

[18] Bowman was brought up on a farm in Michigan, where his mother stimulated his early interest in natural history. When his first effort at drawing a map in school was graded *A*, he had had the "success experience" that turned him toward a career in geography. Teaching in a country school and attending summer institutes to improve his knowledge, he was greatly inspired by hearing one of McFarlane's lectures. Bowman was at the Michigan Normal College in 1901–1902. He then studied with Davis at Harvard in 1902–1903 and returned in the next academic year to teach at Ypsilanti under Jefferson's supervision. In 1904–1905 he returned to Harvard and completed his B.S. degree there in 1905. He was then appointed an instructor at Yale under H. E. Gregory and received his Ph.D. from Yale in 1909. He remained at Yale until 1915 and from 1915 to 1935 was the director of the American Geographical Society. From 1935 to 1948 he was president of Johns Hopkins University (Carter, 1950; Martin, 1977, 1980; Wrigley, 1951).

scholars in geography. Bowman was called on to teach a course in geography at the Yale Forestry School; out of his notes concerning surface features and soils related to forest cover, he wrote *Forest Physiography*, the first book to give systematic coverage of the physical characteristics of the regions of the United States (Bowman, 1911).

Bowman spent three field seasons studying the Andes of Peru, Bolivia, and northern Chile. In 1907 he landed at Iquique, Chile, and made his way across the Atacama Desert to the Bolivian Altiplano and thence to the forested eastern slopes of the Andes. He returned by way of Peru, through Cuzco, Arequipa, and Mollendo. From this field expedition he wrote his doctoral dissertation, "The Geography of the Central Andes." In 1911 he was the geologist-geographer on the Yale Peruvian Expedition headed by Hiram Bingham, during which Bingman rediscovered the lost Inca fortress of Machu Picchu. Bowman described his role in the 1911 expedition as follows:

> The geographic work of the Yale Peruvian Expedition of 1911 was essentially a reconnaissance of the Peruvian Andes along the 73rd meridian. The route led from the tropical plains of the lower Urubamba southward over lofty snow-covered passes to the desert coast at Camaná. The strong climatic and topographic contrasts and the varied human life which the region contains are of geographic interest chiefly because they present so many and such clear cases of environmental control within short distances.
>
> ... My division of the Expedition undertook to make a contour map of the two-hundred mile stretch of mountain country between Abancay and the Pacific Coast ... (Bowman, 1916:vii).

In 1913 Bowman received a grant from the American Geographical Society to permit him to return to Peru for a third time. The results were published in two important books (Bowman, 1916, 1924). Clearly, at this time in his life he was still following closely the paradigm of geographical study formulated by Davis, but in the course of time and field experience he became more cautious about so-called environmental controls. While he was at Paris (1918–19) as a member of the American Commission to Negotiate Peace, he wrote, "the Semple bubble ... is forever punctured so far as I am concerned.... I do not believe in that type of geography." Pursuant to his three expeditions to South America, he attempted to draw his findings together. He sought an effective way to generalize the many detailed observations he had gathered regarding the terrain. His imaginative innovation, which he used in *The Andes of Southern Peru* (1916) and *Desert Trails of Atacama* (1924), was his use of "regional diagrams." He recognized in the high Andes of southern Peru six kinds of what he called topographic types:

1. An extensive system of high-level, well-graded, mature slopes ... below which are:
2. Deep canyons with steep, in places cliffed, sides and narrow floors, above which are:
3. Lofty residual mountains composed of resistant, highly deformed rock, now sculptured into a maze of serrate ridges and sharp commanding peaks.

4. Among the forms of high importance, yet causally unrelated to the other closely associated types, are volcanic cones and plateaus of the Western Cordillera.

5. At the valley heads are a full complement of glacial features, such as cirques, hanging valleys, reversed slopes, terminal moraines, and valley trains.

6. Finally there is in all the valley bottoms a deep alluvial fill formed during the glacial period and now in process of dissection (Bowman, 1916:185–186).

The topographic maps show all these features in their complex arrangement and with many variations that make each view unique. The regional diagram, on the other hand, shows these various types in their characteristic arrangement, simplified and compressed within small rectangles (Figs. 36 and 37). These diagrams are what we would now call "empirical generalizations." Bowman wrote of them:

> This compression, though great, respects all essential relations. For example, every location on these diagrams has a concrete illustration but the accidental relations of the field have been omitted; the essential relations are preserved. Each diagram is, therefore, a kind of generalized type map (Bowman, 1916:15).

Some of Bowman's most important work was in the application of geographical methods to the study of practical problems during and after World War I. We will return to this aspect of his career in Chapter 18.

Ellsworth Huntington

Another scholar who studied with Davis at Harvard and was associated with Bowman at Yale was Ellsworth Huntington.[19] While initially a Davisian-inspired

---

[19]Ellsworth Huntington graduated from Beloit College in 1897 and was given an appointment as assistant to the president of Euphrates College in Harput (Turkey). He was in Turkey from 1897 to 1901. He took advantage of every opportunity to travel in different parts of the country, including a trip through the gorge of the Euphrates River during which he kept copious notes on the character of the land, the climate, and the people. In 1901 he received a scholarship to study at Harvard with Davis, and in 1902 he completed the work for the M.A. degree. He started further graduate study leading toward the doctorate, but when the opportunity came to return to Asia for field study he left Harvard. In 1903–1904 he was a member of Raphael Pumpelly's expedition to Central Asia. In 1905–1906 he visited northern India and then went across the Tarim Basin to the Lop Nor, returning by way of Siberia. In 1907 he joined the Yale faculty as instructor in geology. Yale granted him the Ph.D. degree in 1909 on the basis of some of his published works, and he was promoted to assistant professor in 1910. In 1916, however, he resigned from Yale when his request for promotion to professor was turned down. The reason given was his lack of success as a teacher of undergraduate courses. He had to support himself by writing textbooks. After serving in the army in the field of military intelligence in 1918 and 1919, he returned to Yale as a research associate with the rank of professor, but with a token salary (at the start) of $200 a year. He continued as a research scholar at Yale, supervising dissertations and offering graduate courses until his retirement in 1945. He was president of the Ecological Society of America in 1917, president of the Association of American Geographers in 1923, and president of the American Eugenics Society from 1934 to 1938. He was author or coauthor of 28 books and part author of 30 others. He published approximately 240 professional and popular articles (Martin, 1973).

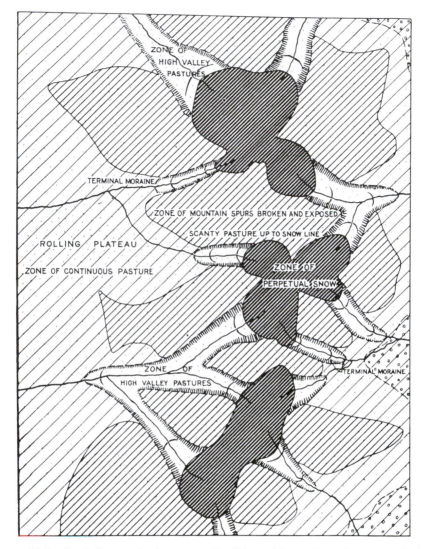

Figure 36 Regional diagram in the eastern Cordillera of Peru. (From Bowman, 1916:65. Reprinted by permission of the American Geographical Society.)

geographer, the locus of Huntington's life's inquiry was investigation of the origin, distribution, longevity, and accomplishment of civilization. He posited climate, "the quality of people," and culture as a triadic causation of human progress. He sought to reveal the environmental platform through space and time whereon mankind had been presented with its climatic circumstance, which, in turn, brought about migration, hastened the processes of selection, and facilitated or denied the advance of culture. This juggling with a thesis of multiple causation was an epic undertaking.

Unfortunately, geographers read only a little of what he wrote, and that usually included *The Pulse of Asia* (1907) and *Civilization and Climate* (1915). In fact,

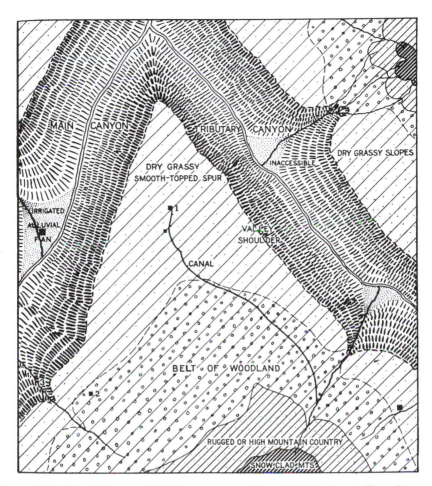

Figure 37 Regional diagram of the deep canyon regions of the Apurimac. (From Bowman, 1916:58. Reprinted by permission of the American Geographical Society.)

Huntington wrote 28 books, parts of 29 other books, more than 240 articles, and more than 100 articles that he did not submit for publication. One can see that it was hardly possible for geographers and others to read a substantial proportion of what he wrote. In consequence, he was rather quickly branded a "determinist" based on a limited reading of his work. Huntington recognized the danger and wrote to a number of geographers pointing this out. As a corrective to this predicament, he wrote *The Pulse of Progress* (1926), which summarized his thought to that time, and then in the clatter of time's hurrying chariot in the 1940s he set about writing a book that would explain his life's work. The first half of this manuscript was published as *The Mainsprings of Civilization* (1945). A sequel to this book, "The Pace of History," was planned, but death prevented completion of the task. However, three chapters have been made available (Martin, 1971).

The fact is that Huntington has been dismissed as a determinist. It must be agreed that early in his career he tied human performance to climate in a way now considered simplistic. And conclusions were drawn from a vast array of data that did not constitute proof. Nevertheless, much of his work involving climate was measured statistically, albeit by cumbersome measurements. This involved matters such as climatic pulsation, climatic variability, climatic optima, the qualities of ozone, the impact of sunspots the counting of tree rings (Fig. 38),[20] and much more.

His feet remained on the ground. He traveled far in native mode, endured hardship, saw much, and recorded same in notebooks and with camera. By the age of 28 he had spent more than seven years in Asia and was the first American geographer to do so. With Davisian training he had gone forth and recorded what he saw and contributed much to the geography of the time.

He was one of the geographers best known to other disciplinarians, and at one time he was one of the geographers best known to the public. (*The Mainsprings of Civilization* sold in paperback to the public for many years after his death.) A careful study of Huntington's lifework reveals that there is very much more to his thought than climate placing limits upon human (and national) capacity.[21] D. H. K. Lee wrote of Huntington's work as follows:

> Huntington's brilliant generalizations covering such a wide range of relationships to climate are worth reading for two reasons: first they are thought-provoking and not all of them have been disproved; and second, as a demonstration of effective presentation they are unequalled (Lee, 1954:473).

Regional Geography and the Delimitation of Regions

In the early years of the twentieth century, those geographers who followed Davis in seeking to define a cohesive field of study including physical and human elements were attracted by the British approach to the delimitation of regions. Many concluded that the highest expression of geographic research was regional geography. Here, within limited confines, the student could go back to causes and forward to consequences without loss of confidence in the results (Fenneman, 1919). But how should one identify and delimit a region?

The first attempt to divide the territory of the 48 contiguous states and territories into physiographic regions was made by John Wesley Powell in 1896 (Powell, 1896). He divided the national territory into 16 regions, some with several subdivisions. Davis himself made a similar map in 1899 (Davis, 1899b:719). This was published in H. R. Mill's *The International Geography* (1899), which was adopted as a textbook in a number of university courses straddling geology, geomorphology and physical geography. In 1911 Bowman wrote *Forest Physiography: Physiography of the United States and Principles of Soils in Relation to Forestry*. This

---

[20] The Ellsworth Huntington collection, one of the largest collections attributable to one geographer, is maintained in the manuscript and archival library, Yale University.

[21] The critics of Huntington are well known; less well known are his supporters. During his life-time geographers finding inspiration in his work included S. B. Jones, A. G. Price, O. H. K. Spate, T. G. Taylor, S. van Valkenburg, and S. S. Visher.

Figure 38 Ellsworth Huntington counting Sequoia rings with Henry Canby

included a chapter entitled "Physiographic Regions, Climatic Regions, Forest Regions," which enlarged upon the physiographic regions of Davis. By 1914 a number of regional divisions of the United States and of North America had been published, each differing in minor ways from the others. In that year W. L. G. Joerg reviewed 21 such maps and drew one of his own that combined the best features of the others. Joerg used the term *natural region* and defined it as "any portion of the earth's surface whose physical conditions are homogeneous." The concept of scale or degree of generalization in delimiting regions had not at that time been widely appreciated. Joerg's regions were highly generalized and, of course, were homogeneous only with respect to the criteria used to define them. N. M. Fenneman also published a study of physiographic boundaries in the United States in 1914; his resulting map incorporated new features not found on the 21 maps that Joerg compared—for example, the separation of the Southern Rockies from the Northern Rockies by the Wyoming Basin (Fenneman, 1914). At the Chicago meeting of the Association of American Geographers in December 1914, one session was devoted to a conference on regions. A committee was appointed to draw a map of physiographic regions on the basis of the criteria accepted by the conference, and Fenneman was named chairman. The map of the physiographic regions with a detailed statement of the characteristics of each region was published in 1916 and included a folded map on the scale of 1/7,000,000 (Fenneman, 1916).

The concept of physical control and human response entered into the drawing of regions with Charles R. Dryer's paper in 1915 (Dryer, 1915). Dryer proposed that the best way to identify natural regions was to measure the economic functions of each. He called his divisions of the United States "natural economic regions." The "solid, liquid, gaseous, and biological phenomena fit into workable living combinations," he wrote (in a surprising return to the natural elements of Aristotle). Economic activities, according to Dryer, fit into these same regions, and a study of the economic activities provides the best guide to delimiting natural regions. This reflection of John F. Unstead's synthetic regions appeared again and again in later years in school and college textbooks that defined a region as homogeneous in its natural characteristics and, therefore, homogeneous in its economic functions. Once the map was drawn, the environmental control was demonstrated. Later, we will see what Albert P. Brigham had to say about this kind of reasoning.

Ellen Churchill Semple

Another pioneer geographer of this period was Ellen Churchill Semple. After graduating from Vassar in 1882, she taught for a few years in her native Louisville, Kentucky. She received a master's degree from Vassar in 1891, earned externally on the basis of a two-year program of readings in sociology and economics, a final written exam, and a thesis, "Slavery: A Study in Sociology." From friends she heard enthusiastic reports about a professor in Germany, at Leipzig, whose lectures were bringing new worlds into view. She went to Leipzig, and in spite of difficulties placed in the way of women who wanted to undertake graduate work, she studied with Ratzel in 1891–92 and again in 1895. She returned to the United States greatly stimulated by Ratzel's new approach to anthropogeography and his interpretations of so-called geographic influences on the course of history. She rejected his ideas about the state as an organism, which he had derived from Herbert Spencer. Semple's aim then was to present Ratzel's ideas in English, but clarified and reorganized, with many new illustrations drawn from different parts of the world. In 1897 she published her first professional article dealing with the role of the Appalachians as a barrier in American history (Semple, 1897), and in 1901 she published a paper based on her own field observations on the highlands of eastern Kentucky regarding the results of isolation on the settlers of that area (Semple, 1901). This second paper started her along the road to fame, but her professional status was confirmed in 1903 with the publication of her first book, *American History and Its Geographic Conditions* (Semple, 1903).

She presented her version of the first volume of Ratzel's *Anthropogeographie* in her book *Influences of Geographic Environment*, which was published in 1911. Here is what she wrote about her method:

> The writer's own method of research has been to compare typical peoples of all stages of cultural development, living under similar geographic conditions. If these peoples of different ethnic stocks but similar environments manifested similar or related social, economic, or historical development, it was reasonable to infer that such similarities were due to environment and not to race. Thus by extensive comparison, the race factor in these problems of two unknown quantities was eliminated for certain large classes of social and historical phenomena (Semple, 1911:vii).

Here is another quotation from the opening paragraph of her book:

> Man is a product of the earth's surface. This means not merely that he is child of the earth, dust of her dust; but that the earth has mothered him, fed him, set him tasks, directed his thoughts, confronted him with difficulties that have strengthened his body and sharpened his wits, given him his problems of navigation or irrigation, and at the same time whispered hints for their solution. . . . On the mountains she has given him leg muscles of iron to climb the slope; along the coast she has left these weak and flabby, but given him instead vigorous development of chest and arm to handle his paddle or oar. In the river valley she attaches him to the fertile soil, circumscribes his ideas and ambitions by a dull round of calm, exacting duties, narrows his outlook to the cramped horizons of his farm. Up on the wind-swept plateaus, in the boundless stretch of grasslands and the waterless tracts of the desert, where he roams with his flocks from pasture to pasture and oasis to oasis, where life knows much hardship but escapes the grind of drudgery, where the watching of the grazing herd gives him leisure for contemplation, and the wide-ranging life of a big horizon, his ideas take on a certain gigantic simplicity; religion becomes monotheism, God becomes one, unrivalled like the sand of the desert and the grass of the steppe, stretching on and on without break or change. Chewing over and over the cud of his simple belief as the one food of his unfed mind, his faith becomes fanaticism; his big spacial ideas, born of that ceaseless regular wandering, outgrow the land that bred them and bear their legitimate fruit in wide imperial conquests (Semple, 1911:1–2).

The quotations suggest two things: first, that her style of writing has a certain literary quality that makes reading it a delight, yet which might—and sometimes does—carry the theme beyond what sober judgment would permit; second, that the concept of the earth as the controlling factor in human life is carried beyond the possibility of objective verification. It is true that in combing the writings of all nations for examples to illustrate her principles, she fell into an error not uncommon when deductive reasoning is followed—she failed to look carefully for examples that contradicted her principles. Is it not possible to find examples of people who worship one God and yet are not pastoral nomads? And are there no examples of inhabitants of the boundless steppes who are pantheists? People who live in pass routes, she wrote, tend to become robbers of passing travelers. Then she presented case after case of people in pass routes who rob for a living. But she did not look for people in pass routes who do not rob, nor did she seek an explanation for robbers who do not live in pass routes.

Nevertheless, two other observations must be made. First, she was very careful to make the point that the environment does not control human action, only that under certain circumstances people tend to behave in predictable ways—which is the first small step toward probability theory. Second, there are some brilliant passages in which her insight is even now thoroughly relevant. Her "islands of ethnic expansion and islands of ethnic retreat" offer an important modification for the contemporary theory of innovation dispersal (Semple, 1911:204–228).

Ellen Semple was an enormously persuasive teacher, and generations of American geographers were brought up to believe these teachings. During the time that she lectured at Chicago and later at Clark University, a large number of future geographers passed through her classrooms. It is easy to condemn her for presenting concepts that have not withstood the test of time, but she must be appreciated for kindling among her students an enthusiasm for the largeness of viewpoint which was hers.[22] Her writing, beautiful in composition, was doubtless a product of the logic of her thought. By the end of the first decade of the 1900s, her work had so permeated American geography that one is able to notice the development of courses around that contribution. For the most part that was to be found in the smaller colleges that preferred this variety of geography to Davisian physiography.

In 1911 Semple started to work on the geography of the Mediterranean region. Over a period of 20 years she was a frequent visitor in the countries bordering the Mediterranean, including the parts of Asia to the east. She did a vast amount of reading in the literature—both ancient and modern—concerning the Mediterranean countries. In 1915 she published the first of numerous articles on different aspects of the region—this one on the mountain barriers and the breaches through them as factors in the history of the region (Semple, 1915). From 1917 to 1918 her knowledge of the Mediterranean world was employed by The Inquiry (see Chapter 19), where she made contributions with special reference to Greece and Turkey. This aside did not interrupt her researches, and she wrote papers dealing with Mediterranean agriculture, the relation of forests to climate, the relation of climate to religion, and the geographic basis of Mediterranean trade. One of the most delightful of these papers described the "Templed Promontories" where the gods were asked to watch over the seafarers who had to sail around dangerous capes (Semple, 1927). All of this work was brought together in her last great book, on which she was at work when stricken in 1929. With great courage she persisted, able to work no more than two hours each day, until the book was completed and published only a few months before her death (Semple, 1931).[22] It is an intellectual curiosity that we do not have a book-length assessment of this remarkable geography (Bushong, 1984).

---

[22]Ellen Churchill Semple was a visiting lecturer at Chicago between 1906 and 1924. In 1912 and again in 1922 she lectured at Oxford in England. She was visiting lecturer at Wellesley College in the fall of 1914, at the University of Colorado in the summer of

Albert Perry Brigham

One of the "outstanding graduate students" who was at Harvard working with Shaler and Davis in 1891–92 was Albert Perry Brigham (James, 1978).[23] Brigham, who was on the faculty of Colgate University for many years, was one of the major supporters of W. M. Davis in the effort to establish geography as a professional field in the United States. Brigham's book *Geographic Influences in American History* was published in 1903, the same year in which Ellen Semple's book on the same subject appeared. Brigham's book (Brigham, 1903) placed heavy emphasis on the origin of what he called geographic conditions but was relatively light on history, as the historians were not slow to point out. Although somewhat different in emphasis, Semple and Brigham's points of view were very similar.

In the course of time Brigham took an increasingly vigorous stand against the way geologically trained geographers asserted the existence of "responses" to environmental conditions without attaching precise meaning to their words and without showing what influences really do and how they do it. He was familiar with the work of the European geographers, especially Ratzel and Brunhes. He insisted that Ratzel was a pioneer and could not be blamed for not exploring all the parameters of man-land relations. In his presidential address before the Association of American Geographers in 1914 (Brigham, 1915), he specified that it was the geographer's task to provide a careful and scientific description of the physical environment, but that geographers should use caution and common sense in asserting the existence of influences and that every possible test should be made to ascertain the validity of any general principles that were suggested. He wrote of the relation of specific factual information to general concepts:

> Our goal is broad generalization. But the formulation of general laws is difficult, and the results insecure until we have a body of concrete and detailed observations. . . . Detailed investigation of single problems, in small and seemingly unimportant fields, must for a long time prepare the way for the formulation of richer and more fundamental conclusions and general principles than we have yet been able to achieve (Brigham, 1915:24–25).

Brigham was especially critical of generalizations concerning the influence of climate. He observed that "perhaps there is no subject, unless it be politics, on which

---

1915, at Western Kentucky University in the summer of 1917, at Columbia University in the summer of 1918, and at the University of California at Los Angeles in 1925. From 1921 until she was stricken with a heart attack in 1929, she was a member of the staff of the Graduate School of Geography at Clark University. In 1921 she was the president of the Association of American Geographers (Colby, 1933).

[23]A. P. Brigham was the minister of a large church in Utica, New York, when he took a summer field course in geology at Harvard in 1889. Two years later he resigned his ministry and went to Harvard for advanced study. He received the M.A. degree at Harvard in 1892, and from then until his retirement in 1925 he was on the faculty of Colgate University. In 1904 he was one of the charter members of the Association of American Geographers and was its secretary-treasurer from 1904 to 1913. In 1914 he was president of the Association (see the *Annals AAG* 20(1930):55–104; and 23(1933):27–32).

men say so much and know so little as about climate." He was especially disturbed by vague and unproved assertions of climatic influence on racial character, skin color, or human institutions. The infinitely variable factors of the total environment, he insisted, produce infinitely diverse results on body and mind.

During the period from the 1890s to World War I, the point of view toward geography that was widely accepted in the United States was that of the search for environmental influences. Furthermore, the meaning was the same whether or not the word "influences" was replaced by "responses" or "adjustments." And in spite of Brigham's warning concerning the need for careful use of words and patient testing of alleged influences, many geographers continued to draw plausible but unverified conclusions from their studies (Hartshorne, 1939:23).

Emory R. Johnson and J. Russell Smith

While anthropogeography was being developed by people trained for the ministry or in history, and physical geography was being developed by people trained in geology, the advanced study in economic and commercial geography was being started by two scholars trained in economics. The individuals who led the way in this aspect of geography were Emory R. Johnson and J. Russell Smith, both in the Wharton School of Finance and Commerce at the University of Pennsylvania.[24] In 1899, when Congress set up the Isthmian Canal Commission to recommend the best route for an interocean canal, Johnson, who was a specialist on the geography of transportation, was appointed to make a cost-benefit study of alternative routes. He selected his graduate student J. Russell Smith as his assistant. Although he had been teaching courses on "The Theory and Geography of Commerce" and on "Physical and Economic Geography," he was not experienced in the actual methods of geographical study. He and Smith had to work out their own procedures. Here is the way Smith described the situation: "We came out of that job with a firm conviction that the American educational field needed geography in the colleges—quick. Because

---

[24]Emory R. Johnson received the Ph.D. in economics for a dissertation, "Inland Waterways: Their Relation to Transportation," at the Wharton School in 1893. Whittlesey lists this as the first doctoral dissertation in geography in the United States (Whittlesey, 1935:213). On the staff of the Wharton School from 1893 until his retirement in 1933, Johnson offered courses in economic geography. He was dean of the Wharton School from 1919 to 1933. Students of Johnson who received the Ph.D. from the Wharton School include J. Paul Goode (1901), J. Russell Smith (1903), and Walter S. Tower (1906). Johnson was a charter member of the Association of American Geographers.

J. Russell Smith graduated from the University of Pennsylvania in economics in 1898 and started graduate work with E. R. Johnson. Johnson and Smith worked on the Isthmian Canal Commission in 1899–1901. In 1901–1902 Smith studied anthropogeography with Ratzel at Leipzig and then returned to the Wharton School, where he completed the Ph.D. in 1903. On the faculty of the Wharton School from 1903 to 1919, he became chairman of the new Department of Geography and Industry, which he founded. From 1919 until his retirement in 1944, he was chairman of the Department of Geography in the School of Business at Columbia University. He was president of the Association of American Geographers in 1943 (Martin, 2001).

of our helplessness in the face of a concrete problem, it convinced us that it was extremely important" (Rowley, 1964:22–23).

In 1919 J. Russell Smith was invited to the newly established School of Business at Columbia University to develop a curriculum in geography leading to advanced degrees. He set himself to remedy what he felt was a serious lack of good textbooks in economic geography and in 1913 published his influential *Industrial and Commercial Geography* (Smith, 1913). The book was revised several times and provided the basic text for courses in this kind of geography for perhaps 50 years. The last edition was published in 1955 with the aid of two coauthors, one of whom was his son, Thomas R. Smith. In 1925 Smith published *North America*. He adopted the concept of human use regions as the framework for his book. While teaching the North America course for 20 years he had perceived 44 of these regions. The book was published at a time when the Roorbach questionnaire leading to the report "The trend of modern geography, a symposium" (1914) established that American geographers believed that one of the most pressing needs within the profession was for a regional geography of North America similar to the regional studies accomplished in Europe. Joerg later wrote (1936) in *The Geography of North America: A History of Its Regional Exposition*: "Here was an account, organized sufficiently minutely by regions and of sufficient length to deal adequately with all the essentials." In addition to this book, Smith published 29 other books for use at various grade levels. He was active in the conservation movement, strongly recommending that steeply sloping land should be planted with tree crops to protect it from erosion. He had developed a tree farm of 2000 acres in Virginia and had become an authority on the oak, chestnut, and honey-locust trees. Here is the way Virginia M. Rowley summarized his career:

> J. Russell Smith is thus a unique man, of unusual energy and versatility. His restless, probing, creative mind caused him to go beyond the narrow subject bounds of a single academic discipline and to view knowledge as a related whole. Some may criticize Smith for his occasional inaccuracies, his untested theories, or, at times, his deemphasis of specific details. These criticisms are sometimes justified and when detrimental to truth and objectivity, reflect definite weaknesses which, as a professional geographer, Smith should have eliminated. On the other hand, it must be remembered that Smith's goal was different from that of the pure research specialist. To him an idea had worth only if it were put to work. We must see Smith the academic geographer, the generalist, as well as synthesizer and experimenter, as many others have seen Shaler, Smith's "unseen master," as more concerned with awakening minds than with imparting specific information (Rowley, 1964:200–201).

GEOGRAPHY AT THE UNIVERSITY OF CHICAGO

The pioneering works of Davis and his disciples, of Ellen C. Semple, and of Emory R. Johnson and J. Russell Smith were of major importance in the introduction of a revitalized geography into the United States. But geography as a professional field could not emerge until there were university departments staffed by scholars trained in geography. This process began at the University of Chicago in 1903, when the first separate department offering advanced graduate study was formed.

The University of Chicago was founded on July 1, 1891. John D. Rockefeller had provided the financial support to organize a new university to be staffed by professionally qualified scholars. William Rainey Harper took office as the first president. Harper was building another university similar to Johns Hopkins, which had been modeled on the German concept of a society of scholars by Daniel Coit Gilman. By the time the first classes were held at Chicago on October 1, 1892, Harper had assembled a faculty of high quality. There were 103 members on the new faculty, and of these 8 were former university presidents. Outstanding scholars included Thorstein Veblen in economics, Albert Michelson in physics, and Thomas C. Chamberlin in geology. Chamberlin, who resigned as president of the University of Wisconsin to accept the appointment of Chicago, brought with him his associate Rollin D. Salisbury.[25] In 1892 H. J. Mackinder visited Chicago and was urged by Harper to join the faculty, but Mackinder did not pursue the matter.

Rollin D. Salisbury

Rollin D. Salisbury, who was the chairman of the Department of Geography at Chicago from 1903 to 1919, was a major force in the development of professional geography in the United States (Pattison, 1981, 1982). He influenced a large number of students and was generally recognized as the best teacher in the university. His freshman course in physiography was always filled. Whereas Davis was noted for his polished lectures, Salisbury was a master of the art of stimulating and directing class discussions. By skillful questioning he ensured student participation. But when any student attempted to obscure a lack of preparation behind a screen of generalities, Salisbury would remark: "perfectly true, perfectly general, perfectly meaningless." He insisted that a student learn to express himself clearly—"not so that he could be understood, but so that he could not be misunderstood" (Chamberlin, 1931:128). His classroom was no place for the dull student, but the better ones were enormously stimulated, especially by participation in his advanced seminars. Students who

---

[25]T. C. Chamberlin was the son of a pioneer farmer in Wisconsin. He graduated from Beloit College in 1866 and returned there as professor of natural history in 1873. He became professor of geology in 1880. From 1882 to 1887 he worked full time on the U. S. Geological Survey. As a geologist he is noted for his demonstration of the occurrence of several advances and retreats of the ice in Wisconsin and for his formulation of the planetesimal hypothesis regarding the origin of the earth. In the field of scientific methodology, he is also known for his paper on the need to use multiple working hypotheses (Chamberlin, 1897). From 1887 to 1892 he was president of the University of Wisconsin. From 1892 until his retirement in 1919, he was chairman of the Department of Geology at the University of Chicago.

R. D. Salisbury graduated from Beloit College in 1881 and succeeded Chamberlin there in 1882. In 1891 he came to the University of Wisconsin but resigned the next year to accept an appointment with Chamberlin as professor of geographic geology at Chicago. In 1894 he was named dean of the Ogden School of Science at Chicago, which position he held until his death in 1922. Salisbury was named chairman of the new Department of Geography in 1903. In 1919, when Chamberlin retired, Salisbury became chairman of the Department of Geology; Harlan H. Barrows replaced him in geography.

attended one of his seminars, even for a short period in the summer session, went out to establish similar seminars in other schools and colleges. In 1913 he started a regular weekly meeting of staff and graduate students in which geographical questions and problems were discussed.[26] In the give and take of scholarly discussion, those in attendance not only clarified in their own minds the methodology and scope of geography, but they also learned a valuable lesson about the way to carry on such a discussion with their peers, to accept criticism without emotional reaction, and to respect the words of others even if they were in disagreement. Charles Colby said that these seminars did more to establish high standards of work and thought at Chicago and elsewhere than any other single part of the program (Colby, 1955).

Salisbury was himself concerned with that part of geography that he called physiography (physical geography). He was hopeful but skeptical that workers in anthropogeography might also develop that part of geography on a scientific basis. He is not properly described as "a follower of Davis," although he did make use of the terminology that Davis had proposed. But in Salisbury's teaching the cycle of erosion was given a minor place. Salisbury also rejected the idea of a simple cause and effect relation between the physical earth and the human response. Physiography, for Salisbury, was the scientific study of the stage setting on which the human drama unfolded. But the relation of the stage setting to human action was not a causal one.

Salisbury published his ideas in his *Physiography* (Salisbury, 1907), which was widely used throughout the United States and went through several editions. With Wallace W. Atwood[27] he made a selection of topographic maps from the U.S. Geological Survey, parts of which were reproduced along with notes interpreting the origin of the landforms (Salisbury and Atwood, 1908). With other members of the geography staff, Salisbury collaborated in a college text that presented a basic course in geography, which for many years was a standard college text in the United States (Salisbury, Barrows, and Tower, 1912). He also demonstrated his method and point of view in studying physical geography in a monograph on the stage setting of Chicago (Salisbury and Alden, 1899).

Building the Department

When the new Department of Geography was established in 1903, Salisbury immediately recruited two young scholars to form his staff. One was J. Paul Goode, who had studied geology at Chicago in 1896–97 but who had gone to the University

---

[26]The first such seminar was held in the fall of 1913 and was attended by Salisbury, Barrows, Tower, and Goode from the staff and by graduate students Charles C. Colby, Wellington D. Jones, A. E. Parkins, William Haas, Stephen S. Visher, and Mary Lanier (Colby, 1955).

[27]Wallace W. Atwood received his B.A. degree in geology at Chicago in 1897. On a field trip to the Devil's Lake Driftless Area of Wisconsin with Salisbury, he developed a keen interest in the interpretation of landforms. He received the Ph.D. in geology at Chicago in 1903, and from 1903 to 1913 he was on the staff of the Department of Geology there.

of Pennsylvania to study economic geography with Johnson.[28] When Salisbury found himself responsible for developing a program of study in geography, he remembered the young student who had so impressed him with sound ideas about geography some six years earlier. He asked Goode to prepare a proposed program of courses for the new department. Goode answered in detail, and most of his suggestions were adopted. In 1903 Salisbury invited Goode to become a member of the geography staff (Martin, 1984).

The other young man recruited at that time was Harlan H. Barrows, who had just received the B.S. degree in geology at Chicago but who had already taught geography for more than five years.[29] Barrows was promoted rapidly, and in 1919, when Salisbury moved back to geology, he was named chairman of the Department of Geography.

Geology and geography at Chicago remained very closely associated during this period. When Chamberlin was away, Salisbury acted as chairman of both departments. Students took courses in both departments. Geology and geography shared space in the same building and made use of the map collection. Salisbury and Goode undertook to set up no new courses that would duplicate work already being offered in geology. But meteorology and climatology, which had been taught in the Department of Geology, were transferred to geography, where they were taught by Goode. The new courses in geography were planned to occupy the great uncultivated field between geology and climatology on the one hand, and biology, history, sociology, economics, anthropology, and political science on the other. The first Ph.D. granted by the new department went to Frederick V. Emerson in 1907 for what was one of the first American urban studies by a geographer—"A Geographic Interpretation of New York City" (Emerson, 1908–1909).[30]

---

[28] J. Paul Goode received the B.S. degree from the University of Minnesota in 1889 and from then until 1898 held the position of professor of natural science at the Minnesota State Normal College at Moorhead, Minnesota. In 1894 he attended a summer session at Harvard and worked with Davis. In 1896–1897 he was a "fellow in geology" at Chicago. In 1899 he was appointed professor of physical science and geography at Eastern Illinois State Normal College. Finding no place where he could pursue advanced graduate study in geography, he went to the University of Pennsylvania to study with E. R. Johnson and received the Ph.D. in economics in 1901. He was an instructor in geography at Pennsylvania when, in 1903, he was offered a position at Chicago (Martin, 1984).

[29] Harlan H. Barrows completed an undergraduate program at Michigan State Normal College under Charles T. McFarlane in 1896. He had been teaching at the Ferris Industrial School (later Ferris Institute) in Big Rapids, Michigan, when he decided to take additional undergraduate work in geology at Chicago. He received the B.S. degree in geology in 1903. He was such an outstanding student that he was immediately appointed as Salisbury's assistant. He was promoted to instructor in geography in 1907, to assistant professor in 1908, to associate professor in 1910, and to professor in 1914. From 1919 until the time of his retirement in 1942, he was chairman of the department. (For other aspects of his career, see Chapter 16.)

[30] F. V. Emerson studied at Edinboro (Pennsylvania), Colgate, Cornell, and Harvard before completing his graduate studies at Chicago. He then worked with C. F. Marbut at the University of Missouri and from 1913 to his death in 1919 was professor of geology at Louisiana State University and also director of the Soil Survey of Louisiana.

Within a decade the new department had already established its preeminent position in the training of the younger generation of geographers. In addition to Salisbury, whose chief concern was with the physical earth, and Goode, who offered work not only in meteorology and climatology but also in the economic and commercial geography of Europe and the tropics, cartography, and the history of geographic thought. Barrows began to develop his ideas on the historical geography of the United States, which came into full flower after World War I (Barrows, 1962). His course was very popular with the undergraduates. A questionnaire circulated among Chicago alumni some years later listed Barrows's course on historical geography as one of the most worthwhile courses in the whole undergraduate program of the university (Colby, 1955). During his career Barrows offered some 25 different courses (Koelsch, 1969, 1976).

Walter S. Tower was added to the department in 1911 to offer courses in the economic geography of South America and in political geography (Tower, 1910).[31] Tower was one of the earliest "regional specialists" in the Latin American field to be appointed to a university post in the United States. Others who taught occasionally at Chicago were Ellen C. Semple, in alternate years between 1906 and 1924, and Bailey Willis, a geologist from Stanford University.

The graduate students who worked in the department during the first decade included many who became leaders of the profession in the period after World War I.[32] When the government of Argentina planned the construction of railroads westward across Patagonia after 1902, it looked to the example of railroad surveys undertaken earlier in the American West and requested assistance from the U.S. Geological Survey. Bailey Willis was appointed to organize and operate this survey. When he found that he needed someone trained in economic geography, he turned to Salisbury for a recommendation. The young man who worked with Bailey Willis in Patagonia in 1912 was Wellington D. Jones. This experience in Patagonia inspired Jones, along with his fellow graduate student Carl Sauer, to suggest the importance of detailed field mapping of agricultural areas. For the first time it was suggested that maps of land use should be prepared at the same scale and the same degree of detail as the maps of the physical environment. The paper that Jones and Sauer published in 1915 had been thoroughly discussed at the Chicago seminar (Jones and Sauer, 1915).

Colby identifies three factors in pre-1917 America that caused the rapid increase in the teaching and writing of geography. We have already mentioned the critical importance of the surveys of the American West and especially the pioneer work of Gilbert, Powell, Wheeler, and Hayden. Second, also in the years prior to World War I, a great increase in overseas commerce was causing the public to demand the teaching of commercial and economic geography—just as the traders of Amsterdam had made similar demands in the seventeenth century and were answered by Varenius. The University of Pennsylvania responded to this demand in the United States through the activities of the economists and economic

---

[31] For a biographical sketch of Walter S. Tower, see Chapter 18.

[32] These students included Charles C. Colby, Wellington D. Jones, Stephen S. Visher, V. C. Finch, Carl O. Sauer, Mary Lanier, Mary Dopp, Mabel C. Stark, and L. P. Denoyer.

geographers in the Wharton School. The third influence that Colby listed was the rapid opening up of new natural resources, including oil. This drew the attention of educators to the need for teaching the geography of resources and the methods of conserving them. All these matters were discussed at length by the participants in the Chicago seminar.[33]

Another distinctive characteristic of the Chicago group was the emphasis it placed on field studies. In the tradition of the exploring expeditions of the West, all graduate students were expected to examine the character of the landscapes and to identify geographical problems from direct observation. In September 1913 Tower led a party of six students in a traverse of the northern Appalachians from Pittsburgh to Harrisburg. In September 1914 Barrows led a much larger group on foot across the southern Appalachians. In 1915 Goode conducted a trip to the West, visiting ranches, mines, and irrigated areas and including a visit to the Panama Pacific Exposition in San Francisco. This kind of field course became a distinctive feature of Chicago before World War I. But, as we will see, students such as Wellington D. Jones, Carl O. Sauer, and K. C. McMurry began to visualize a quite different kind of field experience in which students would not be taken on a conducted tour but would be set to work in a restricted area to identify problems and demonstrate their ability to find answers.

In their field studies the geographers learned to cooperate with scholars in other disciplines. For example, Henry C. Cowles, professor of botany at Chicago, was interested in plant ecology. He involved geographers in his studies of plant succession on the Indiana Dunes. Cowles was one of the founders of the Association of American Geographers (1904) and its president in 1910.

MODERN GEOGRAPHY IN 1914

In 1914 George B. Roorbach, who was assistant professor of economic geography at the University of Pennsylvania, published the results of a questionnaire he had sent out to people who called themselves geographers. He found that in a seminar discussion of the scope and method of geography almost everyone had his or her own definition. As in Germany four decades earlier, very few people who taught geography had been formally educated as geographers. Therefore, each new geographer felt impelled to answer the question: "What is geography?" And true to the nature of most scholars, it would not do to accept any other scholar's definition of the field. Consequently, there was little resolve concerning the nature of geography. Roorbach asked for a listing of the most important tasks to be undertaken by geographers. He received 29 replies, all but four of which were from scholars

---

[33]The four members of the department staff made distinctive contributions to the seminar discussions. Salisbury contributed the sure touch of the master in directing these discussions. Barrows had the keenest mind: He had a prodigious memory and was a strict logician. Tower introduced challenging and original ideas. And Goode never failed to insist that the best way to communicate geographical ideas was through the expert use of maps (Colby, 1955).

in the United States. The four others were well-known British geographers (Roorbach, 1914).

Roorbach found an almost unanimous agreement that geography was the study of the relationship between the earth and life—which was essentially the idea proposed by Davis. The respondents then listed the following tasks as important in the order given:

1. The exact determination of the influence of geographic environment. This was placed first by 22 of the 29 respondents.

2. Regional studies of selected areas. There were some British suggestions that a major task would be to divide the world into its major natural regions.

3. The definition and organization of geographical material.

4. The improvement of the teaching of geography.

5. The study of the influence of geographic factors on history.

6. The exploration of unknown or little-known places (suggested by the British geographer John Scott Keltie and by Robert E. Peary).

7. The study of physical geography.

So it was in 1914. Geographers did not realize that culture, not nature, determined the significance of environment, site, and natural resources, in spite of the critique of environmentalism advanced by ethnologists since before 1900. Moreover, two interuniversity conferences between Columbia-trained ethnologists and Yale geographers, arranged by Franz Boas in 1913 for exploring the problem of environmental conditioning of society, failed to inoculate geographers against the naturalistic assumption. The principle that culture is the fundamental extraenvironmental factor in the derivation of human activities did not penetrate geography more generally until after World War II (Speth, 1978:10–11; 1999).

REFERENCES: CHAPTER 15

Aay, H. 1981. "Textbook Chronicles: Disciplinary History and The Growth of Geographic Knowledge." In B. Blouet, ed., *The Evolution of Academic Geography in the United States.* Pp. 291–301. Hamden, Conn.: Shoe String Press.

Adams, C. C. 1907. "Some Phases of Future Geographical Work in America." *Bulletin of the American Geographical Society* 39:1–12.

Barrows, H. H. 1962. *Lectures on the Historical Geography of the United States, as Given in 1933.* Ed. W. A. Koelsch. Chicago: University of Chicago, Department of Geography.

Baulig, H. 1950. "William Morris Davis: Master of Method." *Annals AAG* 40:188–195.

Beckinsale, R. P. 1981. "W. M. Davis and American Geography: 1880–1930." In B. Blouet, ed., *The Evolution of Academic Geography in the United States.* Pp. 107–122. Hamden, Conn.: Shoe String Press.

Beckinsale, R. P., and Chorley, R. J. 1991. *The History of the Study of Landforms or the Development of Geomorphology.* Vol. 3: *Historical and Regional Geomorphology 1890–1950.* London and New York: Routledge.

Bladen W. A., and Karan, P. P. 1983. *The Evolution of Geographic Thought in America: A Kentucky Root.* Dubuque, Iowa: Kendall/Hunt Publishing Co., Iowa.

Block, R. H. 1984. "Henry Gannett, 1846–1914." *Geographers: Biobibliographical Studies.* Vol. 8, pp. 45–49. London: Mansell.

Blouet, B. W., ed. 1981. *The Origins of Academic Geography in the United States.* Hamden, Conn.: Archon Books.

Bowman, I. 1911. *Forest Physiography: Physiography of the United States and Principles of Soils in Relation to Forestry.* New York: John Wiley & Sons.

———. 1916. *The Andes of Southern Peru: Geographical Reconnaissance Along the Seventy-third Meridian.* New York: Henry Holt.

———. 1924. *Desert Trails of Atacama.* New York: American Geographical Society.

Brigham, A. P. 1903. *Geographic Influences in American History.* Boston: Ginn & Co.

———. 1915. "Problems of Geographic Influence." *Annals AAG* 5:3–25.

———. 1924. "The Association of American Geographers." *Annals AAG* 14:109–116.

Bryan, K. 1935. "William Morris Davis—Leader in Geomorphology and Geography." *Annals AAG* 25:23–31.

Bushong, A. 1984. "Ellen Churchill Semples 1863–1932." *Geographers* 8:87–94.

Butzer, K. W. 1964. *Environment and Archeology.* Chicago: Aldine.

Carter, G. F. 1950. "Isaiah Bowman, 1878–1950." *Annals AAG* 40:335–350.

Chamberlin, R. T. 1931. "Memorial to Rollin D Salisbury." *Bulletin of the Geological Society of America* 42:126–138.

Chamberlin, T. C. 1897. "The Method of the Multiple Working Hypotheses." *Journal of Geology* 5:837–848.

Champlin, M. D. 1992. "Raphael Pumpelly 1837–1923." *Geographers: Biobibliographical Studies.* Vol. 14, pp. 83–92. London: Mansell.

Chappell, J. E., Jr. 1970. "Climatic Change Reconsidered: Another Look at 'The Pulse of Asia.'" *Geographical Review* 60:347–373.

Chorley, R. J. 1965. "A Re-evaluation of the Geomorphic System of W. M. Davis." In R. J. Chorley and P. Haggett, eds., *Frontiers in Geographic Teaching.* Pp. 21–38. London: Methuen.

Chorley, R. J., Beckinsale, R. P., and Dunn, A. J. 1973. *The History of the Study of Landforms, or the Development of Geomorphology.* Vol. 2: *The Life and Work of William Morris Davis.* London: Methuen.

Chorley, R. J., Dunn, A. J., and Beckinsale, R. P. 1964. *The History of the Study of Landforms, or the Development of Geomorphology.* Vol. 1: *Geomorphology Before Davis.* New York: John Wiley & Sons.

Colby, C. C. 1929. "Twenty-Five Years of the Association of American Geographers: A Secretarial Review." *Annals AAG* 19:59–61.

———. 1933. "Ellen Churchill Semple." *Annals AAG* 23:229–240.

———. 1936. "Changing Currents of Geographic Thought in America." *Annals AAG* 26:1–37.

———. 1955. "Narrative of Five Decades." In *A Half Century of Geography—What Next?* (Papers presented at the alumni reunion, June 5, 1954.) Chicago: University of Chicago, Department of Geography.

Cole, D. 1999. *Franz Boas: The Early Years, 1858–1906.* Seattle and London: University of Washington Press.

Darrah, W. C. 1951. *Powell of the Colorado.* Princeton, N.J.: Princeton University Press.

Davis, W. M. 1899a. "The Geographical Cycle." *Geographical Journal* 14:481–504.

———. 1899b. "The United States of America." In H. R. Mill, ed., *The International Geography.* New York: D. Appleton.

———. 1905. "The Opportunity for the Association of American Geographers." *Bulletin of the American Geographical Society* 37:84–86.

————. 1906. "An Inductive Study of the Content of Geography." *Bulletin of the American Geographical Society* 38:67–84 (reprinted in Davis, 1909).

————. 1909. *Geographical Essays*. Ed. D. W. Johnson. Boston: Ginn & Co.

————. 1910. "Experiments in Geographical Description." *Bulletin of the American Geographical Society* 42:401–435.

————. 1911. "The Colorado Front Range, A Study in Physiographic Presentation." *Annals AAG* 1:21–84.

————. 1912. *Die erklärende Beschreibung der Landformen*. Trans. A. Rühl. Leipzig: Teubner.

————. 1915. "The Principles of Geographic Description." *Annals AAG* 5:61–105.

————. 1919. "Passarge's Principles of Landscape Description." *Geographical Review* 8:266–273.

————. 1922. "Peneplains and the Geographical Cycle." *Bulletin of the Geological Society of America* 33:587–598.

————. 1924. "The Progress of Geography in the United States." *Annals AAG* 14:159–215.

————. 1928. *The Coral Reef Problem*. New York: American Geographical Society.

————. 1930a. "The Origin of Limestone Caverns." *Bulletin of the Geological Society of America* 41:475–628.

————. 1930b. "Rock Floors in Arid and Humid Climates." *Journal of Geology* 38:1–27, 136–158.

————. 1932. "A Retrospect of Geography." *Annals AAG* 22:211–230.

Davis, W. M., and Daly, R. A. 1930. "Geology and Geography, 1858–1928." In S. E. Morison, ed., *The Development of Harvard University, 1869–1929*. Pp. 307–328. Cambridge, Mass., Harvard University Press.

de Martonne, E. 1909. *Traité de géographie physique*. Paris: Armand Colin.

Dryer, C. R. 1915. "Natural Economic Regions." *Annals AAG* 5:121–125.

————. 1920. "Genetic Geography: The Development of Geographic Sense and Concept." *Annals AAG* 10:3–16.

————. 1924. "A Century of Geographic Education in the United States." *Annals AAG* 14:117–149.

Dunbar, G. S. 1978. "George Davidson: 1825–1911." In *Geographers: Biobibliographical Studies*. Vol. 2, pp. 33–37. London: Mansell.

Emerson, F. V. 1908–1909. "A Geographic Interpretation of New York City." *Bulletin of the American Geographical Society* 40:587–612, 726–738; 41:3–20.

Fenneman, N. M. 1914. "Physiographic Boundaries within the United States." *Annals AAG* 4:84–134.

————. 1916. "Physiographic Divisions of the United States." *Annals AAG* 6:19–98 (with folded map on a scale of 1/7,000,000).

————. 1919. "The Circumference of Geography." *Geographical Review* 7:168–175.

Friis, H. R. 1981. "The Role of Geographers and Geography in the Federal Government: 1774–1905." In B. Blouet, ed., *The Evolution of Academic Geography in the United States*. Pp. 37–56. Hamden, Conn.: Shoe String Press.

Gilbert, G. K. 1878. *Report on the Geology of the Henry Mountains*. Washington, D.C.: Department of the Interior.

————. 1886. "The Inculcation of the Scientific Method by Example." *American Journal of Science*, 3rd ser., 31:284–299.

Gross, W. E. 1972. "The American Philosophical Society and the Growth of Meteorology in the United States: 1835–1850." *Annals of Science* 29, no. 4:321–338.

Hann, J. 1903. *Handbook of Climatology*, Part I. Trans. R. DeC. Ward. New York: Macmillan.

Hartshorne, R. 1939. *The Nature of Geography, A Critical Survey of Current Thought in the Light of the Past*. Lancaster, Pa.: Association of American Geographers.

Heyman, R. 2001. "Libraries as Armouries: Daniel Coit Gilman, Geography, and the Uses of a University." *Environment and Planning D: Society and Space* 19:1–6.

Huntington, E. 1907. *The Pulse of Asia*. Boston: Houghton Mifflin.

———. 1915. *Civilization and Climate*. New Haven, Conn.: Yale University Press.

———. 1924. *The Character of Races as Influenced by Physical Environment, Natural Selection, and Historical Development*. New York: Charles Scribner.

———. 1945. *Mainsprings of Civilization*. New York: John Wiley & Sons.

Huntington, E., and Cushing, S. W. 1920. *Principles of Human Geography*. New York: John Wiley & Sons.

James, P. E. 1977. "Grove Karl Gilbert, 1843–1918." *Geographers: Biobibliographical Studies*. Vol. 1, pp. 25–33. London: Mansell.

———. 1978. "Albert Perry Brigham: 1855–1932." In *Geographers: Biobibliographical Studies*. Vol. 2, pp. 13–19. London: Mansell.

———. 1979. "John Wesley Powell, 1834–1902." *Geographers: Biobibliographical Studies* 3:117–124.

James, P. E., and Ehrenberg, Ralph E. 1975. "The Original Members of the Association of American Geographers." *The Professional Geographer* 27:327–335.

James, P. E., and Martin, G. J. 1979. "On AAG History." *The Professional Geographer* 31:353–357.

Jefferson, M. 1909. "The Anthropography of Some Great Cities; A Study in Distribution of Population." *Bulletin of the American Geographical Society* 41:537–566.

———. 1915. "How American Cities Grow." *Bulletin of the American Geographical Society* 47:19–37.

———. 1928. "The Civilizing Rails." *Economic Geography* 4:217–231.

———. 1939. "The Law of the Primate City." *Geographical Review* 29:226–232.

Joerg, W. L. G. 1914. "The Subdivision of North America into Natural Regions: A Preliminary Inquiry." *Annals AAG* 4:55–83.

Jones, W. D., and Sauer, C. O. 1915. "Outline for Field Work in Geography." *Bulletin of the American Geographical Society* 47:520–525.

Koelsch, W. A. 1962. *Lectures on the Historical Geography of the United States as Given in 1933* [by Harlan H. Barrows]. Chicago: University of Chicago, Department of Geography.

———. 1969. "The Historical Geography of Harlan H. Barrows." *Annals AAG* 59:632–651.

———. 1979a. "Nathaniel Southgate Shaler: 1841–1906." In *Geographers: Biobibliographical Studies*. Vol. 3, pp. 133–139. London: Mansell.

———. 1979b. "Wallace Walter Atwood: 1872–1949." In *Geographers: Biobibliographical Studies*. Vol. 3, pp. 13–18. London: Mansell.

———. 1981. "The New England Meteorological Society: 1884–1896. A Study in Professionalization." In B. Blouet, ed., *The Evolution of Academic Geography in the United States*. Pp. 89–104. Hamden, Conn.: Shoe String Press.

Koelsch, W. A. 2004. "Franz Boas, Geographer, and the Problem of Disciplinary Identity." *Journal of the History of the Behavioural Sciences* 40, 1:1–22.

Krug-Genthe, M. 1903. "Die Geographie in die Vereinigten Staaten." *Geographische Zeitschrift* 9:626–637, 666–685.

Lee, D. H. K. 1954. "Physiological Climatology." In P. E. James and C. F. Jones, eds., *American Geography, Inventory and Prospect*. Pp. 470–483. Syracuse, N.Y.: Syracuse University Press.

Leighly, J. December 1958. "John Muir's Image of the West." *Annals AAG* 48, no. 4:309–318.

———. 1977. "Matthew Fontaine Maury, 1806–1873." *Geographers: Biobibliographical Studies.* Vol. 1, pp. 59–63. London: Mansell.

Lewis, G. M. 1981. "Amerindian Antecedents of American Academic Geography." In B. Blouet, ed. *The Evolution of Academic Geography in the United States.* Pp. 19–35. Hamden, Conn.: Shoe String Press.

Libbey, J., Jr. 1884. "The Life and Scientific Work of Arnold Guyot." *Bulletin of the American Geographical Society* 16:194–221.

Livingstone, D. N. 1987. *Nathaniel Southgate Shaler and the Culture of American Science.* Tuscaloosa and London: University of Alabama Press.

Lowenthal, D. 1953. "George Perkins Marsh and the American Geographical Tradition." *Geographical Review* 43:207–213.

———. 1958. *George Perkins Marsh, Versatile Vermonter.* New York: Columbia University Press.

———. 2000. *George Perkins Marsh: Prophet of Conservation.* Seattle and London: University of Washington Press.

Martin, G. J. 1968. *Mark Jefferson, Geographer.* Ypsilanti: Eastern Michigan University Press.

———. 1973. *Ellsworth Huntington: His Life and Thought.* Hamden, Conn.: Shoe String Press.

———. 1974. "A Fragment on the Penck(s)-Davis Conflict." *Special Libraries Association, Geography and Map Division Bulletin* 98:11–27.

———. 1977. "Isaiah Bowman: 1878–1950." In *Geographers: Biobibliographical Studies.* Vol. 1, pp. 9–18. London: Mansell.

———. 1980. *The Life and Thought of Isaiah Bowman.* Hamden, Conn.: Shoe String Press.

———. 1984. "John Paul Goode 1862–1932." *Geographers: Biobibliographical Studies.* Vol. 8, pp. 51–55. London: Mansell.

———. 1985. "Paradigm Change: A History of Geography in the United States, 1892–1925." *National Geographic Research* Spring 217–235.

———. 1988. "On American and German Geography circa 1850–1940." *Proceedings of the New England-St. Lawrence Valley Geographical Society* 18:15–25.

———. 2001. "J. Russell Smith 1874–1966." *Geographers: Biobibliographical Studies.* 21, 97–113. Continuum: New York: London.

———. 2003. "From the Cycle of Erosion to 'The Morphology of Landscape': Or Some Thought Concerning Geography as It Was in the Early Years of Carl Sauer." In *Culture, Land and Legacy: Perspectives on Carl O. Sauer and the Berkeley School of Geography.* Edited by Kent Mathewson and Martin S. Kenzer. Pp. 19–53. Baton Rouge: Geoscience Publications, Louisiana State University.

Mayo, W. L. 1965. *The Development and Status of Secondary School Geography in the United States and Canada.* Ann Arbor, Mich.: University Publishers.

Parkins, A. E. 1934. "The Geography of American Geographers." *Journal of Geography* 33:221–230.

Pattison, W. D. 1981. "Rollin Salisbury." In B. Blouet, ed., *The Evolution of Academic Geography in the United States.* Pp. 151–163. Hamden, Conn.: Shoe String Press.

———. 1982. "Rollin D. Salisbury 1858–1922." *Geographers: Biobibliographical Studies.* Pp. 105–113. London: Mansell.

Powell, J. W. 1896. "Physiographic Regions of the United States." In *Physiography of the United States.* Washington, D.C.: National Geographic Society, Monograph No. 1.

Pyne, S. E. 1980. *Grove Karl Gilbert: A Great Engine of Research.* University of Texas Press.

Roorbach, G. B. 1914. "The Trend of Modern Geography—A Symposium." *Bulletin of the American Geographical Society* 46:801–816.

Rowley, V. M. 1964. *J. Russell Smith: Geographer, Educator, and Conservationist.* Philadelphia: University of Pennsylvania Press.

Salisbury, R. D. 1907. *Physiography.* New York: Henry Holt.

Salisbury, R. D., and Alden, W. C. 1899. *The Geography of Chicago and Its Environs.* Chicago: University of Chicago Press.

Salisbury, R. D., and Atwood, W. W. 1908. *The Interpretation of Topographic Maps.* Washington, D.C.: U.S. Geological Survey, Professional Paper 60.

Salisbury, R. D., Barrows, H. H., and Tower, W. S. 1912. *The Elements of Geography.* New York: Henry Holt.

Sauer, C. O. 1966. "On the Background of Geography in the United States." In *Heidelberger Studien zur Kulturgeographie,* Festgabe zum 65, Geburtstag von Gottfried Pfeifer. Pp. 59–70. Weisbaden: Franz Steiner.

———. 1971. "The Formative Years of Ratzel in the United States." *Annals AAG* 61:245–256.

Semple, E. C. 1897. "The Influence of the Appalachian Barrier upon Colonial History." *Journal of School Geography* 1:33–41.

———. 1901. "The Anglo-Saxons of the Kentucky Mountains." *Geographical Journal* 17:588–623; reprinted in the *Bulletin of the American Geographical Society* 42(1910):561–594.

———. 1903. *American History and Its Geographic Conditions.* Boston: Houghton Mifflin. Revised by the author with C. F. Jones, 1933.

———. 1911. *Influences of Geographic Environment.* New York: Henry Holt.

———. 1915. "The Barrier Boundary of the Mediterranean Basin and Its Northern Breaches as Factors in History." *Annals AAG* 5:27–59.

———. 1927. "Templed Promontories of the Ancient Mediterranean." *Geographical Review* 17:353–386.

———. 1931. *The Geography of the Mediterranean Region, Its Relation to Ancient History.* New York: Henry Holt.

Shaler, N. 1909. *The Autobiography of Nathaniel Southgate Shaler, with a Supplementary Memoir by His Wife.* Boston and New York: Houghton Mifflin.

Sherwood, M. 1977. "Alfred Hulse Brooks: 1871–1924." In *Geographers: Biobibliographical Studies.* Vol. 1, pp. 19–23. London: Mansell.

Smith, J. R. 1913. *Industrial and Commercial Geography.* New York: Henry Holt.

———. 1952. "American Geography: 1900–1904," *The Professional Geographer* 4:4–7.

Spate, O. H. K. 1958. "The End of an Old Song? The Determinism-Possibilism Problem." *Geographical Review* 48:280–282.

Speth, W. W. 1978. "The Anthropogeographic Theory of Franz Boas." *Anthropos* 73:1–31.

———. 1999. *How It Came to Be: Carl O Sauer, Franz Boas and the Meanings of Anthropogeography.* Ellensburg, Wash.: Ephemera Press.

Strahler, A. N. 1950. "Davis' Concepts of Slope Development Viewed in the Light of Recent Quantitative Investigations." *Annals AAG* 40:209–213.

Tinkler, K. J. 1985. *A Short History of Geomorphology.* Totowa, N.J.: Barnes and Noble.

Tower, W. S. 1910. "Scientific Geography: The Relation of Its Contents to Its Subdivisions." *Bulletin of the American Geographical Society* 42:801–825.

Visher, S. S. 1948. "Memoir to Ellsworth Huntington, 1876–1947." *Annals AAG* 38:38–50.

Ward, R. DeC. 1908. *Climate, Considered Especially in Relation to Man.* New York: G. P. Putnam's Sons.

———. 1925. *Climates of the United States.* Boston: Ginn & Co.

Ward, R. DeC., and Brooks, C. F. 1936. *The Climates of North America*. In W. Köppen and R. Geiger, eds., *Handbuch der Klimatologie*, Vol. 2, Part J. Berlin: Borntraeger.

Warntz, W. 1981. "*Geographia Generalis* and the Earliest Development of Academic Geography in the United States." In B. Blouet ed., *The Evolution of Academic Geography in the United States*. Pp. 245–263. Hamden, Conn.: Shoe String Press.

Whittlesey, D. S. 1935. "Dissertations in Geography Accepted by Universities in the United States for the Degree of Ph.D. as of May, 1935." *Annals AAG* 25:211–237.

Williams, F. L. 1963. *Matthew Fontaine Maury, Scientist of the Sea*. New Brunswick, N.J.: Rutgers University Press.

Wright, J. K. 1952. *Geography in the Making, the American Geographical Society 1851–1951*. New York: American Geographical Society.

———. 1961. "Daniel Coit Gilman: Geographer and Historian." *Geographical Review* 51:381–399.

Wrigley, G. M. 1951. "Isaiah Bowman." *Geographical Review* 14:7–65.

16

THE NEW GEOGRAPHY IN THE UNITED STATES

World War I to Midcentury

> *Scarcely was physical geography established, or perhaps I should say rejuvenated and reestablished, before an insistent demand arose that it be "humanized." This demand met with prompt response, and the center of gravity within the geographic field has shifted steadily from the extreme physical side toward the human side, until geographers in increasing numbers define their subject as dealing solely with the mutual relations between man and his natural environment. By "natural environment" they of course mean the combined physical and biological environments.*
>
> —Harlan H. Barrows, in his presidential address to the
> Association of American Geographers, 1922

The period from World War I to the 1950s witnessed an ongoing search for a new formulation of acceptable geographical study. Trained geographers began to emerge from graduate departments of geography and to enter the profession, with the result that the traditionally close ties with geology were gradually loosened (Harris, 1979; Trewartha, 1979). Combined departments of geology and geography were gradually separated. In the course of time, the focus of geographical inquiry shifted toward social science and away from exclusive concern with earth science. Indeed, many were deeply disturbed by the growing neglect of the methods and concepts derived from geology and by the tendency to relinquish the study of physical geography to other disciplines. The period has been incorrectly described as one in which geographers devoted themselves to the "mere description of unique places" without any effort to formulate general concepts. Such a characterization seems unwarranted. Much attention was given to the information and use of concepts and models, and many principles and ideas now current can be traced to their early appearance in the 1920s and 1930s.

From the 1880s to the 1910s, a number of geographical societies had been founded. With the entry of America into the war in 1917, geographers became involved in war-related work. The Inquiry was established in New York; it was so titled to retain anonymity while preparing for the fashioning of a peace treaty. Bowman advanced political geography, Johnson wrote of military geography, Whitbeck introduced the term *mental map*, and J. Paul Goode was about to publish the first American world atlas with which geographers could be satisfied (Martin, 1985:230). A number of geographers attained high military rank, bringing more attention to the discipline. This helped its advance on many campuses when "Johnny came marching home."

At this time the ideas of William Morris Davis were little challenged in geomorphology and were only beginning to be challenged in human geography. With the benefit of hindsight, we can now see that the careful observation and measurement of physical processes were neglected in favor of qualitative studies of natural history. In the field of human geography, social Darwinism was under attack; indeed, most of the historians and other social scientists had already rejected it (Barnes, 1925; Hayes, 1908). Many geographers, too, were ready to follow A. P. Brigham in rejecting strict environmental determinism and R. D. Salisbury in avoiding simple cause and effect explanations for complex associations of things on the earth's surface. But not all the geographers were aware of the validity of the criticisms of Davis's scheme of human response to physical controls. The persuasive teaching of Ellen Semple, the creative work of Ellsworth Huntington, and to a lesser extent the work of Ray H. Whitbeck (1926) continued to support some kind of environmental control of human behavior (Huntington, 1924). Long after the physical cause and human response thesis had been dropped, some geographers continued to use the language of "geographic factor" and "environmental control" (Atwood, 1935; Baker, 1921; Lewthwaite, 1966; Martin, 1951; Peattie, 1929, 1940; Whitbeck and Thomas, 1932).

The tradition established at Harvard was carried on after Davis's retirement in 1912 by Wallace W. Atwood (Bushong, 1981).[1] As professor of physiography at Harvard, Atwood attracted many students who were excited by his teaching and by his leadership in field studies. After 1921, when the Clark Graduate School of Geography was established with Atwood as director, students came not only from the United States but also from many foreign countries. Atwood's school texts were very popular, departing from the traditional organization by political units and adopting one based on natural regions. It has been said that "no American has ever brought geography to so many people." Unfortunately, the geographical ideas he taught were already disputed by his colleagues when he reached the peak of his influence, much as Davis's ideas of the causal notion were already outmoded when he used them as the organizing principle of the "new geography."[2]

---

[1]Wallace W. Atwood was on the staff of the Department of Geology at Chicago when he was selected to succeed D. W. Johnson at Harvard. At Harvard Atwood continued his interest in field studies in geomorphology and in the teaching of geography in elementary and secondary schools. His study of the San Juan Mountains of Colorado (Atwood and Mather, 1932) is a classic of its kind. The last chapter deals with "The Utilization of the San Juan Region by Man." In 1920 he became president of Clark University and in 1921 the director of the Graduate School of Geography. In 1925 he founded the periodical *Economic Geography*. He was president of the Association of American Geographers in 1939. He retired in 1946.

[2]Another brilliant teacher who supported the ideas of environmental determinism was T. Griffith Taylor. He was on the staff of the Department of Geography at Chicago from 1928 to 1935. Taylor's work in Canada is discussed in Chapter 12.

CHANGING CONCEPTS

The period after World War I witnessed the gradual erosion of concepts of physical controls and human responses and a vigorous competition among proposals for new approaches to geographical inquiry (Brunhes, 1925). There is always a certain lag in such changes, a regrettable persistence of traditional error (James, 1967; Jastrow, 1936). But such a period of change is an exciting one because a variety of new ideas are used experimentally (Popper, 1959; Wright, 1966).

There were four main currents of geographic thought to examine. One proposal was that the scope of geographical study should be narrowed to focus on the adjustments made by humans to both their physical and biotic environment. This was a removal from the Davisian proposition in centering the adjustment of mankind on both the physical and biotic environment—whereas the Davisian formulation centered mankind more simplistically in an equation of physiographic controls and organic responses. This was the proposal that geography should be described as *human ecology*. A second proposal was that geographers should focus on the identification and explanation of observed differences from place to place on the face of the earth. Such studies are included in *chorology*, or the study of places or regions. But chorology was to be more than descriptive. The third and fourth currents included the search for explanations that would make sense out of observed diversity. These took two chief directions: one was to seek genetic explanation in terms of processes of change acting through time, leading to *historical geography* and its specialized offshoot *sequent occupance*; the other was to seek functional explanations, leading to the concept of the *functional organization of space*. These explanatory procedures were applied in various topical fields.[3] Meanwhile, the decade after World War I also saw a notable shift of professional attention from academic studies to the use of geographic concepts and methods in the study of practical economic, social, and political questions. *Applied geography*, as it developed in the period between World War I and the decade of the 1950s, is the subject of Chapter 18.

Human Ecology

That geography should focus on the study of human ecology, or the adjustment of humans to their natural surroundings, was presented by Harlan H. Barrows in his presidential address before the Association of American Geographers in 1922 (Barrows, 1923).[4] Adjustment, as Barrows used the word, was not caused by the physical environment but was a matter of human choice. Barrows felt, however, that, although the subject matter of geography had been partly lost to other disciplines, it was still too broad and that such specialties as geomorphology, climatology,

---

[3] For full summaries of the contributions made in the various fields of geography in the United States up to 1954 together with extensive references to published materials, see James and Jones (1954). See also Colby (1936) and Whitaker (1954).

[4] Previous American workers in the emerging field of ecology include C. E. Bessey, F. E. Clements, R. Pound, and H. C. Cowles (Armstrong and Martin, 1998). It seems that J. P. Goode was the first to adopt the term "human ecology" in 1907.

and biogeography should be relinquished. Like others before him, he sought a unifying theme that would bring coherence to the study of geography. The unifying theme, he argued, could be provided by restricting attention to human ecology. He continued:

> I believe that those relationships between man and the earth which result from his efforts to get a living are in general the most direct and intimate; that most other relationships are established through these; that, accordingly, the further development of economic regional geography should be promoted assiduously, and that upon economic geography for the most part other divisions of the subject must be based. . . . I believe that geography has been too much a library subject, and too little a field subject. I hold that the field is the geographer's laboratory. I believe that we have made only a beginning in the development of rigorous, scientific methods of field work in physiography and geology, and that the development of a thoroughly effective technique in field work is perhaps our greatest immediate need. Since most of us are "rebuilt geologists" do we not, in general, study the geological items and merely observe, in more or less haphazard fashion, the geographical items? Precisely how should one study in the field those relationships which are truly geographic? . . .
>
> I believe that much of our so-called geographical exposition is something else, that to be truly geographic a discussion must involve from beginning to end an explanatory treatment in orderly sequence of human relationships, and that the development of a satisfactory technique of exposition is only less important than the perfection of field methods (Barrows, 1923:13–14).

But geographers still had to examine skillfully two or more different sets of factors. To be sure, Barrows insisted that the physical conditions should only be studied in relation to humans, but this proved to be more easily said than done. Although Barrows' paper has often been quoted and assigned as reading for graduate students, it did not provide guidelines for a new orientation of the field (Hartshorne, 1939:123).

Chorology

Some sturdy chorologic inquiry was rendered by Mark Jefferson (1917) and W. L. Joerg (1914, 1936), but a much greater impact on the development of geography in the United States resulted from Carl O. Sauer's study entitled "The Morphology of Landscape" (1925). (Also see Leighly, 1976; and Stanislawski, 1975.) This article was written shortly after Sauer became chairman of the Department of Geography at the University of California (Berkeley) in 1923. It was intended as a kind of inaugural lecture—a declaration outlining his concept of the field of geography to his colleagues in other departments of the university. Such a declaration was deemed necessary because of the common and uncritical acceptance of the earlier definitions of geography solely in terms of environmental influences. Sauer insisted that no field of study could be defined in terms of a single causal hypothesis that would commit the student to a particular outcome of an investigation in advance (Sauer, 1927:173). To go into the field to look for influences or evidence of control exerted by the physical conditions is to accept a single dogma. Sauer referred to Siegfried Passarge, who recommended that the first step in any geographic study must be to determine the facts by describing the visible characteristics of an area without attempting to explain them in advance.

Sauer went back to the writings of Humboldt and Hettner, who supported the so-called chorological concept of the nature of geography. And he read numerous other German geographers (Kenzer, 1987; Mathewson and Kenzer, eds., 2003). Geography, Sauer pointed out, is concerned with the study of things associated in area on the earth's surface and with the differences from place to place, both physical and cultural. Man, behaving in accordance with the norms of his culture, performs work on the physical and biotic features of his natural surroundings and transforms them into the cultural landscape.

> The design of the landscape includes (1) the features of the natural area and (2) the forms superimposed on the physical landscape by the activities of man, the cultural landscape. Man is the latest agent in fashioning of the landscape. The study of geography begins therefore with physical geography, but—coasts are marked by ports; mountains have flung over them the trails and workings of man. A phrase that has been much used in German literature, unknown to me as to origin, characterizes the purpose perfectly: "the development of the cultural out of the natural landscape." This is the newer orientation that continues the traditional position (Sauer, 1927:186–187).

This, Sauer suggested, is what geography is all about. It is the study of areas, not to describe them as unique occurrences—for there is no such thing as an idiographic science—but rather to identify the regularities and recurrences from place to place that permit the formulation of generalizations. To understand the changes human beings have made on the face of the earth, it is necessary to go back far enough in time to establish the nature of the processes. Geography as chorology, or the study of the associations and interconnections of things in areas or regions, is what Sauer calls a "naïvely given section of reality"—that is, a division of knowledge that is accepted as axiomatic. He concludes his paper with these remarks:

> Our naïvely given section of reality, the landscape, is undergoing manifold change. This contact of man with his changeful home, as expressed through the cultural landscape, is our field of work. We are concerned with the importance of site to man, and also with his transformation of the site. Altogether we deal with the interrelation of group, or culture, and site, as expressed in the various landscapes of the world (Sauer, 1925:53).[5]

Sauer's purpose was to make a clean break with the traditional (American) geography inherited from the period before World War I. He might have discussed "the morphology of regions, or areas." But the word *region* in 1925 was encrusted with more ambiguities than the word *landscape*, including the notion of the uniform physiographic region that was also uniform in human response. The word *area* was even more ambiguous. As a result of these difficulties with confused word meanings, discussions of the nature of geography that followed not infrequently descended to controversy over the meanings of words.

---

[5]For Hartshorne's criticism of the use of the word *landscape* and the justification for its use presented by Josef Schmithüsen, see also Broek (1938).

Sauer's paper won widespread acceptance among the younger members of the profession, most of whom had completed their graduate training since 1920 and had recently been appointed to one of the several new geography staffs then being formed.[6] The new geographers had been raised on the search for geographic influences, but by 1925 there was enough skepticism concerning the content or method they had been taught to make the younger generation ready to accept a change of paradigm. With enthusiasm they turned to the study of landscapes, or regions, seeking the kind of interacting systems among diverse phenomena that gave character to particular places and tracing the changes introduced by the human settlement back to origins (Broek, 1938; Dodge, 1932). Here is what Norton Ginsburg wrote many years later in a position paper for the Commission on College Geography:

> Theirs was above all a "scientific" geography, concerned with regions as systems, and with the comparative method as a device for developing hypotheses concerning areal relations and processes. The use of statistics was simple and even primitive, to be sure, but their concerns were far from trivial, and the problems with which they dealt were of—to use a somewhat abused word—"overriding importance," at least to the development of geographic discipline.

The younger generation developed new jargon, including the use of the symbols of the Köppen classification of climates, and proceeded to reject the older generation of seekers after environmental influences. Since most of these younger geographers had taken at least some of their graduate work at Chicago, where they had been participants in Salisbury's famous seminar, they spread to other universities the idea of regular staff-student discussions of philosophical or methodological questions.

Yet there is a curious fact about the impact of Sauer's paper. These things had all been said before. In 1924 Sauer himself had published a paper in the *Annals AAG* attacking the study of influences and advocating the field survey of the "areal expression of man's activities" (Sauer, 1924). Instead of going into the field with a set of a priori principles concerning the effect of the physical environment on man, one should seek to observe the facts and then draw conclusions from them. This part of Sauer's proposal drew immediate criticism from some of the older generation. As Dryer pointed out, no one could actually observe anything or describe anything without some kind of working hypothesis, conscious or subconscious. There would be no way to select things to record and describe. If anyone does try to do

---

[6]Sauer had been appointed to the newly renamed Department of Geology and Geography at Michigan in 1915. The chairman was the geologist William H. Hobbs. In 1921 Wallace W. Atwood became chairman of the newly founded Graduate School of Geography at Clark University. In 1923, when the Department of Geography was established in the Social Science Division at Michigan, K. C. McMurry became its chairman. Sauer became chairman of the department at the University of California in Berkeley. New separate departments were established in 1925 at Minnesota and in 1928 at Wisconsin. Meanwhile, there were many positions to be filled in departments of geology and geography. The number of new Ph.D.'s increased rapidly: 10 in 1916–1920; 32 in 1921–1925; 66 in 1926–1930; and 51 in 1931–1935 (the period of the Great Depression) (Whittlesey, 1935; see Browning, 1970, and Hewes, 1946).

what Passarge and Sauer recommend, he wrote, "the result is likely to be a catalogue half rubbish, like a child's collection from a dump heap, and wholly unscientific."[7]

Dryer himself, in his presidential address to the Association of American Geographers in 1919, had presented the chorological concept, but not by that name:

> It seems clear and beyond question that the psychological foundation of the geographic concept is the sense of distribution in terrestrial space. We must concede the pertinence of the doctrine of Kant that "geography is a narration of occurrences which are coexistent in space." The idea, more sharply put by Bain in the statement that "the foundation of geography is the conception of occupied space," fits and includes every work generally recognized as geography from Strabo to Ritter and Reclus. With various additions and qualifications, it forms the essence of most of the current and accepted definitions of geography, of which quotation is unnecessary (Dryer, 1920:5–6).

N. M. Fenneman had made almost the same point in "The Circumference of Geography," his presidential address to the Association of American Geographers in 1918 (Fenneman, 1919).

Dryer's paper seems to have had slight impact on his fellow geographers. Nor, for that matter, was Alexander Bain's[8] very modern-sounding idea of geography as dealing with "the conception of occupied space" (1879) given any attention. Sauer, who was present at the St. Louis meeting of the association in 1919 when Dryer gave his paper, makes no reference to it in "The Morphology of Landscape." The report on the St. Louis meeting in the *Geographical Review* has the following to say about Dryer's address:

> President Dryer's address on "Genetic Geography: The Development of the Geographic Sense and Concept" was scholarly to a high degree and will rank among the finest presidential addresses that have been presented before the Association. It ought to be given a much wider circulation than it will receive if its publication is confined to the Association's *Annals* [*Geographical Review* 9(1920):139].

The report on the meeting goes on to say that the average attendance at the sessions was about 35, half of whom were members, and that only three of the members were from eastern colleges. The large number of younger people about to enter the profession had not started in 1919.

After 1925, when a new generation of younger geographers began to emerge, it became common for geographers to report on situations where the physical features of an area were not of major importance. While some of the older geographers and a few of the younger ones (Fig. 39) continued to report on responses or influences, many of the younger ones took delight in describing cases where other factors were more significant than the physical ones. Richard Hartshorne presented a paper to the association in 1926 concerning the location factor in geography with special reference to manufacturing industries (Hartshorne, 1927). Location relative to the sources of raw materials, markets, power, and labor was more important than

---

[7]See Dryer in *Geographical Review* 16 (1926):348–350.

[8]Alexander Bain (1818–1903), a Scottish philosopher, professor of logic and English at Aberdeen from 1860 to 1880, in *Education as a Science* (London, 1879), p. 272.

Figure 39 William Morris Davis and Preston E. James attend a meeting at Clark University in the 1920s

location relative to such features as relief, drainage, soil, or climate. For those who had been "explaining" the concentration of cotton textile factories in New England by the humidity of the climate (which permitted the spinning of thread without snarling due to static electricity), this reference to relative location with no mention of the elements of the physical environment came as an innovation. People who came to such conclusions were accused of leaving the "ge" out of geography.

Historical Geography

Those who adopted the chorological theme were never satisfied merely to describe the content of an area in static terms. Attention was necessarily focused on the processes, or sequences of events, that provided an explanation of the observed landscapes. To explain is to make sense out of apparently endless diversity. Of course, the study of sequences of events gave a dynamic quality to regional studies that purely contemporary description could not provide. Andrew H. Clark explains it as follows:

The genetic approach to geographical study inevitably leads to an examination of the past. This does not mean that one is to seek simple causes in the past to account for contemporary conditions, but rather that the conditions observed at any period of time are to be understood as momentary states in continuing and complex processes of change. Simple cause and effect relations are elusive, for no matter how far back a scholar may penetrate there is always a more distant past calling for further investigation. The genetic approach focuses attention on processes, for whatever interests us in the contemporary scene is to be understood only in terms of the processes at work to produce it. It is not, therefore, a search for origins in any ultimate sense, but rather views the present, or any particular time, as a point in a long continuum (Clark, 1954:71).

It is important to understand that the new approach to historical geography that appeared in America in the 1920s was not at all like that of Brigham and Semple in 1903 or like Barrows's course on "The Influence of Geography on American History." Barrows had been greatly influenced in his early years by Ellen Semple's interpretation of Ratzel and by the historian Frederick Jackson Turner, who in 1893 gave his famous lecture, "The Significance of the Frontier in American History" (Koelsch, 1969:634). Turner was an eloquent speaker for geographical influences on history. But at some time between 1920 and 1922 Barrows changed his basic approach to this topic. In 1923 he renamed his course "Historical Geography of the United States," and he focused his attention on examples of "creative human adjustments to a passive natural environment" (Koelsch, 1969:637). We should also note that Ellen Semple's book on the Mediterranean, published in 1931, offers outstanding examples of the method of historical geography.

The course Barrows gave at Chicago made a lasting impression on his students. Many of the graduate students wrote dissertations that can be classified as historical geography (e.g., Parkins, 1918); dissertations in other universities in this period were also contributions to historical geography (Clark, 1954:84–85). Yet it seems that not many of the new geographers trained in this way continued to produce studies in historical geography as such, although most of them made use of genetic explanations that involved some attention to the time dimension. Some of the most important studies in historical geography were written by nongeographers.[9] And one study that described in detail the movement of a former hill town down into the valley to locate at a new site on the railroad was written by the geologist, J. W. Goldthwait (the story of Lyme, New Hampshire, in Goldthwait, 1927). This paper, which was published in the *Geographical Review*, was regarded for many years as a model of its kind.

During the period we are discussing, two American geographers became the chief innovators in historical geography (Clark, 1954). One was Ralph H. Brown, the author of *Mirror for Americans* (Brown, 1943). In this study Brown undertook to write a

---

[9]For example: Allan C. Bogue, *From Prairie to Corn Belt* (Chicago, 1963); Bernard DeVoto, *The Course of Empire* (Boston, 1950); H. A. Innis, *The Fur Trade in Canada* (New Haven, Conn., 1930); J. C. Malin, *The Grassland of North America: Prolegomena to Its History* (Lawrence, Kans., 1947); and W. P. Webb, *The Great Plains* (New York, 1927).

geography of the eastern seaboard of North America as portrayed in about 1810 in the writings of the previous two decades or so. This imaginative approach to the re-creation of a past geography as perceived by scholars of the time foreshadows the modern attention to environmental perception. Brown then published a second book, *Historical Geography of the United States* (Brown, 1948), in which he traced the geographical changes during the course of settlement. Unfortunately, the career of this outstanding innovator was cut short by his untimely death at the age of 50 (Dodge, 1948; Miles, 1985).

The other major source of inspiration in historical geography was Carl O. Sauer (Speth, 1981). At Berkeley Sauer formed close intellectual ties with two other workers in allied fields: Herbert E. Bolton, historian, and Alfred L. Kroeber, anthropologist. These outstanding scholars, each bringing a different background to his studies, came together on the problems of interpreting Latin America. The combination proved enormously stimulating not only to these three men, but also to the many graduate students in all three fields. The first of many monographic studies that Sauer wrote with a graduate student as coauthor described the prehistoric Indian frontier of settlement on the Pacific coast of Mexico (Sauer and Brand, 1932). Sauer himself undertook to locate the old colonial highway from Guadalajara to Tucson on the basis of field study (Sauer, 1932). Additional samples from these studies in historical geography include: Broek, 1932; Carter, 1945; Clark, 1949; Hewes, 1950; Kniffen, 1931; Meigs, 1935; Parsons, 1949; Spencer, 1939; and West, 1952. Sauer extended his works on historical geography to cover a wide variety of topics (Sauer, 1952, 1956, 1966b, and his presidential address before the Association of American Geographers, 1941).

Out of these studies of sequences of settlement certain principles began to emerge. One was the principle that the same physical conditions of the land could have quite different meanings for people with different attitudes toward their environment, different objectives in making use of it, and different levels of technological skills. In agricultural areas it was clear that slope had one meaning for the man with a hoe and quite another for the man with a tractor-drawn plow. It might be that the introduction of machinery could reduce the arable area of a country or change the kind of soil considered desirable. People with one kind of culture might concentrate their settlements on flattish uplands, whereas another people in the same area might concentrate in the valleys. Water power sites that were useful for the location of industries before the advent of steam lost that attraction when power came from other sources.

In the mid-1920s Preston James traced the changing significance of the land through a sequence of periods with different cultures in an arbitrarily outlined area on either side of the Blackstone Valley, extending from the outskirts of Worcester, Massachusetts, to the environs of Providence, Rhode Island. The study, published in 1929, was summarized as follows:

> Thus the landscapes of the Blackstone area are made up of a complex of cultural impressions set one upon the other. The three chief cultures, the native Indian, the rural European, and the urban manufacturing, have each modified the natural setting in a unique and characteristic way. Forms developed by the Indian culture are visible, even today, in the shell mounds, the deposits of chipped stones and

broken utensils, or the scarcely discernible trails. The forms of the rural European culture are visible on every side, some of them continuing without change of function to the present, others significantly modified in their use, and others remaining as weather-beaten ruins or brush-entangled fields to tell of a period which exists no more. . . . Finally the urban landscape, in spite of its relatively small area, has come to occupy the position of commanding importance around which the economy of the region is oriented (James, 1929:108).

In that same year Derwent Whittlesey gave a name to this kind of study. He described the studies of the stages of change in the occupance of an area as *sequent occupance*. Referring especially to New England, he wrote:

. . . each generation of human occupance is linked to its forbear and to its offspring, and each exhibits an individuality expressive of mutations in some elements of its natural and cultural characteristics. Moreover, the life history of each discloses the inevitability of the transformation from stage to stage (Whittlesey, 1929:163).

Studies in sequent occupance represent the antithesis of environmental determinism. In a sense they represent a form of cultural determinism, for it is recognized that with any significant change in the attitudes, objectives, or technical skills of the inhabitants of a region, the significance of the resource base must be reappraised. A large number of studies published during the 1920s and 1930s made use of the method of sequent occupance, whether or not that term was adopted (e.g., Colby, 1924; James, 1927, 1931; Platt, 1928, 1933).

This was not a period when general concepts were neglected. Oliver E. Baker made effective use of economic principles to explain the development of American agriculture (Baker, 1921, 1923), and Harold H. McCarty used general concepts to interpret economic conditions and population regions in America (McCarty, 1940, 1942). Both Whittlesey and Hartshorne formulated theoretical structures to enlarge the reach of political geography (Hartshorne, 1950; Whittlesey, 1939). In 1939 Mark Jefferson wrote, "A country's leading city is always disproportionately large and exceptionally expressive of national capacity and feeling"—which he called the "Law of the Primate City" (Jefferson, 1939:231). As early as 1921 Marcel Aurousseau, an Australian geographer working in Washington, D.C., investigated the world distribution of population and sought to quantify the "expansion ratios" of already occupied regions (Aurousseau, 1921). In 1932 Stanley D. Dodge proposed that studies of population could be related to a statistically normal growth curve, a portion of a sine curve (Dodge, 1933). Studying population changes in Vermont and New Hampshire, he classified each minor civil division in terms of its position on the curve of growth or decline. Applying the concept to the whole of New England, he revealed a new pattern of population regions by plotting his results on a map (Dodge, 1935).

One of the most imaginative geographers of the period was Robert S. Platt,[10] who was a member of the Department of Geography at Chicago from 1919 to 1957.

---

[10]Robert S. Platt graduated from Yale in 1914 in history and philosophy. He taught for a year in the Yale Collegiate School at Changsha, China, and then returned for graduate

On his first field trip to the Antilles in 1922, he discarded the ideas of environmental determinism and became one of the most eloquent adversaries of those who continued to speak of responses or controls (Platt, 1946, 1948). In 1928 in a report on a field study of a small Wisconsin community, Platt first formulated the concept of a hierarchy of central places. "The radius of the community," he wrote, "is measured by the reach of the village institutions" (Platt, 1928:92). He noted that the individual farmer looked to the smallest village for those immediate services that had to be close enough for daily contacts. For his larger needs, which did not require such frequent visits, the farmer looked to the larger towns. The hierarchy that Platt identified started with the individual farmer, proceeded to the village of Newport, then Ellison Bay and Sister Bay, to Sturgeon Bay (the county seat), to Green Bay (the regional center), and to Chicago (the metropolis). If Platt were studying Ellison Bay today, he would have described it as a spatial system and he would have used quantitative techniques to make his observations of the functional relationships more precise. But the ideas were all there in 1928 (Thoman, 1979).

STUDIES OF SCOPE AND METHOD

Geographers in North America, like their colleagues in other parts of the world, had to clarify their own ideas regarding scope and method. Geography in America was nurtured in its early years by geologists, and most of the first generation of scholars in geography had a common background in geology. The few who were not geologists were mostly meteorologists or botanists. The direction offered by Davis in the 1880s and 1890s had led to the causal notion becoming the first paradigm in American geography soon after the turn of the century. As departments of geography were established in the universities and as graduate students trained in geography began to enter the profession, the spirit and purpose of geography were examined closely (Finch, 1939; Johnson, 1929). Initially, the main objective was to establish geography as an independent field of study. There was concern to define limits that would separate geography from other fields. This drive toward disciplinal independence probably retarded the development of geographic ideas because the workers in any field of learning must be in close contact with ideas being generated in other fields.

The habit of discussing philosophical and methodological questions was supported in at least three ways. One was the nature of Salisbury's seminars and many others that were patterned on the Chicago example. Then there were the many opportunities for such discussion offered by the annual meetings of the Association

---

study in geography at Chicago. He was appointed instructor in geography in 1919 and received the Ph.D. in 1920. He was a professor by 1939, and from 1949 until his retirement in 1957 he was chairman of the Department of Geography. When he returned to Chicago after teaching in China, he found that Wellington Jones was already teaching a course on Asia and that Walter S. Tower, who had taught a course on South America, had left the university, so he turned his attention to Latin America as his region of special interest. He was president of the Association of American Geographers in 1945 (Thoman, 1979).

Wallace W. Atwood

Harlan H. Barrows

Isaiah Bowman

J. Paul Goode

Richard Hartshorne

Ellsworth Huntington

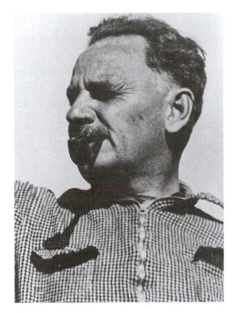

Carl O. Sauer

John K. Wright

of American Geographers. By longstanding tradition the presidents of the association make use of their presidential addresses to set forth their own ideas concerning
the scope of geography. And of no small importance were the foreign visiting
lecturers who were invited to American universities for periods ranging from a
summer session to a whole academic year.[11]

Moreover, there was a widely based tradition in America of studying geography
out of doors, and discussions concerning geographical ideas and procedures were
more vigorous in the field than within the walls of the seminar rooms. Few indeed
were the young geographers of that period who were content to give only a verbal
definition of geography in logical terms. In the field there was no difficulty in
moving the discussion promptly to the search for an operational definition—that is,
in reaching an agreement about what must be done to identify a geographic idea
or how a geographic idea could be used to increase knowledge of the earth as the
home of man. In the field the concept of the region came alive. Symptomatic of
this operational approach to the definition of geography was the oft-quoted remark
that "geography is what geographers do."[12]

These two traditions—the habit of discussion and familarity with field study—
resulted in the organization early in the 1920s of an annual spring field conference
(James and Mather, 1977). The first such conference was held in 1923 in the Indiana
Dunes south of Lake Michigan. The participants were former students at Chicago
who held positions in several midwestern universities.[13] A somewhat larger group
met in May 1924 at Bagley, Wisconsin, and in May 1925 at Hennepin, Illinois.
In 1925 a report representing the joint conclusions reached at these conferences
was published, entitled "Detailed Field Mapping in the Study of the Economic
Geography of an Agricultural Area" (Jones and Finch, 1925). Thereafter field
conferences were held almost every spring and included approximately the same
people. In 1926 a second conference was organized consisting of junior scholars.
In 1935 both groups were combined, meeting that spring in Menominee, Michigan,
and in the spring of 1936 in Pokagan State Park in Indiana. In 1938 the

---

[11]For example, the French geographer Raoul Blanchard lectured at Harvard in 1917, at
Columbia in 1922, at Chicago in 1927, and at Berkeley in 1932. He returned to give a
lecture series at Harvard for one semester every year between 1928 and 1936. In addition
to Blanchard in 1927, other Europeans invited to Chicago included James Fairgrieve in
1920, Sten De Geer in 1922, Ernest Young in 1924, L. Rodwell Jones in 1925, Helge
Nelson in 1926, and Patrick Bryan in the summers of 1928 and 1929.

[12]Credited to A. E. Parkins (see J. Russell Whitaker in the *Annals AAG* 31 [1941]:48).

[13]The following geographers participated in these conferences: from Chicago:
C. C. Colby, W. D. Jones, R. S. Platt, D. S. Whittlesey, and C. O. Sauer in 1923 only
(Whittlesey continued to attend after he went to Harvard in 1928); from Wisconsin:
V. C. Finch, A. K. Lobeck (until he went to Columbia in 1929); from Minnesota:
D. H. Davis; from Northwestern: W. H. Haas; from Michigan: K. C. McMurry; from
George Peabody: A. E. Parkins. The junior group that met first in 1926 included: from
Michigan: S. D. Dodge, R. B. Hall, P. E. James; from Minnesota: R. H. Brown, Samuel
N. Dicken, R. Hartshorne; from Wisconsin: L. Durand, G. T. Trewartha, J. R. Whitaker;
from Chicago: H. M. Leppard.

geographers held their conference in the Muskingum watershed in Ohio, where the Soil Conservation Service was carrying out a program of erosion control based on detailed studies of rainfall and runoff. In 1940 at Pokagan State Park proposed field studies were discussed. During the 1920s and 1930s many papers that originated in conference discussions were published (Finch, 1933; Hall, 1934; Hartshorne, 1932; James, 1931; James, Jones, and Finch, 1934; Jones, 1930; Platt, 1931, 1935; and Whittlesey, 1925, 1927).[14]

Experiments in Method

The field conferences helped focus attention on problems of methodology. How were geographical problems to be identified in the field? How were the necessary data to be collected in useful form? W. D. Jones had returned from Patagonia many years earlier with an appreciation of the need to prepare maps of land use on the same scale and degree of generalization as the traditional maps of the physical features. Jones and Sauer had jointly published a paper on this subject in 1915. The field conference discussions of field methods began where the earlier discussions had ceased.

The basic problem involved in the study of agricultural areas was how to identify and plot on maps the significant units of area relevant to the understanding and guidance of land use. At first it was proposed that several separate maps, all on the same scale, should be prepared to show the critical elements in land use problems. The next step was to reduce the number of such maps to two: one to show the conditions of the land, including units of soil, slope, drainage, and cover of wild plants; the other to show the categories of land use. When these two maps were superimposed, the precise covariance of the physical features and the land use could be examined in detail. Later, the suggestion was made that all this information could be plotted on one map by making use of a fractional code symbol. The denominator of the fraction was made up of digits representing categories of soil, slope, and drainage; the numerator comprised digits representing types of land use or wild vegetation. In the course of field mapping, when the observer noted any difference

---

[14]The members of these two field conference groups shaped the policies of the profession in the 1920s and 1930s. In those days membership in the Association of American Geographers was by election only, based on "an original contribution to some branch of geography beyond the doctoral dissertation." A nominating committee each year selected a single slate of officers. The president and vice president held office for one year only, but the officers whose terms ran for several years naturally had the greatest influence on association policy. These were the secretary, the treasurer, and the editor of the *Annals*. At least two of these were always members of the conference group. Often at the spring conferences one evening was devoted to a discussion of association problems. Actually, the conference membership was not exclusive because the group was always looking for younger men with both energy and ability. Nevertheless, when the new constitution of the association was adopted after World War II, which opened the membership to any interested person, the existence of a "clique" was severely criticized by the flood of new geographers then emerging from the graduate schools. The association is now democratically operated under the management of a paid executive director. In 1941 the membership was 167; in 2003 it was approximately 7000.

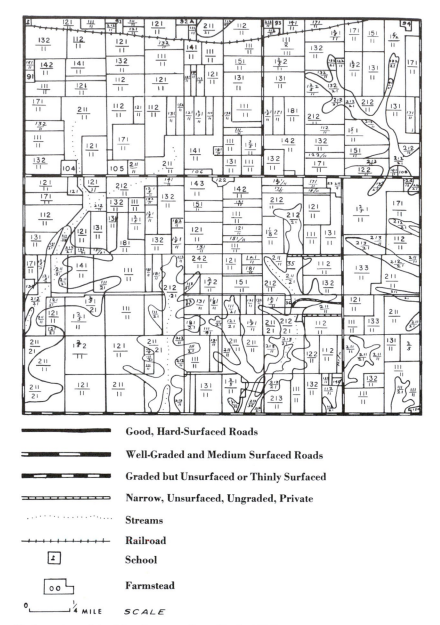

Good, Hard-Surfaced Roads

Well-Graded and Medium Surfaced Roads

Graded but Unsurfaced or Thinly Surfaced

Narrow, Unsurfaced, Ungraded, Private

Streams

Railroad

School

Farmstead

SCALE

Figure 40 A portion of the Montfort area (from Finch, 1933)

of the physical land or the land use, a boundary had to be drawn and a new fractional code symbol shown. In each unit area, represented by one fraction, there was a single, uniform association of land quality and land use. Since every microscopic point on the face of the earth differs from all other points, these unit areas were, in fact, generalizations. Within each unit area there was a certain range of diversity

NUMERATOR

Left-hand Digit: Major Use Type	Second Digit: Specific Crop or Use Type	Third Digit: Condition of Crop
	1. Corn (maize)	1. Good
	2. Oats	2. Medium
1. Tilled land	3. Hay in rotation	3. Poor
	4. Pasture in rotation	
	5. Barley	
	6. Wheat	
	7. Peas (mainly for canning)	
	8. Soy beans	
	9. Potatoes	
	T. Tobacco	
	X Sudan grass	
	2/5 Oats and barley mixed	
	1. Open grass pasture	1. Good
2. Permanent grassland	2. Pasture with scattered trees or brush	2. Medium
	3. Wooded pasture	3. Poor
	4. Permanent grass cut for hay	
3. Timber land	1. Pastured	1. Good
	2. Not pastured	2. Medium
		3. Poor
4. Idle land	1. Is capable of use	

DENOMINATOR

Left-hand Digit: Slope of Land	Second Digit: Soil Type (Wis. Soil Survey Terminology)	Letter x: Condition of Drainage
1. Level, 0°–3°	1. Marshall silt loam	X Poor
2. Rolling, 3°–9°	2. Knox silt loam	XX Very poor
3. Rough, 9°–15°	3. Knox silt loam (steep phase)	
4. Steep, over 15°	4. Lintonia silt loam	
	5. Wabash silt loam	
	6. Rough, stony land	

that had to be small enough to be acceptable—and this, of course, depended on the proposed use for the resulting map.

The fractional code system was tested by V. C. Finch with a group of graduate students from the University of Wisconsin, and the results were published in the so-called Montfort Study in 1933 (Finch, 1933). The map was published on a scale of approximately 1/15,000 (Fig. 40) and provided an extraordinary amount of detailed information. From this map it was possible to prepare a number of thematic maps of individual elements, such as slope, soil, or land used for specific purposes. But the field mapping had required so much time that Finch was unable to recommend the method as useful. In Chapter 18 we will look at a new method of random sampling applied experimentally to a part of the Montfort area.

The number of books and articles by American geographers increased so greatly after the 1920s that it is impossible to do more here than offer a few selected examples. Work was done in all the topical fields of geography: in population and settlement studies; in urban geography, transportation, and other aspects of economic geography; and also in the various fields of physical geography and biogeography. All this work is reported at length in *American Geography: Inventory and Prospect* (James and Jones, 1954).

The innovative papers in the field of population geography may be singled out as examples of the many studies in the 1930s, 1940s, and 1950s. The experiments with the mapping of population were greatly stimulated by the work of the Swedish geographer Sten De Geer. Various kinds of dot maps and density maps were prepared. But one basic problem continued to bother the workers in this field. The census data are summed up within enumeration areas that are seldom relevant to the kinds of problems geographers want to study. In 1936 John K. Wright published a short paper on the mapping of population, using Cape Cod as an example.[15] He showed that quite different patterns of population are brought out by using enumeration areas of different sizes and shapes. He presented a quantitative method for distributing the densities within a large enumeration area, such as a township. He described his method as follows:

> Assume, for example, a township with a known average density of 100 persons to the square mile. Assume, further, that examination of topographic maps and consideration of other evidence have shown that this township may be divided into two parts, *m*, comprising 0.8 of the entire area of the township and having a relatively sparse population, and *n*, comprising the remaining 0.2 of the township and having a relatively dense population. If, then, we estimate that the density of population in *m* is 10 persons to the square mile, a density of 460 to the square mile must be assigned to *n* in order that the estimated densities of *m* and *n* may be consistent with 100, the average density for the township as a whole (Wright, 1936:107).

Wright provided a table to make estimates of density consistent with average figures for whole enumeration areas.

Another innovation in the study of population was offered in 1954 by Lester E. Klimm of the University of Pennsylvania (Klimm, 1954). He pointed to the existence of large empty areas within such a long-settled region as the northeastern states. He described empty areas as follows:

---

[15] John K. Wright graduated from Harvard in history in 1913 and received the M.A. in 1914 and the Ph.D. in 1922. After serving in World War I, he was appointed librarian at the American Geographical Society. At the society he devised a new research catalog for use by geographers, with books classified both topically and regionally. He edited many of the society publications. From 1938 to 1949 he was the director of the American Geographical Society, and in 1946 he was president of the Association of American Geographers. Some of his more important writings are included in a book entitled *Human Nature in Geography* (Wright, 1952; 1966) (see also Koelsch, 2003).

These empty areas are not used for farming, no one lives in them, they contain virtually no recreational or commercial structures. Most of the surface is in woods or brush, ranging from "barrens," burnt-over land, or bog to large tracts of managed commercial forest in Maine, New Hampshire, and New York and extensive areas of state and national forest. Forestry and recreation are the principal present uses. Where they have resulted in structures, the areas occupied have been classified as not being empty (Klimm, 1954:325).

Klimm's map on a scale of 1/2,000,000 shows large, continuous empty areas, especially in northern Maine and New Hampshire, in the Adirondacks, the Allegheny Plateau, the Catskills and Poconos, and in the New Jersey Pine Barrens. He also shows a patchwork of smaller empty areas scattered in many parts of the region. It is clear from an examination of this study that the first step in preparing a map of population (by whatever method population is to be shown) must be to mark off the empty areas. Of course, the population data by census districts entirely obscures this kind of information.

These are only a few examples of the many experiments with method that were tried and then discussed in seminars and at meetings of the association.

Definition of the Field

With the retirement of Davis from Harvard, the arrival of muted criticism from R. Tarr and W. S. Tower, and the posture of students who had returned from advanced study in Germany with a non-Davisian viewpoint, there arose a desire to define the field. In 1910 Tower published a detailed essay on the subject (Tower, 1910). This was followed by publication of the results of the Roorbach questionnaire (1914); a 50 page essay by Jean Brunhes, "Human Geography," in Harry Elmer Barnes' *The History and Prospects of the Social Sciences* (1925); and Sauer's *Recent Developments in the Social Sciences* (1927). Yet even an approximate definition of the field was still lacking.

In 1930 the Commission on the Social Studies in the Schools appointed by the American Historical Association invited Isaiah Bowman to represent geography and subsequently to write a book on the relation of that branch of science to the social studies. The book Bowman wrote was *Geography in Relation to the Social Sciences* (1934). Its content includes: "By Way of Definition," "Measurement in Geography," "Population and Land Studies," "Technique in Geographical Analysis," "Regional Geography," "Economic and Political Bearings," and "Conclusions." Bowman had for some years been predisposed toward undertaking such a book. Schoolteachers, fellow geographers, and university administrators had frequently asked him to define geography. There was a void in the literature concerning the scope and nature of the field. Points of view conflicted, and there was little philosophical cohesion to geography. Bowman revealed his thoughts and feelings on the scope of geography:

This world is made up of regions and each region has its own personality, its own set of significant conditions. A Tibetan yak driver, an Egyptian fellah, an Uros fisherman, an Argentine hacendado, a Kansas farmer, a Peace River pioneer— each lives in a world whose conditions and outlook are almost completely unlike the others. To apprehend those earth qualities, conditions, outlines, measured components, and interactions that enable us to look understandingly at man in relation to the pervasive elements of his complex regional environment—these are the most distinctive as they are the culminating purposes of geographical research (1934:4).

Bowman did not wish to impose an orthodoxy on workers in the field, but he did suppose that a consensus was desirable. The book represented his own geographical point of view. Five years later another book was published that synthesized many geographers' viewpoints concerning the nature of geography. Occasionally, a book is published that stands as a landmark in the history of geographic thought. Such a one is *The Nature of Geography* (Hartshorne, 1939). Hartshorne, then at the University of Minnesota, had been a graduate student at Chicago and a participant in the spring conferences. His published studies during the 1920s and 1930s ranged widely over the field, including studies of agricultural regions, transportation and urban development, climate, and studies of the factors in the location of manufacturing industries. He also published a paper on racial distributions in the United States and on some fundamental concepts in political geography (Hartshorne, 1927, 1932, 1938, 1950). Field studies on the boundary problems of the upper Silesian industrial district (Hartshorne, 1934) excited his curiosity about boundary questions in general. When he was granted a sabbatical leave in 1938–39 together with financial aid from the Social Science Research Fund of the University of Minnesota, he planned to make a field survey of European boundary problems. But 1938 was no time for an American geographer to be examining European boundaries with notebooks, maps, and camera, for the events leading up to World War II had already begun. Before leaving for Europe, he had submitted a paper to the *Annals* regarding certain methodological questions. In Vienna, hoping that he might yet be able to carry on field studies, he received several letters from Derwent Whittlesey, then editor of the *Annals*, suggesting additional materials that could be added to his paper. He made use of the library at the University of Vienna to consult new sources of information. But as time went on and conditions grew worse rather than better, he focused his attention on the many documentary materials available in European libraries and also carried on interviews with leading geographers. The result was a book of nearly 500 pages (Hartshorne, 1979).

Hartshorne describes his purpose as follows:

The detailed examination of the nature of geography which this paper endeavors to present is not based on any assumption that geography is or ought to be a science—or that it ought to be anything other than it is. Assuming only that geography is some kind of knowledge concerned with the earth, we will endeavor to discover exactly what kind of knowledge it is. Whether science or an art, or in what particular sense a science or an art, or both, are questions which we must face free of any value concepts of titles. . . .

The writer's concern . . . is to present geography as other geographers see it —or have seen it in the past. If we wish to keep on the track—or return to the proper track . . . we must first look back of us to see in what direction that track has led. Our first task will be to learn what geography has been in its historical development (Hartshorne, 1939:205, 207).

Hartshorne's book was widely read and is generally accepted as an authoritative account of the points of view of many of the contributors of geographical ideas. It is a product of careful scholarship. But in the course of time colleagues and graduate students in seminars all across the country attempted to identify the positive conclusions regarding the nature of the field, and this proved to be very difficult. Hartshorne had either quoted or paraphrased some 300 methodological writings, some of which departed from what he identified as the mainstream of geographic scholarship. The continued discussion of these fundamental questions raised certain doubts and challenges that required clear answers (Hartshorne, 1955, 1958; Schaefer, 1953). Hartshorne undertook to provide a restatement of the positive conclusions to be drawn about the nature of geography. His *Perspective on the Nature of Geography* (1959) is organized around ten topics, each of which is the subject of a chapter:

1. What is meant by "geography as the study of areal differentiation"?
2. What is meant by the earth's surface?
3. Is the integration of heterogeneous phenomena a peculiarity of geography?
4. What is the measure of significance in geography?
5. Must we distinguish between human and natural factors?
6. The division of geography by topical fields—the dualism of physical and human geography.
7. Time and genesis in geography.
8. Is geography divided between systematic and regional geography?
9. Does geography seek to formulate scientific laws or to describe individual cases?
10. The place of geography in a classification of the sciences.[16]

The conclusions that Hartshorne reached (1959) are summarized in the following statements (note that the earlier ones are modified somewhat by the later ones):

Geography is concerned to provide accurate, orderly, and rational description and interpretation of the variable character of the earth surface (p. 21).

The earth surface is the outer shell of the earth where lithosphere, hydrosphere, atmosphere, biosphere, and anthroposphere are intermingled. This is the geographer's universe (pp. 22–25).

The goal of geography, the comprehension of the earth surface, involves therefore the analysis and synthesis of integrations composed of interrelated phenomena of the greatest degree of heterogeneity of perhaps any field of science (p. 35).

---

[16]Hartshorne sidesteps the question: Is geography an art or a science? "Whether such a field is to be called 'science' is a semantic question, depending on what particular definition is given to a word on which there is much disagreement" (1959:11). He also insists that writers on methodology should read carefully the writings of others of whom they are critical.

Any phenomenon, whether of nature or of man, is significant in geography to the extent and degree to which its interrelations with other phenomena in the same place or its interconnections with phenomena in other places determines the areal variations of those phenomena, and hence the totality of areal variation, measured in respect to significance to man (p. 46).

In describing and analyzing individual features and elements, we are free to utilize whatever categories of classification are empirically significant to the study of their interrelationships, without concern for the abstract distinction between those of human origin and those of natural origin (p. 64).

The traditional organization of geography by topics into two halves, "physical" and "human," and the division of each half into sectors based on similarity of the dominant phenomena in each, is of relatively recent origin and has proven detrimental to the purpose of geography—the comprehension of the integrations of phenomena of diverse character which fill areas in varying ways over the earth (p. 79).

. . . historical studies of changing integrations are essentially geography rather than history as long as the focus of attention is maintained on the character of areas, changing in consequence of certain processes, in contrast to the historical interest in the processes themselves (p. 107).

Geographic studies do not fall into two groups (topical and regional) but are distributed along a gradual continuum from topical studies of the most elementary integration at one end to regional studies of a most complete integration at the other (p. 144).

[Regarding the question of approach:] We start with "observation"—sensory description, often assumed to be the sole meaning of "description." We proceed to "analysis"—the description of the several parts of what has been observed as they appear to be related to each other. Next we state a hypothesis of relationships among the elements and processes. If sound, we have arrived at a higher level of knowledge—"cognitive description" of the elements and interrelationships among them (p. 171).

[Regarding Hettner's concept of geography as a chorological science:] Acceptance of the concept is in no way essential to geographic work. But students who cannot accept the particular characteristics empirically demonstrated as essential to geography, because they cannot understand that necessity, repeatedly attempt to change the subject to fit their view of what a science should be. The long history of such attempts demonstrates that their only effects are the personal frustration and professional unhappiness of those who try to fit a square peg into a round hole (p. 181).

To comprehend these areal variations fully we must dip back into past relationships of the factors involved, and those whose interest so directs them may reach as far back into history as the availability of data may permit. Release from the necessity of focusing our attention on the relations between two particular groups of features, human and nonhuman, permits a wider expansion of interest and at the same time a more effective coherence of the entire field. The opportunity to develop generic studies leading to scientific principles is present in the many forms of topical geography. Likewise, the unlimited number of unique places in the world, each of which is important and intellectually significant at least to those who live there, provides an inexhaustible field for those most interested in this type of research (p. 183).

American Geography: Inventory and Prospect

As the fiftieth anniversary of the founding of the Association of American Geographers approached, some geographers felt that this was a time for stock-taking. World War II had created an unprecedented demand for trained geographers, and those engaged in any of the numerous branches of war work had to devise new methods and make use of new and often unfamiliar materials. In 1949 at a meeting in Evanston, Illinois, attended by the chairmen of several committees that had been appointed by the National Research Council to discuss different aspects of geography, it was decided to undertake a series of symposia for the discussion of geographical questions and eventually to publish a book on the results. Preston E. James and Clarence F. Jones were appointed to direct the project. With funds from the Social Science Research Council and the National Research Council a number of conferences were set up, each one to consider the controversial problems of the various parts of the field (Fig. 41). The whole program called for wide discussion throughout the profession. The original drafts of chapters were read critically by members of the committees and were also presented to sessions of the association and to university seminars throughout the country. The resulting book, therefore, represents the combined thoughts of between 100 and 200 geographers. In fact, a major accomplishment of the project was its stimulation of widespread discussion throughout the profession of the objectives, methods, and concepts of geography (James and Jones, 1954). The following is a list of the chapters and the principal author of each:

1. The Field of Geography, Preston E. James
2. The Regional Concept and the Regional Method, Derwent Whittlesey
3. Historical Geography, Andrew H. Clark
4. The Geographic Study of Population, Preston E. James
5. Settlement Geography, Clyde F. Kohn
6. Urban Geography, Harold H. Mayer
7. Political Geography, Richard Hartshorne
8. The Geography of Resources, J. Russell Whitaker
9. The Fields of Economic Geography, Raymond E. Murphy
    Marketing Geography, William Applebaum
    Recreational Geography, K. C. McMurry
10. Agricultural Geography, Harold H. McCarty
11. The Geography of Mineral Production, Raymond E. Murphy
12. The Geography of Manufacturing, Chauncy D. Harris
13. Transportation Geography, Edward L. Ullman
14. Climatology, John Leighly
15. Geomorphology, Louis C. Peltier
16. The Geographic Study of Soils, Carleton P. Barnes
17. The Geographic Study of Water on the Land, Peveril Meigs III

Figure 41 The committee appointed to design *American Geography: Inventory and Prospect*. (*Bottom row, left to right*): C. F. Kohn (Iowa), H. Lemons (U.S. Army Research), L. O. Quam (ONR), M. F. Burrill (Interior), G. D. Hudson (Northwestern), A. H. Clark (Wisconsin). (*Top row, left to right*): J. K. Wright (AGS), D. S. Whittlesey (Harvard), P. E. James (Syracuse), J. R. Whitaker (George Peabody Teachers College), C. F. Jones (Northwestern), R. Hartshorne (Wisconsin), R. E. Murphy (Clark), C. M. Davis (Michigan), J. A. Russell (Illinois), and H. H. Mayer (Chicago).

18. The Geographic Study of the Oceans, C. J. Burke and Francis E. Elliott

19. Plant Geography, A. W. Küchler

20. Animal Geography, L. C. Stuart

21. Medical Geography, Jacques M. May

22. Physiological Climatology, D. H. K. Lee

23. Military Geography, Joseph A. Russell

24. Field Techniques, Charles M. Davis

25. The Interpretation of Air Photographs, Hibberd V. B. Kline, Jr.

26. Geographic Cartography, Arthur H. Robinson

It is important that no one answer was to be expected regarding controversial issues. The purpose was to identify differences of opinion and, so far as possible, to eliminate differences among the definitions of words. The various chapters did not necessarily record an accepted version of geography, but they did express the variety of interests and points of view that were actually held by members of the profession at midcentury. Not the least important was the extensive bibliography of geographical writings included in each chapter (Wooldridge, 1956; Wooldridge and East, 1958).

REFERENCES: CHAPTER 16

Armstrong, P. H., and G. J. Martin. 1998. "Frederic Edward Clements, 1874–1945." *Geographers: Biobibliographical Studies* 18:36–46.

Atwood, W. W. 1935. "The Increasing Significance of Geographic Conditions in the Growth of Nation-States." *Annals AAG* 25:1–16.

Atwood, W. W., and Mather, K. F. 1932. *Physiography and Quaternary Geology of the San Juan Mountains, Colorado.* Washington, D.C.: U.S. Geological Survey, Professional Paper 166.

Aurousseau, M. 1921. "The Distribution of Population: A Constructive Problem." *Geographical Review* 11:563–592.

Baker, O. E. 1921. "The Increasing Importance of the Physical Conditions Determining the Utilization of Land for Agricultural and Forest Production in the United States." *Annals AAG* 11:17–46.

———. 1923. "Land Utilization in the United States: Geographic Aspects of the Problem." *Geographical Review* 13:1–26.

Barnes, H. E. 1925. *History and Prospects of the Social Sciences.* New York: Alfred A. Knopf.

Barrows, H. H. 1923. "Geography as Human Ecology." *Annals AAG* 13:1–14.

———. 1962. *Lectures on the Historical Geography of the United States, as Given in 1933.* Ed. W. A. Koelsch. Chicago. University of Chicago, Department of Geography.

Barton, Thomas Frank. 1964. "Leadership in the Early Years of the National Council of Geography Teachers, 1916–1935." *Journal of Geography* 63, no. 8.

Beckinsale, R. P. 1981. "W. M. Davis and American Geography (1880–1934)." Pp. 107–122. In Blouet, ed. *The Origins of Academic Geography in the United States.* Hamden, Ct.: Archon Books.

Billington, Ray Allen. 1973. *Frederick Jackson Turner: Historian, Scholar, Teacher.* New York: Oxford University Press.

Block, Robert H. 1980. "Frederick Jackson Turner and American Geography." *Annals of the Association of American Geographers* 70, 1:31–42.

Bowman, I. 1934. *Geography in Relation to the Social Sciences.* New York: Charles Scribner.

Broek, J. O. M. 1932. *Santa Clara Valley, California: A Study in Landscape Change.* Utrecht, Netherlands: Osthoek.

Broek, J. O. M. 1938. "The Concept of Landscape in Human Geography." *Comptes rendus de la congrés internationale de géographie* 2, Sec. 3a:103–109.

Browning, C. E. 1970. *A Bibliography of Dissertations in Geography 1901–1969.* Browning, C. E. 1983. *A Bibliography of Dissertations in Geography, 1969–1982.* Chapel Hill: University of North Carolina, Department of Geography.

Brunhes, J. 1925. "Human Geography." In H. E. Barnes, ed., *History and Prospects of the Social Sciences.* New York: Alfred A. Knopf.

Bushong, A. 1981. "Geographers and Their Mentors: A Genealogical View of American Academic Geography." In B. Blouet, ed., *The Evolution of Academic Geography in the United States.* Pp. 193–219. Hamden, Conn.: Shoe String Press.

Calef, W. 1982. "Charles Carlyle Colby 1884–1965." In *Geographers: Biobibliographical Studies.* Vol. 6, pp. 17–22. London: Mansell.

Carter, G. F. 1945. *Plant Geography and Culture History in the American Southwest.* New York: Viking Fund Publications in Anthropology, No. 5.

Clark, A. H. 1949. *The Invasion of New Zealand by People, Plants, and Animals: The South Island.* New Brunswick, N.J.: Rutgers University Press.

———. 1954. "Historical Geography." In P. E. James and C. F. Jones, eds., *American Geography, Inventory and Prospect.* Pp. 70–105. Syracuse, N.Y.: Syracuse University Press.

Colby, C. C. 1924. "The California Raisin Industry." *Annals AAG* 14:49–108.

———. 1936. "Changing Currents of Geographic Thought in America." *Annals AAG* 26:1–37.

Coppens, L. M. 1985. 'Ralph Hall Brown 1898–1948." *Geographers: Biobibliographical Studies* 9:15–20.

Davis, W. M. 1928. *The Coral Reef Problem.* American Geographical Society, Special Publication, no. 9. New York: American Geographical Society.

Dodge, S. D. 1932. "The Vermont Valley: A Chorographical Study." *Papers of the Michigan Academy of Science, Arts and Letters* 17:241–274.

———. 1933. "A Study of Population in Vermont and New Hampshire." *Papers of the Michigan Academy of Science, Arts and Letters* 18:131–136.

———. 1935. "A Study of Population Regions in New England on a New Basis." *Annals AAG* 25:197–210.

———. 1948. "Ralph Hall Brown, 1898–1948." *Annals AAG* 38:305–309.

Dryer, C. R. 1920. "Genetic Geography: The Development of the Geographic Sense and Concept." *Annals AAG* 10:3–16.

Fairchild, W. B. 1979. " 'The Geographical Review' and the American Geographical Society." *Annals AAG* 69:33–38.

Fenneman, N. M. 1919. "The Circumference of Geography." *Annals AAG* 9:3–11.

———. 1931. *Physiography of Western United States.* New York: McGraw Hill.

———. 1938. *Physiography of Eastern United States.* New York: McGraw Hill.

———. 1939. "The Rise of Physiography." *Bulletin of the Geological Society of America* 50, no. 3:349–360.

Finch, V. C. 1933. *Montfort: A Study in Landscape Types in Southwestern Wisconsin.* Chicago: Geographical Society of Chicago, Bulletin 9.

———. 1939. "Geographical Science and Social Philosophy." *Annals AAG* 29:1–28.

Goldthwait, J. W. 1927. "A Town That Has Gone Downhill." *Geographical Review* 17:527–552.

Goode, J. P. 1903. "Geographical Societies of America." *Journal of Geography* 2, no. 7:343–350.

Grosvenor, G. H. 1948. "The National Geographic Society and Its Magazine." In *National Geographic Magazine: Cumulative Index, 1899–1946.* Foreword, pp. 1–115. Washington, D.C.: National Geographic Society.

———. 1957. *The National Geographic Society and Its Magazine.* Washington, D.C.: National Geographic Society.

Hall, R. B. 1934. "The Cities of Japan: Notes on Distribution and Inherited Forms." *Annals AAG* 24:175–200.

Harris, C. D. 1979. "Geography at Chicago in the 1930s and 1940s." *Annals AAG* 69:21–32.

Hartshorne, R. 1927. "Location as a Factor in Geography." *Annals AAG* 17:92–99.

———. 1932. "The Twin City District: A Unique Form of Urban Landscape." *Geographical Review* 22:431–442.

———. 1934. "Upper Silesian Industrial District." *Geographical Review* 24:423–438.

————. 1935. "Recent Developments in Political Geography." *American Political Science Review* 29:785–804, 943–966.

————. 1938. "Racial Maps of the United States." *Geographical Review* 28:276–288.

————. 1939. *The Nature of Geography, a Critical Survey of Current Thought in the Light of the Past.* Lancaster, Pa.: Association of American Geographers.

————. 1948. "On the Mores of Methodological Discussion." *Annals AAG* 38:113–125.

————. 1950. "Functional Approach to Political Geography." *Annals AAG* 40:95–130.

————. 1955. " 'Exceptionalism in Geography' Re-examined." *Annals AAG* 45:205–244.

————. 1958. "The Concept of Geography as a Science of Space, from Kant and Humboldt to Hettner." *Annals AAG* 48:97–108.

————. 1959. *Perspective on the Nature of Geography.* Chicago: Rand McNally.

————. 1979. "Notes Toward a Bibliobiography of 'the Nature of Geography.' " *Annals AAG* 69:63–76.

Hayes, E. C. 1908. "Sociology and Psychology; Sociology and Geography." *American Journal of Sociology* 14:371–407.

Hewes, L. 1946. "Dissertations in Geography Accepted by Universities in the United States and Canada for the Degree of Ph.D., June, 1935, to June, 1946, and Those Currently in Progress." *Annals AAG* 36:215–247.

————. 1950. "Some Features of Early Woodland and Prairie Settlement in a Central Iowa County." *Annals AAG* 40:40–57.

Heyman, R. 2001. "Libraries as Armouries: Daniel Coit Gilman, Geography, and the Uses of a University." *Environment and Planning D: Society and Space* 19:1–6.

Hobbs, W. H. 1952. *The Autobiography of William Herbert Hobbs.* Ann Arbor, Mich.: J. W. Edwards.

Hooson, David J. M. 1981. "Carl O. Sauer." In Brian W. Blouet, ed., *The Origins of Academic Geography in the United States.* Pp. 165–174. Hamden, Conn.: Archon Books.

Huntington, E. 1924. "Geography and Natural Selection." *Annals AAG* 14:1–16.

James, P. E. 1927. "A Geographic Reconnaissance of Trinidad." *Economic Geography* 3:87–109.

————. 1929. "The Blackstone Valley, a Study in Chorography in Southern New England." *Annals AAG* 19:67–109.

————. 1931. "Vicksburg, a Study in Urban Geography." *Geographical Review* 21:234–243.

————. 1967. "On the Origin and Persistence of Error in Geography." *Annals AAG* 57:1–24.

James, P. E., and Jones, C. F., eds. 1954. *American Geography, Inventory and Prospect.* Syracuse, N.Y.: Syracuse University Press.

James, P. E., Jones, W. D., and Finch, V. C. 1934. "Conventionalizing Geographic Investigation and Presentation." *Annals AAG* 24:77–122.

James, P. E., and Martin, G. J. 1979. *The Association of American Geographers: The First Seventy-Five Years, 1904–1979.* Washington, D.C.: Association of American Geographers.

James, P. E., and Mather, E. C. 1977. "The Role of Periodic Field Conferences in the Development of Geographical Ideas in the United States." *Geographical Review* 67:446–461.

Jastrow, J., ed. 1936. *The Story of Human Error.* New York: Appleton-Century.

Jefferson, M. 1917. "Some Considerations on the Geographical Provinces of the United States." *Annals AAG* 7:3–15.

————. 1939. "The Law of the Primate City." *Geographical Review* 29:226–232.

Joerg, W. L. G. 1914. "The Subdivisions of North America into Natural Regions: A Preliminary Inquiry." *Annals AAG* 4:55–83.

————. 1936. "The Geography of North America: A History of Its Regional Exposition." *Geographical Review* 26:640–663.

Johnson, D. W. 1929. "The Geographic Prospect." *Annals AAG* 19:167–231.

Jones, W. D. 1930. "Ratios and Isopleth Maps in Regional Investigation of Agricultural Land Occupance." *Annals AAG* 20:177–195.

Jones, W. D. 1930. "Ratios and Isopleth Maps in Regional Investigation of Agricultural Land Occupance." *Annals AAG* 20:177–195.

Jones, W. D. and V. C. Finch. 1925. *Annals of the Association of American Geographers* 15:148–157.

Kenzer, M. Ed. 1987. *Carl O. Sauer: A Tribute*. Corvallis: Oregon State University Press.

King, P. B. and S. A. Schumm, eds. 1980. *The Physical Geography (Geomorphology) of William Morris Davis*. Norwich, England: Geo Abstracts Limited.

Klimm, L. E. 1954. "The Empty Areas of the Northeastern United States." *Geographical Review* 33:325–345.

Kniffen, F. B. 1931. "Lower California Studies III: The Primitive Cultural Landscape of the Colorado Delta." *University of California Publications in Geography* 5:43–66.

————. 1932. "Lower California Studies IV: The Natural Landscape of the Colorado Delta." *University of California Publications in Geography* 5:149–244.

Koelsch, W. A. 1969. "The Historical Geography of Harlan H. Barrows." *Annals AAG* 59:632–651.

————. 1979. "Wallace Walter Atwood, 1872–1949." In *Geographers: Biobibliographical Studies*. Vol. 3, pp. 13–18. London: Mansell.

————. 1983. "Robert DeCourcy Ward, 1867–1931." In *Geographers: Biobibliographical Studies*. Vol. 7. London: Mansell.

————. 2003. "John Kirtland Wright, 1891–1969." *Geographers: Biobibliographical Studies* 22:169–181.

Kuhn, T. S. 1962. *The Structure of Scientific Revolutions*. Chicago: University of Chicago Press.

Leighly, J., ed. 1963. *Land and Life, a Selection from the Writings of Carl Ortwin Sauer*. Berkeley and Los Angeles: University of California Press.

————. 1976. "Carl Ortwin Sauer, 1889–1975." *Annals AAG* 66:337–348.

Lewthwaite, G. R. 1966. "Environmentalism and Determinism: A Search for Clarification." *Annals AAG* 56:1–23.

Martin, A. F. 1951. "The Necessity for Determinism." *Transactions and Papers, Institute of British Geographers* 17:1–12.

Martin, G. J. 1971. "Ellsworth Huntington and 'The Pace of History.'" *The Connecticut Review* 5:83–123.

————. 1987. "Preston Everett James 1899–1986." *Geographers: Biobibliographical Studies* 11:63–70.

————. 1994. "Richard Hartshorne, 1899–1992." *Annals of the Association of American Geographers* 84:480–492.

Mathewson, K., and M. S. Kenzer eds. 2003. *Culture, Land, and Legacy*. Baton Rouge, La: Geoscience Publications, Louisiana State University.

McCarty, H. H. 1940. *The Geographic Basis of American Economic Life*. New York: Harper Bros.

————. 1942. "A Functional Analysis of Population Distribution." *Geographical Review* 32:282–293.

Meigs, P. 1935. "The Dominican Mission Frontier of Lower California." *University of California Publications in Geography* 7:1–192.

Mikesell, M. W. 1975. "The Rise and Decline of 'Sequent Occupance': A Chapter in the History of American Geography." In David Lowenthal and Martyn J. Bowden, eds., *Geographies of the Mind: Essays in Historical Geosophy in Honor of John Kirtland Wright*. Pp. 149–169. New York: Oxford University Press.

———. 1978. "Tradition and Innovation in Cultural Geography." *Annals AAG* 68 no. 1:1–16.

Moulton, Benjamin. 1987. "Charles Redaway Dryer 1850–1927." In *Geographers: Bio-bibliographical Studies*. Vol. 11, pp. 23–26. London: Mansell.

Parkins, A. E. 1918. *The Historical Geography of Detroit*. Lansing: Michigan Historical Commission.

———. 1934. "The Geography of American Geographers." *Journal of Geography* 33:221–230.

Parsons, J. J. 1949. *Antioqueño Colonization in Western Columbia*. Berkeley and Los Angeles: University of California Press, Ibero-Americana 32.

Peattie, R. 1929. "Andorra: A Study in Mountain Geography." *Geographical Review* 19:218–233.

———. 1940. *Geography in Human Destiny*. New York: George W. Stewart.

Penn, Mischa, and F. Lukermann. 2003. "Chorology and Landscape: An Internalist Reading of 'The Morphology of Landscape.'" In Mathewson and Kenzer, 2003. *Culture, Land and Legacy: Perspectives on Carl O. Sauer and Berkeley School Geography*. Pp. 233–259. Baton Rouge, La: Geoscience publications, Louisiana State University.

Pfeifer, G. "Regional Geography in the United States Since the War; a Review of Trends in Theory and Method." Trans. J. B. Leighly, 1938. American Geographical Society, Mimeographed Pub. No. 2.

Platt, R. S. 1928. "A Detail of Regional Geography: Ellison Bay Community as an Industrial Organism." *Annals AAG* 18:81–126.

———. 1931. "An Urban Field Study: Marquette, Michigan." *Annals AAG* 21:52–73.

———. 1933. "Magdalena altipac: A Study in Terrene Occupancy in Mexico." *Bulletin of the Geographical Society of Chicago* 9:45–75.

———. 1935. "Field Approach to Regions." *Annals AAG* 25:153–174.

———. 1946. "Problems of Our Times." *Annals AAG* 36:1–43.

———. 1948. "Environmentalism Versus Geography." *American Journal of Sociology* 53:351–358.

———. 1959. *Field Study in American Geography, The Development of Theory and Method Exemplified by Selections*. Chicago: University of Chicago, Department of Geography, Research Paper No. 61.

Popper, K. R. 1959. *The Logic of Scientific Discovery*. London: Hutchinson.

Ryan, B. 1986. "Nevin Melancthon Fenneman, 1865–1945." In *Geographers: Biobibliographical Studies*. Vol. 10, pp. 57–68. London: Mansell.

Sauer, C. O. 1924. "The Survey Method in Geography and Its Objectives." *Annals AAG* 14:17–33.

———. 1925. "The Morphology of Landscape." *University of California Publications in Geography* 2:19–53.

———. 1927. "Recent Developments in Cultural Geography." In E. C. Hayes, ed., *Recent Developments in the Social Sciences*. Pp. 154–212. Philadelphia: J. B. Lippincott.

———. 1931. "Cultural Geography." In *Encyclopedia of the Social Sciences*. Vol. 6, pp. 621–623. New York: Macmillan.

———. 1932. *The Road to Cibola*. Berkeley and Los Angeles: University of California Press, Ibero-Americana 3.

————. 1941. "Foreword to Historical Geography." *Annals AAG* 31:1–24.

————. 1952. *Agricultural Origins and Dispersals.* New York: American Geographical Society, Bowman Memorial Lectures, Ser. 2.

————. 1956. "The Agency of Man on the Earth." In W. L. Thomas, ed., *Man's Role in Changing the Face of the Earth.* Pp. 46–69. Chicago: University of Chicago Press.

————. 1966a. "On the Background of Geography in the United States." *Heidelberger Studien zur Kulturgeographie, Festgabe für Gottfried Pfeiffer* 15:59–71.

————. 1966b. *The Early Spanish Main.* Berkeley and Los Angeles: University of California Press.

Sauer, C. O., and Brand, D. D. 1932. *Aztatlán, Prehistoric Mexican Frontier on the Pacific Coast.* Berkeley and Los Angeles: University of California Press, Ibero-Americana 1.

Schaefer, F. K. 1953. "Exceptionalism in Geography: A Methodological Examination." *Annals AAG* 43:226–249.

Sherwood, M. 1977. "Alfred Hulse Brooks, 1871–1924." In *Geographers: Biobibliographical Studies.* Vol. 1. Pp. 19–23. London: Mansell.

Spencer, J. E. 1939. "Changing Chungking: The Rebuilding of an Old Chinese City." *Geographical Review* 29:46–60.

Speth, W. W. 1977. "Carl Ortwin Sauer on Destructive Expoitation." *Biological Conservation* 11, no. 2:145–160.

————. 1981. "Berkeley Geography, 1923–1933." In B. W. Blouet, ed., *The Origins of Academic Geography in the United States.* Pp. 221–244. Hamden, Conn.: Shoestring Press.

Stanislawski, D. 1946. "The Origin and Spread of the Grid-Pattern Town." *Geographical Review* 36:105–120.

————. 1947. "Early Spanish Town-Planning in the New World." *Geographical Review* 37:94–105.

————. 1975. "Carl Ortwin Sauer, 1889–1975." *The Journal of Geography* 74:548–554.

Taylor, T. G., ed. 1951. *Geography in the Twentieth Century.* New York: Philosophical Library.

Trewartha, G. T. 1979. "Geography at Wisconsin." *Annals AAG* 69:16–21.

Thoman, R. S. 1979. "Robert Swanton Platt, 1891–1964." In *Geographers: Biobibliographical Studies.* Vol. 3, pp. 107–116. London: Mansell.

Walker, H. Jesse. 1980. "Richard Joel Russell, 1895–1971." In *Geographers: Biobibliographical Studies.* Vol. 4, pp. 127–138. London: Mansell.

West, R. C. 1952. *Colonial Placer Mining in Colombia.* Baton Rouge: Louisiana State University Press.

————. 1979. *Carl Sauer's Fieldwork in Latin America.* Dellplain Latin American Studies, 3. Ann Arbor, Mich.: University Microfilms International.

Whitaker, J. R. 1954. "The Way Lies Open." *Annals AAG* 44:231–244.

Whitbeck, R. H. 1926. "Adjustments to Environment in South America: An Interplay of Influences." *Annals AAG* 16:1–11.

Whitbeck, R. H., and Thomas, O. J. 1932. *The Geographic Factor.* New York: Century.

Whittemore, K. T. 1972. "Celebrating Seventy-Five Years of the 'Journal of Geography' 1897–1972." *Journal of Geography* 71:7–18.

Whittlesey, D. S. 1925. "Field Maps for the Geography of an Agricultural Area." *Annals AAG* 15:187–191.

————. 1927. "Devices for Accumulating Geographic Data in the Field." *Annals AAG* 17:72–78.

————. 1929. "Sequent Occupance." *Annals AAG* 19:162–165.

————. 1935. "Dissertations in Geography Accepted by Universities in the United States for the Degree of Ph.D. as of May, 1935." *Annals AAG* 25:211–237.

————. 1939. *The Earth and the State.* New York: Henry Holt.

Williams, M. 1983. "'The Apple of My Eye': Carl Sauer and Historical Geography." *Journal of Historical Geography* 9, no. 1:1–28.

Wooldridge, S. W. 1956. *The Geographer as Scientist*. London: Thomas Nelson.

Wooldridge, S. W., and East, W. G. 1958. *The Spirit and Purpose of Geography*. London: Hutchinson.

Wright, J. K. 1952. *Geography in the Making: The American Geographical Society, 1851–1951*. New York: American Geographical Society.

———. 1966. *Human Nature in Geography: Fourteen Papers, 1925–1965*. Cambridge, Mass.: Harvard University Press.

17

THE NEW GEOGRAPHY IN THE UNITED STATES

Midcentury to the Present

> *False facts are highly injurious to the progress of science, for they often endure long; but false views, if supported by some evidence, do little harm, for every one takes a salutary pleasure in proving their falseness.*
> —*Charles R. Darwin,* The Descent of Man

It is often said that "war teaches geography," although the missing qualifier is "belatedly." Immediately following World War II undergraduate and graduate enrollments in geography departments in the United States increased substantially. Prominent geographers had played significant wartime and postwar roles in intelligence and reconstruction, setting examples students wanted to follow. War veterans sought academic degrees in a society that had begun to demand credentials. Many students who had served in foreign areas returned with a heightened interest in geographical studies.

Most undergraduate students who enrolled in geography departments were introduced to the discipline by way of Finch and Trewartha's *Elements of Geography* (Mikesell, 2002) or James's *Outline of Geography*, although the number of available basic texts was growing rapidly. Regional geography was a strong curricular component. Introductory world regional courses were followed at higher levels by courses focused on such geographic realms as East Asia, Europe, North America, South America, and Africa south of the Sahara. Many leading geographers were regional specialists known for their field research and publications. Some were renowned for the elegant prose and strong substance of their textbooks: P. E. James and C. F. Jones (Middle and South America), M. Jefferson (Europe), W. A. Hance (Africa), G. B. Cressey (China), and E. Huntington and N. Ginsburg (Asia).

In 1945 only 28 universities in the United States were offering full programs at graduate level leading to the Ph.D. degree in geography, and only one (the University of Toronto) in Canada. These numbers increased rapidly in the years that followed, and the existing departments expanded significantly. Martin Kenzer has detailed the growth of geography at Northwestern University during this period (Kenzer, 1983), a story repeated in various forms at numerous other institutions in the immediate postwar period. This expansion created a professional job market in universities, and regional geographers took leading roles in promoting that development. At Northwestern University, for example, the senior-level Geography of

Africa course was required of all graduate students entering the African Studies Center, headed by the redoubtable anthropologist Melville Herskovits.

Field work also was ascendant. Field research in foreign areas was encouraged, if not required. Many geography departments operated their own field stations in the United States and had links with departments abroad. The University of Michigan ran a field station at Mill Springs in Kentucky; Northwestern University located a station at Platteville, Wisconsin. In such settings graduate students learned to analyze and map soils, identify crops, record weather changes, recognize and assess human impacts on natural environments, and map urban functions while acquiring many other skills that would serve them well in later field endeavors. Returning to the same field setting year after year, the accumulating data provided valuable insights into the changing physical and cultural geographies of the field station areas.

Geographers faced an old problem in such situations: the problem of scale. Usually, the small area was studied at a large scale; the large region was taught at a small scale. The transferability of small-area definition to large-region generalization was a difficulty geographers addressed in various ways. Platt (1942) had devised the case-study method; others, too, used the small area as representative of the larger region. In the process a massive quantity of local as well as global patterns, mapped at scales ranging from macro to micro, accumulated and became an invaluable inventory that informed school systems as well as the public. Mapping at all scales had improved dramatically during this heyday of regional geography.

Hartshorne's *The Nature of Geography* was required reading during the 1940s and 1950s in nearly all graduate departments (with the notable exception of Berkeley). This work did not prescribe, but it revealed much of what had been taught, thought, and said about the nature of geography. It therefore dealt with the concept of the region, its history, its development, and its contemporary role in several countries. It certainly helped maintain the position of regional analysis as central to the mission of American geography.

New developments began to be seen in human geography during the 1950s. For university teaching purposes most geographers had begun to specialize in one topical and one regional subject matter. Specializations became more frequent, narrower, and, with the expansion of departments, more feasible. The systematic specialization might be in urban, economic, physical, political, or historical geography. (Many other systematic specializations were already being practiced.) In addition, the marriage of the systematic and the regional enterprise was becoming more frequent, as, for example, in the historical geography of the British Isles, the economic geography of Japan, and the urban geography of Southeast Asia.

FOUR BOOKS OF NOTE

The 1950s witnessed an explosion of geographic writing ranging from field studies to philosophical works. Four volumes published during this decade had an especially significant impact on the discipline. *Geography in the Twentieth Century* (Taylor, 1951) brought together what had been accomplished in geography with special reference to North America and Europe. This collection of essays summarized a variety of geographic undertakings, as well as presenting several lucid national histories of geography. Fine essays by Taylor's Toronto colleagues G. Tatham and D. F. Putnam also reminded a U.S. audience that geography was prospering in Canada.

American Geography: Inventory and Prospect (James and Jones, 1954), published on the 50th anniversary of the founding of the Association of American Geographers, was a systematic assessment of the discipline; all chapters were written by committee. The result was thorough, if traditional, coverage of what had been accomplished, significant for its valuable bibliography, although some complained that it was "all inventory and no prospect." Derwent Whittlesey, then teaching geography at Harvard, chaired the group that prepared what may be called the dominant chapter in the volume, dealing with regional geography.

Two years later another important work, *Man's Role in Changing the Face of the Earth* (Thomas, 1956), arose from a symposium on the topic organized by the author and editor. The book was inspired by the work of George Perkins Marsh and was much influenced by Sauer. Its intent was to reveal details about conservation objectives and destructive realities in a world dominated by an ever-growing human population. In some respects Thomas's volume marked a high point for American geography, suggesting a confluence and fruition of the physical and human geographies that had by then begun to drift apart. But though it had an important impact on the development of postwar cultural geography, it could not heal a rift that would become one of the discipline's major dilemmas.

The appearance of *Perspective on The Nature of Geography* (Hartshorne, 1959) marked another high point—the culmination of interest during the 1950s in questions involving "geographic thought," the theory and substance that made geography what it was. Appearing two decades after *The Nature of Geography* (Hartshorne, 1939), this book with its predecessor was required reading in many graduate schools in America. Many students described the reading of this pair of books as exhaustive as well as exhausting given their dense prose, comprehensive coverage, theoretical content, and bibliographic detail.

What these four books (among many of the period) signified was a discipline that had come of age, was thriving at an unprecedented number of universities and colleges, and had a strong scholarly representation and effective intellectual representatives. Geography in America had three growing professional associations with complementary missions, and the Association of American Geographers (AAG) could contemplate a second half-century of progress.

SIGNS OF CHANGE

It was also during the 1950s, however, that new developments began to mark several fields of human geography, for example, model building in urban geography and mathematical approaches to economic geography. Specialization in topical (systematic) as well as regional areas became the norm. Such systematic specializations as urban, economic, political, cultural, and historical geography on the one hand, and climatology, pedology, geomorphology, and biogeography on the other signaled a fragmentation that would herald the transformation of the discipline.

Specialization of the kind just described would, according to many critics, lead away from what F. K. Schaefer called "Exceptionalism in Geography" (Schaefer,

1953). He argued that Hartshorne had declared that geography was essentially idiographic, that is, focused on the unique and specific, whereas geography should be nomothetic in mode, seeking generalizations and universal laws. Schaefer's article was flawed by his use of erroneous data, and he misrepresented Hartshorne's position, but his attack on an establishment figure encouraged a younger generation of geographers to seek a new law-seeking geography (Martin, 1989).

Yet other contributions argued for change. Torsten Hägerstrand's work concerning innovation diffusion in Sweden had entered into American thought with J. Leighly's article "Innovation and Area" (1954). Hägerstrand visited the University of Washington for an extended period in 1959, lending inspiration to the movement that was to become "spatial science". The work of German geographers and cryptogeographers such as Walter Christaller, J. H. von Thünen, August Lösch, and Alfred Weber had begun to enter U.S. geography via economic, commercial, and industrial geography. As these works were translated into English in the 1950s and 1960s, their adoption and exploitation surged (Weber had been translated in 1929.) Pursuant to his World War II work in Washington, D.C., Edward Ackerman had urged more systematic work (and less emphasis on regional geography) as the direction that was needed and that could produce a nomothetic product (Ackerman, 1945, 1958). An economist, Walter Isard, who published *Location and Space Economy* in 1956, was developing what he called regional science. Many geographers believed that Isard was offering a newly minted and more precise approach to regional geography and joined with him in a new Regional Science Association. Urban geography had displayed great analytical rigor throughout the 1950s, which gave it a leading role during the measurement-focused period. Certainly, there was a desire for more accuracy, and numeracy was thought to be the route to this increased accuracy. But the application of mathematics to geography was not new; it had already been practiced decades earlier by geographers including Henry C. Carey, Ellsworth Huntington, and Robert E. Horton. What was new was the strength of the movement toward quantitative analysis throughout human geography.

QUANTITATIVE AND THEORETICAL GEOGRAPHY

The transformation that took place in geographical methodology beginning in the middle 1950s accelerated in the early 1960s. By 1963 Ian Burton, in an article that coined the phrase "The Quantitative Revolution and Theoretical Geography," could write that the quantitative revolution was over. That "revolution" was part of a wider movement that emphasized the construction of theory, the building of mathematical models, and the extensive use of statistical analysis throughout the social and behavioral sciences (Burton, 1963; Morrill, 1984).

In geography a spontaneous movement had arisen at the universities of Washington and Iowa under the leadership of William Garrison and Edward Ullman (at Washington) (Eyre, 1978) and Harold McCarty (at Iowa). Further activity in this context emerged at the University of Wisconsin under the cartographer Arthur Robinson and with the work of the Social Physics Project of J. Q. Stewart at Princeton and William Warntz of the American Geographical Society (in this project an attempt was made to apply Newtonian physics to the distribution of phenomena in space). The graduate students at these schools sought a nomothetic approach to geography based on spatial and econometric theory. The group was inspired by the article of Fred Schaefer at Iowa, by the work of Torsten Hägerstrand at Lund (Hägerstrand, 1967), early developments in regional science, and theoretical and statistical work in economics. In some cases the new ideas came through cartography.

As noted, the University of Washington students were led into this new geography by pathfinder W. Garrison, supported by E. Ullman (the philosopher of the enterprise) and by G. Donald Hudson as chairman. Two works in particular steered the movement toward a more theoretical geography: the mathematical foundation of A. Scheidegger's *Theoretical Geomorphology* (1961) and the attempt by W. Bunge to establish geography as a science in his *Theoretical Geography* (1962). (Perhaps, in this context, S. Gregory's *Statistical Methods and the Geographer* might be mentioned as a work that facilitated these developments.) The initial group of students at Washington included Brian Berry, Duane Marble, William Bunge, Michael Dacey, Richard Morrill, and Edwin Thomas (Berry, 1993). With doctorates in hand, these students began their careers in several of the major geography departments, especially at Chicago, Northwestern, and Michigan. A series of discussion papers was initiated and spread across much of the United States. By 1959 the National Science Foundation began to sponsor workshops and symposia on the new methodology. Thus, the new methodology was diffused to Canada, Britain, other parts of Europe, Australia, and New Zealand. Another generation of new methodologists emerged (Morrill, 1991), which included John Cole, Richard Chorley, Stanley Gregory, Peter Haggett, Ronald Johnson, and David Harvey. These geographers propelled a change in the discipline from a descriptive regional and integrative undertaking to one with a spatial and theoretical approach. The transformation was quantitative in that geographical concepts were frequently expressed in statistical or mathematical terms.

By the mid 1960s the "space cadets" were ready to spread their viewpoint throughout the discipline (and not restrict their viewpoints to a few specialties). In this respect the publication of *The Science of Geography* (1965) was noteworthy. This was a report of the National Academy of Sciences and the National Research Council, which formulated research priorities for geography. More theoretical work was urged, and a distinction was made between theoretical deductive thinkers and empirical inductive workers. This comparison between "thinkers" and "workers" (the latter being the empiricists) aroused resentment. But such etymology was not surprising given the preponderance of theory-inclined members on the committee.

Five years later another report was prepared for the Committee on Science and Public Policy of the National Academy of Sciences and the Problems and Policy Committee of the Social Science Research Council (Taaffe, 1970). This report also focused on human geography as the study of spatial organization expressed as patterns and processes. Meanwhile, David Harvey's *Explanation in Geography* (1969) endorsed scientific methodology: It urged the development of geographic theory, but it did not offer a critique of the positivism upon which the quantitative movement was built.

By this time more than a decade of effort had been invested in geography's new direction. Whatever their appellation, this first group of determined scholars did transform (and divide) the discipline. As David Harvey states, "along with many others

(Chorley, Haggett, Ullman, Garrison, Berry, Morrill, and Hägerstrand in leading roles), we helped bend the structure of formal geography, against considerable opposition, to our collective will" (Harvey, 2002). A group of these young modernizers—as they saw themselves—from Michigan State University, Wayne State University, and the University of Michigan met regularly outside their departments to plot strategy and to exchange knowledge and ideas. Similar initiatives were taking place elsewhere. In his obituary of Peter Gould, Peter Haggett (2003; see also Stoddart, 2000) cites a letter dated November 30, 1965, in which Gould invites him to a "quantitative seminar" he had formed at Penn State: "In the beginning of February [1966] we are holding with Ohio State (Ned Taaffe, Howard Gauthier, Les King), Pittsburgh, and Penn (Alan Scott and Julian Wolpert) an informal get together. Four sessions: one on graph theory; one on theory of search; one on spectral analysis; and one on learning and behavioral models. Do come over and join us. It should be fun."

Such technical terminology had never been part of geography's lexicon, and soon the professional literature became a blend of the traditional and this "avant-garde" language. Curricula, too, were changing: "Statistics" or "quantitative methods" became part of undergraduate requirements, and regional studies, foreign languages, and field research declined proportionately. Geography was experiencing the first of several revolutionary transformations that would mark its course during the second half of the twentieth century.

Proponents of the new geography argued that traditional geography was intellectually weak and that academics in other disciplines viewed the subject merely as an interpretation of unique places (Gould, 1979). They also believed that through the new methodology geography could join the mainstream of science, pursuing and attaining an understanding of the organization and evolution of landscape. This newly designed geography sought to analyze what gave the landscape its face (physical as well as cultural). To a considerable extent the division between the new and the old geography was a division between younger and older geographers. Unfortunately, the division became personal and sometimes bitter. It has been claimed that traditional editors would not give the new geography's manuscripts a fair reading. Nevertheless, it was not long before the new wave began to occupy senior positions in the field, including department chairs and other significant professional offices. Books and manuscripts representing this new geography were published in large numbers.

It is pointless to speculate on the issue of which was the better geography. The new geography promised greater accuracy, could be verified, would lead to generalizations, and could be cumulative in its buildup of scientific knowledge. It also occupied significant growth areas of the discipline. Such matters as the size and location of cities and the location of businesses had been studied by urban and economic geographers, two of the most developed areas in American geography for decades. This geography was procedurally new and received much attention. Initially it added less to content than to method: It attempted to reorganize the way geographers made measurements, it changed the language used from narrative to numerate, and it sought laws from the data sorted and sifted by computers. The goal was to make geography more scientific. An introductory textbook literature emerged to carry this new geography into academic departments. Foremost among these, perhaps, was *Spatial Organization: The Geographer's View of the World* (1971) by R. F. Abler, J. S. Adams, and P. R. Gould. A new era in U.S. geography was underway in which much excellent work was produced.

Yet the established geography, centered around the theme "man on the land," had won its way as the end product of Davisian thought and had established itself in the literature, course offerings, examination systems, dissertations, and as the mindset for a large percentage of American geographers. The collision of the two geographies, conservative and innovative, was without equal in the history of the field. Their articulation, characterized by the terms *idiographic* and *nomothetic*, was provided in a brief and terse statement by W. R. Siddall (1961), who had been a graduate student at Seattle in the late 1950s. The two geographies were character-ized in the publications of C. P. Snow in his explication of the two cultures—the humanities and the sciences—commencing in the 1950s and thereafter published in several editions and several languages. It was a stage in the emergence of scientific accomplishment that made such a collision inevitable. Rocketry, the jumbo jet, Sputnik, and the computer all suggested speed, accuracy, and emergence from the previous unknowing age. Yet previous workers in geography had scoured the landscapes about which they wrote, coming to know them intimately and ranking among the country's leading area specialists familiar with foreign languages and cultural settings. The new workers urging a new geography did much of their field work in the computer laboratory. Hence, there was another consequence to the decline of regional geography: that part of regional study dealing with the physical environ-ment also declined. Departments of Earth Science and Departments of Geology took over many of the courses abandoned or neglected by geographers. Departments of Geography and specialist periodical literature focused on human geography, dividing what had previously been considered indivisible.

In the 1950s Senator Joseph McCarthy's committees began to investigate sci-entists' foreign field research, area study programs, foreign language study, and other "suspect" activities. Owen Lattimore of Johns Hopkins University was the first of McCarthy's geographer victims (Dunbar, 2000). Thereafter a number of other geographers were accused of communist sympathies. While none of the geographers was ever found guilty of overtly supporting communism or being involved in communist activities, the investigative process could take years and cost the victim money (legal assistance), anxiety, and professional standing. All this discouraged the earlier enthusiasm for the field-based study of distant regions. And since many of those attacked for un-American activity were Asian specialists, the number of geographers specializing in Asian studies sharply declined, even though this was one area of the world that would inevitably grow in significance. Approximately one decade after the cessation of these investigations the U.S. military was heavily committed in Vietnam.

The issue of "falling behind" in the science race had permeated American society since the "wake-up call" of 1957, when the Russians launched the world's first artificial satellite. An obsession with winning, with becoming number one, had developed. The threat of placing second to the Russians during the cold war proved traumatic to the national psyche. At once resources were poured into science. Geography benefited (though not to the extent of physics). Summer institutes were financed to develop quantitative techniques in geography and to emphasize the scientific properties of geography. Given these opportunities and the promise of government-inspired direction, more geographers studied statistics, mathematics, and computer technology.

While this work was in progress, danger was looming. One of the logical casualties of the quantitative-theoretical movement was what remained of regional geography, the discipline's center of gravity just a decade earlier. Measurable accuracy of the kind required by the theoretical constraints of quantitative analysis made large-scale (small-area) study the norm; a reluctance to generalize from such research to larger regions diminished the utility of smaller-scale frameworks. Introductory world regional geography courses, to which first-year students had flocked in large numbers, lost their constituency. Upper-level regional courses, centering as they did on large regions, were deemed inappropriate for a discipline in search of universal laws. While universities such as Northwestern continued to offer such courses as African Political Systems (political science), Peoples of Africa (anthropology), and History of Subsaharan Africa, the geography department dropped its once-required Geography of Africa course as inconsistent with the new geography.

In an attempt to reverse this decline, H. J. de Blij in 1971 published a textbook that sought to combine the large-scale theoretical with the small-scale substantive, placing such constructs as Christaller's central-place theory and neo-Thunian schemas in their larger regional settings (de Blij, 1971). This effort was only partially successful; although it revived the world-regional–based introductory course as the optimal introduction to college geography, it did not stimulate a revival of follow-up, upper-level regional courses. By the last decade of the twentieth century, some major departments of geography in the United States offered neither regional nor physical courses as part of their curricula.

TWO EXPERIMENTAL PROJECTS

The expansive spirit of the 1960s inspired two large undertakings by the Association of American Geographers: the Commission on College Geography (CCG) (1963–74) and the High School Geography Project (1961–70). These were good times to launch such ventures. A spirit of optimism prevailed in the United States, professional growth was rapid, and diffusion of innovative thinking was swifter than ever before. It is difficult to assess the overall impact of the projects. The CCG, directed by John F. Lounsbury, published three series of publications: the general series, the resource papers, and the technical papers. Some geographers at the time argued that the commission's publications were individualistic and only sometimes relevant to what students were studying in college-level geography. Yet they were detailed, most were well written and much more specialized than textbooks, and they could be adopted for use in standard courses. Nevertheless, only modest use was made of them countrywide. The CCG also sponsored several summer institutes for college teachers of geography and developed a consulting service. The CCG was terminated in 1974.

The High School Geography Project was a larger undertaking than the Commission on College Geography project. The Association of American Geographers and the National Council for Geographic Education formed a joint committee on education to develop and implement a plan. Gilbert F. White and Clyde Kohn became cochairmen. Initially (1961) a grant was provided from the Ford Foundation's Fund for the Advancement of Education to facilitate the development of a proposal for a high school geography program to be applied throughout the United States. The proposal, when developed, was funded by the National Science Foundation in 1963 and continued until 1970. Directors were W. D. Pattison (1961–64), N. F. Helburn (1964–69), and D. G. Kurfman (1969–70). Authors were carefully selected to write essays that were intended to be parts of a course in progress. Large numbers of these were written by people with diverse viewpoints. Conferences, workshops, and institutes were held across the country that examined and discussed the essays—a course in the making. Out of this rigorous process of sorting and sifting a compromise was reached between the traditional geography and the more recently developed spatial science (McNee, 1973). At its high point in the middle 1960s, the High School Geography Project was a large enterprise that employed many people throughout North America, especially at its Boulder headquarters. The published course book, *Geography in an Urban Age* (1970), was the result; it was revised in 1974. It was not a textbook in the traditional mode. Furthermore, the work was aimed at grades 10 to 12, while most geography was taught in grades 7 to 9. Pilot projects were launched across the country, but there was little enthusiasm for this geography among the professional educators. Social studies courses continued to dominate the curriculum, with history at the forefront, a condition little changed since the 1910s. The outcome of this large investment of activity was not as successful as had been hoped (Pattison, 1970). Cognitive dissonance had developed between what the public, teachers, school boards, and educators thought geography was and the salient features of the newly emerging (quantitative) geography, which loomed large in the 1970 volume. Even so, with the membership of the National Council for Geographic Education (dominated by teachers) fully aware of this project, the cause of geography was given emphasis in the high schools of the country (Stoltman, 1980).

NEW PHILOSOPHIES

Paralleling the development of a spatial science was the assumption of a *positivist* philosophy (which limits knowledge to facts that can be observed and to the relationships among these facts). This, it was thought, would win acceptance in the larger scientific community and would provide both explanation and prediction. Yet it became apparent that peoples and environmental settings are always unalike. Both have individuality and thereby variation, which meant that modeling in a positivist mode was problematic. By the very early 1970s spatial science and positivism were losing some support.

The United States had been involved in the Vietnam War for several years, and half a million men were involved in the fighting. At home the peace movement, civil rights and related race riots, the assassination of Martin Luther King, Jr., a slowing economy, and new modes of behavior among the younger generation all conspired to raise questions about the assumptions underlying a positivist philosophy. A surge of critical thinking arose in the 1970s as a result of dissatis-faction with the mechanistic models and assumptions of the quantitative revolution. New foundations for geographical inquiry, especially directed toward bringing the whole into juxtaposition with social justice, were sought. Positivism did not appear to address the needs of the human condition.

One group of geographers began to turn their attention to humanism, and a second group associated itself with radical and "from the left" postures. This remained the majority philosophical position of U.S. geography for approximately the next 20 years. Humanist geography bestows a major role to human awareness, human agency, and human existence. Initially, humanistic geography shared much with behavioral geography but then distanced itself from the behavioral enterprise as it recognized more and more the subjectivity of the investigator and the inves-tigated. Humanism has by its nature been associated with phenomenology and existentialism. Its trend more recently has been to seek growth and fulfillment via connection with the humanities (Buttimer, 1990).

Humanism and science frequently find themselves in conflict, although at other times they beneficially coexist; at still other times one leads the other only to be over-taken in a never ending series of actions and reactions. Those geographers who turned to humanism pursued the hermeneutic tradition (usually associated with Dilthey) that offered *understanding*. Positivism had instead sought *explanation*. Since posit-ivist explanation had failed somewhat in its aim, geographers such as Saarinen (1969), Cox and Golledge (1969, 1981) and others turned their attention to psychology, whereby they might gain some understanding of human perception and behavior. This was not a new mode for geographers. P. Vidal de La Blache and C. O. Sauer shared the notion that people fashioned the landscape from their culture. J. K. Wright had encouraged the investigation of imagination (1947), a matter later developed by D. Lowenthal (1961), who emphasized separate personal worlds of experience as essential to larger geographical comprehension. Meanwhile, in 1952 W. Kirk had published an article that seems to have functioned as the seminal essay concerning the behavioral environment. In thinking of the position of historical geography with regard to the human-environment conundrum, he proposed Gestalt psychology as a source for the integration of mind and nature. And at the University of Chicago from 1960 to 1965 G. F. White organized a number of studies concerning the perception of natural hazards as a guide to human behavior, which in turn led to R. Kates's (1971) study concerning natural hazards, hypotheses, and models. Behavioral geographers sought to measure and comprehend responses from indi-viduals to particular circumstances.

In contradistinction to humanistic geographers, an increasing percentage of geographers have sought involvement in changing some aspect of society or, in the more radical case, in changing society itself, if necessary by revolutionary means. This involves developing theories concerning the processes underlying social conditions that can be subjected to programmatic analysis. Marxist geographers dominate this group and argue that processes can and should be changed by political action. Yet there were a number of other variants of radical geography. In 1969 R. Peet and a number of graduate students at Clark University initiated the journal *Antipode*. Harvey's shift to Marxist analysis began to set a tone for *Antipode* in the

1970s. His *Social Justice and the City* (1973) created a critique and model for this new radical geography.

As a counterpart to Harvey's theoretical contribution, W. Bunge organized populist, community-based "expeditions" in Detroit (Fitzgerald) and later Toronto. These were organized demonstrations and are perhaps unique in the history of American geographical field work. In 1974 a Union of Socialist Geographers was created, permitting group action of a kind that might have brought questions by authority had such activity been carried out by individuals acting alone.

By then a state of eclectic pluralism characterized the geographers' undertaking. Problematically, geography now seemed to have become a discipline without a core. Human geography dominated but was quickly divided into a multiplicity of specializations. These exhibited centrifugal tendencies; economic geographers consorted with economists, historical geographers with historians, political geographers with political scientists, cultural geographers with anthropologists, and so on. These were divisive forces; there were no centripetal forces to draw the whole back to a core. In 1989 G. Gaile and C. Wilmott could write "the core of geography is the set of assumptions, concepts, models and theories that geographers bring to their research and teaching." That was a fair assessment of "the core" in the late 1980s, but it was not a substantive core that could demand allegiance from the fragmented specialties. When in 1982 J. F. Hart made a reasoned case for regional geography "as the highest form of the geographer's art," the cycle had come full circle. D. W. Meinig's first volume of *The Shaping of America*, a major series of books on historical geography, was published in 1986, then came John Borchert's *America's Northern Heartland: An Economic and Historical Geography of the Upper Midwest* (1987), John Agnew's *The United States in the World Economy: A Regional Geography* (1987), and Wilbur Zelinsky's *Nation into State: The Shifting Symbolic Foundations of American Nationalism* (1988).

With the arrival of postmodernism the approval of general theory was not possible, and a newly constructed form of regional geography seemed possible, even probable. P. Haggett (1996) has written "I expect a new regional geography to evolve which will supplement traditional skills by integrating quantitative models from economic geography, regional economies and regional science."

Many more books and articles, regional by intent, theme, and inspiration, were published through the 1990s. A large number of high school and college geography textbooks continue to be published with a regional theme into the present. Meanwhile, C. Salter (1999) wrote about the enduring nature of evocative regional geography, A. Murphy (2003) wrote of rethinking the place of regional geography, and M. Lewis and K. Wigen (1997) examined the regional process in studying traditional divisions of the world, both real and imagined. While much of this is referred to as the new "regional geography"—a term minted by Nigel Thrift (1983)—it is a version of regional geography as it has evolved over time, incorporating new methods but with a similar objective. The regional geography thought to be passé had been hard won over a period of seven or eight decades. Its significance may have been lost on a generation that had not practiced it firsthand and that was not very familiar with its literature. Although regional geography was largely dropped from campus course offerings during the 1970s and 1980s, it had a peculiar ability to survive in both the media and the culture. And it would have been invaluable to have had detailed regional geographies available of Vietnam, North Korea, Afghanistan, Iraq, and the Middle East over the last 40 years.

With the decline of regional geography came the further separation of physical and human geography. The two latter geographies had found integration in the context of region. With the reduction of the regional idea, both human and physical geography no longer had a bond and became two different subjects that often were not taught in the same department.

During the 1990s geography was subject to yet another philosophical shift, the cultural turn. Marxist and radical approaches began to fade, and a new theoretically grounded cultural geography (compare to a theory-free Berkeley cultural geography) was in the ascendant. Dominant in this theoretical turbulence was postmodernism, a philosophy that is difficult to define, but one that has made itself known to the human and social sciences. It derived from architecture and other artistic and cultural experiences. Then came deconstruction, a search for difference with a rejection of grand theory. (Soja, 1987; Harvey, 1989; Relph, 1991; D. Gregory, 1989).

With newly minted philosophies, methodologies, and epistemologies came newly formed Association of American Geographer specialty groups. Geography would seem to have had an unusual opportunity to present itself successfully to a large audience. However, differences had emerged between academic and public notions of what was geographical. Academic geography had become ever more remote from the public notion of the discipline.

Geographic philosophies have changed frequently during the last 50 years. Philosophies multiply and coexist, although the newer offerings have more following and support than the older. Yet how necessary are philosophies that frequently change, are too sophisticated for the comprehension of many, and are continually being subdivided or replaced? The nagging question remains: In what ways do these philosophies help advance the geographic objective?

WHAT WERE THE NEW GEOGRAPHIES?

With the passing of regional geography as the core of the discipline, a process that began in the 1950s, specialization began to increase. At first this specialization assumed the character of systematic categories such as economic geography, medical geography, and population geography. With increasing specialization—an inevitable part of the growth process associated with investigation—areas of study were further subdivided. Informal groups began to emerge, such as correspondence groups, special sessions at meetings, local groups of geographers sharing similar interests as well as new newsletters and journals. The International Geographical Union had provided the example with the formation of commissions and working groups much earlier in the century, and many other disciplines had provided the example of similarly specialized groups.

The formation of specialty groups within the Association of American Geographers (AAG) began in 1978. Prior to that time, however, special sessions at the national and regional levels had heralded this development (Conzen, 1978). In early

2004 52 specialty groups functioned; in previous years the number had occasionally been higher. The numbers of such groups and the titles of the groups provide a notion of the direction geography is taking in the United States. (In a typical year most AAG members opt to join approximately two specialty groups each.) In 2003–04 the groups with the largest memberships included geographic information systems (1182), urban geography (737), and remote sensing (485). The groups with the smallest numbers included disability (21), aging (43), and Canadian (59). Regional geography specialty groups posted only modest support: Latin America (282), Africa (170), China (153), Asia (144), Europe (134), Russia (131), Middle East (82), and Canada (59). Australia-New Zealand is not included in the regions, and Asia is not subdivided, as was once standard practice. Incidentally, these low figures for global regions came at a time when the public need for geographical knowledge of many parts of the world was at a premium. When matters relating to Israel were so vexed and misunderstood in the United States, considerably less than 1 percent of AAG membership belonged to the Middle East specialty group.

This increased institutionalization of specialized groups led to renewed proclamations of fragmentation. When Gaile and Willmott proposed to the AAG a collection of essays to summarize the work of the specialty groups, the association declined to support it. Gaile and Willmott published the study independently (1989). The association felt that "the fission evident within geography . . . may have gone as far as it can or should go in American geography". (Abler, Marcus, and Olson, 1992, p. 384). A companion volume intended to embrace the next decade of American geography, with essays provided by selected author(s) from each specialty group, was also published (Gaile and Willmott, 2003). In an attempt to demonstrate unity amid diversity, the association, at the request of its council, planned a volume to be entitled *American Geography: Survey and Synthesis*. This title was revised to *Geography's Inner Worlds: Pervasive Themes in Contemporary American Geography* (1992) and was designated by the AAG as a contribution to the 1992 International Geographical Congress held in Washington, D.C. The book was concerned with "cross-cutting and unifying concepts and methods" and included four themes: "what geography is about," "what geographers do," "how geographers think," and "why geographers think that way." The book goes a long way toward explaining the geography of geographers, and the editors hoped that synthesis would be added to analysis and that understanding would complement explanation as the future of geography as a discipline unfolds. The persistent tone of the work is evocative of a desire to integrate and to show what is held in common among a variety of contending geographies suggested by 52 specialty groups.

The reality in the early twenty-first century, however, is that American geography seems to have a periphery without a core. Whereas once the core fed the periphery and was sustained by it in reciprocity, there is now a periphery on which much good work is accomplished, but it is eclectic, of great variety and not unidirectional as it once was. This centrifugal process moving the discipline (whose folk and professional models are already dissonant) away from a core of subject matter and toward ways of reasoning dates from the 1950s. The academic and intellectual mission has become obscure. Suggestive of a potential core is Koelsch's endorsement of Hanson's (1999) assertion that space, place, and environment are complementary rather than competitive notions within geography (Koelsch, 2001, p. 275). Perhaps this is so, but there are others who reason that a core is not essential and that from the periphery will emerge by way of hybrid vigour and evolutionary strife a new disciplinal arrangement. That, too, is possible. Science advances in a variety of ways.

Some of the 52 geographies have been much affected by the new methods, while others have been little affected. The latter were typified by the newly wrought genre of feminist geography and studies in the history of geographical thought. These are briefly introduced below. Geographies much affected by the new methods are numerous and include examples such as global information systems (GIS), cartography, urban planning, and applications of geographic thinking. These and other specialty group areas may have entered into a golden age. The two chapters that follow (Chapters 18 and 19) examine this development.

FEMINISM AND THE HISTORY OF GEOGRAPHIC THOUGHT AS EXAMPLES OF SUBJECT AREAS LITTLE AFFECTED BY NUMERACY

Feminist geography derived encouragement from the women's movement of the 1960s and was further encouraged by such movements as the radical, Marxist, and welfare groups that concerned themselves especially with the structures of social inequality. It was fundamentally an academic expression of concern that a pervasive inequality of the sexes existed in the field of geography, both as actors and as subjects of research, and that the dimensions of this matter should be examined.

The first serious study to be published on the status of women within the geographic profession was by Wilbur Zelinsky. Entitled "Women in Geography: A Brief Factual Account" (1973), it was a searching analysis in disparity and initiated professional concern and further investigation. Other studies were soon to emphasize the differences between male and female spatial behavior in an attempt to establish the significance of gender in geographic research. A review of this work was published by W. Zelinsky, J. Monk, and S. Hanson in 1982. Human geography was also criticized by J. Monk and S. Hanson (1982) for failing to evaluate the role of women and the issue of gender.

The movement became more international by the 1985–90 period, by which time contributions had come from the United Kingdom, the Netherlands, France, and Spain, with some measure of subject emphasis on women in the developing world. Studies were made with special reference to disadvantages imposed on women by capitalist and patriarchal structures. The "women and gender study group" developed this thought considerably and then published *Geography and Gender* (1984), which was designed to function as a primary text for undergraduate work, especially in the American system.

Further investigations concerned gendered space, the invisible contributions of women, reproduction as production, and other matters that are a product of female and male difference. There developed the formulation that men—deriving authority as father figures and heads of households—are associated with rationality, domination of a "female" earth, and power over women. Women, by this same formulation, have symbolic power by being linked with nature and by being mothers and keepers of homes. The crux of the matter was succinctly expressed by G. Rose (1993) in explaining that there has been a male bias in structuring the discipline. Membership figures for the AAG indicate that in the constructive phase of discipline building (of American geography), women rarely constituted 10 percent of that membership.

Knowledge and disciplinal structure are social constructs, which is why Western societies often suffer ethnocentric bias when studying other parts of the world (E. Said, 1978). This is one of many notions that gender geography presents to the field as a whole. As with other specializations, feminist geography has produced its own groups and its own journals. Within the International Geographical Union,[1] symposia on gender were organized by 1981, and a Commission on Gender was established in 1992. *Gender, Place and Culture* was inaugurated in 1994, and reviews of feminist literature are published regularly in *Progress in Human Geography* (Monk, 2001). This led to further fragmentation of human geography, both substantively and epistemologically, yet this divisiveness is a price that must be paid for ongoing inquiry of this order of magnitude.

A second area little affected by the new methods is the history of geographical thought, the essence of this book. This study of the history of geography should not be (but is occasionally) confused with historical geography. The task of the history of geography is to learn something of the contribution of our forebears, be they predisciplinary or postdisciplinary in chronology, to study something of their methods and philosophies, and ultimately to construct histories of the evolution of the field at different times in different places. Both knowing what "geography" has meant in varying temporal contexts and knowing how we have arrived at our present comprehension and purposes are the objectives of this area of scholarship.

In the first half of the twentieth century there was little scholarly work in the history of geography by American geographers. (The work of J. K. Wright, R. Hartshorne, and some of the AAG presidential addresses were the exception.) Biographic and prosopographic dissertations concerning geographers were sponsored by departments of education or history. This was to change in the years to come, although dissertations in the history of geographical thought remained very few. Specialists remained rare, and the principal scholar in this area, J. K. Wright of the American Geographical Society (AGS), found few helpers in his attempt to develop the field. Courses and seminars on the subject were a mix of philosophy and methodology, with little history of thought included.

The AAG established an archives and a history committee in 1971. From this development emerged a formal AAG archive and the *History of Geography Newsletter* (1981). In 1988 the title was changed to *History of Geography Journal*. It was the only publication of the period that dealt exclusively with the history of geography. The formation of the archive and history committee also led to increasing awareness of the need to form a proper archival deposit for the association.

---

[1]Noteworthy is the election of Anne Buttimer to the presidency of the International Geographical Union, 2000–2004. She is the first female elected to this office.

Eventually, this archival collection was located with the AGS special collections at the University of Wisconsin at Milwaukee. It was helpful to James and Martin in writing and editing *The Association of American Geographers: The First Seventy-Five Years, 1904–1979* (1978). On the same occasion, in 1979, a special issue of the *Annals*, "Seventy-Five Years in American Geography," was published. It constituted an eclectic set of essays and reminded an American audience that one could look back as well as forward.

Meanwhile, in 1977 *Geographers: Biobibliographical Studies* was founded as one of the functions of the commission on the history of geographical thought of the International Geographical Union. A number of American geographers and geographers from more than 30 other countries contributed essays to this ongoing series concerning the lives and thought of deceased geographers. More than 50 American geographers have been made the subject of these biobibliographies. The commission also sponsored symposia on the occasions of the International Geographical Congresses and the regional congresses. Much geographic thought has been generated and shared in this way. In this series of meetings, the gathering in Lincoln, Nebraska, in 1979 generated *The Origins of Academic Geography in the United States* (Blouet, ed., 1981). This anthology contained a series of valuable essays concerning the history of American geography. Maynard Weston Dow, in a remarkable series entitled "Geographers on Film," beginning in 1970 filmed interviews of geographers and has also filmed geographic occasions of note, such as seminars and banquet addresses. He has amassed a treasury of visual and oral lore of the field in a unique medium (Dow, 1974, 1981). These films also contain footage of a number of now deceased U.S. (and other national) geographers (Fig. 42).

Many investigations of the history of American geography have been conducted since the 1950s. There are a number of biographies that embrace varying portions of the development of American geography. Examples include J. R. Smith (Rowley, 1964), M. Jefferson, E. Huntington, and I. Bowman (Martin, 1968, 1972, 1981), W. M. Davis (Chorley, Beckinsale, and Dunn, 1973), G. K. Gilbert (Pyne, 1980), P. E. James (D. Robinson, 1980), N. S. Shaler (Livingstone, 1987), and T. G. Taylor (M. Sanderson, 1988). R. J. Moss wrote *The Life of Jedidiah Morse* (1995), and in 1999 W. W. Speth provided *How It Came To Be: Carl O. Sauer, Franz Boas and the Meaning of Anthropogeography*. D. Lowenthal wrote *George Perkins Marsh, Versatile Vermonter* (1958) and several articles concerning Marsh, then set about a revision of *Versatile Vermonter*. The resultant book, much more than a revision, was published in 2000. Subsequently, "The Lowenthal Papers," including works by four authors, was published (Olwig, 2003). P. R. Gould (1985), S. M. Dicken (1986), Y. Tuan (1999), and H. J. de Blij (2000) offered works autobiographical in character. P. Gould and F. Pitts (2002) introduced 14 contemporary autobiographical accounts of American geographers in *Geographical Voices*. Two collections of papers relating to the work and accomplishment of Carl Sauer must also be mentioned (Kenzer, 1987; Mathewson and Kenzer, 2003).

Histories of a number of university departments of geography have been written. Among the more significant of these were perhaps California (G. S. Dunbar, 1981), Cincinnati (B. Ryan, 1983), Clark (W. Koelsch, 1980, 1987, 1988), Harvard (N. Smith, 1987; T. Glick, 1988), Wisconsin (C. W. Olmstead, 1987), and Yale

Figure 42 M. W. Dow building "Geographers on Film" (1979)

(G. Martin, 1988). Reaching further back in time was C. J. Glacken's (1967) *Traces on the Rhodian Shore . . .*, K. W. Butzer's (1992) "The Americans Before and After 1492: Current Geographical Research," and studies by W. Warntz concerning Newton and Varenius (1981, 1989). In G. S. Dunbar (ed., 2001), W. Koelsch offered "Academic Geography, American Style: An Institutional Perspective," a summary of the history of American geography from the mid-1800's to the present. G. Martin presented a summary of work undertaken by American geographers from 1975 to 1999 in Gaile and Willmott (2003). There remain many gaps in our knowledge, not the least of which relates to the significant contributions of female geographers, but a start has been made (e.g., M. Berman [1974] and J. Monk [1998]).

Interesting projects are sometimes inspired, at least in part, by previous works in the history of geography. One such project developed by Clark University faculty led by B. L. Turner II stands revealed as task and accomplishment in *The Earth as Transformed by Human Action: Global and Regional Changes in the Biosphere over the Past 300 Years*. The inspiration for this project derived by linear descent, in the first instance from George Perkins Marsh who inverted the determinist thesis to study the impact of humans at the Earth's surface. Secondly, Marsh was celebrated by a symposium held at Princeton (1955) which led to publication of many of the papers presented on that occasion in the book *Man's Role in Changing the Face of the Earth* (ed. W. L. Thomas, 1956). The 1990 volume mentioned above goes far beyond the detail of Marsh and (ed.) Thomas, but was inspired by them. Other notions once developed have seemingly been ignored for a time, then later have been re-examined with newly kindled enthusiasm. Examples include "mental maps," "central place," and geosophy.

There is no mainstream to this genre, but a variety of individual efforts that do not of themselves form a philosophic whole. While origins of ideas, explanations of the evolution of thought, and precursors of laws may be identified, the genre is idiographic in tone and idiosyncratic in choice of topic. The history of geography may become a theoretically self-contained enterprise, no longer considered a prerequisite to larger geographic comprehension, and a geography devoid of its history is emerging. Yet as a foundation to disciplinary comprehension the study remains of much value. This large undertaking has not been subject to revision by the post-1950s theoretical enterprise, although it has attempted to record it in part.

THE PROFESSION: ITS INSTITUTIONS AND SIZE

How many geographers are there in the United States today? Indeed, what makes a geographer? Membership numbers for the large geographic organizations provide some guidance. The Association of American Geographers (AAG), founded in 1904, had a membership of only hundreds in 1954; it now has a varying annual membership typically between 7500 to 8400, of whom approximately 75 percent are college-level geography faculty, the remainder being largely applied geographers. Minorities constitute a small percentage of the membership of the AAG. African American, Hispanics, and Native American members combined total less that 4 percent of the membership. Women make up a varying percentage but typically less than a third of total membership. The association has a permanent central office staff in Washington, D.C. (Fig. 43).

The AAG publishes *The Annals* (initiated in 1911), *The Professional Geographer* (1946), and the *AAG Newsletter*. Conjointly this threesome is essential for the geographer. The AAG also organizes an annual meeting, typically attended by some 3000 to 4000 geographers. The association is divided into nine regional divisions—an arrangement that provides opportunity for graduate student and novitiate faculty involvement. Divisions may publish newsletters or selected papers from the annual meeting and provide the opportunity to talk with fellow geographers.

The American Geographical Society (founded in 1851) is now decentralized, with its official headquarters (plus its archival collection) in New York City, its library, map collections, and staff in the Golda Meir Library at the University of Wisconsin, Milwaukee (since 1978), and the editorial staff of *The Geographical Review* currently in the geography department of Clark University, Worcester, Massachusetts (Figs. 44 and 45). The Society has memberships and subscriptions approximating 5000. Its imposing *Current Geographical Publications* continues to be published 10 times a year from Milwaukee. *Focus* and *Ubique* are published from the New York office.

Figure 43 Executive Directors of the AAG: (*right*) Ronald F. Abler (1989–2002) and (*left*) Douglas Richardson (2003–)

Figure 44 The American Geographical Society staff, June 1956

Figure 45 Sesquicentennial of The American Geographical Society: Director, Mary Lynne Bird, 2002

The National Council for Geographic Education is devoted to the improvement of the teaching of geography at all levels, currently headquartered in Jacksonville, Alabama. It was founded in 1915, and the *Journal of Geography* (founded 1902) became its official publication. It holds annual meetings that are generously attended.

Finally, there is the The National Geographic Society, founded in 1888. It has approximately 6.5 million US members who automatically become subscribers to the *National Geographic Magazine*. In this respect it stands alone. Also rendering it unique is the fact that it has been a product of the Bell—Grosvenor family for the length of its life. While its natural history style of geography has not won much academic support, it has provided the public with a "magic carpet" (a term adopted by Gilbert H. Grosvenor) by introducing the public to many parts of the world they would never otherwise visit. Now, with the realization that academic geography needs public support that merges with support for geography in the schools, there is a new-found enthusiasm for the Society's explorations, geographic alliances, the geography bee, and other activities (discussed later in the chapter).

These are the discipline's main institutions in the United States. What of its academic personnel? Of approximately 2200 accredited four-year institutions in the United States, only some 250 offer undergraduate or graduate degrees in geography. Between 1945 and 1990 100,000 bachelor's degrees were awarded in geography.

In 1990 586,000 college-level students took geography courses at more than 600 institutions in the United States (Walker, 1991). Approximately 3000 students graduate with bachelor's degrees in geography each year. Approximately 550 students graduate with master's degrees in geography annually, and 120 to 130 with Ph.D. degrees (from a total of 52 departments offering Ph.D. degrees). To be noted, however, is a substantial inflow of English-speaking geographers from abroad who undertake doctoral work in the United States or who come as faculty members (after 1989). Few of those with bachelor's or master's degrees retain a relationship with their departments of origin or communities. This too frequently means that roots are lost and input into policy, advising leaders, and otherwise helping thoughtfully in communities is foregone. This is particularly unfortunate at a time when geographers do not write for the public as they once did.

The profession has shown a marked growth since the 1950s in journals, newsletters, reference works, and atlases, and not withstanding some severe losses of eminent departments of geography (e.g., Harvard, Yale, Michigan, Pittsburgh, Columbia and Northwestern), other departments are emerging with doctoral granting status that with proper nurture may go some distance toward redressing that loss. Nevertheless, closure of departments at eminent universities suggests that geography is not well thought of by those institutions that have traditionally contributed the nation's educational, economic, cultural, and political leaders.

SOME INTERNATIONAL DIMENSIONS TO THE DEVELOPMENT OF GEOGRAPHY IN THE UNITED STATES

External influences are brought into American geography by way of books, periodicals, correspondence, visiting lecturers, professorships, and so on. There are also specific international activities that enrich the content of the discipline. These activities have included participation by American geographers in the International Geographical Congresses held in the United States in 1904 (Fig. 46), 1952 (Fig. 47), and 1992 (Fig. 48). During the congresses large numbers of American geographers have access to the different viewpoints of foreign geographers, while some distinguished visitors may remain in the United States giving lectures and seminars and assisting with projects. The same benefits accrue from participation in the international congresses held abroad as well as the regional conferences. American geographers also are members of study groups and commissions of the International Geographical Union, which organizes congresses and conferences.[2] Other

---

[2]The United States provided four presidents of the International Geographical Congress series who delivered addresses as follows: Robert E. Peary, 1904 (Washington DC); Isaiah Bowman, 1934 (Warsaw); George B. Cressey, 1952 (Washington DC); Roland J. Fuchs, 1992 (Washington DC).

Figure 46 Delegates participating in the Eighth International Geographical Congress aboard a Chicago horse-drawn sightseeing coach. Gilbert H. Grosvenor sits on the top deck (hat on knees) facing his wife, Elsie. (Copyright NGS)

Figure 47 Delegates at the 17th Congress of the International Geographical Union, 1952, Washington, D.C. (*From left to right*): Dr. Hans Bobek, Austria; Mrs. Carolyn Patterson, USA; Dr. Cheng-Slang Chen, China; Mr. Nafis Ahmad, Pakistan, and Dr. Carl Troll, Germany. (Copyright NGS)

Figure 48 The International Geographical Congress, 1992, Washington, D.C.: Initiating an institutional history of the Congresses

international exchanges occur in meetings of other bodies, including the International Cartographic Association, the International Geosphere-Biosphere Programme, the Scientific Committee on Problems of the Environment, the Intergovernmental Panel on Climatic Change, the International Human Dimensions of Global Environmental Change Programme, the International Council for Scientific Unions, and many additional global organizations and field work activities.

At these and other congresses the main language is invariably English. Other languages may be used (traditionally French with the International Geographical Congresses), but it is clearly English that dominates. This constitutes a large advantage to geographers from the United Kingdom, Australia, New Zealand, Canada, and other parts of the one-time British Empire, but especially to American geographers, of whom a small percentage are fluent in a foreign language. It is of even greater advantage given the amount of literature published in English. Here again the American geographers have an advantage in negotiating the ever increasing body of literature. As Garcia-Ramon (2003) has said, "the growing hegemony of English as a global language privileges the geographical discourse of the Anglophone world. . . . Linguistic hegemony is a form of power that empowers some while disempowering others." Gutierrez and Lopez-Nieva (2001) write in a similar vein in "Are international journals of human geography really international?" Chauncy Harris has made a comprehensive study of the use of English with regard to the International Geographical Congresses (1871–2000), and with regard to the use of English in geographical periodicals and serials from 1882 to 2000 (2001).

Works in foreign languages that are considered significant additions to the literature are invariably translated into English, such as Von Thünen, Weber, Christaller, Lösch. These works were little known and hardly cited by American geographers until they were translated. Prior to 1954 this was perhaps less true, as a higher percentage of American geographers were fluent in French or German or both, and geographers and geographical ideas predominantly traveled from Europe to the United States. In the last 50 years this order of things has been reversed, and now American geographers and geographical literature are exported from the United States to Europe.

In Washington, D.C., the Office of the Geographer of the Department of State was led by a series of geographers including George J. Demko (1984–89), a geographer from Ohio State University.[3] He established a more rigorous geographical program in the State Department. This meant bringing in more cartographic computer specialists and more and geographically better trained personnel. The result was greater efficiency and an ability to swiftly produce required data in an appropriate format. (Much of the work centered on boundary disputes.) This was a major accomplishment as many of the problems in the State Department are uniquely geographic. In his capacity as head of the International Research and Exchange Board's Geography Commission for the U.S. Academy of Sciences and the Soviet Academy of Sciences, Demko arranged travel, joint seminars, and publications. American publications had been prepared in earlier years at the request of the Soviets, who had also translated American books such as *American Geography: Inventory and Prospect* (1954; published in Russian, 1957: *All Possible Worlds* was published in Russian, 1988). Soviet geographers prepared a book for the 1960 International Geographical Congress in Stockholm that was published in English (1962) as *Soviet Geography: Accomplishments and Tasks.*

Exchanges of geographers took place between the two countries, and Demko, R. Fuchs, and R. Taaffe were among the early American graduate students to visit the faculty of geography of Moscow State University. C. D. Harris attended the third Congress of the Geographical Society of the U.S.S.R. in Kiev in 1960. In that same year T. Shabad, a journalist on the staff of the *New York Times*, initiated *Soviet Geography: Review and Translation*, published initially by the American Geographical Society. This was a significant development, as it provided access to Russian geographical thinking for an American audience. Then began a series of U.S.–U.S.S.R. exchanges. The Soviet Commission met in Moscow and Siberia in

---

[3]Geographers who have held this post include the following: Lawrence Martin (1920–1923), S. Whittemore Boggs (1924–1954), Sophia A. Saucerman [Assistant Geographer] (1954–1957), G. Etzel Pearcy (1957–1969), Robert D. Hodgson (1971–1979), Lewis M. Alexander (1980–1987), J. Millard Burr (Acting, 1982–1984), George J. Demko (1984–1989), and William B. Wood (1989–).

Figure 49 U.S.–U.S.S.R. field trip, Lake Baikal, Siberia, 1983

1983 (Fig. 49), in Vilnius in 1985, and starting in 1987 in Washington, D.C., and Milwaukee.[4] At the Twenty-third International Geographical Congress held in Moscow in 1976, which put Soviet geography on display, a number of American geographers were present, especially those interested in Soviet-U.S. exchange. Demko arranged further exchanges with Soviet, Hungarian, Czech, and Bulgarian geographers in the years 1979–87.

It is difficult to assess the significance of these activities, but they did bring new viewpoints into American geography, created cooperation at a time when relations between the United States and the U.S.S.R. were under strain, and provided opportunity for further exchange. It is interesting to note that Yale Richmond makes special mention of the Soviet-U.S. Geography Commission in *Cultural Exchanges and the Cold War* (2003, pp. 53–44, 79).

GEOGRAPHIC ILLITERACY

During the 1980s a well-grounded concern developed throughout the United States that the citizenry, and especially the school-age population, had little knowledge of the world or even the country in which they lived. Academic studies as well as media polls revealed the extent of American ignorance regarding even the location of places and countries. When a U.S. president arrived in the capital of Brazil, Brasilia, and pronounced himself pleased to be in Bolivia, his gaffe appeared on the front pages of all the major newspapers, and editorials asked the obvious question: Did the country's leaders know much more about the world than the geographically illiterate public?

---

[4]Resultant to these meetings, books were published: *Urban Geography in the United States and the Soviet Union* (Lappo et al., 1992), *The Art and Science of Geography: Soviet and American Perspectives* (Demko et al., 1992), and *Global Change: A Geographical Approach* (Mather et al., 1991). An account of these activities was published by Annenkov and Demko (1984) as "Development of Relations between Geographers of the United States and the U.S.S.R. from the 1950s to the 1980s."

Calls for a revival of geography as a school subject came from numerous quarters, notably including the president of the National Geographic Society, Gilbert M. Grosvenor. The society had long had a difficult relationship with the profession; to many professional geographers the popularization of its magazine under the rubric of geography was inappropriate and misleading. Many geographers felt that there was rather little geography in *National Geographic*. To the society and its leadership, however, professional geographers seemed snobbish, insulated, and often unimaginative (de Blij, 1995). But on the matter of resurrecting geography in the schools, all parties could agree, and Grosvenor, encouraged by the small cadre of professional geographers then working on the staff of the National Geographic Society, vigorously took the lead. In a memorable address before the AAG during its 1983 meeting in Washington, D.C., Grosvenor laid out his plans to enhance collaboration between the society and the association, proposed a campaign to promote the teaching of geography in American schools, and announced that a geographer had been appointed editor of a new scholarly journal that was to reflect the society's support for scientific research as well as high school education (Grosvenor, 1983). That journal was *National Geographic Research* (later to be retitled *Research and Exploration*).

The society also launched what was to become known as the Geographic Alliance Program, funded by many millions of dollars provided by the society. Its primary objective was the training of geography teachers from across the country, groups of whom were invited to the society headquarters in Washington, D.C., for intensive training sessions, following which it was their responsibility to transmit this information to their colleagues in their respective states. The program involved a total of 160,000 teachers from all 50 states by 1995. It soon began to reverse the decline of geography in the high schools, although the road back proved challenging. Meanwhile, the society introduced other incentives, such as an annual National Geography Bee (which draws up to 6 million participants) and supported Senator Bill Bradley of New Jersey in establishing a Geography Awareness Week each November to publicize the ongoing revival of the subject.

Grosvenor also consulted with the National Governors' Association in an attempt to improve geographical education in the school system. Resulting from this activity, in 1989 President George Bush joined with state governors to request "national standards" in a number of fields, including geography. (The other subjects were science, mathematics, English, and history.) The National Geographic Society funded a coalition of representatives from the four major geography organizations to publish *Geography for Life: National Geography Standards*, (1994). This book includes standards for grades K–4, 5–8, and 9–12 that "specify the essential subject matter, skills, and perspectives that all students should have in order to attain high levels of competency" (DeSouza, 1994). When geography was included as a core subject in "Goals 2000: Educate America Act", Section 102, it was the culmination of more than a decade of work and reform in geographic education.

In the mid-1980s, when these initiatives began, public awareness of geography rose, and professional geographers frequently found themselves having to explain just what it is that geographers do. Surely the nation's leaders were geographically literate, and just what were the "big ideas" in geography that could compare to, say, research on the moon or on the origins of humanity? On the first point, the answers were not reassuring. In his memoirs, former Secretary of State Henry Kissinger describes how his "vaunted" national security staff had briefed President Nixon for a visit by the president of Mauritius—but had confused Mauritius with Mauritania, producing an embarrassing exchange in the Oval Office (Kissinger, 1999).

The point created consternation among professional geographers. The discipline was undergoing continuing transformations, and its fragmentation, reflected by the growing number of AAG specialty groups, continued. Gone were the days when such shorthand as "spatial organization" or "regional specialization" or "environmental study" could begin to answer questions about geography's core concerns. R. F. Abler raised the issue in the title of his 1987 AAG presidential address "What Shall We Say? To Whom Shall We Speak?" (Abler, 1987). "We would be wise," Abler writes, "to refocus our attention on the core of our discipline—on the things only geographers can do and do well—and to reduce our emphasis on our too-numerous specialties and diverse interests" (p. 516). But what are these things? "I have often been impressed," Abler adds, "with the way traditional regional geographers with bad reputations among their junior colleagues can fill classrooms with enviable numbers of interested, enthusiastic students . . . if we do not move quickly to reclaim regional geography, that part of our intellectual birthright will be irretrievably lost."

Clearly, however, no single answer to a question regarding geography's core would satisfy all, or even a majority, of professional geographers in an era of post-structuralism, postmodernism, feminism, neo-Marxism, and other diverse pursuits. In his 1987 address Abler remarked that it occurred to him that "we influence far more people—students, colleagues inside and outside the discipline, and people in the public and private sectors—by talking rather than writing." This notion was underscored when ABC Television in 1988 appointed Harm de Blij as geography editor on its program *Good Morning America*. Reaching more than 5 million viewers with each appearance, using maps, and demonstrating the utility of geographic perspectives rather than trying to define the discipline, he elicited thousands of written responses, many from high school students interested in majoring in geography at college (de Blij, 1990).

Large as this audience was, it did not include some opinion makers, such as the columnist Richard Cohen of the *Washington Post*, who did not get the message. "Geography is such a frumpy science (have you ever seen a geographer interviewed on television?)," he wrote on June 27, 1995. Among responses to his column was one entitled "Geography in the Public Eye" (Grosvenor, 1995). "Geography is destiny; location really does matter, whether buying real estate or making global decisions. Ever since a Gallup survey showed that one in seven Americans couldn't locate their country on a blank world map, we've been working on this problem."

PUBLIC DEBATE

With geography now in the public arena, the debate among geographers intensified. At the 2001 AAG meeting in New York, the opening session talk given by the *New York Times* journalist John N. Wilford, "A Science Writer's View of Geography," aroused much comment (Wilford, 2001). Wilford suggested that geographers have done a poor job of speaking the popular language, of conveying in simple and direct terms what is important about their work. He argued that geographers lack a "big-picture focus." In other sciences, he said, big-picture elements are clear: the age and accelerating expansion of the universe, the formation of galaxies, the nature of extrasolar planets. "But I don't know what the comparable questions are in geography," Wilford said. As a result, comparatively little geography appears in the popular press. Why should this be so? Is it, he asked, because geography is too fragmented for anyone to speak with authority on its big picture? In fact, is there a big picture at all? Wilford wondered if geographers are "actually discouraged from thinking and speaking out in the popular language." Wilford proclaimed himself to be unaware of a single person in geography with some fluency in the popular language, but cited the late Peter Gould as someone who did. But "without some Goulds, how is geography to be understood by the public?"

R. F. Abler, then AAG executive director, wrote in the April 21, 2001, *AAG Newsletter* that geographers would benefit greatly if they cultivated an ability to speak to the public about geography's big picture in plain language. He conducted an informal survey following Wilford's appearance that indicated that "the majority thought [Wilford's] impressions to be ill-informed and his advice arrogant and unhelpful" (Abler, 2001). Perhaps one-fourth of those he consulted, however, felt that Wilford had been perceptive about the discipline and correct in his recommendations. But how useful to the discipline is the exposure Wilford and his colleagues provide in the popular press? A similar majority probably would answer that the link is unproven and that the required "popularization" of geography would do it more harm than good.

There can be no doubt, however, about the paucity of geographic perspectives and geographers' voices in the public debates over major issues of the day. Electronic and print media are replete with commentaries by political scientists, economists, historians, and others on issues ranging from global warming to globalization and from terrorism to trade agreements. During the run-up to the 2003 war against Iraq's Saddam Hussein regime and its aftermath, not a single geographer was among the horde of network and cable commentators, although several dimensions of this crisis were patently geographic. Certainly there was no dearth of informed opinion among geographers, as was evident during the 2003 New Orleans AAG meeting. But why was all this activism not translated for public consumption?

One consequence of this situation is that a good deal of "big-issue" geography is being written today by nongeographers, some of it less than well informed. Jean Grove's magisterial and highly technical book *The Little Ice Age* (Grove, 1988) was followed by a book of identical title written for the general public by an archaeologist (Fagan, 2000). The Harvard history professor D. Landes begins his *Wealth and Poverty of Nations* (certainly a big-picture issue) with an attack on geography as a discipline and then proceeds to commit the very intellectual offenses that did so much damage to geography in the first half of the twentieth century (Landes, 1998). The Columbia University economist J. D. Sachs, in an article based on an address at the United States Naval War College, discovers that "virtually all of the rich countries of the world are outside the tropics, and virtually all of the poor countries are in them . . . climate, then, accounts for a quite significant proportion of the cross-national and cross-regional disparities of world income" (Sachs, 2000). Wilford describes *Guns, Germs and Steel* by physiologist J. Diamond as "the best book on geography in recent years," and indeed it is a tour de force—but the weakest link in its multidisciplinary chain is the geographic (Diamond, 1997).

The dilemma continues, in part because many geographers have not learned the "plain language" to which Abler refers and also because geography's "big pictures" are not easily constructed. Geography does not offer spectacular missions to Mars or dramatic tomb-openings in Maya country. Geography's relevance is clear and demonstrable, its key questions are crucial, but its major goals are achieved incrementally, not sensationally. To articulate that relevance, those questions and goals remain principal and critical tasks for the next generation of professional geographers. Having made the case for securing public support, geographers may reflect on their last half-century of accomplishment with considerable satisfaction— not necessarily with the multidirectional thrust of their undertaking, or with the end product—but in the larger reaches of multiple investigations that promise both further commitment and discovery.

REFERENCES: CHAPTER 17

Abler, R. F., Adams, J. S., and Gould, P. R. 1971. *Spatial Organization: The Geographer's View of the World*. Englewood Cliffs, N.J.: Prentice Hall.

Abler, R. F. 1987. "What Shall We Say? To Whom Shall We Speak?" *Annals AAG*, 77(4):521 (first quote), 520 (second quote).

Abler, R. F. 2001. "From the Meridian: Wilford's Science Writer's View of Geography" *AAG Newsletter* 36(4):1.

Abler, R. F., Marcus, M. G., and Olson, J. M. 1992. *Geography's Inner Worlds: Pervasive Themes in Contemporary Geography*. New Brunswick, N.J.: Rutgers University Press.

Ackerman, E. A. 1945. "Geographic Training, Wartime Research, and Immediate Professional Objectives." *Annals AAG* 35:121–143.

———. 1958. *Geography as a Fundamental Research Discipline*. University of Chicago Department of Geography Research Paper No. 53. Chicago: University of Chicago.

Ackerman, E. A., et al. 1965. *The Science of Geography*. Washington, D.C.: National Academy of Science and National Research Council.

Agnew, J. A. 1987. *The United States in the World-Economy: A Regional Geography*. Cambridge: Cambridge University Press.

Annenkov, V. V. and Demko, G. J. 1984. *Soviet Geography* 25:749–757.

Barnes, T. 2001. "Lives Lived and Lives Told: Biographies of Geography's Quantitative Revolution". *Environment and Planning D: Society and Space* 19:409–429.

Berman, M. 1974. "Millicent Todd Bingham." *Geographers: Biobibliographical Studies* 11:7–12.

Berry, B. J. L. 1993. "Geography's Quantitative Revolution: Initial Conditions, 1954–1960 —A Personal Memoir." *Urban Geography* 14:434–441.

Blaut, J. 1979. A Radical Critique of Cultural Geography: *Antipode* 11:25–29.

Blouet, B. W., ed. 1981. *The Origins of Academic Geography in the United States*. Hamden, Conn.: Archon Books.

Borchert, J. R. 1987. *America's Northern Heartland*. Minneapolis: University of Minnesota Press.

Bunge, W. 1966. *Theoretical Geography*. Lund Studies in Geography, Series C1. Lund: C. W. K. Gleerup.

Bunge, W. 1979. "Perspective on *Theoretical Geography*" *Annals AAG* 69:169–174.

Burton, I. 1963. "The Quantitative Revolution and Theoretical Geography." *Canadian Geographer* 7:151–162.

Buttimer, A. 1990. Geography, Humanism and Global Concern." *Annals AAG* 80:1–33.

Butzer, K. W. 1992. "The Americans Before and After 1492: Current Geographical Research." *Annals AAG* 82:3.

Chorley, R. J., R. P. Beckinsale, and A. J. Dunn. 1973. *The History of the Study of Landforms, or the Development of Geomorphology. Vol. 2: The Life and Work of William Morris Davis*. London: Methuen.

Conzen, M. P. 1978. "The Formation of Specialty Groups within the AAG." *The Professional Geographer* 20:309–314.

Cox, K. R. and R. G. Golledge, eds. 1969. *Behavioral Problems in Geography: A Symposium*. Studies in Geography 17. Evanston Ill.: Northwestern University.

———, eds. 1981. *Behavioral Problems in Geography Revisited*. London: Methuen.

De Blij, H. J. 1971. *Geography: Realms, Regions and Concepts*. New York: John Wiley and Sons.

———. 1990. "Geography on 'Good Morning America.'" *Focus* 40, 4:32.

———. 1995. *Harm de Blij's Geography Book*. New York: John Wiley and Sons.

———. 2000. *Wartime Encounter with Geography*. Lewes, Sussex, U.K.: The Book Guild.

———. 2001. "The Big Picture." *The Geographical Bulletin* 43, 2:69.

Diamond, J. 1997. *Guns, Germs and Steel*. New York: Norton.

DeSouza, A. R. 1994. *Geography for Life: National Geography Standards, 1994*. Washington, D.C.: National Geographic Research and Exploration.

Dicken, S. M. 1986. *The Education of a Hillbilly: Sixty Years in Six Colleges: The Memoirs of Samuel Newton Dicken with the Assistance of Emily Fry Dicken*. Privately printed for the Lane County Geographical Society.

Dow, M. W. 1974. "The Oral History of Geography." *The Professional Geographer* 26:430–435.

———. 1981. "Geographers on Film: The First Seven Years." *History of Geography Newsletter* 1:21–28.

Duncan, N. 1996. "Postmodernism in Human Geography." Pp. 429–458. In *Concepts in Human Geography*. Eds. C. Earle, K. Mathewson, and M. S. Kenzer. Maryland: Rowman & Littlefield.

Dunbar, G. S. 1981. *Geography in the University of California (Berkeley and Los Angeles) 1868–1941*. Published by the author, printed by De Vorss, Marina del Rey, California. Revised ed. 1996.

———. 2000. "Owen Lattimore, 1900–1989." *Geographers: Biobibliographical Studies* 20:24–42.

Eyre, J. D. 1978. *A Man for All Regions: The Contributions of Edward L. Ullman to Geography*. University of North Carolina at Chapel Hill, Department of Geography, Studies in Geography No. 11.

Fagan, B. 2000. *The Little Ice Age*. New York: Basic Books.

Gaile, G. L. and C. J. Willmott. 1989. *Geography in America*. Columbus, Ohio: Merrill.

———. 2003. *Geography in America at the Dawn of the 21st Century*. Oxford: Oxford University Press.

Garcia-Ramon, M. D. 2003. "Globalization and International Geography: The Questions of Languages and Scholarly Traditions." *Progress in Human Geography* 25:1–5.

Glick, T. F. 1988. "Before the Revolution: Edward Ullman and the Crisis of Geography at Harvard, 1949–1950." Pp. 49–62. In Harmon and Rickard, 1988. *Geography in New England.* A special publication of the New England/St. Lawrence Valley Geographical Society.

Gould, P. R. 1979. "Geography, 1957–1977: The Augean Period." *Annals AAG* 69:139–150.

———. 1985. *The Geographer at Work.* London: Routledge and Kegan Paul.

Gould, P. R. and F. R. Pitts. 2002. *Geographical Voices: Fourteen Autobiographical Essays.* Syracuse: Syracuse University Press.

Gregory, D. 1989. "Areal Differentiation and Post-Modern Human Geography." Pp. 67–96. In D. Gregory and R. Walford, eds. *Horizons in Human Geography.* London: MacMillan.

Gregory, S. 1963. *Statistical Methods and the Geographer.* London: Longman.

Grosvenor, G. M. 1995. "Geography in the Public Eye." *The Washington Post.* July 28:A26.

Grove, J. M. 1988. *The Little Ice Age.* London: Routledge.

Hägerstrand, T. 1967. *Innovation Diffusion as a Spatial Process.* Chicago: University of Chicago Press.

Haggett, P. 1996. "Geographical Futures: Some Personal Speculations." P. 70. In *Companion Encyclopedia of Geography: The Environment and Humankind.* Eds. I. Douglas, R. Huggett, and M. Robinson. London: Routledge.

———. 2003. "Peter Robin Gould, 1932–2000." *Annals AAG* 93, 4:925–934.

Hanson, S. 1999. "Isms and Schisms: Healing the Rift Between the Nature-Society and Space-Society Traditions in Human Geography." *Annals AAG* 89:133–143.

Harris, C. D. 2001. "English as International Language in Geography: Development and Limitations." *The Geographical Review* 91:675–689.

Hart, J. F. ed. 1972. *Regions of the United States.* New York: Harper and Row.

———. 1975. *The Look of the Land.* Englewood Cliffs, N.J.: Prentice-Hall.

———. 1982. "The Highest Form of the Geographer's Art." *Annals AAG* 72:1–29.

Hartshorne, Richard. 1939. *The Nature of Geography, a Critical Survey of Current Thought in the Light of the Past.* Lancaster, Penn.: Association of American Geographers.

———. 1959. *Perspective on the Nature of Geography.* Chicago: Rand McNally.

Harvey, D. 1969. *Explanation in Geography.* New York: St Martin's Press.

———. 1973. *Social Justice and the City.* London: Edward Arnold.

———. 1989. *The Condition of Postmodernity.* Oxford: Basil Blackwell.

———. 2002. "Memories and Desires." In P. Gould and F. R. Pitts, eds., *Geographical Voices.* Syracuse, N.Y.: Syracuse University Press, 166.

Isard, W. 1956. *Location and Space Economy.* New York: John Wiley and Sons.

———. 1975. *An Introduction to Regional Science.* Englewood Cliffs, N.J.: Prentice-Hall.

James, P. E. 1935. *An Outline of Geography.* Boston: Ginn and Co.

James, P. E. and C. F. Jones, eds. 1954. *American Geography: Inventory and Prospect.* Syracuse, N.Y.: Syracuse University Press.

James, P. E. and G. J. Martin. 1978. *The Association of American Geographers: The First Seventy-Five Years, 1904–1979.* Washington, D.C.: Association of American Geographers.

Kates, R. W. 1971. "Natural Hazard in Human Ecological Perspective: Hypotheses and Models." *Economic Geography* 47:438–451.

Kenzer, M. S. 1983. "The Formation of an Independent Department of Geography at Northwestern University: A Chapter in the Growth of American Geography Following World War II." *History of Geography Newsletter* 3:30–37.

———, ed. 1987. *Carl O. Sauer: A Tribute.* Corvallis: Oregon State University Press.

Kirk, W. 1952. "Historical Geography and the Concept of the Behavioural Environment." *Indian Geographical Journal* 25:152–160.

Kissinger, H. 1999. *Years of Renewal*. New York: Simon & Schuster.

Koelsch, W. A. 1987. *Clark University: A narrative History, 1887–1987*. Worcester, Mass.: Clark University Press.

———. 1988. "Geography at Clark: The First Fifty Years, 1921–1971." Pp. 40–48. In Harmon and Rickard, 1988. *Geography in New England*. A special publication of the New England/St. Lawrence Valley Geographical Society.

———. 2001. "Academic Geography, American Style: An Institutional Perspective." Pp. 245–279. In *Geography: Discipline, Profession and Subject since 1870*. Ed. G. S. Dunbar. Dordrecht, Netherlands: Kluwer Academic Publishers.

Landes, D. 1998. *The Wealth and Poverty of Nations*. New York: Norton.

Leighly, J. 1954. "Innovation and Area." *Geographical Review* 44:439–441.

Lewis, M. W. and K. Wigen. 1997. *The Myth of Continents: A Critique of Metageography*. Berkeley: University of California Press.

Livingstone, D. N. 1981. *Nathaniel Southgate Shaler and the Culture of American Science*. Tuscaloosa, AL: University of Alabama Press.

Lowenthal, D. 1958. *George Perkins Marsh, Versatile Vermonter*. New York: Columbia University Press.

———. 1961. "Geography, Experience and Imagination: Towards a Geographical Epistemology." *Annals AAG* 51:241–260.

———. 2000. *George Perkins Marsh: Prophet of Conservation*. Seattle: University of Washington Press.

McNee, R. B. 1973. "Does Geography Have a Structure? Can It Be 'Discovered'? The Case of the High School Geography Project." Pp. 285–313. In R. J. Chorley, ed. *Directions in Geography*. London: Methuen.

Martin, G. J. 1988. "Geography, Geographers and Yale University, c. 1770–1970." Pp. 2–9. In Harmon and Rickard, 1988. *Geography in New England*. A special publication of the New England/St. Lawrence Valley Geographical Society.

———. 1989. "*The Nature of Geography* and the Schaefer-Hartshorne Debate." Pp. 69–90. In *Reflections on Richard Hartshorne's The Nature of Geography*. Occasional publication of the Association of American Geographers.

———. 2003. "The History of Geography." Pp. 550–561. In Gaile and Willmott, eds. 2003. *Geography in America at the Dawn of the 21st Century*. Oxford: Oxford University Press.

Mathewson, K. and M. S. Kenzer, eds. 2003. *Culture, Land and Legacy: Perspectives on Carl Sauer and Berkeley School Geography*. Baton Rouge, La.: Geoscience Publications, Louisiana State University.

Meinig, D. W., ed. 1971. *On Geography: Selected Writings of Preston E. James*. Syracuse, N.Y.: Syracuse University Press.

———. 1987. *The Shaping of America: A Geographical Perspective on 500 Years of History. Vol. 1. Atlantic America, 1492–1800*. New Haven, Conn.: Yale University.

Mikesell, M. 2002. "Finch, V. C. and Trewartha, G. T. 1949. *Elements of Geography: Physical and Cultural*." 3rd ed. *Progress in Human Geography* 26, 3:401–404.

Monk, J. 1998. "The Women Were Always Welcome at Clark." *Economic Geography* 78 (extra issue): 14–30.

———. 2001. "Gender and Feminist Studies in Geography." In N. J. Smelser and P. B. Baltes, eds. *International Encyclopedia of Social and Behavioral Sciences*. Amsterdam: Elsevier Science. 9:5924–5929.

Morrill, R. L. 1984. "Recollections of the 'Quantitative Revolution's Early Years: The University of Washington 1955–1965.'" Pp. 57–72. In M. Billinge, D. Gregory and R. Martin, eds. *Recollections of a Revolution*. London: Macmillan.

————. 1991. "Quantitative Revolution." Pp. 142–143. In G. S. Dunbar, ed. *Modern Geography: An Encyclopedic Survey*. New York: Garland Publishing.

Moss, R. J. 1995. *The Life of Jedidiah Morse: A Station of Peculiar Exposure*. Knoxville: University of Tennessee Press.

Murphy, A. B. 2003. "Rethinking the Place of Regional Geography." (President's Column). *AAG Newsletter* 9:3.

Olmstead, C. W. 1987. *Science Hall: The First Century*. Madison, Wis.: Department of Geography, University of Wisconsin.

Olwig, K. 2003. "Forum: The Lowenthal Papers." *Annals AAG* 93, 4:851–885.

Pattison, W. D. 1970. "The Producers: A Social History." From *Geographic Discipline to Inquiring Student, Final Report on the High School Geography Project*. Washington, D.C.: 57–169.

Peet, R., ed. 1977. *Radical Geography: Alternative Viewpoints on Contemporary Social Issues*. Chicago: Maaroufa.

Platt, R. S. 1942. *Latin America, Countrysides and United Regions*. New York: McGraw-Hill.

Pyne, S. E. 1980. *Grove Karl Gilbert: A Great Engine of Research*. Austin: University of Texas Press.

Rediscovering Geography Committee. 1997. *Rediscovering Geography: New Relevance for Science and Society*. Washington, D.C.: National Academy Press.

Relph, E. 1991. "Post-Modern Geography." Review Essay. *The Canadian Geographer* 35, 1:98–105.

Robinson, D. J. 1980. *Studying Latin America: Essays in Honor of Preston James*. Syracuse, N.Y.: Department of Geography, Syracuse University.

Rose, G. 1993. *Feminism and Geography*. Cambridge: Polity Press.

Rowley, V. M. 1964. *J. Russell Smith: Geographer, Educator, and Conservationist*. Philadelphia: University of Pennsylvania Press.

Ryan, B. 1983. *Seventy-Five Years of Geography at the University of Cincinnati*. Cincinnati: The University of Cincinnati, Department of Geography.

Saarinen, T. F. 1969. *Perception of Environment*. Washington, D.C.: Association of American Geographers.

Sachs, J. D. 2000. "The Geography of Economic Development." United States Naval War College: Jerome E. Levy Occasional Paper in Economic Geography and World Order 1:9.

Sack, R. D. 1980. *Conceptions of Space in Social Thought*. London: Macmillan.

————. 1986. *Human Territoriality: Its Theory and History*. Cambridge: Cambridge University Press.

Said, E. 1978. *Orientalism*. New York: Harper.

Salter, C. L. 1999. "The Enduring Nature of Evocative Regional Geography." *The North American Geographer* 1, 1:4–22.

Sauer, C. O. 1956. "Retrospect." In W. L. Thomas, ed. *Man's Role in Changing the Face of the Earth*. Pp. 1131–1135. Chicago: University of Chicago Press.

Schaefer, F. 1953. "Exceptionalism in Geography. A Methodological Examination." *Annals AAG* 43:226–249.

Scheidegger, A. E. 1961. *Theoretical Geomorphology* (revised 1970). Berlin: Springer-Verlag.

Siddall, W. R. 1961. "Two Kinds of Geography." *Economic Geography* 37:189.

Smith, N. 1987. "Academic War over the Field of Geography: The Elimination of Geography at Harvard, 1947–1951." *Annals AAG* 77, 2:155–172.

Soja, E. 1987. "The Postmodernization of Geography: A Review Essay." *Annals AAG* 77:289–296.

Stoddart, D. R. 2000. "Becoming—and being—Peter Gould." *The Geographical Review* 90, 2:238–247.

Stoltman, J. P. 1980. "Round One for HSGP: A Report on Acceptance and Diffusion." *Professional Geographer* 23:209–215.

Stone, K. H. 1979. "Geography's War-Time Service." *Annals AAG* 69, 1:95.

Taafe, E. J., ed. 1970. *Geography: Report of the Behavioral and Social Sciences Survey.* Englewood Cliffs, N.J.: Prentice Hall.

Taylor, T. G., ed. 1951. *Geography in the Twentieth Century.* London: Methuen.

Thomas, W. L., ed. 1956. *Man's Role in Changing the Face of the Earth.* Chicago: University of Chicago Press.

Thrift, N. 1983. "On the Determination of Social Action in Space and Time." *Environment and Planning D: Society and Space* 1:23–57.

Tuan, Yi-fu. 1999. *Who Am I?: An Autobiography of Emotion, Mind, and Spirit.* Madison: University of Wisconsin Press.

Turner, B. L., II, Clark, W. C., Kates, R. W., Richards, J. F., Mathews, J. T., and Meyer, W. B., eds. 1990. *The Earth as Transformed by Human Action: Global and Regional Changes in the Biosphere over the Past 300 Years.* Cambridge: Cambridge University Press.

U.S. Department of Education. 1992. *America 2000: An Education Strategy.* Washington, D.C.: Department of Education.

Walker, W. J., ed. 1991. *Schwendeman's Directory of College Geography of the United States.* Richmond: Geographical Studies and Research Center, Eastern Kentucky University.

Warntz, W. 1981. "*Geographia Generalis* and the Earliest Development of American Academic Geography." Pp. 245–263. In Blouet, ed. 1981. *The Origins of Academic Geography in the United States.* Hamden, Conn.: Archon Books.

———. 1989. "Newton, the Newtonians, and the *Geographia Generalis Varenii.*" *Annals AAG* 79:165–191.

White, G. F., ed. 1974. *Natural Hazards: Local, National, Global.* New York: Oxford University Press.

White, G. F. and J. E. Haas. 1975. *Assessment of Research on Natural Hazards.* Cambridge, Mass.: Massachusetts Institute of Technology Press.

Whittlesey, D. 1954. "The Regional Concept and Regional Method." Pp. 19–69. In P. E. James and C. F. Jones, eds. *American Geography: Inventory and Prospect.* Syracuse, N.Y.: Syracuse University Press.

Wright, J. K. 1947. "Terrae Incognitae: The Place of the Imagination in Geography." *Annals AAG* 37, 1:15.

Zelinsky, W. 1973. *The Cultural Geography of the United States.* Englewood Cliffs, N.J.: Prentice Hall.

———. 1973. "Women in Geography: A brief factual account." *The Professional Geographer* 25, 2:151–165.

18

APPLIED GEOGRAPHY

Interest in planning for the classification and use of the land and other natural resources is not new to our science; in fact, it is one of the most persistent interests in American geography.

—Charles C. Colby, in his presidential address to the
Association of American Geographers, 1936

The period from World War I to the decade of the 1950s witnessed a notable increase in the application of geographical knowledge and skills to the study of practical problems of government and business.[1] Of course, there has never been a time when the search for knowledge about the earth as the home of man has not been undertaken for practical purposes as well as for the satisfaction of intellectual curiosity. There are few fields of learning in which the relevance of concepts and specialized methods to practical needs can be more clearly demonstrated. We have many examples of such applications, as when Strabo wrote his geography for the use of Roman administrators, or when Varenius wrote his special studies of Japan and Siam for the merchants of Amsterdam, or when Maury made use of his wind and current charts to formulate improved sailing directions. One of the distinctive characteristics of American geography has been the tradition of the resource inventory—studies of this sort were made long before Thomas Jefferson gave Lewis and Clark specific instructions about the kinds of information needed; such practical purposes motivated the Great Surveys of the American West and the numerous surveys of resources along proposed rail lines. But the men who led these surveys were not trained as geographers—they had to work out their own objectives and methods as the work proceeded.

In addition to the work of the multiple geological surveys was the work of the Coast Survey, the General Land Office, the Lake Survey, Pacific Railroad Survey, and many more. When the Association of American Geographers was formed in 1904, fully one-third of its membership (of 48 persons) were applied geographers. Then, at a time when it was unfashionable to grant women university posts, a number of women distinguished themselves in the applied sector, including Gladys Wrigley, Helen Strong, Clara Le Gear, Evelyn Pruitt, Betty Didcoct Burrill, and Dorothy A. Muncy. Of course, many more did excellent work, but they are not as well known.

---

[1]Two of the earliest books to adopt the title "applied geography" are J. Scott Keltie, *Applied Geography; A Preliminary Sketch*, 1890 (second edition 1908) and Alexander Stevens, *An Introduction to Applied Geography*, 1921.

As the number of professionally trained geographers gradually increased, a certain proportion of the younger generation—then as now—expressed impatience with theoretical studies and demanded that geographic investigations be clearly relevant to the practical problems involving public or private policy. In the 1920s and 1930s some geographers were disenchanted with the experiments with field methods applied to small areas, and a demand arose for some tangible connection with the "overriding" economic, social, or political problems of the day. There was no clear answer to the question of how one might demonstrate a tangible connection, so each scholar who decided to use geographical studies for practical ends had to formulate his own justification. The result was the appearance of a wide variety of studies in what might be called applied geography, in the sense that the purpose was to provide the basis for planning remedial action.

The first large breakthrough in the use of professionally trained geographers to study practical problems came during World War I and its aftermath. During the 1920s and 1930s not only were certain wartime projects continued and completed but also new kinds of applied research were undertaken. A large number of geographers were called into both military and civilian service during World War II, and since then opportunities for employment in various branches of government as well as in private business firms have grown rapidly. Many studies involving the geographical analysis of location or of areal spread are still made by non-geographers, but in the contemporary period there is a growing appreciation of the value of the professional training that geographers have.

IN WORLD WAR I

With U.S. participation in World War I and the peace conference that followed, 51 association members contributed their talents. Five members (W. W. Atwood, H. A. Gleason, H. E. Gregory, A. E. Parkins, and R. H. Whitbeck) offered instruction to groups of soldiers and the Students Army Training Corps (SATC) program; five members (C. F. Brooks, A. J. Henry, A. McAdie, J. Warren Smith, and R. DeC. Ward) contributed research findings on aerography, meteorology, and climatology; 10 members (N. A. Bengtson, A. H. Brooks, R. M. Brown, G. E. Condra, N. H. Darton, G. R. Mansfield, F. E. Matthes, O. E. Meinzer, P. S. Smith, and T. W. Vaughan) pursued research concerning the location of minerals; four members (S. W. Cushing, J. W. Goldthwait, E. Huntington, and L. Martin) were attached to intelligence; H. Bingham advised on aviation; W. Bowie advised on map projections; E. E. Free advised on gases; W. Churchill and D. W. Johnson wrote propaganda tracts; and G. E. Nichols functioned as botanical adviser on sphagnum moss for the American Red Cross (much used as a substitute for cotton in absorbent surgical dressing). At this time the international significance of commodities began to make itself felt.

Commodity Studies

One of the first uses of the knowledge and skills of trained geographers was in connection with certain commodity studies. In the fall of 1917 the Division of Planning and Statistics of the U.S. Shipping Board was asked to make a survey of imports. The problem was that there were not enough ships to carry the supplies of war in addition to normal trade. The board was asked to classify all imports into one of three categories: (1) those commodities so necessary to the war effort that all available supplies must be imported; (2) those commodities essential to the war effort, but not in such short supply that all available production had to be imported; and (3) those commodities not required for military or civilian needs. As usual in such wartime agencies, the answer was expected the next morning. The chairman of the research group was Dean Edwin F. Gay of the Harvard Business School. By chance Walter S. Tower, the Chicago geographer, was in Washington, D.C., at the time, and he was asked to help in this rush job because he had been teaching a course at Chicago in the geography of commerce. It is reported that "by midnight Gay realized that Tower knew more about commodities and where they could be obtained than any of the others" (Colby, 1955:13). The next day Tower was appointed head of the Commodities Section of the Shipping Board and was asked to bring together a staff of experts in this field.[2] Among others he selected Vernor C. Finch of Wisconsin, coauthor with Oliver E. Baker of *Atlas of World Agriculture* (Finch and Baker, 1917); William H. Haas of Northwestern University, a former student at Chicago; and George B. Roorbach, an assistant professor of economic geography at the University of Pennsylvania. Later Charles C. Colby of Chicago also joined the staff. These university people found what so many have had to rediscover in later years—that intelligence reports must be summarized in simple, unambiguous language on page one. Colby turned out to be especially adept at extracting the "meat" from a report and writing useful summaries. Those who remember his presence at many later meetings of geographers recall his amazing ability to summarize long and involved discussions in terse sentences so clear that discussants wondered what they had been arguing about. Colby became a very effective writer of proposals to be submitted for financing to government agencies, an ability that he no doubt possessed before 1917, but it certainly improved with his wartime experience.

The discovery that geographers had useful skills for commodity studies led to the establishment of another Division of Planning and Statistics, this time in the War Trade Board, also under Dean Gay. The geographer selected to head the research unit was Harlan H. Barrows of Chicago. For his staff Barrows recruited J. Russell

---

[2]Walter S. Tower received the B.A. degree from Harvard in 1903 and the M.A. in 1904. In 1906 he completed the Ph.D. at the University of Pennsylvania with a dissertation, "A Regional and Economic Geography of Pennsylvania." He taught economic geography at the University of Pennsylvania from 1906 to 1911, at which time he became a member of the Department of Geography at the University of Chicago. He was promoted to professor in 1916, but in 1917 he was granted leave of absence to work for the War Shipping Board in Washington. He never returned to a university post. From 1919 to 1921 he was trade adviser to the Consolidated Steel Company; from 1921 to 1924 he was the U.S. commercial attaché in London; from 1924 to 1933 he was with the Bethlehem Steel Company; from 1933 to 1940 he was executive secretary of the American Iron and Steel Institute; and from 1940 to 1952 he was president of the institute. He retired in 1952 and lived in Carmel, California, until his death at age 88 in 1969.

Smith of the University of Pennsylvania, Ray H. Whitbeck of Wisconsin, and Nels
A. Bengtson of Nebraska. Other work, too, was undertaken in conjunction with the
war effort.[3]

The Inquiry

In September 1917 President Wilson directed his close friend and adviser, Col. Edward
M. House, to set up an organization that would gather the most complete collection of information possible and prepare it for use at the coming Paris Peace
Conference. Under the direction of President S. E. Mezes of the College of the City
of New York, some 150 persons were recruited, including historians, economists,
geographers, journalists, and others with special knowledge of particular areas. The
group, which became known as The Inquiry, carried on its research in the building
of the American Geographical Society in New York City. The work of The Inquiry
was made possible by its access to the library and map collections of the society.[4]

The subjects studied by The Inquiry included the political and diplomatic history of Europe; international law, including the geographic interpretation of problems of territorial waters and of interconnections across frontier zones; economics
and economic geography; physiography in relation to strategic boundaries; and many
other more detailed investigations of major problem areas where plebiscites were
to be carried out. A major part of the work was in the field of cartography, in which
a map-making program of unprecedented size and detail was undertaken. First, a
set of new base maps was made, showing prewar political boundaries, the complete
drainage system, the roads and railroads, and the cities and towns. Some of the maps
were on scales of 1/1,000,000 or 1/3,000,000. These were the general maps of Europe
as a whole and the somewhat more detailed maps of the Balkans. But there were
also a great number of very large-scale maps, such as the map of Alsace-Lorraine
on a scale of 1/250,000. For several places of critical importance, block diagrams
were drawn showing the geological structure on the sides of the block and the
terrain on its surface. All these maps were available at the Paris Peace Conference
to be used for the study of various boundary proposals. On these maps information
was plotted to show population density, ethnic composition, agriculture, industrial
centers, mineral resources, and many other things needed by the Paris Peace

---

[3] W. M. Davis prepared *A Handbook of Northern France* (1918) describing the terrain
features of the war zones, illustrated with his incomparable pen and ink sketches. The
book was printed in pocket size. Some 400 copies were supplied to infantry officers at
the front, while an additional 5000 copies were supplied to YMCA libraries for the use
of troops. Ellsworth Huntington and H. E. Gregory wrote and edited *The Geography of
Europe* (New Haven, Conn.: Yale University Press, 1918), to which 17 other geographers
made contributions and which was intended to serve as an up-to-date textbook for courses
in the SATC in the universities. The work was sponsored by the National Research
Council.

[4] "The American Geographical Society's Contribution to the Peace Conference,"
Geographical Review 7(1919):1–10. See also Lawrence E. Gelfand (1963) and Arthur
Walworth (1976).

Conference (Wright, 1952:200). The maps were also made available to universities in the United States, where courses in war aims were being offered.

Isaiah Bowman, who had been the director of the American Geographical Society since 1915, supervised the geographical studies. He brought together a number of geographers to work with him. Mark Jefferson, his former teacher at Ypsilanti, became chief cartographer.[5] Most of the geographers were assigned to study various topics and regions of Europe. Fortunately, the society had already published the results of the relation between language and the division of that continent into separate states, which had been done by Leon Dominian (1917). But much more detailed information would be needed when new political boundaries were to be drawn. A few members of The Inquiry worked on the collection of material concerning other parts of the world: Bailey Willis reported on the problem areas of Latin America, especially on the background of the Tacna-Arica dispute between Peru and Chile; H. L. Shantz sought information about the plant resources of Africa; C. F. Marbut was assigned the task of compiling a soil map of Africa; and J. Warren Smith (of the U.S. Weather Bureau) prepared maps of climatic elements.

The Paris Peace Conference

On December 4, 1918, The Inquiry specialists, their assistants (together with the materials they had gathered), and numerous officials, including President Woodrow Wilson, sailed for France on U.S.S. *George Washington*. At Paris Bowman was given the title of chief territorial specialist of the American Commission to Negotiate Peace; Mark Jefferson was appointed chief cartographer (C. Stratton and A. K. Lobeck were his assistants). At Paris the map became everything, and it is said that there was coined the aphorism "One map is worth ten thousand words." The map became the international language of the conference, and the Americans were best prepared to make these maps. Copies of the American-made maps concerning European matters were reduced in size and entered in the Black Book, and maps of colonial matters were entered into the Red Book of the American delegation. These books were constantly consulted by leaders and diplomats of many of the delegations at the conference (Martin, 1966). With the maps and the abstract principle of Wilsonian justice, some 3000 miles of new boundaries were created for the former states of Central Europe (Rhoads, 1954) (Fig. 50).

Wartime Projects Published Later

The information gathered and digested during the war and its aftermath was not entirely buried in government documents. Isaiah Bowman wrote a book concerning the problem areas of the world that remained for many years the most authoritative study in political geography—*The New World* (1921). In this book

---

[5]For biographical data on Bowman and Jefferson, see Chapter 15. Other geographers who worked on The Inquiry were O. E. Baker, N. M. Fenneman, W. L. G. Joerg, C. F. Marbut, E. C. Semple, H. L. Shantz, and B. Willis. Army officers assigned to The Inquiry were Major D. W. Johnson and Major Lawrence Martin as well as two nongeographers, Captain W. C. Farabee and Captain S. K. Hornbeck.

Figure 50 Territorial specialists (and others) meet with General Le Ronde (*extreme right*) at the Paris Peace Conference, 1918–1919. (*Front, second from left*) Isaiah Bowman, (*second from right*) Cambon, talking with General Le Ronde

Bowman did not attempt to build a theory of political geography, but rather to analyze in informative detail the particular problems of particular regions, with adequate description of the local setting and the historical background so that the reader understood what was going on in the postwar world. It served its purpose admirably.[6] H. L. Shantz and C. F. Marbut also published a monograph on the vegetation and soils of Africa, which for many years remained a definitive statement on that subject (Shantz and Marbut, 1923). D. W. Johnson's report on the relation of military strategy and tactics to the terrain features was published in 1921. This was the first substantial contribution by an American to military geography (Johnson, 1921).

For many years after the war Bowman and the American Geographical Society were involved in boundary studies and a program of mapping. While the Paris Peace Conference settled the affairs of Europe, the Latin American countries suffered poorly marked boundaries. Guatemala and Honduras asked the United States to settle their dispute over their common boundary; Robert Lansing, the secretary of state, turned to the man with whom he had dealt on such matters in Paris. He asked Bowman to

---

[6]The Department of State placed a copy of the book in each of the U.S. consular offices around the world, and the Carnegie Endowment for International Peace distributed copies to the leading centers of teaching and research in international affairs (Wright, 1952:255).

arrange a study of the Guatemala-Honduras boundary and suggest a solution. The American Geographical Society organized a research team under the direction of Major Percy H. Ashmead to make a map of the area, showing not only the details of the terrain but also the distribution of people and their ways of using the land. The survey was made in 1919, and a suggested solution was submitted to Mr. Lansing. The negotiations took 14 years, but the settlement (which was accepted in 1933) was based on the maps and recommendations of the society's report.

The Millionth Map of Hispanic America

Bowman's work in Peru as well as his experience with the Guatemala-Honduras problem brought to light a real lack of geographical information on Latin America as a whole. There was no reliable map of the region. With the methods of surveying then in use, it would have taken many decades and vast sums of money to produce a useful map. But Bowman knew that a large number of original surveys in manuscript form had been made by private companies for a variety of purposes. He proposed that the American Geographical Society undertake a major research program leading to the compilation of a map of Hispanic America on a scale of 1/1,000,000, conforming to the standards and format of the International Map of the World that was originally proposed by Albrecht Penck. Raye R. Platt, reviewing the completion of the Millionth Map in 1946, of which he had been director for 15 years, quoted the annual report of the council of the American Geographical Society for 1920 as follows:

> The first step in the development of this program aims at the review and classification of all available scientific data of a geographical nature that pertains to Hispanic America. . . . The work will involve the compilation of maps— topographic and distributional—on various scales, but always including sheets on the scale of 1/1,000,000 which will conform to the scheme of the International Map. . . . The undertaking is an ambitious one, but the Society is happy to say that assurances of cooperation have been given by the whole group of Hispanic American countries in a cordial spirit that augurs well not only for the immediate scientific results but also for the fostering of mutual understanding and sympathetic relations toward which the field of geography offers a peculiarly fortunate approach (Platt, 1946:2).

As part of the map of Hispanic America project, the society supported the publication of a series of research studies. Some were based on field surveys, such as the reports on European colonies in Chile, Argentina, and Brazil by Mark Jefferson (1921, 1926); the studies of land settlement problems in Mexico and Chile by George M. McBride (1923, 1936); or the additional studies of Peru carried out by O. M. Miller (1929). The study of the central Andes by Alan G. Ogilvie, the Scottish geographer, was compiled from the drawings of the Millionth Map and from Bowman's copious notes (Ogilvie, 1922).[7] The map was used to help adjudicate disputes between Chile and Peru in 1925, Bolivia and Paraguay in 1929, Colombia and Peru in 1932, and Colombia and Venezuela in 1933. And the Hispanic America Map, completed

---

[7]Also based on field study was Bowman's *Desert Trails of Atacama* (Bowman, 1924).

in 1946 in 107 sheets (more than 300 square feet in extent), was a contribution to the Millionth Map of the World, then in progress.

LAND CLASSIFICATION STUDIES

Another quite different application of geography to the solution of practical problems has to do with studies of land quality and land use. It had long been recognized that detailed information concerning land resources was needed if plans for better resource use were to be properly guided. The destruction of the land through improper use, which the public began to talk about in the early 1970s, had been reported by geographers a century before—notably by George Perkins Marsh and Nathaniel Southgate Shaler. Ratzel used the expressive German term *Raubbau*, or robber economy, to describe a form of land use that destroys the land base. As noted in Chapter 7, efforts to classify land in terms of its potential use began in the early days of the independent United States and were carried to new levels of utility in the Great Surveys of the West and in the work of the U.S. Geological Survey in the latter part of the nineteenth century. Major steps were taken, however, to enlarge the scope of land classification studies and to improve methods during the 1920s and 1930s. The discussions of method in the annual field conferences led directly to practical applications, especially in the program of resource inventory carried out in the state of Michigan. So important was the work of the Michigan Land Economic Survey that the record needs to be presented in some detail.

The Michigan Land Economic Survey

To understand the problems of public policy that made the Land Economic Survey important we must review the conditions of land and land use in Michigan at the end of World War I. From the point of view of its natural features, the state of Michigan can be divided into two quite different parts. To the south of a line drawn roughly from Saginaw Bay on the east to Muskegon on the west (Fig. 51), Michigan is a part of the productive agricultural plains of the Middle West. The soils are mostly loams, in only a few small spots is the surface too steep for cultivation, and the growing season is long enough to permit the ripening of grain crops. But north of this line the physical character of Michigan is very different. Here the land is made up of a deep accumulation of glacial deposits—moraines, till plains, and sandy outwash plains. The Upper Peninsula is similar to the northern part of the Lower Peninsula as far west as Marquette. West of this city the knobby crystalline rock hills of the Canadian Shield form the surface. Whereas the southern part of Michigan was once covered by a broadleaf forest similar to that of Ohio and eastern Indiana, the north country was covered by one of the finest stands of white pine to be found in America. The white pine, intermingled with broadleaf species, formed a dense and almost unbroken forest cover. When settlers came into southern Michigan, they cleared the broadleaf forests and established farms, creating what became a part of the Hay and Dairy Belt. But the first penetration of the country north of Saginaw Bay, which came after the Civil War, was based on

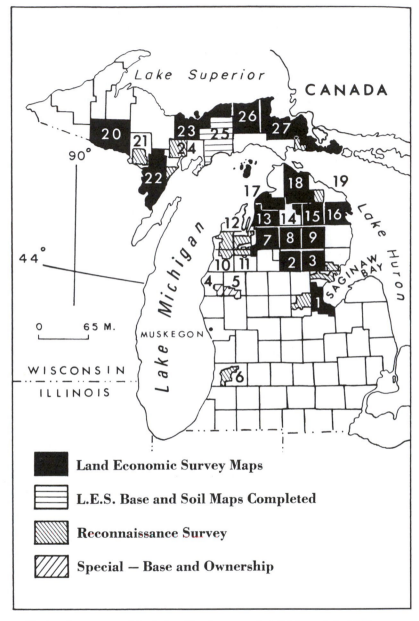

Figure 51 Areal coverage of land classification maps in Michigan, July 1939

1. Bay	8. Crawford	15. Montmorency	22. Menominee
2. Roscommon	9. Oscoda	16. Alpena	23. Alger
3. Ogemaw	10. Benzie	17. Emmet	24. Delta
4. Mason	11. Grand Traverse	18. Cheboygan	25. Schoolcraft
5. Lake	12. Leelanau	19. Presque Isle	26. Luce
6. Alligan	13. Antrim	20. Iron	27. Chippewa
7. Kalkaska	14. Otsego	21. Dickinson	

lumbering rather than farming. From the stands of white pine came the wood that was used to build most of the houses and other structures of the prairie states.

In those days no questions were raised about the way private interests made use of natural resources. The lumberjack was there to cut down trees and transport the logs to the sawmills. No one thought of requiring the replanting of cutover lands, and the lumberjack removed all the trees, leaving none to provide for reseeding. The slashings left after the tree trunks were cleared of branches were left lying on the ground, and the wood-burning locomotives were not equipped with screens to catch sparks. Forest fires were frequent and destructive, especially when there were high winds after a long dry period. On October 8, 1871, on the same day that the famous Chicago fire started, a forest fire started near Petoskey on Lake Michigan. The same conditions that made the Chicago fire so destructive—high winds after a long period of drought—made the forests almost ready to explode. During the following days the fire swept on a broad front all the way across the Lower Peninsula to the shores of Lake Huron. It has been estimated that more trees were destroyed by fire than were cut by the lumberjacks' axes.

When forest fires and lumbering had completed their work, no trees were left on the completely denuded land. A scrubby second growth of brush appeared in some places; in others there was only a growth of low plants that barely covered the charred remains of the forest. The large lumber companies tried to sell their land to farmers, and it was actually offered for sale at $2 an acre. Some farmers thought this was too good a bargain to miss. The result was a thin scattering of isolated farms and a few small towns. But large parts of the lumber properties could not be sold and were simply abandoned. In Michigan, when an owner fails to pay his taxes for seven years, the land reverts to state ownership. By 1910 not only had the lumber companies moved away, but even some of the farmers had found nothing but disillusionment in the poor sandy soils and the short growing season. Only in widely scattered localities were there small groups of rural people struggling to hold on. Already there were some people who believed that the greatest values to be found in the north country were in the wild game animals—fish, game birds, and deer. The problem of idle land was already critical because the people of the southern part of the state were forced to carry the tax burden of supporting the scattered settlements of the north.

Two men provided the leadership that resulted in effective action. One was Carl O. Sauer, a member of the staff of the Department of Geology and Geography at the University of Michigan. Sauer's interest in the field survey of land quality and use dated back to his days as a graduate student at Chicago and his association there with Wellington D. Jones, recently returned from a survey of northern Patagonia with Bailey Willis. When Sauer joined the Michigan faculty in 1915, he found the state facing a serious practical problem but lacking the kind of specific information needed to formulate a remedial policy. The agricultural experts wanted only to find ways to make farming pay; the forestry experts wanted only to plant trees; and the hunting and fishing clubs wanted to keep everyone else out. People who were familiar with the character of the north country knew that it contained a great diversity of physical conditions that made the adoption of any one general policy impossible. There was need for just the kind of attention to the mapping of significant differences from place to place that geographers in that period had been talking about. Sauer's plan for a land classification survey represented the application of ideas generated in professional discussion to real practical problems (Sauer, 1919, 1921).

At Ann Arbor Sauer met a forester and naturalist named Parish S. Lovejoy, who was also deeply concerned about the problem of the cutover lands. Lovejoy had the knowledge and the commitment to become a kind of gadfly to stir various groups into action—the meetings of scientists, the hunting clubs, the members of the state legislature. Speaking before such groups, Lovejoy pointed out eloquently that a third of the state of Michigan was bankrupt and that the situation was spreading. Some kind of public policy, he insisted, must be adopted, and quickly. But no policy will be worth anything, he continued, unless it is based on an accurate and detailed knowledge of the relevant facts. The gist of what Lovejoy had to say is contained in the annual report of the Michigan Academy of Sciences for 1921.[8] At a special session of the Academy on "Michigan's Idle Lands," held in 1920, the nature of the critical situation in the north country was spelled out, and a program of action was presented. A resolution was adopted and sent to the State Department of Conservation recommending immediate action on setting up a land survey.

The newly formed Department of Conservation took the recommended action. Securing the cooperation of the U.S. Department of Agriculture, the University of Michigan, and the Michigan Agricultural College (now Michigan State University), funds were made available for an experimental field study of Charlevoix County to be carried out in the summer of 1922. The results proved the value of the information to be obtained, and the Michigan Land Economic Survey was created. The survey was directed to make a detailed inventory of the cutover lands, together with reports on the current economic situation by counties. Because the inventory of land and land use was to be used not only as a basis for developing some kind of policy but also to guide programs of land management, the mapping had to be done in great detail; yet, since there was no previous experience to indicate the best method to use, much of the work had to be improvised to meet clear and specific needs. One of the field surveyors observed about the procedures:

> Field operations in the first year were experimental, and many changes [in categories to be mapped] were later made. . . . Field crews consisted of a cover and base mapper together with a soil and slope mapper. Most of the field time was taken up with boundary delineation and the determination of soil types. One section [one square mile] per day was considered a good field accomplishment. The instruments were a compass together with a soil augur; the only basic maps were the General Land Office plats, more than half a century old, showing the section corners of which little evidence remained, and drainage features where these crossed the section lines. . . . The original field scale was eight inches to the mile, subsequently changed to four inches (quoting Horace Clark in Davis, 1969:18–19).

After the first few field sessions a more-or-less established procedure was followed (Barnes, 1929). The work was done by soil specialists, foresters, and

---

[8]P. S. Lovejoy, "The Need for a Policy for the Cut-Over Lands of Michigan," *22nd Annual Report of the Michigan Academy of Sciences* (1921):5–7.

graduate students in geography from the University of Michigan. After a summer in the field with compass and notebook, most of the survey teams were able to pace a straight line through brush and swamps and could usually find the weather-beaten stakes that had been set out to mark the section corners by the General Land Office surveyors 50 years before. Students of geography who experienced this kind of practical training had no trouble with the identification of areal associations or the concept of the unit area.

In fact, the concepts of the unit area and of the land type were the major professional contributions of the survey. Remember that in the 1920s the experimental studies of small areas, such as the study of Montfort by V. C. Finch, had not yet been made. The fractional code system, which implies the existence of a unit area—uniform with respect to the physical land and the land use or cover— had not been devised. The field operations of the survey resulted in a series of maps of individual elements: lay of the land (showing five categories of slope); soil types (based on the standard definitions of the U.S. Soil Survey); drainage features; cover (including wild vegetation, crops, or planted pastures and also abandoned farms); population; political organization; assessed valuation; tax delinquency; landownership and, where taxes were still being paid, the intention of the owner in maintaining possession of the land; and trade areas.

Each of these items was plotted on a separate map. But when the maps were compared in the office, it became clear that certain conditions were found repeatedly to form the same associations. Not only could certain natural land types be identified (repeated associations of slope, soil, drainage, and wild cover), but also certain economic conditions were found to have a high correlation with certain land types. Wade DeVries, a land economist, may have been the first to call attention to these associations (DeVries, 1927, 1928). But it remained for J. O. Veatch, a soil specialist, to define the types and to prepare a statewide scheme of the regions (Veatch, 1930, 1933, 1953).

Meanwhile, Lee Roy Schoenmann provided an example of how the information gathered by the survey could be used to establish local zoning rules for rural areas. He took the maps for Alger County and presented them at a series of meetings with local businesspeople and farmers. As a result, the community set up its own restrictions on land use, designating certain areas for farming and pasture, other areas for fish and wildlife preserves, and still other areas for reforestation (Schoenmann, 1931).

Only about half the counties of the original cutover area had been mapped when the survey was brought to an end in 1933 during the Great Depression. Enough work had been done in different parts of the north country to demonstrate that much of the area could not be managed by private owners. The completed maps (Fig. 51) provided enough data so that land management planning could be extended to neighboring counties. Furthermore, by 1933 the method of field mapping by pacing with a compass was outmoded through the use of vertical air photographs (Chapter 19). In addition, certain pressure groups in the state wanted to see the survey cease publishing this kind of information. For example, real estate interests were attempting to sell land on the shores of the numerous inland lakes, but not all the lake shores were sandy and suitable for summer homes. The survey classified the lake shores, showing where they were sandy and where boggy. Pressures similar to those used against Powell in his surveys of the West reappeared in Michigan in the 1930s. Enough had been accomplished, however, to make it clear that the geographers, with their methods of identifying areal associations and with their experience in the analysis of the interplay of diverse processes (which we would now describe as spatial systems), were in a position to make a distinctive contribution to the expanding field of land classification and land use planning (McMurry, 1936).

Further Development of the Land Classification Idea

The land classification idea was picked up by many individuals and many agencies of the federal and state governments. To reduce costs, mapping was done by air photographs, and continued experimentation was done with different scales and categories. Several previously unmapped Michigan counties were examined by these new methods. By 1940 the Subcommittee on Land Classification (Charles C. Colby, chairman) of the Land Committee of the National Resources Planning Board reported the existence of 72 separate land classification projects then being carried on by 46 agencies of the federal government and by 28 state agencies (Colby, 1941).

Also in the planning field was Harlan H. Barrows on the Water Resources Committee of the National Resources Planning Board. Barrows made three major contributions to water planning studies. First, he insisted on the use of clear, simple English in the writing of reports for publication. Second, this insistence on clear writing also meant clear thinking. He played an important role in formulating policy regarding multiple-purpose river development projects that were developed during the 1930s. Third, he designed the procedure for integrated regional studies as an essential basis for policy planning. Between 1935 and 1938 he drew up the plan that was accepted by the states of Colorado, New Mexico, and Texas for the allocation of the water of the upper Rio Grande. He was successful in working out similar cooperative solutions for the water of the Pecos River, the Red River of the North, and the Columbia Basin. The impact of his efforts in these river basin projects is summarized as follows:

> In subsequent years, the report of those investigations shaped technical analysis of water projects over many other areas. Moreover, the list of questions posed for investigation still is an incisive classification of resource-use problems in an irrigated area. His outline for the investigations became the springboard for studies and policy discussions cutting across all the relevant disciplines and levels of jurisdiction. He continued the same type of analysis as a consultant to the Department of the Interior concerned with resource development problems in Alaska and in the Central Valley of California (Colby and White, 1961:398–399).

In connection with the planning for economic development in the Tennessee Valley by the Tennessee Valley Authority, geographers were employed in the Division of Land Planning and Housing. G. Donald Hudson was named Chief Geographer of the Land Classification Section, one of several sections in the Division. Hudson had six geographers under his jurisdiction, most of whom came from the University of Chicago. Their first undertaking was to make a land use inventory of the Tennessee Valley. To accomplish this, they created and then adopted the "Unit Area Method."

They mapped approximately one half of the Valley before the entire task was terminated owing to political differences of opinion on the project. Meanwhile the geographers had developed an inventory of scenic resources, and had made studies of town economic bases and trade areas. A special group was assigned to study the proposed reservoir sites to determine the extent of lands to be purchased by the Authority—the so-called taking line studies i.e. to determine which farms would be made marginal by the reservoir. Hudson described the method for defining and mapping unit areas in 1936 (Hudson, 1936).

In 1941 the Land Committee was searching for less costly methods of surveying the land and land use of an area as a basis for improving the economy of depressed areas. Charles C. Colby and Victor Roterus were appointed as consultants to prepare an inexpensive method for gathering the necessary information. Their proposal was published in 1943, along with a Livelihood Area map of the United States (Colby and Roterus, 1943). The so-called area analysis method was based on an outline to be filled in by field observation and was summarized under four main headings: (1) the employment pattern, (2) the conditions affecting employment and income (natural resources, economic activities, and institutions), (3) directions of desirable readjustment, and (4) the proposed program of remedial action. Between 1941 and 1943 the method was applied to many small areas in different parts of the United States and proved useful in the guidance of efforts to rebuild depressed economies (Fig. 52).

Since World War II the whole approach to studies of this kind has been revolutionized through the use of computer programs, with data supplied from new remote sensing devices. These changes are discussed in Chapter 19.

Figure 52 Areas of the United States reported by area analysis method

Land Classification Studies in Latin America

Land classification studies are of special importance in countries with developing economies, such as those of Latin America. But before the 1950s few Latin American geographers were trained in the methods of the field survey. The first major attempt to apply land classification methods to a Latin American country was directed by geographers from the United States or trained in the United States. This was in Puerto Rico.

The Puerto Rico Rural Land Classification Program was carried out between June 1949 and August 1951. Here was a small island, some 3435 square miles in area, which in 1950 had a population density of 642 people per square mile. With a purely agricultural economy such a density could not be adequately supported. Furthermore, the greater part of the land of high potential productivity was used to grow sugar cane, while basic foods had to be imported. But the island government adopted a policy of providing for the rapid improvement of the economy by making better use of the land to produce crops and by investing heavily in new manufacturing industries to be scattered throughout the island and connected to ports by new all-weather roads.[9] Rafael Picó (who holds a Ph.D. in geography from Clark University) was chairman of the Puerto Rico Planning Board. He understood better than most Latin Americans at that time that economic planning had to be based on detailed knowledge of the resource base. Picó asked G. Donald Hudson, then chairman of the Department of Geography at Northwestern University, to help in planning such a survey. In March 1949 Hudson and his Latin American specialist at Northwestern, Clarence F. Jones, went to Puerto Rico to work out plans. It was decided that the island should be covered by maps on the very large scale of 1/10,000. Each year Northwestern University would supply some advanced graduate students, either from their own university or from other graduate departments around the United States. Each of these graduate students would be joined by a Puerto Rican student to form a field team, and each team would be assigned a section of the island to survey. Hudson, Jones, and Picó made an experimental traverse across the island to test the categories of land and land use. Mapping was done on vertical air photographs, and unit areas were identified by the fractional code method. The first field team started work in July 1949.[10]

---

[9]Governor Muñoz Marin was Puerto Rico's first elected governor in 1948. Puerto Rico in 1952 became the Commonwealth of Puerto Rico, freely associated with the United States. Governor Muñoz Marin undertook to develop the island's economy in what has been called Operation Bootstrap.

[10]The mapping was done on a scale large enough so that every plot of land, used or unused, could be shown. Eight categories of land use were identified: (1) cropped land, (2) pasture and harvested forage, (3) forest brush, (4) nonproductive land, (5) rural public community service land, (6) land used for quarrying or mining, (7) urban and manufacturing, and (8) miscellaneous, such as canals, water storage tanks, roads, railroads, and farm buildings. The physical characteristics of the land, which appeared as the denominator of the fraction, included soil types (as defined by the U.S. Soil Survey), degree of slope, conditions of drainage, rate of erosion, and amount of stoniness and rock exposure. See *Rural Land Classification Program of Puerto Rico* (Evanston, Ill.: Northwestern University Studies in Geography, 1952). See also Jones and Berrios (1956) and Jones and Picó (1955).

The Puerto Rico survey demonstrated the value of this kind of inventory. Land redistribution in some other parts of Latin America (where it was done with no mapped information) has brought disastrous results. In Puerto Rico the information gathered by the survey was used as the basis for replanning land use; crops were matched to favorable land types, or they were removed from unsuited lands, such as steep slopes, where they brought destructive erosion. The survey information was used to plan the routes of new roads and to locate the numerous small manufacturing plants in relation to population and accessibility. The success of Operation Bootstrap was in no small measure based on the existence of reliable, detailed knowledge about the land quality and existing land use.

Since that time, many somewhat similar surveys have been undertaken in Latin America, some by the Organization of American States, some by the Agency for International Development, and others by Latin American government agencies, as in Brazil. Some surveys, as in Chile, were done by private agencies in the United States on contract.

THE PIONEER BELT STUDIES

Also intended to provide knowledge on which to formulate policy were the studies of pioneer settlement initiated by Isaiah Bowman. In 1925 he turned his attention to problems relating to the thinly populated areas on the margins of settlement. He submitted a proposal to the National Research Council (NRC) to obtain funds for studies of pioneer areas. After two years of consideration by a special committee and by the Division of Geology and Geography of the NRC, the project was recommended to the Social Science Research Council. In 1931 both councils endorsed Bowman's program, and the Council of the American Geographical Society also gave its support.

Bowman outlined the nature of the problem (Bowman, 1932). Pioneering in the 1930s, he wrote, was not at all like the pioneering of the past century when new settlers depended almost entirely on their own muscles. Now pioneers wanted the latest machinery, the best medical services, and well-developed facilities to connect them with markets. Yet pioneer zones are always experimental. When prices for farm products are low, pioneers may seek new lands where the price per acre is lower than in settled communities, but these same pioneers may be driven back from low-priced land by droughts. Pioneer belt studies are not solely concerned with the possibilities of new settlement; they may also be concerned with the need to withdraw from less favorable places. In Malthus's time more food could be produced by moving farmers onto new lands and creating new agricultural communities. But by the 1930s increasing the supply of food without increasing the prices was done by reducing the number of farmers and withdrawing them from the marginal lands. Productivity in the modern world is increased by concentration in the more accessible and better suited areas and withdrawal from the remote and marginal areas. But all such changes must be applied in particular places. Bowman proposed to study pioneer movements around the world and to identify certain general conditions: not only the kinds of physical conditions considered favorable but also the attitudes and objectives that led people to become pioneers and the economic, social, and political institutions that could best support pioneers. But Bowman also proposed to investigate the particular and unique conditions in specific pioneer areas, knowledge of which would be essential to the formulation of policy. His proposal ranged widely over all the fields of the social sciences and was essentially inter-disciplinary in character.

Several studies of pioneer belts in general (and also of specific ones) were published during the 1930s. Bowman's book *The Pioneer Fringe* (1931) stated the nature of the problem and offered examples from the western United States, Canada, Australia, southern Africa, Siberia, Mongolia, Manchuria, and South America. A volume containing 27 cooperative studies of particular pioneer regions followed the next year (Joerg, 1932). Finally, Bowman, aided substantially by Karl Pelzer, summarized the results of the whole undertaking in a report on the world's potential pioneer areas (Bowman, 1937). Meanwhile, pioneer studies were vigorously pur-sued in Canada under the direction of the Canadian Pioneer Problems Committee headed by William A. Mackintosh of Queens University. Under the general title *Canadian Frontiers of Settlement*, edited by Mackintosh and Joerg, eight separate volumes were published, starting in 1934 (Innis, 1935).

Bowman gave a convocation address at the University of Western Ontario in 1937 in which he summarized his own point of view toward geography in the 1930s:

> Within limits that have varied widely in time, geography has set itself the task of understanding man's relation to the earth, and I shall presently attempt to explain that phrase with some precision. Always there must be food and clothing, toler-able if not optimum temperature ranges for both man and the things he requires, transport needs and desires, and, unhappily, wars and famines, for a time at least, as well as great conquests and conditional conquests of at least the local and immediate in the environment. Out of this play of forces—by no means either infinite or hopelessly complex—man is progressively creating and experimenting, and the chief experiment is himself. He is changing himself as well as the world as he goes along (Bowman, 1938:2).

For a more complete statement of Bowman's point of view, see his *Geography in Relation to the Social Sciences* (Bowman, 1934).

Geography in World War II

The demand for the services of geographers in World War II far exceeded the supply of experienced and properly trained professionals. Geographers were needed in all the kinds of work performed during World War I and also in many research studies. Geographers worked as commissioned officers or as noncommissioned draftees assigned to intelligence agencies. Geographers in large numbers came as civilians either for full-time positions in war agencies or for short-term specific studies.

By 1943 there were more than 300 geographers working in Washington, D.C. These included 75 geographers in the Research and Analysis Branch of the Office of Strategic Services (initially known as the Coordinator of Information), 46 in the War Department, 23 in the Intelligence Division (G2), and an additional 23 in the Army Map Service. The office of the Geographer, Department of State, had 13 geo-graphers; 15 were employed in the Board on Geographic Names, 12 in the Office of Economic Warfare, and 12 within the Department of Agriculture. In addition, 8 geographers were employed by the Geological Survey and 6 by the Coast and Geodetic

Survey. There were 5 geographers in the Weather Bureau, 4 in the Map Division of the Library of Congress, and 18 others were scattered among a variety of agencies. (These figures do not include cartographers or others engaged in producing maps and charts.) Approximately 25 other geographers were employed in posts overseas.

Some of these geographers helped to prepare the compilations of information about countries or parts of countries, either as a basis for planning military operations or as a guide to military government after the war. The Joint Army-Navy Intelligence Studies (JANIS) brought together many kinds of data with which geographers had no previous experience. But a very important part of the JANIS program consisted of the compilation and publication of many detailed maps of special features. Large numbers of geographers in the Office of Strategic Services were assigned to the cartographic work, while others worked on the various countries where JANIS handbooks were needed. Many geographers also worked on special problems and prepared background reports for the guidance of those who were responsible for decisions.

A few examples of the kinds of work geographers did can be offered. One had to do with the kinds of uniforms and equipment needed in different environments. In 1940 the quartermaster general of the army had three sets of uniforms for military use: temperate, torrid, and frigid. When troops occupied the Aleutian Islands, they were equipped with temperate zone uniforms; when these proved to be quite inadequate, however, it was clear that Aristotle's climatic zones were no longer useful. The quartermaster general established a research laboratory in Natick, Massachusetts, to test different kinds of equipment under a great variety of artificially produced climatic conditions. The problem was to identify the important differences of climate and other environmental conditions, not only to find the kinds of equipment best suited to them, but also to find out in detail where such environments would be encountered all over the earth. The result was the so-called clothing atlas. This atlas specifies in detail by means of a complicated key the variety of equipment necessary to carry on field operations in the world's many kinds of environment. This work was continued and expanded after the war.

Before the landings in Normandy many geographers, including those in the intelligence branch of the army, were busy making detailed studies of the beaches and the terrain behind beaches. Johnson's *Battlefields of the World War* had become a historical document, for the changed technology of warfare rendered his interpretation of the significance of terrain obsolete. When warfare became mechanized, the pattern of paved roads became more significant than the arrangement of hills and valleys. Small villages where paved roads came together became more important than cuestas. The basic point was that an army operating on foot can move as rapidly off the road as on it, but a mechanized army moves with great speed on a road, regardless of slope, and very much more slowly off a road. This is another example of the general principle suggested earlier that the significance of the physical and biotic features of the earth changes with changes in the attitudes, objectives, and technical skills of man himself.

In the Pacific theater there was a serious lack of any reliable information about the character of beaches or the terrain. Geographers were set to work combing the literature to find descriptions or old photographs. Missionaries and tourists had provided some information, but it was scattered and hidden in much irrelevant detail.

Yet maps were compiled and published showing in amazing detail the arrangement of coral reefs, cliffs, roads, caves, and other features of military importance.

One group of geographers received special training in the study of transportation facilities. What were the essential items of equipment in a port that would determine its capacity to handle traffic? A trained port engineer had to explain such matters. Geographers learned that in some of the ports of western Europe the tidal range was so great that gates had to be provided to keep the water from draining out at low tide. The condition of the gates was critical. Then, what about the conditions of roads and railroads? This group of geographers, together with expert photographers from the motion picture studios of California, undertook to provide descriptions, photographs, and maps showing the condition of transportation facilities immediately after the armies started their advance eastward. The materials gathered proved to have great practical utility for those commanders in charge of logistics.

A very important function that could be performed only by an experienced regional specialist was the interpretation of capabilities and intentions of foreign countries. Unfortunately, the number of geographers who had specialized in the study of foreign areas before World War II was quite inadequate for the demand. Those geographers who had called themselves regional specialists had focused on parts of the United States or Latin America. The number of geographers who had specialized in European, Asian, or African countries was very small. As a result, the work of foreign area interpretation was done by language specialists, historians, and others who happened to have a familiarity with areas in question. The further result was that many persons—geographers and others—who were assigned to such positions proved inept and unreliable. E. A. Ackerman, pointing a critical finger at what he described as inadequate professional training in the systematic aspects of geography, summarized the role of geographers in the war effort:

> In the three years from 1941 to 1944 American geographers dealt almost constantly with a series of difficult professional problems. The profession as a whole may take pride in the manner in which these situations were met. Both the well-known and the previously obscure showed skill, imagination, energy, and unselfishness as they perspired over wartime tasks. Our techniques advanced, and the prestige of the profession increased notably during those years. Scholars and administrators who had scarcely heard of geography before Pearl Harbor are now familiar with its methods and its results. Geography unquestionably has wider recognition than ever before in this country.
>
> However, an assessment would hardly be honest if one were to stop with praise of our recent performance. Wartime experience has high-lighted a number of flaws in theoretical approach and in the past methods of training men for the profession. It is no exaggeration to say that geography's wartime achievements are based more on individual ingenuity than on thorough, foresighted training. The geographer perfectly or even adequately trained for the specialty into which he was thrown has generally been an exception. The unfamiliarity of most young American geographers with foreign geographic literature; their almost universal ignorance of foreign languages; their bibliographic ineptness; and their general lack of systematic specialties are but a few points which may be cited in proof. All these were just as regular a source of difficulty as the strangeness of the problems and the pressure under which we worked (Ackerman, 1945:121–122).

Ackerman arrived at the important conclusion that a major source of difficulty in the preparation of geographers before World War II was a widespread belief in the essential duality of the subject. In many places it was felt that a geographer might become either a regional specialist without any training in a systematic field or that a geographer might become a specialist in any one of a number of systematic fields. This is the duality that the German geographers had resolved and that Hartshorne had attacked in 1939. It was the duality that participants in the annual field conferences deplored. But wartime experience with the employment of many poorly trained or partly trained people proved that the conceptual structure of geography was still not widely understood.

Geography in the White House

President Franklin D. Roosevelt had a keen interest in geography. In 1921 he had been elected to the council of the American Geographical Society. Since that time he had remained fascinated by the subject and had developed a substantial knowledge of atlases. Pursuant to the German annexation of Austria in March 1938 and increased anti-Semitic activity by the Nazis, Roosevelt began to think about resettlement of European Jewry—and other refugees—on a large scale. He held private meetings with Isaiah Bowman, exploring possibilities as to where several million such people might be relocated. As a result of these meetings, Bowman arranged for a team of workers (including geographers) to make feasibility studies of refugee settlement in different parts of the world. In addition, President Roosevelt initiated the M Project (M for anonymity), which, under Bowman's direction and through the person of Henry Field, provided some 666 studies in 20,000 pages and an *Atlas of Population and Migration Trends* (Martin, 1980).

Bowman worked in the Department of State three days a week and was frequently called on to advise Sumner Welles, Cordell Hull, and the president. He was made a member of the Stettinius Mission to London (1944), the Dumbarton Oaks Conference (1944), and the San Francisco Conference (1945). At all stages leading to the creation of the United Nations Charter, the geographical point of view was found to be of value (Martin, 1980).

OTHER APPLICATIONS OF GEOGRAPHY

The applications of geography to practical problems took place in many other sectors before the 1950s. One of these was in marketing research for private business firms. In 1931 William Applebaum was working on a thesis on secondary commercial centers of Cincinnati. In Cincinnati Applebaum began to focus his attention on the location factor in the development of outlying retail market centers. He also learned that the Kroger Company was looking for the best places to locate planned supermarkets. Applebaum turned his attention to the selection of sites for Kroger, and his results proved so useful that the company became interested in his methods. He went to work for Kroger to apply his method to the selection of other supermarket locations. Since 1931 and especially since World War II, most business firms engaged in selling to the public have added market research departments; the demand for persons with geographic training to work in the field of market research has increased rapidly.

What do geographers do when they work on a problem of retail store location? They make maps of the distribution of potential customers. But the population maps that are made by counting the number of people in census districts lack the relevant detail needed for such studies. It is necessary to plot on a map the arrangement of people along specific streets and to know the patterns of their daily trips to work or to retail stores. Often a store location proves much better on one side of the street than on the other, depending on the customers' routes to work. It is also necessary to map the areas from which other competing stores draw their customers. Because geographers are usually familiar with the making and use of detailed maps and with the map analysis of location problems, their contribution to the study of market areas has become widely recognized and appreciated (Applebaum, 1952).

Since World War II the use of geographers in marketing research problems has been greatly extended. In 1961 a whole issue of *Economic Geography* was devoted to a series of papers detailing examples of this kind of applied research.[11] The latest mathematical procedures have been applied to this kind of investigation (Applebaum and Cohen, 1961). Applebaum continued to publish on matters relating to marketing research until 1974. Other imaginative applications of geographic methods have been made to the study of the operations of large corporations (McNee, 1961). Meanwhile, economic geographers have been conducting studies on the location of economic activity ranging from iron and steel plants to flour milling.

Geographers such as Gilbert White have made studies of natural hazards, including major river floods (Kates, 1962; White, 1973). Studies of military matters from the geographical point of view have been made by Joseph A. Russell and others. Yet other geographers have worked with agencies of the United Nations, have made environmental studies in the wake of ever-increasing industrial pollution (F. W. McBride formed his own agency), and have studied the effects of weather and climate on humans (physiological climatology). Especially noteworthy in the last-named case is the work of E. Huntington and D. H. K. Lee (Martin, 1974). C. W. Thornthwaite applied his remarkable knowledge of climatology and other aspects of the physical environment to the dairy industry in New Jersey with remarkable success (1931, 1933). L. D. Stamp demonstrated the value of land utilization study (1931, 1952). H. H. Bennett studied soil erosion in relation to the productivity of the land (1928). E. L. Ullman was a member of the board of directors of Amtrak. M. I. Glassner has advised the government of Nepal in negotiating a transit treaty with India and has functioned as consultant to the UN Development Program for

---

[11]*Economic Geography* 37(1961): Saul B. Cohen, "Location Research Programming for Voluntary Food Chains," 1–11; Bart J. Epstein, "Evaluation of an Established Planned Shopping Center," 12–21; Howard L. Green, "Planning a National Retail Growth Program," 22–32; Harold R. Imus, "Projecting Sales Potentials for Department Stores in Regional Shopping Centers," 33–41; Jack C. Ransome, "The Organization of Location Research in a Large Supermarket Chain," 42–47; William Applebaum, "Teaching Marketing Geography by the Case Method," 48–60.

the land-locked countries of Asia. Professional geographers have been employed by the Bureau of the Census since the 1920s.

In more recent years an increasing number of geographers in the United States have turned their attention to the applied movement. To cater to the need for a more relevant geography, applied courses and programs were introduced at a variety of academic institutions. This development was encouraged by a decreasing number of academic posts available and a constant, if not increasing, number of geographers to fill them. It was thought that the applied movement would increase job opportunities, and this is indeed what happened. In 1976 *The Geographical Review* announced that it would introduce a new section on applied research. J. D. Harrison (1977) sampled the national scene and noted that nearly 90 percent of the departments approached were expecting to emphasize more applied geography in the immediate future (Dunbar, 1978). The *AAG Newsletter* and the *Professional Geographer* were prepared to include sections on applied geography.[12]

In 1978 J. Frazier and B. J. Epstein cofounded the Applied Geography Conference at Binghamton State University in New York. From 1928 to 2002 25 volumes of *Papers and Proceedings of the Applied Geography Conference* were published. Beginning in 2003 *Papers of the Applied Geography Conference* superceded the previous publication. This conference now draws approximately 200 participants from business, government, and the academic community. It has also generated a number of publications (Frazier, 1982; Frazier, Epstein, and Schoolmaster, 1995). Several universities in North America now have concentrations and masters degree programs in applied geography and are placing their students in meaningful and relevant employment such as with supermarket chains.

Specialty groups of the AAG were founded at the same time as the Binghamton conference. While applied geographer might join planning, urban transport, and other specialty groups, it is worth noting that applied geography is in the top third of specialty groups by body count and in 1996 had a total of 456 members and ranked 5th of 47 groups. In 1983 the first annual James R. Anderson Medal of Honor was awarded to an applied geographer and has continued to the present. By the turn of the twenty-first century applied geography had become a study group of the International Geographical Union and by 2003 was a commission. This commission has provided *Applied Geography: A World Perspective* (editors A. Bailly and L. J. Gibson). The first of three parts of this book is entitled "History and Epistemological Foundations"; this provides an insightful perspective.

Meanwhile, the *Applied Geography Newsletter* and the international journal *Applied Geography* began publication in 1980 (the latter has now ceased publication). A *Directory of Applied Geographers* was produced by the Association of American Geographers in 1981, and in 1983 Canada began publication of the *Operational Geographer*. Academic debate ensued over the role of "applied" and what was presumably a "pure" geography. By the 1980s it had become clear that more and

---

[12] Special mention should be made of the work of L. D. Stamp, an English geographer. See especially his *Applied Geography* (1960), a primer on the subject for that period. See also "Laurence Dudley Stamp, 1898–1966" by M. J. Wise, *Geographers: Biobibliographical Studies* 1988, 12: 175–187.

more academic appointments were being made in techniques and fewer consequently in the traditional academic areas (Green, 1983). Growth was swift in geographic information systems (GIS). In 1985 the AAG did not recognize it as a specialty. By 1987 it was the third-largest specialty group of the association, and by 1989 the second-largest specialty group by only nine members.[13] Meanwhile, a minor debate had surfaced concerning the distinction between applied and nonapplied geographers. Few attempted a definition of applied geography.

The applied posture led to increasingly specialized techniques, and some felt that it took the applied geographer farther away from the traditional geographer. R. A. Rundstrom and M. S. Kenzer have voiced concern that the academic core of geography is being eroded and that rather than being trained to think critically, students are learning to serve business and government (Rundstrom and Kenzer, 1989). F. G. Wynn has added that "to tailor our graduates to the specific skill requirements of employers is to abrogate our responsibilities as educators and to undermine one of the central attractions of our discipline as a field of inquiry" (Wynn, 1983). Torrierri and Ratcliffe (2004) suggest that over the last 15 years there has developed a need for funding leading to cost-cutting measures. A department's ability to secure funded research, preferably with "overheads" going to the institution, is regarded favorably and may well lead to departmental growth.

The applied research practiced in North America during the 1980s does not seem to have had any structure other than that created by its capacity to satisfy societal needs. A problem besetting geography in the United States has been that during the 1980s (and even prior to that) only a very small percentage (frequently less than 1 percent) of undergraduate students selected geography as a major subject (Wilbanks and Libbee, 1979). The idea of persuading undergraduates that they could be employed in planning agencies and the like, an idea reaffirmed by a past president of the AAG, George Demko (1988), is gained, some argue, at a cost to the discipline. Academicians who work exclusively within the university may see the applied person as one who borrows their materials and sells it in the marketplace. Their concern is that this practice will dominate and change the direction of the discipline. Academic practitioners are anxious about the standing of geography, especially since the demise of some departments at some of the best-known universities in the country is well known. T. G. Gordon has epitomized this state of affairs with two brief phrases: "the intellectual core"[14] and "anti-intellectualism."[15] There seems to be no clear definition of the limits to what is disciplinal and what is applied.

Applied geography is a hybrid whose epistemology evades definition.[16] It does not have a theoretical core nor a coherent structure and is characterized by

---

[13]AAG. "1988 Topical and Areal Proficiencies and Specialty Group Membership," *AAG Newsletter* 24, 3 (1989):10.

[14]*AAG Newsletter* 1988 23(5):1.

[15]*AAG Newsletter* 1988 23(6):2.

[16]Applied geographers were involved in matters including daylight saving time, Nile Waters agreements, the High Aswan Dam project, Zuyder Zee and Everglades reclamation projects, desert irrigation, and innumerable other works.

pragmatism. It is client driven and usually not curiosity driven. Large numbers of problems are posed that lead individuals or teams to adopt ad hoc postures drawing on the geographer's discipline. The work should also draw on neutral, value-free thinking in the resolution of problems, but this state of mind is hard to develop. Applied geographers may function as gatherers of information (a valuable function in the United States in both world wars) or synthesizers, though only occasionally does the geographer participate in the policy-making stage. Possibly this is due to the small number of geographers both in the university administrative structure and the political structure of most developed societies. This state of affairs may, however, be changed as this new geography presents itself as more relevant to contemporary society. Much of applied geography is privately negotiated. In consequence, we do not know the size of the industry. It has become a full-time occupation for considerable numbers and a part-time occupation for many more. Their knowledge is a commodity that can be marketed. This, in turn, has led to a new call for the further application of spatial science and locational analysis based on GIS and model building. The number of problems demanding attention is so large that this genre may come to dominate in a society demanding answers. Torrieri and Ratcliffe (1904) note several broad categories of applied research worthy of enumeration: market and location analysis, medical geography, land-use planning, environmental issues and policy, transportation planning and routing, and the geography of crime. Much excellent work has been accomplished, and membership of this guild continues to increase.

John F. Hart asked the question, "Why Applied Geography" (Frazier, 1982; Hart, 1989). He believes that all geographical knowledge is of value in understanding the world around us and that all of it is applicable. So why, he asks, add the term *applied* geography. In time applied geography may well find itself as part of traditional geography. An applied curriculum has been developed at Southwest Texas State University and is perhaps the largest in the country. There applied geography has been defined as "the use of geographic content, principles and methods in research and other activities designed to aid in the resolution of human problems." Internship programs have led to professional careers. The department began a master's of applied geography degree in 1983. In addition, Ryerson Polytechnical Institute in Toronto has founded a School of Applied Geography. There is some value in demonstrating the utility of the discipline vis-à-vis other disciplines.

Applied geography continues strongly in physical geography, especially in climatology and geomorphology. Recreation, tourism, and sport, sometimes referred to as leisure geography, is one of the fastest growing segments of the applied movement, as are studies in optimal location. While each of these activities exploits the cartographer, cartography itself is one of the most dominant forms of the applied movement. The map probably has never commanded higher esteem than at present, presenting a clarity amidst perplexity. Whether for military purposes, boundary settlement issues, the layout of utility lines in a town, or simply as a tourist aid, the map is of paramount importance (Kenzer, 1989).

REFERENCES: CHAPTER 18

AAG. 1989. "AAG Topical and Areal Proficiencies and Specialty Group Membership." *AAG Newsletter* 24, no. 3:10.

Ackerman, E. A. 1945. "Geographic Training, Wartime Research, and Immediate Professional Objectives." *Annals AAG* 35:121–143.

Applebaum, W. 1952. "A Technique for Constructing a Population and Urban Land Use Map." *Economic Geography* 28:240–243.

Applebaum, W., and Cohen, S. B. 1961. "The Dynamics of Store Trading Areas and Market Equilibrium." *Annals AAG* 51:73–101.

Barnes, C. P. 1929. "Land Resource Inventory in Michigan." *Economic Geography* 5:22–35.

Bennett, H. H. 1928. "The Geographical Relation of Soil Erosion to Land Productivity." *Geographical Review* 18:579–605.

Bowman, I., 1921. *The New World, Problems in Political Geography.* New York: World Book Co.

———. 1924. *Desert Trails of Atacama.* New York: American Geographical Society, Special Publication No. 5.

———. 1931. *The Pioneer Fringe.* New York: American Geographical Society.

———. 1932. "Planning in Pioneer Settlement." *Annals AAG* 22:93–107.

———. 1934. *Geography in Relation to the Social Sciences.* New York: Charles Scribner.

———. 1937. *The Limits of Land Settlement, a Report on Present-Day Possibilities.* New York: Council on Foreign Relations.

———. 1938. "Geography in the Creative Experiment." *Geographical Review* 28:1–19.

Colby, C. C. 1936. "Changing Currents of Geographic Thought in America." *Annals AAG* 26:1–37.

———, ed. 1941. *Land Classification in the United States.* Washington, D.C.: Report of the Land Committee to the National Resources Planning Board.

———. 1955. "Narrative of Five Decades." In *A Half Century of Geography—What Next?* (Papers presented at the alumni reunion, June 5, 1954.) Chicago: University of Chicago, Department of Geography.

Colby, C. C., and Roterus, V. 1943. *Area Analysis—A Method of Public Works Planning.* Washington, D.C.: Technical Paper No. 6 of the Land Committee, National Resources Planning Board.

Colby, C. C., and White, G. F. 1961. "Harlan H. Barrows, 1877–1960." *Annals AAG* 51:395–400.

Davis, C. M. 1969. "A Study of the Land Type." In *The Michigan Land Economic Survey.* Pp. 15–41. Ann Arbor, Mich.: Office of Research Administration, Project No. 08055.

Davis, W. M. 1918. *A Handbook of Northern France.* Cambridge, Mass.: Harvard University Press.

Demko, G. J. 1988. "Geography beyond the Ivory Tower." *Annals AAG* 78:575–579.

DeVries, W. 1927. "An Economic Survey of Chippewa County, Michigan." *Papers of the Michigan Academy of Science, Arts and Letters* 8:255–268.

———. 1928. "Correlation of Physical and Economic Factors as Shown by the Michigan Land Economic Survey Data." *Journal of Land and Public Utility Economics* 4:295–300.

Dominian, L. 1917. *The Frontiers of Language and Nationality in Europe.* New York: American Geographical Society.

Dunbar, G. S. 1978. "What Was Applied Geography?" *Professional Geographer* 30:238–239.

Finch, V. C., and Baker, O. E. 1917. *Atlas of World Agriculture.* Washington, D.C.: U.S. Department of Agriculture.

Frazier, J. W. ed. 1982. *Applied Geography: Selected Perspectives.* New York: Prentice-Hall.

Frazier, J. W., B. J. Epstein, and F. A. Schoolmaster. 1995. "Contributions to Applied Geography: The 1978–1994 Applied Geography Conferences," *Papers and Proceedings of Applied Geography Conferences* 18:1–13.

Gelfand, L. E. 1963. *The Inquiry.* New Haven, Conn.: Yale University Press.

Green, D. B. 1983. "Teaching Positions in Geography in the United States: What Specialties Have Been in Demand?" *AAG Newsletter* 18:14–15.

Haggett, P. 1990. *The Geographer's Art.* Oxford: Basil Blackwell Ltd.

Harrison, J. D. 1977. "What Is Applied Geography?" *Professional Geographer* 29:297–300.

Harrison, J. D. and Larson, R. D. 1977. "Geography and Planning: The Need for an Applied Interface. *Professional Geographer* 29:139–147.

Hudson, G. D. 1936. "The Unit Area Method of Land Classification." *Annals AAG* 26:99–112.

Innis, H. A. 1935. "Canadian Frontiers of Settlement: A Review." *Geographical Review* 25:92–106.

James, P. E. and C. F. Jones. 1954. *American Geography: Inventory and Prospect.* Syracuse, N.Y.: Syracuse University Press.

Jefferson, M. 1921. *Recent Colonization in Chile.* New York: American Geographical Society, Research Series No. 6.

———. 1926. *Peopling the Argentine Pampa.* New York: American Geographical Society, Research Series No. 16.

Joerg, W. L. G., ed. 1932. *Pioneer Settlement, Cooperative Studies by Twenty-Six Authors.* New York: American Geographical Society.

Johnson, D. W. 1921. *Battlefields of the World War, Western and Southern Fronts: A Study in Military Geography.* New York: American Geographical Society, Research Series No. 3.

Jones, C. F., and Picó, R., eds. 1955. *Symposium on the Geography of Puerto Rico.* Rio Piedras: University of Puerto Rico Press.

Kates, R. W. 1962. *Hazard and Choice Perception in Flood Plain Management.* Chicago: University of Chicago, Department of Geography, Research Paper 78.

Kenzer, M. S. 1984. Comments from the Outside: The Sixth Annual Applied Geography Conference, October 12–15, 1983. *Applied Geography* 4:85–86.

———, ed. 1989. *On Becoming a Professional Geographer.* Columbus, Ohio: Merrill.

———, ed. 1989. *Applied Geography: Issues, Questions, and Concerns.* Dordrecht: Kluwer Academic Publishers.

———. 1991. "Applied Geography." In *Modern Geography: An Encyclopedic Survey* (G. S. Dunbar, ed.), pp. 3–4. New York and London: Garland Publishing.

———. 1992. "Comment: Applied and Academic Geography and the Remainder of the Twentieth Century." *Applied Geography* 12:207–210.

Martin, G. J., ed. 1966. *Mark Jefferson: Paris Peace Conference Diary.* Ann Arbor: Edwards.

———. 1974. "'Civilization and Climate,' Revisited." *Geography and Map Division, Special Libraries Association Bulletin,* No. 96, pp. 10–17.

———. 1980. "'The Science of Settlement' and Resettlement Schemes." In *The Life and Thought of Isaiah Bowman.* Pp. 123–139. Hamden, Conn.: Shoe String Press.

McBride, G. M. 1923. *The Land Systems of Mexico.* New York: American Geographical Society, Research Series No. 16.

———. 1936. *Chile: Land and Society.* New York: American Geographical Society, Research Series No. 19.

McMurry, K. C. 1936. "Geographic Contributions to Land-Use Planning." *Annals AAG* 26:91–98.

McNee, R. B. 1961. "Centrifugal-Centripetal Forces in International Petroleum Company Regions." *Annals AAG* 51:124–138.

Miller, O. M. 1929. "The 1927–1928 Peruvian Expedition of the American Geographical Society." *Geographical Review* 19:1–37.

Ogilvie, A. G. 1922. *Geography of the Central Andes*. New York: American Geographical Society, Map of Hispanic America Publication No. 1.

Palm, R. I. and A. J. Brazel. 1992. "Applications of Geographic Concepts and Methods." Pp. 342–362. In R. F. Abler, M. G. Marcus, and J. M. Olson, eds. *Geography's Inner Worlds*. New Brunswick, N.J.: Rutgers University Press.

Platt, R. R. 1946. "The Map of Hispanic America on the Scale of 1:1,000,000." *Geographical Review* 36:1–28.

Rhoads, J. B. 1954. "Preliminary Inventories." *Cartographic Records of the American Commission to Negotiate Peace*, No. 68. U.S. Washington, D.C.: National Archives.

Richardson, D. B. 1989. "Doing Geography: A Perspective on Geography in the Private Sector." In M. S. Kenzer, ed. *On Becoming a Professional Geographer*, pp. 66–74. Columbus, Ohio: Merrill.

Rundstrom, R. A. and Kenzer, M. S. 1989. "The Decline of Fieldwork in Human Geography." *Professional Geographer* 41:294–303.

Russell, J. A. 1983. "Specialty Fields of Applied Geography." *Professional Geographer* 35:471–475.

Sauer, C. O. 1919. "Mapping the Utilization of the Land." *Geographical Review* 8:47–54.

———. 1921. "The Problem of Land Classification." *Annals AAG* 11:3–16.

Schoenmann, L. R. 1931. "Land Inventory for Rural Planning in Alger County, Michigan." *Papers of the Michigan Academy of Science, Arts and Letters* 16:320–361.

Shantz, H. L. and Marbut, C. F. 1923. *The Vegetation and Soils of Africa*. New York: American Geographical Society, Research Series No. 13.

Stamp, L. D. 1931. "The Land Utilization Survey of Britain." *Geographical Journal* 78:40–53.

———. 1952. *Land for Tomorrow: The Underdeveloped World*. Bloomington: Indiana University Press.

———. 1960. *Applied Geography*. London: Penguin Books.

Thornthwaite, C. W. 1931. "The Climates of North America According to a New Classification." *Geographical Review* 21:633–655.

———. 1933. "The Climates of the Earth." *Geographical Review* 23:433–440.

Torrieri, N. K. and M. R. Ratcliffe. 2004. "Applied Geography." Pp. 541–547. In Gaile and Willmott (eds.). *Geography in America at the Dawn of the 21st Century*. Oxford: Oxford University Press.

Veatch, J. O. 1930. "Natural Geographic Divisions of Land." *Papers of the Michigan Academy of Science, Arts and Letters* 14:417–432.

———. 1933. "Classification of Land on a Geographic Basis." *Papers of the Michigan Academy of Science, Arts and Letters* 19:359–365.

———. 1953. *Soils and Land of Michigan*. East Lansing: Michigan State University Press.

Walworth, A. 1976. *America's Moment: 1918—American Diplomacy at the End of World War I*. New York: W. W. Norton.

White, G. F. 1973. "Natural Hazards Research." In R. J. Chorley, ed., *Directions in Geography*. Pp. 193–216. London: Methuen.

Wilbanks, T. J. and Libbee, M. 1979. "Avoiding the Demise of Geography in the United States." *Professional Geographer* 31:1–7.

Wynn, G. 1983. "Human Geography in a Changing World." *New Zealand Geographer* 39:64–69.

Wright, J. K. 1952. *Geography in the Making, The American Geographical Society, 1851–1951*. New York: American Geographical Society.

19

New Methods of Observation and Analysis

The idea that thought is the measure of all things, that there is such a thing as utter logical rigor, that conclusions can be drawn endowed with inescapable necessity, that mathematics has an absolute validity and controls experience— these are not the ideas of a modest animal. Not only do our theories betray the somewhat bumptious traits of self-appreciation, but especially obvious through them all is the thread of incorrigible optimism so characteristic of human beings.

. . . When will we learn that logic, mathematics, physical theory, are all only inventions for formulating in compact and manageable form what we already know, and like all inventions do not achieve complete success in accomplishing what they were designed to do, much less complete success in fields beyond the scope of the original design, and that our only justification for hoping to penetrate at all into the unknown with these inventions is our past experience that sometimes we have been fortunate enough to be able to push on a short distance by acquired momentum?

—*Percy W. Bridgman,* The Nature of Physical Theory *(1964)*

The industrial revolution has finally caught up with the study of geography. For thousands of years the techniques of observation and the methods of analysis remained essentially the same. People on foot, on horseback, in horse-drawn carriages, and in canoes and sailing ships traveled through different parts of the world. They recorded the things they saw by direct observation, analyzing their observations into component parts to clarify their descriptions and testing their hypotheses regarding processes by additional direct observation. A geographer was one who gained more than common satisfaction in visiting unfamiliar and out-of-the-way places and returning to interpret the circumstances observed in the strange new worlds beyond the far horizons. No one did this kind of work better than Alexander von Humboldt, the last great universal scholar. But now, and especially since World War II, there has been a revolution in the technology of observation and analysis.

Certain dates mark these technological advances. In 1950 the U.S. Bureau of the Census installed its first electronic computer, known as UNIVAC. In 1957 the Soviet Union sent up its first successful satellite, *Sputnik I.* In 1962 the first geodetic satellite was placed in orbit, and thereafter the age-old problem regarding the shape of the earth was resolved with a degree of accuracy never possible before. *Nimbus I,* the first weather satellite, was launched in 1964, and *Nimbus II* in 1966. The result was information concerning the state of the earth's atmosphere that brought synoptic meteorology from an art to a science.

The computer came just in time.[1] With this electronic device mathematical computations can be carried out in seconds that would have required a long time for an individual with pencil and paper. Areas of the earth's surface defined as homogeneous by specified and measured criteria can be outlined in minutes from satellite data fed into computers. Furthermore, for every piece of information about the face of the earth that Humboldt could command, the student of geography today has tens of thousands of items of information to overwhelm the traditional methods of analysis. Data banks can store this information, which can be recalled in moments. The transformation of technology was so remarkable that a large number of scholars raised in the older traditions seemed unprepared to move forward into the new world. The fruition of this technology came in July 1969 with the lunar landing of Apollo 11. Television promotion of this accomplishment and postcard, Christmas card, and magazine likenesses of the earth from a distance altered humans' images of the planet on which they lived (Cosgrove, 2001).

THE TECHNOLOGY OF OBSERVATION

Although the ancient Greeks understood the method of calculating the size and shape of the earth, they lacked the instruments to make measurements with sufficient accuracy. Not until the seventeenth century were the methods of a national mapping program worked out by the Cassinis in France. Newton's and Huygen's deduction that the earth must be flattened at the poles was challenged by the Cassinis on the basis of their measurement of the Paris meridian across France. This led to the expeditions of La Condamine and Maupertuis and the confirmation of the polar flattening.

The modern technology of measurement would not have advanced without several international agreements and several technical innovations made in the seventeenth and eighteenth centuries. The pendulum clock was one of these, and another was the development of a chronometer that could be used at sea. And there had to be some kind of international agreement concerning units of measurement—a type of agreement that the Greeks and Romans were never able to reach. It was not until 1791 that the French adopted the standard meter as 1/10,000,000 of the meridian from the equator to the pole passing through Paris. A rod consisting of an alloy of metals that minimizes changes of length owing to temperature changes is kept in an air-conditioned vault in Paris; this rod has been accepted throughout the world as the standard of linear measurement. Even the length of the foot, which is still the popular unit of measurement in the United States, is determined by comparison with the French meter. In 1960 the rod was replaced by an optical method of determination based on the wave properties of light; in 1983 this was superseded by another and more precise method based on the speed of light.

---

[1]The first electronic computer, ENIAC, was made by Eckert and Mauchly at the University of Pennsylvania in 1946. It was designed for the Army Ordnance Corps to calculate ballistic trajectories that involved the numerical solution of very difficult differential equations.

As a result of these agreements and the technical improvements of the contemporary period, the measurement of the shape of the earth has now been done in great detail. The observations of satellite motions every hour or so reveal minute differences in the shape of the earth, and the observation of these minute differences in motion has been made possible by a new kind of satellite tracking camera devised and constructed by the Smithsonian Astrophysical Observatory (King-Hele, 1967). The earth is now described as slightly pear-shaped with a bulge south of the equator. A map of departures from a geoid-shaped sea level shows bulges in western Europe, to the north of New Guinea, and between Africa and Antarctica. The average equatorial diameter of the earth is 12,756.38 km (7926.42 miles), and the average polar diameter is 12,713.56 km (7899.83 miles). The equatorial circumference is now measured as 40,075.51 km (24,902.45 miles).

Mapping the Earth

The method of making a large-scale map of a country that was developed by the Cassinis remained the standard procedure until the 1930s. G. R. Crone describes the steps that had to be taken in making a national survey as follows:

1. Determination of mean sea level, at one point at least, to which all altitudes are referred.

2. A preliminary plane table reconnaissance to select suitable points for the triangulation, and the erection of beacons over them.

3. Determination of initial latitude, longitude, and azimuth (for direction) which will "tie" the map to the earth surface.

4. Careful measurement of the base or bases with tape or wire of a special alloy.

5. Triangulation, the theodolite being used to observe horizontal angles from the base and beaconed points, and to measure altitudes by readings of the vertical angles.

6. Calculation of the triangulation and heights, and the transference of the trig points to the sheets issued to plane tablers.

7. The filling-in on the sheets by plane tablers of the required topographical detail—contour lines, rivers, woods, settlements, routes, and names. (Crone, 1950:152)

One of the finest examples of such an undertaking was provided by the Ordnance Survey of Great Britain. This was begun in 1791 and eventually covered the land on the scale of 1:63,360. Originally inspired by military purposes, it was later to be used in a multiplicity of contexts (Harley, 1975). By the end of the nineteenth century, almost all the countries of Europe had been covered by topographic (large-scale) maps, but each country used its own scale and its own projection, and each included different categories of features. Some 15 different prime meridians were in use. Since the maps were not comparable, no overall topographic coverage for Europe was available. Outside of Europe small parts of North America had been mapped, and there was a survey of India. This was the Great Trigonometrical Survey of India, which was joined by Sir George Everest as assistant superintendent in 1823. By 1830 Everest was appointed Surveyor-General of India. When he retired in 1843, the national survey network had been formed (Smith, 1994). This formidable undertaking, serving as a model of its kind and covering a vast area, has been described in *Mapping an Empire: The Geographical Construction of British India, 1765–1843* (Edney, 1999). In other parts of the world, large-scale maps were few (Robinson, 1956). So much had been learned concerning surveying and cartography that beginning in 1898 national atlases began to make their appearance: Finland, 1898; Sweden, 1900; Canada, 1906.

Two important steps were taken in the late nineteenth century. First, in 1884 a conference was convened in Washington, D.C., to discuss the adoption of a single prime meridian. As a result, 25 nations agreed to use the meridian of Greenwich in England (the astronomical observatory on the Thames, now in the suburbs of London) as the 0° meridian and to measure longitude east and west of Greenwich (Fig. 53). During the nationalistic days of the 1800s, many countries had adopted a prime meridian through their own capital city, but by the end of the century, most of the world's maps made use of the Greenwich meridian. Second, at the Fifth International Geographical Congress held in Bern in 1891 Albrecht Penck proposed that an international map of the world should be made on a scale of 1/1,000,000 using uniform symbols and conforming to agreed standards. The "Millionth Map" was melded into the digital chart of the world (DCW) in the 1980s.

Meanwhile, the production of large-scale maps (on scales of 1/100,000 or larger) continued slowly. Arthur H. Robinson summarized the world mapping situation in 1956:

> The need for such maps is continually increasing, for, as the world's population increases and life becomes more complex, more and more planning becomes necessary to insure adequate food supplies and transportation facilities. This means the extension of soil surveys, land-use surveys, water-supply surveys, erosion surveys, population surveys, and a host of others. None of these can be carried on adequately without the basic topographic map as a point of departure. The use of the topographic map as a base is by no means limited to those activities in which detailed maps are generally conceded to be indispensable; it also serves an important function as a source of information for a wide range of interests, from "marketing" maps to "treasure" maps (Robinson, 1956:296).

A century after Penck suggested the 1/1,000,000 map of the world, the task has not been completed.

Vertical Air Photography

Vertical air photography came into use during World War I. Even in the Civil War, photographs had been taken from balloons for military purposes, but the first vertical photographs for intelligence purposes were used during the trench warfare on the western front. Skilled photo interpreters learned how to identify objects in the pictures when they were viewed from directly overhead and how to penetrate the screen of camouflage intended to hide military objects from overhead viewers.

The applications of vertical air photography to geographical studies were appreciated even during the war. In 1917 the first vertical photograph (of a part of Paris) was published in the *Geographical Review* to demonstrate the utility of such

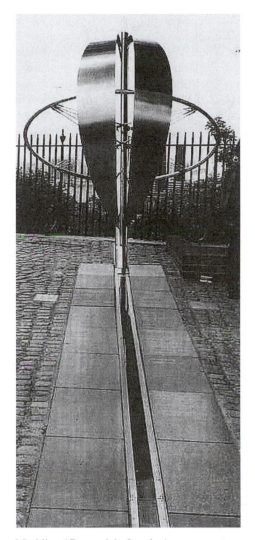

Figure 53 The Prime Meridian (Greenwich, London)

pictures for geographical studies (Woodhouse, 1917:337). After the war the airplane was used to experiment with air photography for geographical purposes, and several pioneer studies were published during the 1920s. In 1920 Willis T. Lee, a geomorphologist on the staff of the U.S. Geological Survey, published a paper demonstrating how air pictures could be used to interpret the landforms and settlement patterns of the Coastal Plain in the eastern United States (Lee, 1920; also Wright, 1952:330–334). In 1921 Jules Blache, a professor of geography at Grenoble, made use of photographs taken from airplanes to interpret the conditions of life in Morocco, even beyond the territory then controlled by France (Blache, 1921). In 1922 the American Geographical Society published a book by Willis T. Lee showing a variety of landforms and settlement types as seen from the air (Lee, 1922).

A pioneer in the development of new techniques for using photographs to make topographic-scale maps was the Scottish cartographer Osborn M. Miller. An artillery officer during World War I, Miller appreciated the potential uses of air photography not only for spotting artillery targets but also for making maps for peaceful purposes. In 1923 Miller joined the staff of the American Geographical Society to offer training in this work in the School of Surveying under the direction of Alexander Hamilton Rice (Wright, 1952:320–322). Miller had completed such a course of study at the Royal Geographical Society in London. The purpose was to train explorers in the techniques of field mapping. In 1926, when Miller was in Zurich, he witnessed a demonstration in the laboratories of the Wild Instrument Company of a device for plotting topographic maps from stereoscopic pairs of air photographs. When adjacent pictures were overlapped by about 60 percent, the eye, viewing the pictures through a specially designed pair of lenses, could see the terrain and settlement features in relief. The stereoscopic plotting instrument permitted the mapmaker to draw contour lines by following the levels viewed in the photographs. Furthermore, it was possible to use oblique air photographs after applying the corrections devised by Miller (Miller, 1931). Thereafter, both vertical and oblique air photographs were used to make topographic maps much more rapidly and at less cost per square mile than had ever been possible before.[2]

The Use of Air Photographs for Plotting Geographical Data

In addition to plotting points and lines for the purpose of making a large-scale base map, geographers were also intrigued by the possibility of plotting the areas occupied by the phenomena they wanted to study. At the Ann Arbor meeting of the Association of American Geographers in 1922, W. L. G. Joerg of the American Geographical Society showed how vertical photographs could be used to make maps of American cities in much greater detail than was provided by the usual topographic map.[3]

Nevertheless, the field mapping of the Michigan Land Economic Survey during the 1920s and 1930s was done in the traditional manner by field surveyors equipped with plane tables, compasses, and alidades. The plane table was set up on a tripod and oriented by the compass (with care to make sure that the compass was not deflected by a pocket knife or a wire fence). The alidade was used to sight at distant objects and to draw the line of sight on the map. Distances were measured by pacing. Although the geographers were aware of the new techniques for field-mapping, there were not enough airplanes properly equipped with cameras to apply these improved methods.

The use of vertical air photographs was a major breakthrough in field mapping techniques. The first experiment in the use of vertical air photography to map the

---

[2]After these pioneering experiments the field of photogrammetry, a branch of engineering, developed more and more precise methods of producing maps from air photographs. It improved both cameras and methods of plotting the maps to overcome errors of perspective, including eventually even the effect of differences in altitude of the earth's surface.

[3]Reported in *Annals AAG* 13 (1923):211.

vegetation and land use of a small area was done by K. C. McMurry, then chairman of the Department of Geography at the University of Michigan.[4] The experimental area was Isle Royale in Lake Superior. The entire island was photographed—after a delay of more than a month because forest fire haze made photography impossible. McMurry mounted the photographs in the form of a mosaic, and from them he identified certain areas he thought were representative of differing vegetation and land use categories. The type areas made up about 25 percent of the total area. He then went into the field to make maps of these areas by traditional methods. The comparison of the photographs with the ground maps made it possible to extend the mapping to the whole island by photo interpretation (Russell, Foster, and McMurry, 1943).

By this time the geographers who met at the annual spring field conferences were ready to try mapping geographical data directly on the photographs. Using the photographs in the field made it possible to identify slope, soil, drainage, vegetation, land use, and settlement features and to plot these features directly on the pictures in relation to the visible objects picked up by the camera. The first major land and land use inventory to use the photographs in this way was the survey of the area administered by the Tennessee Valley Authority, carried out under the direction of G. Donald Hudson (TVA, 1983). With the photographs to provide exact outlines of fields and other features, it was possible to plot information about land use, physical land conditions, crop yields, land value, market areas of towns, and other relevant information much more rapidly and with much greater accuracy than could be done by the traditional methods used in the Michigan survey. After that time mapping geographical information on vertical air photographs became the standard procedure.

Radar and Infrared Imagery

World War II produced another and even more far-reaching breakthrough in the use of new devices for the remote sensing of the face of the earth. In many ways the history of the application of the new devices to geographical research was repeated. In the 1920s the Michigan Land Economic Survey had to make use of already outmoded methods of field mapping, but even today the new devices for mapping all kinds of important features are still not available to the great majority of geographers. Yet it is clear that the profession has been for some years on the threshold of a new era of low-cost field observation, with a new command of detail and a variety of resolution levels. Only two of the new devices are described: the side-looking airborne radar (SLAR) and infrared color film. Both are dependent on the new high-altitude reconnaissance planes.

The airborne radar is capable of scanning large areas rapidly and, if necessary, repeatedly. The plane carries its own source of energy, which permits the emission of thousands of pulses of electromagnetic energy per second. Some of the energy is reflected back to the airplane from the earth's surface. A specially devised camera with a continuous film strip records the radar imagery from a cathode-ray tube. Robert B. Simpson has summarized the capabilities of SLAR as follows:

---

[4]Reported in *Annals AAG* 22 (1932):69.

Assuming the availability of a properly equipped aircraft, almost any task within the capability of SLAR can be accomplished more cheaply by it. A recent study for the U.S. Agency for International Development indicated that the most pressing elements of a topographic map production program for a typical underdeveloped country could be accomplished by SLAR in one-quarter to one-tenth the time, and at one-quarter to one-tenth the cost of a conventional mapping program.

Among the types of material that such a survey could provide an underdeveloped country are the following:

1. An area mosaic. . . .
2. Natural regions overlays, including those showing geomorphology, and other aspects of geology, vegetation, soil, land utilization, and a variety of other critical economic and scientific parameters, such as the size and shape of drainage basins.
3. Information necessary to determine the areas suitable or practices which will upgrade the economy, such as areas suitable for irrigated floodplain agriculture.
4. Suggestions as to location of mineral resources. . . .
5. A basis for the selection of routes and sites, such as for roads, terminals, industrial plants and railways, in detail appropriate to the scale.
6. Data from which to determine areas requiring later large-scale Class-A map coverage. . . .

Meanwhile, SLAR surveys can provide the data for the production of small-scale, Class-A planimetric sheets, as well as that for the interim production of medium-scale sheets (Simpson, 1966:96).

Another source of geographical data is provided by the use of infrared imagery of the 4.5 to 5.5 micron wavelength band. The cameras are carried in high-altitude airplanes that permit the rapid survey of large areas of the land surface. Already such photography has been applied to forest resource inventories; the pictures not only permit estimates of the available lumber in a forest but also indicate the extent of forest damage by disease, for the diseased trees appear in a distinctive shade of red. It is also possible to make studies of agricultural land use, including types and conditions of crops and differences of cultivation practices (Olson, 1967). With adjustments of camera and film it is possible to focus on soil types, differences of soil productivity, or variations in the availability of water.

These new instruments make possible a census of the whole of the world's population, covering any particular populations at frequent intervals. The patterns of metropolitan areas can be exactly delimited, and the movements of goods and people can be plotted either between urban centers or within urban areas.

Satellite Imagery

The year 1957 witnessed the beginning of another major advance in humankind's long and continuous efforts to increase knowledge of the physical character of the earth. Starting on July 1 and continuing for 18 months, 70 nations cooperated in the International Geophysical Year (IGY). Simultaneous observations carried on with standardized procedures were made all around the world for the purpose of finding

Figure 54 Planet Earth

answers to a variety of geophysical questions. Why do magnetic storms disrupt communications? Can the positions of the continents and ocean basins be mapped so
accurately that minute movements can be measured? Are glaciers receding and ice
caps melting? The observations provided new information for meteorology, geodetic surveys, ionospheric physics, glaciology, oceanography, seismology, and other
branches of geophysics as well as for the study of geomagnetism, gravity measurements, aurora and air glow, solar activity, and cosmic rays. The results provided a
spectacular increase in knowledge of physical processes. Then on October 4, 1957,
Soviet engineers put *Sputnik I* in orbit, and on January 31, 1958, the United States
sent up *Explorer I*. Thereafter a large number of satellites have been placed in orbit,
each to perform specific functions. Satellite imagery now provides a wealth of new
data concerning the face of the earth (Fig. 54).

The satellites can do many things of interest to geographers. Satellites of the *Tiros*
and *Nimbus* series are providing new views of the earth's atmosphere. Every day—
or every hour, if necessary—the receiving stations on the earth get TV pictures of
the earth's cloud cover, thus making possible not only a general view of atmospheric
circulation but also a day-to-day synoptic view of the storm patterns. Weather forecasting takes on a new dimension. From these same satellites temperatures can be
recorded all over the earth by the use of infrared energy detectors. The earth, as a
great system of complex parts kept going by radiation from the sun, can now be
observed as a whole and the inputs of energy exactly measured. The cloud patterns,
from which atmospheric circulation can be read, show no signs of Maury's wind

zones, but they do clearly reveal the more or less permanent oceanic whirls and the cold fronts that push into them on the poleward sides (Barrett, 1970).

NEW ANALYTIC PROCEDURES

All this new technology of observation has been matched in the contemporary period by new analytic procedures. The so-called quantitative revolution means that many of the younger generation of geographers have discovered the value of mathematics and especially of mathematical statistics. But an essential part of the revolution consists in the use of the electronic computer as an analytic device.

The Use of Mathematical Concepts and Statistical Procedures

Following the notable success of econometrics in the 1930s in making the description of economic processes more precise and providing for the objective testing of economic hypotheses, the use of mathematics began to spread into other social and behavioral sciences. Geographers, however, lagged behind the others, in part because the statistical procedures developed in other fields were not directly applicable to the analysis of spatial factors. It remained for mathematically minded geographers and geographically minded economists to develop statistical procedures specifically for the study of geographical problems.

The use of mathematics in geography is not really new. Of course, even in the days of Thales and Eratosthenes there was a branch of geography known as *mathematical geography*, but this had to do chiefly with studies of the form of the earth and the relation of the earth to the celestial bodies. In the modern period *mathematical geography* is again a branch of the field, but it refers to the use of mathematical concepts and statistical procedures for the study of occupied space. The two usages of these words should not be confused. Even in the latter sense, however, statistical procedures have been used in almost every generation during the modern period. Ellsworth Huntington made use of statistical analyses to give plausible support for his otherwise verbal hypotheses regarding the effects of climate. In 1937 John K. Wright used mathematics to provide a quantitative measurement of the variations in the intensity of phenomena over the earth and of degrees of correspondence between two or more phenomena (Wright, 1937). This is known as dasymetric mapping (a technique developed by Henry D. Harness in the early 1800s) (Robinson, 1955). In 1939 M. G. Kendall published a paper analyzing the covariance among 10 crops in 48 counties of England in terms of productivity (Kendall, 1939). After World War II papers pointing to the importance of using quantitative procedures appeared more frequently (Weaver, 1954, 1956). Of special influence was the paper by John Q. Stewart, published in 1947, in which he demonstrated the application of mathematics to the study of urban hierarchy and to other problems of population distribution. Here is what Stewart had to say about the utility of mathematics:

> The way of progress is obstructed by the opinion, common among authorities on economics, politics, and sociology, that human relationships never will be described in mathematical terms. There may be some truth in this as regards the

doings of individual persons. Even the physicist has given up the idea that the behavior of individual particles can be precisely described thus and necessarily contents himself with the discussion of averages. But the time to emphasize individual deviations is after the general averages have been established, not before (Stewart, 1947:461).

In spite of these and other individual efforts in the use of statistical procedures, there were no large numbers of followers. It may be said that the quantitative revolution had its beginning when William L. Garrison offered the first seminar for the training of graduate students in geography in the use of mathematical statistics. This was at the University of Washington in 1955.[5]

Numerous benefits can be derived from the use of mathematical concepts and statistical procedures in geographical studies. Mathematics provides a clear way to avoid the old problem of tracing cause and effect relations. The differential equations in calculus may be used to describe the way variables change through time, but they do not specify an antecedent cause and subsequent effect. Given the condition of a phenomenon at any one time, it is possible to describe its condition either at an earlier or later time. The theory of functions may prescribe that where A exists, B also exists, but it does not have anything to say about one being the cause of the other.

A very important benefit to clarity of thought resulting from the use of mathematical concepts is that deterministic models can be replaced by stochastic models (Lewis, 1965). Since 1927, when Werner K. Heisenberg, a German physicist, formulated the principle of indeterminacy in physics, the basic concept has spread to other fields of study. In physics this principle demonstrates that it is impossible to measure with full accuracy at the same time both the position and the velocity of an electron. On the other hand, when large numbers of electrons are measured, the probability that they will occupy certain positions and be moving with certain velocities becomes more and more predictable. The principle has obvious applications to geography.

Probability theory provides the mathematical foundation on which statistical analysis is built (King, 1969:32). Models based on probability are stochastic models. Statistics are also designed to permit drawing interfaces from a set of observations by consideration of a sample. The inferences are then used to develop empirical generalizations, models, and hypotheses.

---

[5]The seminar during the first few years included Brian J. L. Berry, William Bunge, Michael F. Dacey, Arthur Getis, Duane F. Marble, Robert W. Morrill, John D. Nystuen, and Waldo Tobler. Torsten Hägerstrand was a visiting scholar for one semester. From the University of Washington and University of Lund, training in quantitative procedures spread to numerous universities. In Britain the chief center for the new approach was at Bristol University, where Haggett was invited to become a second professor of geography. The new procedures spread rapidly in the Soviet Union and other parts of the world where geography was developed on the Soviet model. The economist Walter Isard established the Department of Regional Science at the University of Pennsylvania. The Regional Science Association now has branches in many countries.

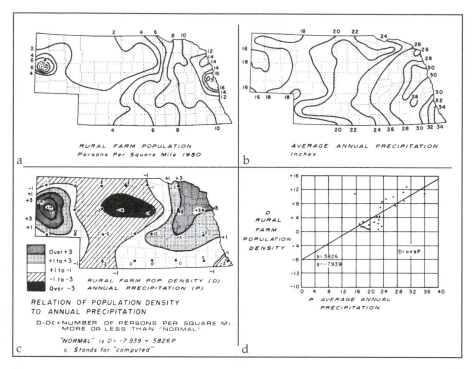

Figure 55 Relation of population to rainfall in Nebraska (from Robinson and Bryson, 1957)

As an example of the statistical approach to an old problem, consider the procedure geographers use in defining the degree of correspondence in the extent of area occupied by two phenomena. This was formerly done by the inspection of superimposed maps from which it was possible to identify in qualitative terms the degree of correspondence. It was also possible to identify the problem areas where the two phenomena did not correspond. The results, however, were not precise (McCarty and Salisbury, 1961). In 1957 Arthur H. Robinson and Reid Bryson demonstrated what might be done to compare the density of rural population and the average annual rainfall in a specific area—Nebraska (Robinson and Bryson, 1957). From an inspection of the two maps (Fig. 55a and b), it is clear that the density of rural population is lower where the average rainfall is less. But there are some exceptions. The two phenomena do not vary exactly in the same way. Therefore, it seems that some other factors besides rainfall are important in explaining the density of population. The statistical comparison of two unlike quantities (population density and rainfall) requires a special procedure to make the figures comparable. After plotting a pattern of random points over the state, readings of the population density and of the rainfall are recorded for each point. These data are then plotted on a scatter diagram where one scale measures population density and the other measures rainfall (Fig. 55d). From the equation that best fits the distribution of dots on the diagram it is possible to discover the density that should be expected for each amount of rainfall. A map showing the departures from the expected densities identifies the problem areas (Fig. 55c). The technique of multiple regression analysis makes possible the measurement of such areal associations (Robinson, 1962).

In this study of Nebraska, two contingent distributions were compared, one contingent on the area of enumeration and the other on the period of observation. But geographers often want to compare the correspondence of two discrete or discontinuous distributions—for example, land use and soil type. The area to be analyzed can be divided into unit areas, similar to those revealed by the fractional code system used in Finch's study of the Montfort area (Fig. 38). The existence of a concentration of certain kinds of land use on certain kinds of soil can be identified by counting the unit areas. By the use of a random sampling of land use and soil at selected points (the chi-square method), the degree to which a hypothetical correlation agrees with the facts can be determined. This method gives a number that increases with the absolute amount of difference between a hypothetical association and an actual, observed distribution. By this method the areal spread of a region can be determined quantitatively.

William Bunge, in his book *Theoretical Geography*, demonstrates the benefits of using sampling procedures by making a partial restudy of Finch's Montfort area (pp. 333–335). A table of random numbers was translated into coordinates to provide a map of random points. At each point the land use was recorded. By noting the occurrence or nonoccurrence of one form of land use (grassland), Bunge arrived at the estimate that this kind of cover was to be found on 21 percent of the area. Finch had calculated that grassland covered 24.1 percent of the area. Which figure is more nearly correct? The use of sampling from random points permits the geographer to calculate his probable error. Furthermore, by using random points plotted on vertical air photographs and then checking the necessary information at each point in the field, it is not necessary to make a complete map of the whole area. Finch spent 120 days in the field, and the preparation of his land use table by planimetering the field map took a year. Bunge estimated that by using the random sampling method the required information could be acquired in three days, and with much greater accuracy. With random sampling techniques it is entirely feasible to obtain good estimates of continental or global land use percentages on a modest budget (Bunge, 1966:104–107).

Geographers have been accustomed to pointing out that the nature of their field of study does not permit the use of controlled experiments. This is no longer true. Statistical procedures offer the equivalent of a geographical laboratory. For example, Waldo Tobler shows how the distorting effects of transportation, terrain, and other conditions can be eliminated on maps to permit effective testing of the concept of central places (Tobler, 1959). When the interaction of several features associated in a spatial system is to be analyzed, a covariance analysis permits the student to keep one element constant while other elements vary in relation to it. It is possible to measure the variations in relation to constant factors so that geographers can now measure functional relations where the distributing effects of "other things" that are not equal are eliminated.[6]

---

[6]For books describing these statistical techniques, see Duncan, Cuzzort, and Duncan, 1961; Haggett, 1966; Cole and King, 1968; King, 1969; and Taaffe, 1970.

The use of statistical methods in geography,[7] as used in computers, was summarized by Greer-Wootten (1972) in a survey of 34 areas of application covering 772 citations. Not all the empirical studies cited therein were without errors of interpretation or misuse of data. Gould (1970) complained that use of statistics in geography might be a "wild goose chase." His warning has been echoed by Clark and Hosking (1986:iii), who say that: "This [methodological revolution] was not without its problems. Examples of ill-conceived analyses, overexuberance in the use of some methods, and gross errors can be found in abundance in the geographic literature."

An updating and expansion of the Greer-Wootten survey would be welcome. The Chorley and Haggett (1967) volume has been widely used in graduate seminars, and Bennett (1981) has documented the progress of spatial analysis in Europe in the 1970s.

The publication in 1968 of *Spatial Analysis*, a book of readings edited by Berry and Marble, gave spatial analysts a firm foundation on which to build their research. For the point-pattern analyst, the reprinting of Matui's (1932) pioneer article was invaluable. *Spatial Analysis* followed on the two volumes of articles and new work in quantitative geography edited by Garrison and Marble (1967). The past quarter-century has witnessed the use of more complicated models (Tobler, 1970), beyond the basic set of programs (Marble, 1967) listed in the second edition of *All Possible Worlds*. An important recent collection of model usage is *Spatial Statistics and Models*, edited by Gaile and Willmott in 1984. Its table of contents is a fine guide to work accomplished by spatial analysts after 1968.

The work by Matui (1932) stimulated Michael Dacey, whose work in point-pattern analysis runs to many dozens of articles. Getis and Boots (1978) and Boots and Getis (1988) have codified the current approach, with many empirical examples. It should be noted that works by two dozen geographers and numerous geologists have been cited in *Spatial Statistics* by Brian D. Ripley (1981). The work by Haining (1982) is in that tradition. Point-pattern analysis studies led to the work on spatial autocorrelation (Dacey, 1968), stimulated by Moran (1948) and Geary (1954). British statisticians and epidemiologists have been more oriented to spatial distributions than have their American colleagues, just as German economists have been more spatially oriented than have their American counterparts. This British tradition continues to be strongly felt by American geographers. One problem in spatial analysis, as pointed out by Mead (1974), is that results change when spatial patterns are viewed at different scales. Problems with the multidimensionality of spatial data are discussed by Nijkamp (1979). Spatial autocorrelation is now routinely taught in introductory quantitative methods textbooks, as in Clark and Hosking (1986), Ebdon (1977), Griffith and Amrhein (1991), Silk (1979) and Taylor (1977). These studies rely on original work by Dacey (1968), Cliff and Ord (1973, 1981), Cliff et al. (1975), and numerous others. A good summary is given in Griffith (1987, 1988).

---

[7]The author wishes to thank Forrest R. Pitts for help with this section of the manuscript.

Diffusion of ideas and techniques through a population of relevant receivers is an old and strong tradition in geography. Modeling of diffusion relies in good part on the pioneer work of Torsten Hägerstrand (1965), who used simulation models whose parameters were based on empirical field work on propensity to communicate over distance. A classical approach to geographical diffusion theory is given by Hudson (1972). Brown (1981) has summarized the diffusion work that he and many students did in the 1970s. Gatrell (1984) has studied the recent history of spatial diffusion modeling. Berry (1972) used a hierarchical diffusion model to explain the flow of information through an urban system of growth centers. Yapa (1977, 1978) has looked at barriers to acceptance of innovations in a third world context. And Gaspar and Gould (1981) have used polyhedral dynamics, also known as q-analysis, to study barriers to the spread of agricultural innovations in Portugal. Gould (1980a, 1980b) introduced the basic ideas of q-analysis to geographers. His recent work with colleagues (Gould et al., 1991) is a diffusion study in "predicting the next AIDS map," using a spatial adaptive filtering technique.

A significant portion of recent diffusion work has been done in medical geography. Cliff et al. (1981) studied the spread of several diseases in Iceland; Cliff and Haggett (1985) studied the spread of measles in the Pacific; and Cliff, Haggett, and Ord (1986) have modeled the spread of influenza epidemics. Pyle (1986) also brought out a book on influenza epidemics. Earlier, Tinline (1970, 1971) had introduced a novel diffusion model in his study of cattle epidemics. R. R. White's (1972) work on probability maps for leukemia echoes the earlier work of Choynowski (1959, reprinted in *Spatial Analysis*) on brain tumors in Poland.

In the study of interaction at a distance, it is necessary to fit curves to the data. The work of Morrill and Pitts (1965) led to use of the expansion method of Casetti (1972), who pioneered its use within geography. It is increasingly useful in many research situations. Eldridge and Jones (1991) have used this technique in a two-dimensional way to study "warped space," where distance-decay functions may vary directionally from a point. Further applications of the expansion method appear in Jones and Casetti (1992).

Simulation and microsimulation have been useful approaches in cultural and economic geography since the early work of Hägerstrand (1952, summarized in 1965). Comments by Pitts (1963) on simulation models, as well as the simulation of Polynesian drift voyages by Levison et al. (1973), were early responses to this technique. More recently, Whitmore (1991) has simulated the population collapse in sixteenth-century Mexico, while Amrhein and MacKinnon (1988) and Pandit and Casetti (1989) have looked at labor forces in this fashion. A simulation of the spread of AIDS in Finland was done by Löytönen (1991). Both simulation and microsimulation have long been used to good effect in physical geography explorations of process.

In their roles as planning consultants, geographers often use Weberian location analysis to suggest new locations (more or fewer) for public and private facilities. Many location–allocation algorithms for computers have been published at the University of Iowa, where this approach is a major concern, an outgrowth of the influence of Harold McCarty. From Iowa have come books and many computer program guides by Ghosh and Rushton (1987), Goodchild and Noronha (1983), Hillsman (1980), and Rushton et al. (1973).

Approaches described by Teitz and Bart (1968) have been used in third world settings by Ayeni et al. (1987) and Rushton (1984, 1988) and have been incorporated in microcomputer interactive contexts as spatial decision support systems by Densham and Rushton (1987) and Honey et al. (1991). Regional forecasting models were introduced into geography by Martin and Oeppen (1975) and Martin (1975).

The influence of mathematical geology as practiced at the University of Kansas has appeared in geographic studies. Tobler (1969a) derived the spectrum of settlements on a highway route, a method questioned by Rayner and Golledge (1973). Tobler (1969b) also produced a topographic replica of West Africa by adding orthogonal sine curves to each other. Another useful technique from geology at Lawrence is kriging, also known as geostatistics. With its "semi-variogram," a curve of autocorrelation over distance, it is now widely applied in soils geography and in trend-surface analysis (Agterberg, 1984).

Networks in geography were introduced by Haggett and Chorley (1969) and were a major part of the Lowe and Moryadas (1975) textbook. The 1965 article by Pitts on the network of early Russian cities used incidence, or presence/absence, data and was based on Garrison (1960). The Garrison-Pitts approach was then used by the anthropologist G. J. Irwin (1974, 1978) in Papua New Guinea. In 1979 Pitts used real distance measures, which produced more relevant stress and minimum access-distance results.

A number of well-written textbooks on statistical use have been published. Among these are King (1969), Lewis (1977), Taylor (1977), Silk (1979), Clark and Hosking (1986), and Griffith and Amrhein (1991). The last two combine univariate and multivariate approaches in a way that earlier texts could not—or did not—do, owing to the need to start from the very lowest levels of introductory mathematics. The Griffith and Amrhein book is very closely argued but rewards the effort put into it. Analytic approaches to behavioral geography are found in Golledge and Stimson (1987), and Golledge and Timmermans (1988) have summarized the use of behavioral models in geography.

A number of the early quantitative people in geography moved on to broad-scale geographic analysis; Brown (1991) on migration and Morrill (1973, 1981) on political reapportionment are examples. Some have joined with the younger generation (often their own students) to question the validity of logical empiricism alone to solve geographic problems. *A Search for Common Ground* (Gould and Olsson, 1982), followed by *A Ground for Common Search* (Golledge et al., 1988) are examples. Hardly a year passes in which a former user of statistics and computers in geography does not publish a *mea culpa* and warn the young to beware. Yet the profession continues to attract imaginative workers, such as Anselin (1988).

GEOGRAPHIC INFORMATION SCIENCE

During the last 15 or so years geographic information science has experienced a remarkable growth in teaching programs, certificate programs, bachelor's and master's degree programs, and now doctoral programs. The surge of this activity has been notable in geography, but it has been adopted substantially and simultaneously in the adjacent social and environmental sciences, in which speed and accuracy of data arrangement and delivery are of significance, spatial relationships increase the complexity of statistical analysis, and heterogeneous behavior makes computer-based modeling essential.

Geographic information science has three major components. The first is geographic information systems (GIS, or geo-informatics) (Longley et al., 2001; DeMers, 2000). This deals with data collection, processing, display, storage of information, and analysis. Geographic information systems automate geographic concepts, facilitate decision making, assist in formulation of hypotheses, and aid prediction. It is a form of computer-based cartography and can create maps both adaptable and purposeful. Operation may be in a multitude of domains, and modern GIS has multiple ramifications.

Global positioning systems (GPS) merged with remotely sensed and survey data has given new power to applied geography. Computerization by local governments of cadastral property characteristics and property taxation data has resulted in merged data sets that underpin modern emergency response systems. Survey data systems have been merged with real-time systems linked to satellite platforms; use of these data has been increasing rapidly. The impetus for this development came . from the research and development branch of the Department of Defense. Only in the last decade did the Department drop its access barriers and make higher-resolution images available to the public, stimulating the new wave of site-specific social and environmental research. The accompanying growth of commercial companies such as Environmental Systems Research Institute (ESRI) also has simplified and streamlined access to and use of GIS.

The second component of geographic information science is spatial analysis. This came to geography from statistical and econometric analysis, providing measures of spatial dependence (i.e., what is observed in one area is dependent on what is in another area). To resolve matters of spatial dependence, a knowledge of mathematics and statistics is needed; examinations in these subjects have now replaced the traditional requirement of a foreign language, commonplace through the 1950s (Anselin, 1988). It was especially during the late 1990s and early 2000s that much progress was made in testing and correcting for spatial dependence in the statistical analysis of geographical patterns. Progress in this regard also was made parallel to geography in spatial econometrics, geophysics, and ecology in what has become a broad multidisciplinary effort. The National Science Foundation made a sequence of two programmatic fundings at the University of California, Santa Barbara. The first was to the National Center for GIS, which focused its attention on GIS software and its diffusion. The second was to the National Center for Spatially Integrated Social Science. The underlying thesis was that these two developments would provide generic capabilities that cut across the social and human-environmental sciences, providing innovative perspectives and new means in integrative thinking.

One other development in geographic information science, yet in its early stages, relates to the emergence of computational geography, a term coined by

S. Openshaw (Openshaw and Abrahart, 2000). It has borrowed largely from computational biology and centers on bioinformatics and the discovery of the underlying biological pattern of DNA; several other disciplines have positioned themselves similarly. Geography has begun to use the power of computation to transcend the received body of location theory, recognizing that heterogeneous human behavior can be modeled in an adaptive agents framework. There is now a land-use change group at the University of Indiana whose primary function is to model the role of interacting actors in land use change. Economist Thomas Schelling in his study of micromotives and macrobehavior has demonstrated from these studies that unexpected results arise from a system of interacting actors; his contribution is the foundation document in the emergence of social complexity theory. Geographic information systems together with spatial analysis and computational geography constitute the three essential parts of geographical information sciences as we enter the twenty-first century.

REFERENCES: CHAPTER 19

Agterberg, F. P. 1984. "Trend Surface Analysis." In Gary L. Gaile and Cort J. Willmott, eds., *Spatial Statistics and Models*. Pp. 147–171. Boston, D. Reidel.

Amrhein, Carl G., and Ross D. MacKinnon. March 1988. "A Microsimulation of a Spatial Labor Market." *Annals AAG* 78, no. 1:112–131.

Anselin, Luc. 1988. *Spatial Econometrics: Methods and Models*. Boston: Kluwer Academic.

Ayeni, B., Rushton, G. and McNulty, M. L. 1987. "Improving the Geographical Accessibility of Health Care in Rural Areas: A Nigerian Case Study." *Social Science and Medicine* 25, no. 10:1083–1094.

Barrett, E. C. 1970. "Rethinking Climatology: An Introduction to the Uses of Weather Satellite Photographic Data in Climatological Studies." *Progress in Geography* 2:153–205.

Bennett, R. J., ed. 1981. *European Progress in Spatial Analysis*. London: Pion.

Berry, B. J. L. 1972. "Hierarchical Diffusion: The Basis of Development Filtering and Spread in a System of Growth Centers." In Niles M. Hansen, ed., *Growth Centers in Regional Economic Development*. Pp. 108–138. New York: Free Press.

Berry, B. J. L., and Marble, Duane F. eds. 1968. *Spatial Analysis: A Reader in Statistical Geography*. Englewood Cliffs, N.J.: Prentice Hall.

Bird, J. B., and Morrison, A. 1964. "Space Photography and Its Geographic Applications." *Geographical Review* 54:463–486.

Blache, J. 1921. "Modes of Life in the Morocco Countryside, Interpretations of Aerial Photographs." *Geographical Review* 11:477–502.

Boots, B. N., and Arthur Getis. 1988. *Point Pattern Analysis*. Newbury Pa,:, Calif.: Sage Publishers.

Brown, L. A. 1981. *Innovation Diffusion: A New Perspective*. London and New York: Methuen.

———. 1991. *Place, Migration, and Development in the Third World*. New York: Routledge.

Brunn, S. D., Cutter, S. L., and Harrington, J. W. Jr. 2004. *Geography and Technology*. Dordrecht Kluwer Academic Publishers.

Bunge, W. 1966. *Theoretical Geography*. Lund: University of Lund. (Lund Studies in Geography.)

Casetti, Emilio. 1972. "Generating Models by the Expansion Method: Applications to Geographic Research." *Geographical Analysis* 4:81–91.

Castells, M. 1996. *The Rise of the Network Society*. Oxford: Blackwell.

Chorley, Richard J., and Haggett, Peter, eds. 1967. *Models in Geography*. London: Methuen.

Choynowski, M. 1959. "Maps Based on Probabilities." *Journal of the American Statistical Association* 54:385–388.

Clark, William A. V., and Hosking, Peter L. 1986. *Statistical Methods for Geographers*. New York: John Wiley & Sons.

Cliff, Andrew D., and Ord, J. Keith. 1973. *Spatial Autocorrelation*. London: Pion.

Cliff, Andrew D., and Ord, J. Keith. 1981. *Spatial Processes: Models and Applications*. London: Pion.

Cliff, Andrew D., and Haggett, Peter. 1985. *The Spread of Measles in Fiji and the Pacific*. Canberra: Australian National University, Department of Human Geography Publication 18.

Cliff, Andrew D., Haggett, Peter, and Ord, J. Keith. 1986. *Spatial Aspects of Influenza Epidemics*. London: Pion.

Cliff, Andrew D., Haggett, Peter, Ord, J. Keith, Bassett, Keith A. and Davies, Richard B. 1975. *Elements of Spatial Structure: A Quantitative Approach*. Cambridge: Cambridge University Press.

Cliff, Andrew D., Haggett, Ord, J. Keith, and Versey, G. R. 1981. *Spatial Diffusion: An Historical Geography of Epidemics in an Island Community*. Cambridge, UK: Cambridge University Press.

Cole, J. P., and King, C. A. M. 1968. *Quantitative Geography. Techniques and Theories in Geography*. London: John Wiley & Sons.

Cosgrove, Denis. 2001. *Apollo's Eye: A Cartographic Genealogy of the Earth in the Western Imagination*. Baltimore: Johns Hopkins University Press.

Crone, G. R. 1950. *Maps and Their Makers, an Introduction to the History of Cartography*. New York: G. P. Putnam's Sons.

Dacey, Michael F. 1968. "A Review on Measures of Contiguity for Two and *k*-color Maps." In Berry and Marble, eds., *Spatial Analysis*. Pp. 479–495.

DeMers, M. N. 2000. *Fundamentals of Geographic Information Systems*. 2nd ed. New York: John Wiley.

Densham, P. J., and Rushton, Gerard. 1987. "Decision Support Systems for Locational Planning," In R. G. Golledge and H. Timmermans, eds., *Behavioral Modeling in Geography*. Pp. 56–96. New York: Croom Helm.

Duncan, O. D., Cuzzort, R. P., and Duncan, B. 1961. *Statistical Geography*. New York: Free Press.

Ebdon, David. 1977. *Statistics in Geography: A Practical Approach*. Oxford: Basil Blackwell.

Eldridge, J. Douglas, and Jones, John Paul, III. November 1991. "Warped Space: A Geography of Distance Decay." *Professional Geographer* 43, no. 4:500–511.

Freeman, L. C. 1965. *Elementary Applied Statistics for Students in Behavioral Science*. New York: John Wiley & Sons.

Gaile, Gary L., and Willmott, Cort J., eds. 1984. *Spatial Statistics and Models*. Boston: D. Reidel.

Garrison, William L. 1960. "Connectivity of the Interstate Highway System." *Papers and Proceedings of the Regional Science Association* 6:121–137.

Garrison, William L., and Duane F. Marble, eds. 1967. *Quantitative Geography*. 2 vols. Studies in Geography, Nos. 13 and 14. Evanston, Ill.: Northwestern University.

Gaspar, G., and Gould, P. 1981. "The Cova de Beira: An Applied Structural Analysis of Agriculture and Communication." In Allan R. Pred, ed., *Space and Time in Geography*. Pp. 183–214. Lund: Gleerup.

Gatrell, Anthony C. 1984. "The Geometry of a Research Specialty: Spatial Diffusion Modelling." *Annals AAG* 74, no. 3:437–453.

Geary, R. C. 1954. "The Contiguity Ratio and Statistical Mapping." *The Incorporated Statistician* 5:115–141.

Getis, Arthur, and Boots, Barry N. 1978. *Models of Spatial Processes: An Approach to the Study of Point, Line and Area Patterns.* Cambridge: Cambridge University Press.

Ghosh, Avijit, and Rushton, Gerard, eds. 1987. *Spatial Analysis and Location-Allocation Models.* New York: Van Nostrand Reinhold.

"*GIS World* Interview—Roger Tomlinson: The Father of GIS." *GIS World* April 1996: 56–60.

Golledge, Reginald G., and Stimson, Robert J. 1987. *Analytical Behavioural Geography.* London: Croom Helm.

Golledge, Reginald G., Couclelis, Helen, and Gould, Peter. 1988. *A Ground for Common Search.* Santa Barbara, Calif.: Santa Barbara Geographical Press.

Golledge, Reginald G., and Timmermans, Harry, eds. 1988. *Behavioural Modelling in Geography and Planning.* London: Croom Helm.

Goodchild, Michael F., and Noronha, V. T. 1983. *Location-Allocation for Small Computers.* Monograph No. 8, Department of Geography, University of Iowa.

Gore, A. 1998. *The Digital Earth: Understanding out Planet in the 21st Century.* Los Angeles: California Science Center.

Gould, Peter. 1970. "Is *Statistix Inferens* the Geographical Name for a Wild Goose?" *Economic Geography* 46 Supplement: 439–448.

———. 1980a. "Q-Analysis, or a Language of Structure: An Introduction for Social Scientists, Geographers, and Planners," *International Journal of Man-Machine Studies* 13:169–199.

———. 1980b. "A Structural Language of Relations." In R. Craig and M. Labovitz, eds., *Future Trends in Geomathematics.* London: Pion.

Gould, Peter, Kabel, Joseph, Gorr, Wilpen, and Golub, Andrew. 1991. "AIDS: Predicting the Next Map." *Interfaces*, 21, no. 3:80–92.

Gould, Peter, and Olsson, Gunnar. 1982. *A Search for Common Ground.* London: Pion.

Greer-Wootten, Bryn. 1972. *A Bibliography of Statistical Applications in Geography.* Washington, D.C.: Association of American Geographers. Technical Paper No. 9.

Griffith, Daniel A. 1987. *Spatial Autocorrelation: A Primer.* Washington, D.C.: Association of American Geographers.

———. 1988. *Advanced Spatial Statistics: Special Topics in the Exploration of Spatial Data Series.* Boston: Kluwer Academic Publishers.

Griffith, Daniel A., and Amrhein, Carl G. 1991. *Statistical Analysis for Geographers.* Englewood Cliffs, N.J.: Prentice Hall.

Hägerstrand, Torsten. 1965. "A Monte Carlo Approach to Diffusion." *Archives Européennes de Sociologie* 6:43–67.

Haggett, P. 1966. *Locational Analysis in Human Geography.* New York: St. Martin's Press.

Haggett, Peter, and Chorley, Richard J. 1969. *Network Analysis in Geography.* New York: St. Martin's Press.

Haggett, Peter, Cliff, Andrew D., and Frey, Allan. 1977. *Locational Models.* New York: John Wiley. (2nd ed. of *Locational Analysis in Human Geography*, 1965)

Haining, Robert. 1982. "Describing and Modeling Rural Settlement Maps." *Annals AAG* 72:211–223.

Hillsman, E. L. 1980. *Heuristic Solutions to Location-Allocation Problems.* Monograph No. 7, Department of Geography, University of Iowa.

Honey, Rex, Rushton, Gerard, Armstrong, M. P., Lolonis, Panos, Dalziel, De, S., and Densham, P. J. 1991. "Stages in the Adoption of a Spatial Decision Support System for Reorganizing Service Delivery Systems." *Environment and Planning* 9:51–63.

Hudson, John. 1972. *Geographic Diffusion Theory.* Evanston, Ill.: Northwestern University, Studies in Geography 19.

Irwin, G. J. 1974. "The Emergence of a Central Place in Coastal Papuan Prehistory: A Theoretical Approach." *Mankind* 9:268–272.

———. 1978. "Pots and Entrepôts: A Study of Settlement, Trade, and the Development of Economic Specialization in Papuan Prehistory." *World Archaeology*: 299–319.

Jensen, J. R. 2000. *Remote Sensing: An Environmental Perspective.* Englewood Cliffs, N.J.: Prentice-Hall.

Jones, John Paul, III, and Casetti, Emilio, eds. 1992. *Applications of the Expansion Method.* London: Routledge.

Kendall, M. G. 1939. "The Geographical Distribution of Crop Productivity in England." *Journal of the Royal Statistical Society* 102:21–62.

King, L. W. 1969. *Statistical Analysis in Geography.* Englewood Cliffs, N.J.: Prentice Hall.

King-Hele, D. 1967. "The Shape of the Earth." *Scientific American* 217:67–76.

Kuhn, H. W., and Kuenne, R. E. 1962. "An Efficient Algorithm for the Numerical Solution of the Generalized Weber Problem in Spatial Economics." *Journal of Regional Science* 4:21–23.

Lee, W. T. 1920. "Airplanes and Geography." *Geographical Review* 10:310–325.

———. 1922. *The Face of the Earth as Seen from the Air: A Study in the Application of Airplane Photography to Geography.* New York: American Geographical Society.

Levison, Michael, Ward, R. Gerard, and Webb, John W. [with the assistance of Trevor I. Fenner and W. Alan Sentance]. 1973. *The Settlement of Polynesia: A Computer Simulation.* Minneapolis: University of Minnesota Press.

Lewis, P. M. 1965. "Three Related Problems in the Formulation of Laws in Geography." *The Professional Geographer* 17:24–27.

Lewis, Peter. 1977. *Maps and Statistics.* London: Methuen.

Lowe, John C., and Moryadas, S. 1975. *The Geography of Movement.* Boston: Houghton Mifflin.

Löytönen, Markku. March 1991. "The Spatial Diffusion of Human Immunodeficiency Virus Type I in Finland." *Annals AAG* 81, no. 1:127–151.

Marble. D. F. 1967. *Some Computer Programs for Geographic Research.* Evanston: Northwestern University, Department of Geography.

Martin, R. L. 1975. "Identification and Estimation of Dynamic Space-Time Forecasting Models for Geographic Data." *Advances in Applied Probability* 7:455–456.

Martin, R. L., and Oeppen, J. E. 1975. "The Identification of Regional Forecasting Models Using Space-Time Correlation Functions." *Institute of British Geographers, Publications.* 66:95–118.

Matui, I. 1932. "Statistical Study of the Distribution of Scattered Villages in Two Regions of the Tonami Plain, Toyama Prefecture." *Japanese Journal of Geology and Geography* 9:251–266.

McCarty, H. H., and Salisbury, N. E. 1961. *Visual Comparison of Isopleth Maps as a Means of Determining Correlations between Spatially Distributed Phenomena.* Iowa City: State University of Iowa, Department of Geography.

Mead, R. 1974. "A Test for Spatial Pattern at Several Scales Using Data from a Grid of Contiguous Quadrats." *Biometrics* 30:295–307.

Miller, O. M. 1931. "Planetabling from the Air. An Approximate Method of Plotting from Oblique Aerial Photography." *Geographical Review* 21:202–212, 660–662.

Moran, P. A. P. 1948. "The Interpretation of Statistical Maps." *Journal of the Royal Statistical Society*, Series B, 19:243–251.

Morrill, Richard L. 1973. "Ideal and Reality in Reapportionment." *Annals AAG* 63:463–477.

———. 1981. *Political Redistricting and Geographic Theory*. Washington, D.C.: Association of American Geographers.

Morrill, Richard L., and Pitts, Forrest R. 1965. "Marriage, Migration, and the Mean Information Field: A Study in Uniqueness and Generality." *Annals AAG* 57, no. 2:401–422.

Nijkamp, Peter. 1979. *Multidimensional Spatial Data and Decision Analysis*. New York: John Wiley & Sons.

Olson, C. E. 1967. "Accuracy of Land-Use Interpretation from Infrared Imagery in the 4.5 to 5.5 Micron Band." *Annals AAG* 57:382–388.

Openshaw, Stan and Robert J. Abrahart, eds. 2000. *Geocomputation*. London and New York: Taylor & Francis.

Pandit, Kavita, and Casetti, Emilio. 1989. "The Shifting Patterns of Sectoral Labor Allocation During Development: Developed Versus Developing Countries." *Annals AAG* 79:329–344.

Pitts, Forrest R. 1963. "Problems in Computer Simulation of Diffusion." *Papers of the Regional Science Association* 11:111–119.

———. 1965. "A Graph Theoretic Approach to Historical Geography." *Professional Geographer* 17, no. 5:15–20.

———. 1979. "The Medieval River Trade Network of Russia Revisited." *Social Networks* 1, no. 3:285–292.

Pyle, Gerard F. 1986. *The Diffusion of Influenza*. Totowa, N.J.: Rowman & Littlefield.

Rayner, John N., and Reginald G. Golledge. 1973. "The Spectrum of U.S. Route 40 Re-examined." *Geographical Analysis* 5:338–350.

Ripley, Brian D. 1981. *Spatial Statistics*. New York: John Wiley & Sons.

Robinson, A. H. 1955. "The 1837 Maps of Henry Drury Harness." *Geographical Journal* 121, part 4, Dec. 1955:440–450.

———. 1956. "Mapping the Land." *Scientific Monthly* 82:294–303.

———. 1962. "Mapping the Correspondence of Isarithmic Maps." *Annals AAG* 52:414–425.

Robinson, A. H., and Bryson, R. A. 1957. "A Method for Describing Quantitatively the Correspondence of Geographical Distributions." *Annals AAG* 47:379–391.

Roszak, T. 1994. *The Cult of Information*. Berkeley: University of California Press.

Rushton, Gerard. 1984. "Use of Location-Allocation Models for Improving the Geographical Accessibility of Rural Services in Developing Countries." *International Regional Science Review* 9:217–240.

———. 1988. "Location Theory, Location-Allocation Models, and Service Development Planning in the Third World." *Economic Geography* 64, no. 2:97–120.

Rushton, Gerard, Goodchild, Michael F., and Ostresh, Lawrence M., Jr. eds. 1973. *Computer Programs for Location-Allocation Problems*. Iowa City: University of Iowa, Department of Geography. Monograph No. 6.

Russell, J. A., Foster, F. W., and McMurry, K. C. 1943. "Some Applications of Aerial Photographs to Geographic Inventory." *Papers of the Michigan Academy of Science, Arts and Letters* 29:315–341.

Schelling, Thomas C. 1978. *Micromotives and Macrobehavior*. New York: Norton.

Silk, John. 1979. *Statistical Concepts in Geography*. London: George Allen & Unwin.

Simpson, R. B. 1966. "Radar, Geographic Tool." *Annals AAG* 56:80–96.

Stewart, J. Q. 1947. "Empirical Mathematical Rules Concerning the Distribution and Equilibrium of Population." *Geographical Review* 37:461–485.

Stewart, J. Q., and Warntz, W. 1958. "Macrogeography and Social Science." *Geographical Review* 48:167–184.

Taaffe, E. J., ed. 1970. *Geography.* Englewood Cliffs, N.J.: Prentice Hall.

Taylor, Peter J. 1977. *Quantitative Methods in Geography: An Introduction to Spatial Analysis.* Boston: Houghton Mifflin.

Teitz, Michael B., and Bart, Polly. 1968. "Heuristic Methods for Estimating the Generalized Vertex Median of a Weighted Graph." *Operations Research* 16: 955–961.

Tinline, Roland R. 1970. "Lee Wave Hypothesis for the Initial Pattern of Spread During the 1967–68 Foot and Mouth Epizootic," *Nature* 227:860–862.

———. 1971. "Linear Operators in Diffusion Research," M. D. I. Chisholm, A. E. Frey, and Peter Haggett, eds. *Regional Forecasting.* Pp. 71–91. London: Butterworth.

Tobler, W. 1959. "Automation and Cartography." *Geographical Review* 49:526–534.

———. 1969a. "The Spectrum of U.S. 40." *Regional Science Association*, Papers, 23:45–52.

———. 1969b. "Geographic Filters and Their Inverses," *Geographical Analysis* 1:234–253.

———, ed. 1970. *Selected Computer Programs.* Ann Arbor: Michigan Geographical Publications.

TVA. *A History of the Tennessee Valley Authority.* 1983 Knoxville, Tenn: TVA. (Information Office)

Weaver, J. C. 1954. "Crop-Combination Regions in the Middle West." *Geographical Review* 44:175–200.

———. 1956. "The County as a Spatial Average in Agricultural Geography." *Geographical Review* 46:536–565.

White, R. R. 1972. "Probability Maps of Leukaemia Mortalities in England and Wales." In N. D. McGlashan, ed., *Medical Geography: Techniques and Field Studies.* London.

Whitmore, Thomas M. 1991. "A Simulation of the Sixteenth-Century Population Collapse in the Basin of Mexico." *Annals AAG* 81:464–487.

Willow Run Laboratories. 1966. *Peaceful Uses of Earth Observation Spacecraft.* 3 vols. Ann Arbor: University of Michigan Press.

Wright, John K. 1937. "Some Measures of Distributions." *Annals, AAG* 27:177–211.

Yapa, Lakshman S. 1977. "The Green Revolution: A Diffusion Model." *Annals AAG* 67, no. 3:350–359.

———. 1978. "Non-Adoption of Innovations: Evidence from Discriminant Analysis." *Economic Geography* 54, no. 2:115–134.

20

Innovation and Tradition

> *The advantages of mathematical models—unambiguity, possibility of strict deduction, verifiability by observed data—are well known. This does not mean that models formulated in ordinary language are to be despised or refused. A verbal model is better than no model at all, or a model which, because it can be formulated mathematically, is forcibly imposed upon and falsifies reality.*
> —*Ludwig von Bertalanffy,* General System Theory, Foundations, Development, Applications

Innovation is the essence of the growth of science. The notion that provides innovation arrives mysteriously, sometimes in seconds and sometimes the product of many years of thought. The unseen forces at work that function as preconditions include the development of scientific thinking, the social conditions surrounding the undertaking, the development of associated sciences, and the individual creativity of geographers. We have shifted our stance from "what" to "how," from form to function, from substance to process, and from matter to energy en route to resolution of the problem. And the geographer is aware of vast numbers of problems ranging from the local to the global; to solve at least some of these is the task. Innovations in thinking might solve some of these problems.

These innovations might come from an individual outside the machinery of science, but they are more likely to be the product of a new scientific posture. These new geographies might present themselves at regular or irregular intervals; they represent the advance of a new viewpoint by way of which new thought may be encouraged. The appearance of a new geography has been loudly and persistently proclaimed by many generations of geographers since ancient times. Usually, the adjective only indicates that some new information is at hand, but occasionally there are genuine innovations of technique or method or in the concepts that provide an enlarged comprehension of some kind of order in earth space. Sometimes a "new geography" means that a whole new world has been brought to light. As long as there is hope for progress in scholarship, the number of all possible worlds awaiting discovery is infinite. But it is always wise to distinguish what is new from what is not new and by examining the record to avoid, as much as possible, the persistence of old error.

In the 1970s and 1980s, when the existence of another "new geography" was loudly proclaimed, there was more need than ever before to raise the question "What is New?" (Dickenson and Clarke, 1972). In the preceding chapter were listed some of the unprecedented changes in the technology of observation and analysis that provide geographers not only with more information than has ever been available before but also with a means of storage and recall and a way to carry out complex analyses. What kinds of questions do geographers ask of these new data? And have the basic purposes of geographic study been given a new direction? The geography of recent years, as in years past, is a mixture of innovation and tradition. That circumstance encourages further study in the history of geographical ideas. A large part of the quest is to understand how it came to be (Chorley, 1973; Chorley and Haggett, 1965; Taylor, 1976; Goodchild and Janelle, 1988).

A review of what geographers have done in the past reveals the persistence of certain kinds of avoidable error (James, 1967). One major source of repeated error seems to be the common failure of too many geographers to read what other geographers, past and present, have written. Strangely, this appears to be a characteristic of all fields of scholarship and is not restricted to the contemporary period. One reason why Strabo's books on geography were found almost intact was that his contemporaries did not read what he had so laboriously written. Again and again we come across examples of general concepts formulated by scholars of one generation that have outlived their usefulness in illuminating the arrangement of things on the earth. Notable is the persistence of Aristotle's ideas concerning the difficulties of life in the so-called torrid zone even after the attack on these ideas began in the thirteenth century. Another example is the continued use of Maury's wind zones. Peter Haggett suggests that progress is marked "by the sound of plummeting hypotheses" (Haggett, 1966:277). The difficulty is that some hypotheses do not plummet soon enough but remain as obstacles to confuse later generations.

Geographers, like scholars in other fields of learning, have been caught in certain semantic traps. Because human beings, alone among animals, possess language in which abstract ideas can be represented by symbols, they very easily confuse the symbol with the "reality" for which it stands (Bertalanffy, 1965). Yet the nature of word symbols, which are not always precisely defined and which carry heavy burdens of connotation, makes possible the development of professional controversy that is largely based on differing interpretations of word meanings. For example, what is the "order" that we seek in our universe? "Order" and "chaos" exist as concepts in the human mind, conjured up by the use of these word symbols. Perhaps what we call chaos is really a kind of natural order not yet comprehended. What do we mean by cause and effect? Actually, the meaning of causality has never been resolved by philosophers. Only recently has the search for functional relationships replaced simplistic cause and effect, and strict determinism has given way to the search for probabilities (Haggett, 1966:23–27).

The acceptance of many "dichotomies" is another example of a semantic trap. A dichotomy exists when two opposites are defined as mutually contradictory, such as good and evil or reason and faith. But a dichotomy does not exist when one of the alleged opposites forms a subordinate part of the other or when one is derived from the other. Furthermore, a dichotomy may exist for some people and not for others, depending on certain basic attitudes of the culture.

A dichotomy that is embedded in our culture is the dualism of the individual and nature, which has long been accepted in geographic thought. From Judeo-Christian teaching comes the directive that humans should establish their conquest over nature. The teleologists had no doubts that the all-wise creator had built the natural world for the special benefit of humankind. This separation of the natural world and the human world was the cornerstone of the conceptual structure developed by the social Darwinists. Yet for a majority of the world's inhabitants, such a dichotomy cannot exist. Among the Buddhists and Hindus, for example, humankind is a part of nature, not separate from it. The individual hopes to be absorbed into the universe after overcoming his or her ignorance, lusts, and angry reactions to frustration.

Among the dichotomies that exist because of the meaning given to word symbols and that have been harmful to the clarity of geographical thought, we may underline five: (1) that geography must be approached either idiographically or nomothetically, but not both; (2) that physical and human geography are separate branches of study with different conceptual structures; (3) that geography must be either topical or regional; (4) that geography must be either deductive or inductive; and (5) that geography as a field of study must be classified either as a science or as an art. The fact that geographic writings may be placed in all these categories destroys the validity of the dichotomies.

A REVIEW OF THE TRADITIONS

As we have seen, methodological discussions involving all these dichotomies and others started in the 1870s, at the time when faculties were being established in universities to offer advanced training in geography. Since the first new appointments were people who had never been trained in a graduate school of geography, each had to define the field to their own satisfaction. In the United States most of the presidents of the Association of American Geographers presented their ideas about the scope and method of geography in their presidential addresses. Before turning to some of the current conceptions, we will review earlier statements, taking guidance from William Pattison, who has suggested that American definitions in general have reflected the history of research work and that this work has exhibited an essential unity attributable to a small number of distinct but affiliated traditions (Pattison, 1964).

Pattison proposes four traditions: (1) an earth science tradition, (2) a man-land tradition, (3) an area studies tradition, and (4) a spatial tradition. He maintains that although all four have found expression throughout the past century of American geography—as continuing parts of a general legacy of Western thought—each has tended to enjoy a time of preference. The earth science tradition, for example, was particularly prominent in the thinking of researchers shortly before the founding of the profession:

> As Mackinder of Oxford has recently expressed it, geography is the study of the present in the light of the past. When thus conceived it forms a fitting complement to geology, which, as defined by the same author, is the study of the past in the light of the present (Davis, 1888).

> The discernment of the meaning of surface features gives soul and sense to that too often soulless and senseless study, geography, for there is significance in every cape and every estuary, in every cataract and every delta. . . . In this phase the new geology is the new geography (Chamberlin, 1892).

The man-land tradition next rose to dominance. Interpretations of it changed during the time of ascendancy as these two definitions show:

Any statement is of geographical quality if it contains . . . some relation between an element of inorganic control and one of organic response (Davis, 1906).

Geographers . . . define their subject as dealing solely with the mutual relations between man and his natural environment. . . . Thus defined geography is the science of human ecology (Barrows, 1923).

The area studies tradition, especially favored in geographic work of the mid-twentieth-century years, is strongly represented in these assertions:

Our fundamental definition of geography [is] the study of the areal differentiation of the world (Hartshorne, 1939:242).

Geography is . . . the field of study that deals with the associations of phenomena that give character to particular places, and with likenesses and differences among places (James and Jones, 1954:6).

Finally, the spatial tradition, vigorously pursued in ensuing years through studies of geometry and movement, is selected for emphasis in these statements:

The main contribution of the geographer is his concern with space and spatial interaction (Ullman, 1953:56).

The contemporary stress is on geography as the study of spatial organization, expressed as patterns and processes (Taaffe, 1970:5–6).

Through the same sequence of decades, many of the important pronouncements on geography drew attention more to the complementarity of the traditions than to their distinctiveness, as these examples demonstrate:

Geography treats the man-environment system primarily from the point of view of space in time. It seeks to explain how subsystems of the physical environment are organized on the earth's surface, and how man distributes himself over the earth in his space relation to physical features and to other men (Ackerman, 1965:1).

Modern geography has continued to give attention to the same questions as did primitive folk, the significance of position, the togetherness of things, the areal distribution of entities and aggregations, the utility of the environment (Sauer, 1966:60).

Geography is the study and science of environmental and societal dynamics and society-environment interactions as they occur in and are conditioned by the real world. Geographic investigations into these are influenced by the character of specific places, as well as by spatial relationships among places and processes at work over a hierarchy of geographic scales (Gaile and Willmott, 2003:1).

Acceptance of any of these positions depends in part on one's own philosophy of geography. Most of them are both innovative and traditional; each expresses a piece of the whole concept of geography as a field of learning (cartography and behavioral geography have been suggested as fifth and sixth traditions [Blaut, 1979]). As fashions in words change, graduate students are presented with new programs of scholarly procedure and invited to abandon old ones. In the meantime frustration with the effort to provide widely accepted logical definitions leads to an increasing attention to operational definitions, which gives added strength to the old quip "Geography is what geographers do." It would seem to be a dissipation of professional energy to find so many printed pages devoted to such questions as whether geography studies areal differentiation or spatial interaction, or whether geography looks for similarities or differences among places, or whether geography is merely descriptive or seeks to explain things. Geography, like all other branches of learning, seeks answers to its questions by all these routes, but none of them exclusively. An individual geographer may claim to devote his or her life to describing the unique character of places, yet it is logically impossible to identify a characteristic as unique without some measure of the general. Out of this maze of word traps, however, it is possible to distinguish innovations and to record progress toward the goals set up in all these statements. It is important to seek what is new and place it in balance with what is traditional.

SYSTEMS

R. J. Chorley (1962) was first to bring systems theory before the geographical community with "Geomorphology and General Systems Theory." In America Edward A. Ackerman was arguably the first geographer to point to the rise of systems research throughout the scientific world after World War II. All science, said Ackerman, is concerned with four "overriding" problems:

1. The particulate structure of energy and matter (physics).
2. The structure and content of the cosmos (astronomy, astrophysics, geophysics).
3. The origin and physical unity of life forms (biological sciences).
4. The functioning of systems of multivariables, such as life systems and social systems (worked cooperatively by all the sciences).

(Ackerman, 1963:434)

Geographers, Ackerman claimed, must find the concept of a system of many different but interdependent variables ready-made for the meaningful study of "all humanity and its natural environment." Is this what Ritter was seeking when he wrote about "coherent relationships," or Guyot when he wrote of "the great life system," or Hartshorne when he wrote about "integrations," or R. S. Platt when he referred to "process-patterns of dynamic social relations"? How did it happen that only after mid-century did the concept of the system as a unit of study receive overwhelming acceptance?

General System Theory

Anatol Rapoport defines a system as "a whole (a person, a state, a culture, a business firm) which functions as a whole because of the interdependence of its parts" (Rapoport in Buckley, 1966:xvii). The words are new, but the mental image of such structures of interconnected parts goes back at least to the Greek philosophers. However, before World War II organized complexity could only be contemplated qualitatively or described as a "balance of nature."

The scholar who is credited with presenting the first outlines of a general theory of systems is Ludwig von Bertalanffy (Bertalanffy, 1951a, 1951b, 1956, 1962, 1968; see also Boulding, 1956). His account of how he developed these ideas is instructive.

When he started his professional career as a biologist in the 1920s, Bertalanffy found his colleagues seeking more knowledge about the nature of organisms by dissecting them into smaller and smaller parts. It struck him that until the organism was examined as a structure of interdependent parts, no real understanding of the "laws" governing organic life could be gained. From thinking about biological organisms, he broadened his views by recognizing the existence of other kinds of systems to which the concept of systems behavior could also be applied. When he presented these ideas at a seminar in philosophy at the University of Chicago in 1937, the scholarly world was not yet ready for the broad approach. In the 1930s the tendency was to make more and more minute analyses and to be skeptical of general theory. Almost the only field devoted to the development of general theory was physics. Most scientists were seeking simple, one-way, cause-and-effect sequences.

This trend toward the isolation of minute problems for investigation was challenged during World War II and has been definitely reversed in the contemporary period. During the war scholars were called on to work on strategic problems and to formulate policy recommendations involving complex issues. They found that they could not find answers to the questions with which they were confronted as long as they remained within the confines of single disciplines. After the war the value of the interdisciplinary approach to a variety of nonmilitary problems was at last fully recognized. The scholarly world was ready for Bertalanffy's ideas.[1]

General System Theory seeks to identify the characteristics that are common to many different kinds of systems. There are three fundamental aspects of all systems: structure, functioning, and evolution (being, acting, and becoming). When systems are isolated in laboratories or symbolically isolated by statistical procedures, they become closed and irreversible, but on the face of the earth systems are open and reversible as they receive inputs of energy or information and send forth outputs. As more and more scholars began to study different kinds of systems, it was discovered that all systems, however defined, behave in certain predictable ways. For example, it was noted that the growth curve of organisms (the S-curve) is mathematically very similar to growth curves for the spread of innovations, for economic development, or for populations.[2] General System Theory searches for the abstract properties that can be applied to all systems. Such *isomorphisms* form the basic structure of General System Theory and can be used to predict the working of drainage systems (Chorley, 1962), ecosystems (DeLaubenfels, 1970:112–120), political systems (Cohen and Rosenthal, 1971), economic systems, and many others.

---

[1] In 1948 Norbert Wiener published his book *Cybernetics* (Cambridge, Mass.: M.I.T. Press). Cybernetics (literally, steersmanship) is a science of communications that seeks to move information and directives through administrative channels more efficiently. From this new approach have come the application of automation to industry and the computerized analysis of complex administrative problems in government. Since the first formulation by Wiener, the objectives of this field of study have been broadened. Cybernetics is a special application of General System Theory.

[2] Note the use of this concept in the studies of population growth and decline by Stanley D. Dodge in the 1930s (Chapter 16).

Consider the isomorphism between the behavior of heat in a thermodynamic system and the movement of information in an economic system. The second law of thermodynamics states that in any heat system that undergoes irreversible change (without inputs or outputs of energy) there must be a loss of the energy available to do work. The mathematical factor that measures the unavailable energy in a thermodynamic system is called *entropy*. In a closed thermodynamic system entropy increases. In an economic system in which information concerning markets or new technology or other matters is being communicated, there is a loss of information because communication is not perfect. The increase of entropy in a thermodynamic system and the loss of information in an economic system may be described by the same mathematical formulas.

Spatial Systems

Geographers are especially concerned with any systems that involve, as functionally important variables, such spatial elements as location, distance, direction, extent, density, succession, or derivatives of these. Any system of which one or more functionally important variables are spatial is a spatial system (Wilbanks and Symanski, 1968). Such systems are not the same as regions, although a spatial system may provide the criteria by which a region is defined and identified.

Systems, like regions, require redefinition when the student moves from one resolution level to another. Harvey points out that the elements of a system that can be defined at one resolution level may become subsystems themselves when the resolution level is raised. Whole new systems come into focus. Or, by lowering the resolution level, systems may be defined at the global scale. By combining the geographer's concern with specifically defined segments of earth space with the system analyst's concern with the operation of functionally interconnected sets of elements, new insights into the nature of order on the face of the earth may be expected. There should also be an important feedback into General System Theory as a result of focusing attention on the spatial aspects of systems. General System Theory has provided opportunity for the study of those complex integrations of things and events that have always stimulated the curiosity of geographically minded students. Now, for the first time in the history of geographical inquiry, the available technology seems to be matched to the potential power of method. It is important at this time to take stock of the spatial theory available in the contemporary period.

THE CUTTING EDGES OF GEOGRAPHIC RESEARCH

In 1963 the National Academy of Sciences/National Research Council appointed an ad hoc committee within the Division of Earth Sciences to consider the potential contribution of geographic research to the general progress of science.[3] The

---

[3]The National Academy of Sciences was established by the federal government in 1863 to advise the government (on request) on matters of science and technology. The National Research Council was set up in 1916 to promote especially critical scientific research. After World War II the Academy and the Council were combined. The ad hoc Committee

committee identified four "problem areas and clusters of research interest"— physical geography, cultural geography, political geography, and a fourth field in which traditional economic, transportation, and urban geography are grouped under "location theory."

Physical geographers study the physiographic-biotic system as the human habitat, or environment (Ackerman, 1965:14–22). The environment, however, can be examined from several different points of view: in terms of what the inhabitants perceive it to be, in terms of the identification of desirable changes to be brought about by human action, or in terms of a static setting in which people momentarily find themselves. The judicious selection of significant parameters of the natural environment is the task of the physical geographer. Particular stress is laid on the system relations among such elements as surface features, air, water, soil, and biota (Kalesnik, 1964).[4]

Cultural geographers seek an understanding of the interactions between human societies and those features of the human habitat that have been produced or modified by human action (Ackerman, 1965:23–31). Attention is focused on the differences from place to place in the ways of life of human communities. Two different methods of study are commonly in use: developmental, focusing attention on the origin and diffusion of cultures and on cultural growth and retrogression; and functional, focusing on the short-term processes of cultural interaction, spatial organization, and flow or movement. Many of these studies are in the literary tradition and use the method of historical geography (Meinig, 1962, 1968, 1969).

Political geography is the study of the interaction between political processes and geographical areas (Ackerman, 1965:31–44). A major theme is the effect of the spatial arrangement of elements relevant to the operation of political processes. The territorial phenomena of political systems are studied over a wide range of resolution levels—from supranational political organizations, to the nation-state, to the political subdivisions of the state, to metropolitan urban communities, and to local or regional special-purpose administrations (Cohen and Rosenthal, 1971; Kasperson and Minghi, 1969; McColl, 1969; Soja, 1971).

The Science of Geography describes location theory as follows:

> Recent development in all three traditional subjects [economic, urban, and transportation geography] has involved extensive application of mathematical methods to facilitate refinement of theory, and a higher level of generalization than existed

---

on Geography, appointed in 1963, was made up of E. A. Ackerman (Carnegie Institution of Washington), chairman; Brian J. L. Berry (University of Chicago); Reid A. Bryson (University of Wisconsin); Saul B. Cohen (then at Boston University); Edward J. Taaffe (Ohio State University); William L. Thomas, Jr. (California State University at Hayward); and M. Gordon Wolman (Johns Hopkins University).

[4]Ackerman, *The Science of Geography* (1965), includes 11 pages of references to writings on these four problem areas. Since World War II and especially since 1960, the number of geographers and the volume of notable contributions to geography have both increased until it is no longer possible in a book of this size to refer even to a representative sample. The student should consult the *Bibliographie géographique* (Paris: Armand Colin), and *Current Geographical Publications* (Milwaukee: American Geographical Society).

before has emerged. As a result, it would appear that the three traditional fields have been joined in a problem area which we entitle . . . location theory studies. It is of special interest in our discussion because these studies include: (a) research on the qualities of space in a theoretical framework; (b) application of formal systems methods to space relations study; and (c) integration of at least three of the spatial subsystems of culture. . . . It is also of interest for the extent to which its methods and concepts have found practical applications in a revitalized "applied geography" directed publicly to problems of urban and regional planning, and privately to marketing analysis for plant and store location (Ackerman, 1965:44).

Still another survey of the role of geography as a behavioral and social science was carried out by the Panel on Geography of the Behavioral and Social Science Survey and published in 1970 (Taaffe, 1970).[5] This report provides illustrative studies in six different geographical fields (spatial distributions and interrelationships, circulation, regionalization, central-place systems, diffusion, and environmental perception). It also discusses research methods and shows how they are applied to studies of locational analysis, cultural geography, urban studies, and environmental and spatial behavior. It discusses geography and public policy and goes into some detail on the status and trends of the profession in personnel, research, and research training.

Both of these reports recognized that it is in the general range of locational analysis that geographers in the contemporary period have made the largest advances toward the formulation of general concepts. Whether these concepts are to be identified as empirical generalizations, laws, or theories may be left to the judgment of professional geographers. To illustrate the kinds of innovation that have been made, four examples are provided: (1) the gravity model, (2) central-place concepts, (3) diffusion concepts, and (4) regional identification.

Social Physics and the Gravity Model

Relationships between distance and interactions, such as movement of goods and migrations, were written about in the eighteenth and nineteenth centuries by Bishop George Berkeley (1713), Henry Carey (1858), Ernst G. Ravenstein (1885), and Herbert Spencer (1892). It may have been that the pioneering work concerning distance-decay and the gravity model was accomplished by other than geographers (Tocalis, 1978).

---

[5]The Behavioral and Social Science Survey was carried out between 1967 and 1969 under the auspices of the Committee on Science and Public Policy of the National Academy of Sciences and the Problems and Policy Committee of the Social Science Research Council. A general volume, *The Behavioral and Social Sciences: Outlook and Needs* (Englewood Cliffs, N.J.: Prentice-Hall, 1969), discusses the relations among the disciplines and broad questions of utilization of the social sciences by society and makes specific recommendations for public and university policy. The Panel on Geography was made up of Edward J. Taaffe (Ohio State University), chairman; Ian Burton (Toronto); Norton Ginsburg (University of Chicago); Peter R. Gould (Pennsylvania State University); Fred Lukermann (University of Minnesota); and Phillip L. Wagner (Simon Fraser University).

In 1929 W. J. Reilly, studying retail marketing problems, postulated that the movement of persons between two urban centers would be proportional to the product of their populations and inversely proportional to the square of the distance between them.[6] In 1949 this empirical generalization was refined by the economist G. K. Zipf, who formulated the principle of least effort in human behavior (Zipf, 1949/1965). But the astrophysicist John Q. Stewart is often credited with being the first to point out the isomorphic relationship of these concepts with Newton's law of gravitation (Stewart, 1947). Thereafter this concept became known as the gravity model.[7] William Warntz, working with Stewart, also borrowed analogy models from physics in his studies of population potential (Warntz, 1959b, 1964). He suggested that the mathematics of population potential is the same as that which describes a gravitational field, a magnetic potential field, and an electrostatic potential field.

The question is sometimes raised whether the use of analogy models is acceptable. There have, of course, been examples of the misuse of isomorphisms; on the other hand, there have been many very successful uses of them. William Bunge offers the following observation:

> It is an observed fact that once theory is produced it often can be applied to a variety of subjects. In this sense, there is unity of knowledge. To give this assertion substance some examples appropriate to geography are offered.
>
> Consider Enke's paper [Stephen Enke in *Econometrica*, 1951] "Equilibrium among Spatially Separated Markets: Solution by Electric Analogue." Can electricity be expected to behave like a spatial economic system, as he insists? Yes, because it has been found that the underlying mathematics can be translated into certain carefully selected aspects of both subjects. A second illustration of the borrowing of theories is available from Beckmann's "A Continuous Model of Transportation" [in *Econometrica*, 1952]. It is suggested by hydrodynamics. Can water be expected to behave like a spatial economic system? Again, it is the mathematics that can be made to fit features of both sets of phenomena. If social scientists are somewhat defensive because they have been borrowing heavily from mathematics and from theory first used in other fields, they can draw some comfort from the knowledge that there is reciprocity. Programming, first applied in social science, is now being used in designing electric networks (Bunge, 1966:4).

The gravity model, however, has to be refined for maximum applicability to studies of location and spatial interchange. Models describing the volume of exchange between two populations are more expressive when the populations are weighted by some factor such as income per capita, which measures the degree of economic activity (Goodchild, 1992). Distance between places is not a matter of simple linear distance but of rather more sophisticated measures of distance in terms of

---

[6]W. J. Reilly, *Methods for the Study of Retail Relationships* (Austin: University of Texas Press, 1929).

[7]Both the notion of the gravity model and the population potential were specifically stated by Henry C. Carey (1793–1879) in 1868. Carey wrote: "Gravitation is here, as everywhere else in the material world, in the direct ratio of the mass, as in the inverse one of the distance" (McKinney, 1968:103). For recent thought on the gravity model, see Kingsley E. Haynes and A. Stuart Fotheringham (1984). *Gravity and Spatial Interaction Models*.

transport routes and facilities; frequency of movement by land, sea, or air; and the costs of transport (Haggett, 1966:35–40).

From workers such as Zipf, and Stewart and others in social physics and yet other disciplines, the properties of distance are receiving ever more attention. Economists such as Weber, Hoover, and Lösch and later Garrison and Walter Isard's regional science followers gave this notion attention, as it operated on the principle of least effort. Alan G. Wilson, a physicist who became concerned with traffic forecasting while a city councillor in Oxford, established the gravity model on a more thorough mathematical footing. R. J. Johnston (1997:113) states:

> Collaboration between academic geographers and practicing planners on these modeling exercises led to developments which paralleled those in regional science in the United States. Two new British journals catered for these research areas— *Regional Studies* and *Environment and Planning*: both attracted contributions and readers from other social sciences.

The gravity model remains the best-known exemplar of social physics. It sought analogy with Newton's law of gravitation, assuming that humans interact on Earth as do heavenly bodies beyond planet Earth. Lukermann (1958) made a significant critique of social physics, claiming that assumptions made in the physical models are not met in the world of humanity. R. J. Johnston (2003, 313–314) has provided a detailed recent history regarding development and exploitation of the gravity model.

Central-Place Studies

Efforts to unravel the complexities regarding the spacing and functions of central places and to formulate some testable generalizations relating thereto have been given considerable attention by geographers in North America. E. L. Ullman arguably made the earliest of these contributions with "A Theory of Location for Cities" (1941). There followed "The Nature of Cities" (1945) by C. D. Harris and Ullman, notably providing three models of intra-urban spatial patterns. Then W. L. Garrison made several contributions including a series of three articles summarizing research to that time concerning location theory. Meanwhile, J. E. Brush (1953) had reviewed the hierarchy of central places in southwestern Wisconsin. He concluded that Christaller's work provided a norm against which to measure observed differences in various parts of the world. He saw no possibility of finding precisely the same hierarchies in different cultural regions. A. K. Philbrick (1957) made a study of rural functional organization. M. F. Dacey (1962) infused method into the analysis of central-place study. B. J. L. Berry studied settlement patterns and retail centers in Seattle and Spokane, joining with Garrison in publishing the results (1958a,b, 1959a) (Barnes, 2000). In 1969 Harvey concluded on the basis of numerous attempts to apply the concept in various locales that "the tests have shown that actual spatial patterns do not conform to theoretical expectation" (Harvey, 1969:138). It may be that the economic theory regarding the range of a commodity—that is, the distance a customer will travel in order to make a purchase—is "inherently untestable." Statistical techniques were advanced aided by computers and large data sets to facilitate large-scale numerical testing (Gould, 1969), and economic geographers began arranging their own mathematical requirements as analytic tools needed in the comprehension of location. Curry (1998) and Cadey (1974) developed advanced mathematical models based on stochastic variation. Cliff and Ord (1973) formulated geographical statistics in an original manner.

In the 1990s Paul Krugman, an American economist developed his own "new economic geography" (Barnes, 2003) embracing the notion that comparative cost advantages may be ignored given specialization and trade that secure increased profit. This is now considered an approximate new trade theory. Krugman plans to make economic geography a branch of economics (Krugman, 1995).

This work by American geographers was inspired by the earlier work of the Germans J. H. Von Thünen, A. Weber, and A. Lösch. The original formulation of what is sometimes called central-place theory was accomplished in 1933 by the German geographer Walter Christaller (1933/1966) in a study of the spatial arrangement of tertiary economic functions in southern Germany. Christaller thought of his work as complementary to von Thünen's model of agricultural land use (von Thünen, 1826/1966) and Alfred Weber's model of industrial locations (Weber, 1909/1966). He noted that "the crystallization of mass about a nucleus is part of the elementary order of things," and human settlement obeys this principle as well as physical elements. The foci or nodes around which settlement tends to cluster are what Christaller called central places, each surrounded by a complementary area with which the central place is functionally related. Based on evidence from Germany, Christaller described a "nested hierarchy" of central places ranging through seven orders. Places of the lower orders provide goods and services that are needed frequently with a minimum of travel. Places of higher orders not only provide these same goods and services but also other, more specialized goods and services that are needed less frequently and for which people are willing to travel greater distances. The higher order places have worldwide connections.

Christaller also attempted to explain the spacing and pattern of arrangement of central places. He postulated a uniform plain evenly settled by an agricultural population as the base of the hierarchy. If access to a market is the chief factor in the development of a settlement pattern, then the minimum average travel distance from the complementary area is gained if the area has the shape of a hexagon. But Christaller recognized that optimum conditions would also be found where as many central places as possible are located along a main traffic route between higher order places. Furthermore, where government administration or other purposes are to be served, a further modification of the simple market-oriented type of pattern would appear. The actual pattern to be observed in any area depends on the interplay of these three principles: marketing, traffic, and administration (Berry and Pred, 1961:15–18).

German geographers in the 1930s paid little attention to Christaller's work. It was tested by Otto Schlier in 1937 on the basis of census data (Schlier, 1937), and it was applied to the pattern of settlement in Estonia by Edgar Kant. In 1940 a German economist named August Lösch had investigated Christaller's ideas in a book on the spatial pattern of the economy; in 1944 Lösch published a revised and much enlarged edition, which was translated into English in 1954 (Lösch, 1940, 1944/1954). Lösch found evidence to support Christaller's concepts, both with regard to the nested hierarchy of central places and also the hexagonal patterns of complementary areas.

The study of settlement patterns and hierarchies was greatly stimulated by Christaller's hypotheses, which led to many studies of specific areas and to the formulation of a number of alternative hypotheses.[8] George K. Zipf proposed a rank-order hypothesis that predicts the population of urban places on the basis of the rank of any one city among all the cities of a country (Zipf, 1949/1965:374–386).

Studies of Diffusion

One traditional concern of geographers (at least since Ratzel) has been the inter-pretation of the spread or retrogression of the things that occupy space on the face of the earth. Ratzel, in the second volume of his *Anthropogeographie*, described the patterns of population and culture that had resulted from the process of diffusion from centers of origin. Nor should we forget the contribution that Ellen C. Semple made to the explanation of culture patterns. She discussed the various ways culture change could be brought about—by conquest, infiltration, influence, and many other ways. Semple proceeded to offer two important empirical generalizations in verbal form:

> In general, however, any piecemeal or marginal location of a people justifies the question as to whether it results from encroachment, dismemberment, and con-sequently national or racial decline. This inference as a rule strikes the truth. The abundance of such ethnic islands and reefs—some scarcely distinguishable above the flood of the surrounding population—is due to the fact that when the area of a distribution of any life form, whether racial or merely animal, is for any cause reduced, it does not merely contract but breaks up into detached fragments. . . . Ethnic or political islands of decline can be distinguished from islands of expan-sion by various marks. When survivals of an inferior people, they are generally characterized by inaccessible or unfavorable geographic location. . . . The scattered islands of an intrusive people, bent upon conquest or colonization, are distinguished by a choice of sites favorable to growth and consolidation, and by the rapid extension of their boundaries until that consolidation is achieved; while the people themselves give signs of the rapid differentiation incident to adaptation to a new environment (1911:164–165).

Geographers have traditionally sought explanations of diffusion through the deciphering of historical processes. By plotting patterns resulting from the diffu-sion process on maps, it has been possible to identify centers of origin and directions of spread. Biogeographers and cultural anthropologists have used the cartographic method to throw light on the subjects they were investigating. Carl Sauer's thought-provoking hypotheses regarding agricultural origins and dispersals use the cartographic method to illuminate prehistoric problems (Sauer, 1952). Fred B. Kniffen uses maps of house types to reveal directions of migration, just as H. Kurath did with word usage and pronunciation in the late 1940s (Kurath, 1949;

---

[8]Brian Berry and Allan Pred published a review of the theory and a bibliography of published materials (Berry and Pred, 1961). The flow of studies has continued since 1961. Among the many, see Curry, 1964; Parr and Denike, 1970; Woldenberg, 1968; Berry, 1988.

Kniffen, 1965). The list of substantial contributions to this kind of geographical study is long and growing.[9]

The use of mathematical models to describe and predict the diffusion of innovations was started by the Swedish geographer Torsten Hägerstrand. He completed his doctoral thesis in 1953. This work concerned the adoption of innovations by farmers in central Sweden that could be mapped and simulated. His work on this subject was published in Swedish in 1953. It was not translated until 1967, when Allan Pred provided *Innovation Diffusion as a Spatial Process*. By this time Hägerstrand had lectured in the United States, and U.S. geographers had visited Lund and worked with Hägerstrand. This itself was a case of diffusion of knowledge. In 1962 Wagner and Mikesell had published an anthology entitled *Readings in Cultural Geography*. This work devoted considerable attention to "cultural origins and dispersals," a term later replaced by "spatial diffusion." The book included Hägerstrand's essay "The Propagation of Innovation Waves," which reached large numbers of American geographers. Hägerstrand investigated how innovation spread and then modeled diffusion as a stochastic process using Monte Carlo methods. Individuals, he postulated, are more likely to be informed about an innovation the closer they are to the source of the innovation. Distance, of course, is not mere linear distance but also a measure of contiguity and contact. In 1965 R. S. Yuill made use of simulation models to show the probable patterns of spread around several kinds of barriers (Yuill, 1966; see also Bunge, 1966:112–132; Haggett, 1966:59–60; Morrill, 1970). Other applications of mathematical procedures to the study of innovation diffusions include Edward Soja's study of the economic and political modernization of Kenya (Soja, 1968) and Lawrence Brown's general discussion of diffusion processes (Brown, 1968, 1981; Brown and Moore, 1969).

These studies of the diffusion process have been done at different resolution levels. Hägerstrand was working at a high resolution level that permitted him to focus on specific individuals and made possible certain general concepts regarding individual human behavior. Yuill and Soja were working at a somewhat lower resolution level that focused on groups rather than individuals. It is also possible to study diffusion processes at a very low resolution level—at a global scale. Interestingly, the concepts and models found useful at one resolution level are not necessarily useful at other levels. Diffusion theory can be enriched by the examination of a variety of levels.

In his book *Innovation Diffusion: A New Perspective*, Brown (1981) provides a different tripartite approach: first, the establishment of outlets whereby innovation will be distributed; second, the implementation of a strategy by each outlet to win acceptance in its service area; and third, individual adoption of the innovations, which was previously the main thrust of this type of research. There followed a radical critique of diffusion theory (Blaut, 1977, 1987). Work on origin and dispersal

---

[9]The rural sociologist Everett M. Rogers (*Diffusion of Innovations*, New York: The Free Press, 1962) makes no mention of contributions by geographers to diffusion studies. He mentions Walter H. Kollmorgen as a rural sociologist (p. 32), and refers to Hägerstrand (pp. 154, 298) for his use of game theory.

introduced into cultural geography by Sauer have now been revealed as studies in diffusion (e.g., Morrill, Gaile, and Thrall, 1988). Geographers and others will doubtless continue to modify, reject, and further develop aspects of diffusion processes, but as a body of thought it is part of tradition.

Studies of Regions

Regional geography functioned as the basis of American geography from about 1920 to 1960, and it was a significant undertaking both before and after those years. Regions were defined, mapped, written about and taught in the universities. However, there was a dichotomy: Studies were usually of small areas capable of being known intimately by slow locomotion (e.g. walking or by bicycle), while regional geography classes were continental in extent, such as the geography of North America, the geography of Asia, and the geography of Africa. Scale was the difference between these two varieties of regional geography. Writing and study was done at one scale, while exposition in the classroom was done at a different scale. Compromise was reached in R. S. Platt's *Latin America: Countrysides and United Regions* (1942). In this book Platt presented 94 studies of places in widely separated latitudes and altitudes in Latin America, each of which he contended were components of larger perspectives. Then came "The Regional Concept and the Regional Method" in *American Geography: Inventory and Prospect* (1954). This extended article represented one of the most searching analyses of what we might now call traditional regional geography.

By the end of the 1950s, spatial science began its trajectory through both the discipline and geography department curricula, but as the years passed traditions of this regional treatment were reduced in number and emphasis. It seems to have remained mostly in the smaller college and university departments. Yet just what was it that was so quintessentially "geographic" about the region? Haggett (1990) identified five properties of regions: regions as exemplars, as anomalies, as analogues, as modulators, and as covering sets. The exemplar could be typified by any one of Platt's 94 places. The region as anomaly, in contrast to the region as exemplar, can be found in the mountainous regions of Switzerland and France. Here the south-facing valley walls receive more sunshine and warmth, more snow and ice melt, and a longer growing season, than do the north-facing walls. Property is more expensive on the south-facing walls. Other anomalies may be found in Trewartha's *Earth's Problem Climates* (1961, 1981). Region as analogue is invoked when the climate of the Mediterranean basin is shown to be replicated in California, west Australia, South Africa, and Chile, each providing the ubiquitous grape and its product, wine. Region as modulator suggests that in areas where economic activity is established and for whose product the demand is inelastic, the level of prosperity attained will be reasonably constant. Conversely, regions dominated by industries of elasticity, such as armaments, are subject to vicissitudes of demand. Finally, we come to regions as jig-saws (or covering sets) that totally cover a country or continent. The size of these regions, of course, depends on the scale adopted. Explication is difficult and depends on the skill of the author. This was accomplished superbly by W. M. Davis and H. Baulig in North America, P. E. James and Pierre Denis in South America, W. Meade in Scandinavia, L. Gallois in France, O. H. K. Spate in India, T. G. Taylor in Australia, and many others. These regional studies were born of intense familiarity with and love of the landscape and the viewing of these areas over a period of many years.

Then came numeracy and theory, which are distant from the humanism that inspired and sustained the earlier work in the regional mode. In the 1960s statistical treatment of regional matters was not well understood; what to measure and how to measure it was problematical. With the passage of time, the quantitative movement in human geography became more sophisticated, modeling reached new levels, and workers in this genre became more experienced. A new level of accuracy was brought to the explication of the region.

Walter Isard, an American economist, thought that regional matters might best be dealt with in an interdisciplinary fashion and in 1954 founded the Regional Science Association. It was made up of a group of economists, geographers, other social scientists, and engineers and was "an international association devoted to the free exchange of ideas and viewpoints with the objective of fostering the development of theory and method in regional analysis and related spatial and areal studies" (Isard, 1956, 1960).

The regional science movement gained support internationally and has been especially attractive in those countries in which geographical studies are applied to practical problems. The countries of the developing world find this approach of great value in the guidance of development programs.

Regional science has embraced both empirical modeling and theoretical analysis of problems of location and regional economies. It has also found itself bonded especially to geography and economics; yet it clearly maintained its identity in periodicals such as *Environment and Planning A, International Regional Science Review*, and others.

What started as an innovation with Shaler in 1885 burgeoned with a vast series of studies and learning ranging from college courses to doctoral dissertations to the published literature. Then came the newer numerate geography that permitted greater precision. Numerical values replaced verbal descriptions. Models and computer work replaced muddy-boots field work, photographs, and sketches. Then came reappraisal and rapprochement. Historical geographers, cultural geographers, and traditionalists continued to appreciate the meaning of region, be it large or small. Theorists, frequently searching for answers to specific questions, continued with their quests. Once again, what was innovation has become tradition in the larger folds and reaches of the discipline.

THE POPULATION PROBLEM

Geographers have given themselves more and more to problem solving—at the local, regional, and global levels—throughout the twentieth century. And there were many problems to be solved, including matters relating to war, peace, boundary location, sharing of fresh water, access to the ocean, desertification, . . . the list is very long indeed. Many of the problems are too large for an individual or local group to attempt to solve. International groups are better suited to attempt solutions, especially of global problems. The first commissions of the International Geographical Congress were established for these purposes at the Bern meeting in 1891. Four commissions were established concerning immigration, global time, geographical bibliography, and the Millionth Map. Since then the number of commissions has grown steadily, and for some decades "working groups" were established as subsidiary to the commissions. All "groups" have now been designated commissions; they work at specified tasks that are invariably global in scope.

The Commission on Population was formed at the Paris meeting in 1931, which merged with the Commission on Rural Habitat in 1934 to form the Commission on Habitat and Population. Since then, a commission concerning population study and problems has been at the front of IGC study concerns. Since the AAG opted for specialty groups in 1978, population has remained a major concern (White et al., 1989; Gober and Tyner, 2003).

In 1970 Wilbur Zelinsky, a specialist in population geography at Pennsylvania State University, pointed to the critical importance of examining the results of the continued growth of the world's population and the continued expansion of industrial production. As this "growth syndrome" looms ahead as perhaps the major problem of the twenty-first century, scholars in a variety of fields have contributed studies of different aspects of the problem (Trewartha, 1969). There have been studies by sociologists, demographers, economists, political scientists, ecologists, agronomists—and geographers and others. It is not a new undertaking, though there is now a new-found urgency concerning the matter. In 1798 Thomas Robert Malthus wrote his "Essay on the Principles of Population," postulating that population tended to increase in a geometric progression and foods in an arithmetic progression. This was perhaps the earliest demographic model constructed. Malthus revised his thought in 1803 and again confronted the issue of world population and adequacy of food supply. Approximately 80 years later Ernst Georg Ravenstein made a further substantial contribution to the study of population. He made special studies of the laws of migration. He also believed that the maximum population the planet could sustain was six billion (Ravenstein, 1876, 1885, 1889). These two thinkers worked in Britain, where both the total and the urban populations were surging. It led to studies by people from many fields of study both there and in other countries.

In America there had been an early interest in population from the time Thomas Hutchins, geographer to Congress in the 1780s, had surveyed land and counted people. Jedidiah Morse wrote geography textbooks in which he was concerned to know the population of at least the eastern seaboard. Later, Francis Walker, a Yale geographer, became superintendent of the US. censuses of 1870 and 1880 and supervised publication of the *Statistical Atlas of the U.S.* (1874). He was emphatic in his focus on people and their distribution throughout the United States. Then came the formation of a profession and intellectual discipline. The first professional geographer to make the study of population his specialty was Mark Jefferson. He developed what he termed anthropography[10] and defined this as the study of the distribution of population. He made such studies of North America, India, Japan, much of Europe, and Argentina. These studies attracted the attention of a visitor

---

[10]Anthropography should not be confused with anthropogeography. For the distinction see M. S. W. Jefferson, "Some Considerations on the Geographical Provinces of the USA, 1916." *Annals of the Association of American Geographers* 1917, 7:3–15.

from Australia named Marcel Aurousseau. He worked with the American Geographical Society in the early 1920s and published two distinguished articles relating to population in the *Geographical Review*. These studies led to the geographer of the U.S. Census Bureau accepting an invitation from Bowman of the American Geographical Society to come for a several-month-long stay to learn what geographers were doing with population studies. C. E. Batschelet arrived in May 1924 and worked with Aurousseau under Bowman's direction for more than three months.

In 1924 Curtis F. Marbut gave his presidential address before the Association of American Geographers—"The Rise, Decline, and Revival of Malthusianism in Relation to Geography and Character of Soils" (Marbut, 1925). Oliver E. Baker, among other geographers, was concerned about foodstuffs and population growth.

Many more studies of population were published in American geography, and in 1953 G. T. Trewartha presented "A Case for Population Geography" to the AAG as his presidential address (Trewartha, 1953). By then innovative population studies had become a disciplinal tradition. During the second half of the twentieth century there was ever-increasing concern for an ever-increasing world population. In 1970 Zelinsky, Kosinski, and Prothero edited *Geography and a Crowding World*, a work that seized hold of the potential problem:

> The geographic approach is one of several needed to understand and treat those troubles everyone now recognizes as stemming from rapid population growth and uneven development in the less advanced regions of the world. Precisely the same is true of the afflictions growing out of the unceasing accumulation of human beings and things in the affluent nations, problems glimpsed only dimly as yet or in terms of isolated facets. Geographic analysis is hardly the single magic nostrum for these ills, but it is hard to imagine any workable therapy that excludes it (Zelinsky, 1970:498).

Zelinsky's paper questions the traditional Western belief in the benevolence of continued growth. The problems that are being faced in the economically underdeveloped countries cannot be solved, he argues, until they have been faced and solved in the more advanced countries—first of all in the United States. Geographers can offer diagnoses describing the nature of the illness with which humankind is afflicted; they can become prophets, predicting the likely results of various remedial policies; or they can join with others as the architects of the utopia that must be built if utter disaster is to be avoided. He concludes as follows:

> A thorough review of the present status of human geography, and of population geography in particular, would reveal how woefully deficient we are in terms of practitioners, in terms of both quantity and quality, how we are still lacking in relevant techniques, but most of all that we are totally at sea in terms of ideology, theory, and proper institutional arrangements. Even if we were to be showered with unlimited funds tomorrow morning with which to initiate research on key aspects of the geographic diagnosis of the Growth Syndrome, it would almost certainly be impossible to put enough people with the right skills and attitudes into the proper settings. Given the hope, a not totally unrealistic one, that this picture will alter radically before the end of this decade, I have stated a number of difficult, but ultimately operational, research themes, practical daydreams that could inflame the imagination (Zelinsky, 1970:529).

The "no-growth" thesis has been abundantly examined in recent years by population geographers, demographers, and ecologists. Today the world's population still may be effectively divided into "haves" and "have nots." All continue to want. Resources are finite, while population continues to increase. There were a few thousand humans on Earth a million years ago; between 5 and 10 million 10,000 years ago; more than a billion in 1850; 2.5 billion by 1950; and 6.5 billion today. World population may become 10 billion by 2030 (Bongaarts and Bulatao, 2000). Increasing production means increasing pollution and other deterioration of the environment. That is what those attending the Earth Summit discussed in Rio de Janeiro in 1992 and in New York in 1997. Human migration, evident since early times, continues to the present. It is now a swifter and better informed process, and it carries ever-increasing political consequences. This, too, has been subject to the investigations of geographers (Raven, 2000).

The quest for a sustainable world now concerns the geographer whose global embrace provides opportunity for resolution of this problem. Curiously, it may be that water supply is the fundamental resource problem confronting humanity. The point is that earlier studies in the geography of population have led geographers to confront the larger issue. From there developed an ever-larger quest to view the whole. The enlargement of viewpoint of population geography is now tradition.

THE CONCEPTUAL STRUCTURE OF GEOGRAPHY

In formulating and finding the answers to questions concerning occupied space, geographers have developed *conceptual structures*—that is, frameworks of ideas about the character of the earth's surface and the methods of deriving geographical understandings from the study of it. The methods of formulating conceptual structures in geography do not differ from the methods followed by scholars in other fields of learning. Furthermore, geography is not the only field that is handicapped by the confusion of word meanings, and helpful perspective is gained by examining some of the discussions of theory and method produced in other disciplines. As a result of this confusion, scholars who deal with the philosophies of their fields must make clear in one way or another how they use their word symbols, and, when they introduce a new word symbol, they must make some effort to inform the reader how the new symbol differs in meaning from symbols already in use (Amedeo and Golledge, 1975; Machamer, 2002).

Percepts and Concepts

In this book we started with percepts and concepts.[11] A *percept* is an empirical observation—an observation made through the senses and based on experience. An empirical observation is sometimes called a factual statement. But as was pointed out in Chapter 1, the features on the face of the earth that are perceived are closely

---

[11] See H. N. Lee. 1973. *Percepts, Concepts, and Theoretic Knowledge: A Study in Epistemology.* Memphis State University Press; and B. Burns. 1992. *Percepts, Concepts, and Categories: The Representation and Processing of Information.* Amsterdam: North-Holland.

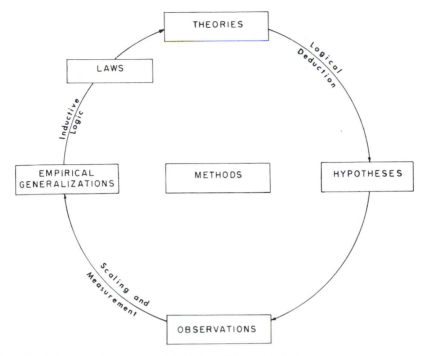

Figure 56 Diagram of concepts (adapted from Wallace, 1980)

influenced, if not determined by, concepts. A *concept*, as we use the word, is a mental image of a thing or event. As such, the word *concept* is a general term that stands for a whole hierarchy of ideas ranging from simple generalizations to abstract theory. Just as a percept is the abstract of sensation, so is a concept the abstract of percepts.

The words used to describe this hierarchy are for the most part poorly defined and overlapping and are often used by the same author in different senses. The hierarchy suggested here is somewhat modified from the diagram (Fig. 56) presented by Walter L. Wallace in *Sociological Theory* (1980:ix). At the base are observations of the kind we call percepts. Proceeding to the right, the circular form of the diagram indicates the important part played by hypotheses in determining observations. But proceeding in the opposite direction, the observations—which must always be of unique things—are generalized in what Wallace calls empirical generalizations. This requires scaling and measurement applied to the basic observations. Then, through inductive logic and no inconsiderable amount of intuition and imagination, the empirical generalizations can be further abstracted into theory. Theory, however, cannot be tested by direct observation. A theory is judged "good" if it permits the formulation of numerous hypotheses by deduction. The hypotheses can then be tested by observation and may yield new information from hitherto unexpected observations.

Geographers (and scholars in other fields) are accustomed to using other terms in various parts of this simple circular diagram. For example, David Harvey writes of empirical laws (Harvey, 1969:31). What exactly is a law, and what can be done with it? Braithwaite defines a law as "a generalization of unrestricted range in space and time"—in other words, a generalization of universal applicability (Braithwaite, 1968:12). In this sense a law would be introduced on the theoretical side of empirical generalizations. The distinction between a generalization and a law would rest on two distinctions: An empirical generalization applies in a particular place or at a particular time, whereas a law is universal, and a law is embraced as one aspect of a still more abstract theory.[12]

There are two major difficulties with the use of the word *law*. In the first place, whether or not we use it in a sophisticated sense, the fact remains that some people, even some scholars, use the word *law* in the anthropomorphic sense: A law governs events, and to break a law is wrong and must result in punishment. Of course, in moments of clarity it is recognized that a scientific law does not govern events—it is governed by events, of which it is a generalization. The word has, and still can, result in obscurity of thought. But even more important is the fact that a law in the strict sense as defined by Braithwaite could scarcely be formulated from geographical evidence. If geography deals with things and events on the face of the earth, then its generalizations cannot be universally true or at least cannot be verified as universals from geographic observations. The only laws that could he defined as universals would be those of physics and chemistry, yet even in physical science there is a certain indeterminacy that makes necessary the use of probability concepts—these are *stochastic* laws. But even when they cannot predict every situation, they are universally probable. In the social and behavioral sciences, however, the only way a law can be identified as universally applicable is to define man's universe as the things and events located on the surface of one medium-sized planet in a very unusual relationship to a medium-sized sun.

Nevertheless, the word *law* is so widely used by geographers and others that to abandon it would not seem possible. As Braithwaite points out, very good results can be derived from treating some more abstract empirical generalizations as if they were laws. In such a case, a law is more widely applicable than a generalization. Nevertheless, it is important to keep these distinctions in mind and not to claim more for geographical study than can be delivered.

Pattern Concepts and Process Concepts

The mental images geographers develop regarding the arrangement of things on the face of the earth are *pattern concepts*. They are empirical generalizations. It is recognized that whatever scale or resolution level one uses to look at the earth's surface, certain kinds of associations of features of diverse origin are found

---

[12]Cole and King recognize six different meanings attached to the word *law* (Cole and King, 1968:522). Golledge and Amedeo also identify six kinds of law, none of which is exactly equivalent to any of Cole and King's laws (Golledge and Amedeo, 1968). Examining other writings on these matters, one finds a great emotional respect for a concept defined as a law but almost no common ground for defining the concept (Minshull, 1970:119–129). See also R. J. Johnston. 1997. "Scientific Method in Human Geography." 73–95. In *Geography and Geographers: Anglo-American Human Geography since 1945*.

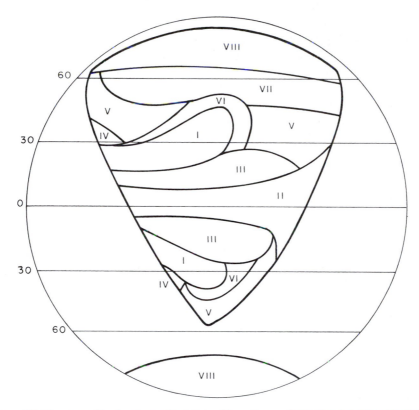

Figure 57 The generalized pattern of habitats. (From James, 1966. Reprinted by permission of the publisher.)

repeatedly in similar situations. Soil associations, for example, are groups of soil types that are found again and again in the same locations relative to each other. Bowman's regional diagrams, which he used to give a generalized picture of the arrangement of surface features in the Andes of southern Peru, offer another example of pattern concepts (Figs. 36 and 37). At a global scale the generalized continent devised by W. Köppen to show the orderly arrangement of climates with reference to latitude and land and water distribution is another example of a pattern concept (Fig. 28). Köppen's generalized continent is shown occupying a hemisphere, with the greatest east-west expanse of land about 70° north and with the smoothed outline of a land mass tapering to the southern extremity at latitude 55° south. On this continent (with differences of altitude removed) Köppen showed the ideal, or generalized, arrangement of his climatic types. This same procedure was used to show the generalized arrangement of habitats, with reference to latitude and to position on the western side, interior, or eastern side of a continental land mass (Fig. 57).[13]

---

[13] W. Köppen, "Versuch einer Klassifikation der Klimate, vorzugsweise nach ihren Beziehungen zur Pflanzenwelt," *Geographische Zeitschrift* 6 (1900):593–611. The eight habitats shown in Fig. 57 are defined in terms of associations of climate, water, vegetation, and soil. (The mountain habitats are omitted because they do not fit on this

Process concepts are those that provide a generalized picture of sequences of events. Examples of such concepts include the model of the cycle of erosion formulated by W. M. Davis. Chorley and Haggett have edited a book in which a variety of kinds of models are illustrated: models of physical systems, models of socio-economic systems, and models of mixed systems (Chorley and Haggett, 1967). They distinguish two varieties of models: those that are deterministic in that if specified conditions exist at a particular moment of time, it is possible to specify what conditions existed at an earlier time or will exist at a later time, and stochastic models, which specify sequences of events within a certain range of probability.[14] In any case they provide an ideal picture of a process against which the actual departures from a norm can be measured. It was concerning such process models that Wellington D. Jones is credited with the remark "Models tell us what would happen if things do not happen the way they do."

Geography: Nomothetic or Idiographic?

The idea that any field of learning can be restricted either to a study of unique things or to the formulation of general concept may date back to 1894, when Wilhelm Windelband used the words *idiographic* and *nomothetic*. Once the word symbols were provided, the existence of a dichotomy could be proclaimed (Siddall, 1961). The dichotomy was turned loose in the literature of geography in 1953 by Fred K. Schaefer. Sauer pointed out in 1925 that although geography was formerly devoted to descriptions of unique places as such, geographers had for a long time been seeking to formulate illuminating generalizations about the earth and man's place on it (Sauer, 1925:27). But in the 1950s, after both Hettner and Hartshorne had been incorrectly quoted as saying that geography is essentially idiographic and is not concerned with general concepts, this characterization gained an amazingly wide acceptance (Harvey, 1969:50–51).[15]

---

generalized continent.) The habitats are: I. Dry lands; II. Tropical forest lands; III. Tropical woodlands and savannas; IV. Mediterranean lands; V. Mid-latitude mixed forests; VI. Mid-latitude grasslands; VII. Boreal forests and woodlands; VIII. Polar lands. Each of these habitats occupies a particular location relative to latitude and the land-water factor.

[14]The term *model* is very difficult to define because it is used in so many different senses. Harvey quotes one opinion that "a model can be a theory, or a law, or a relationship, or a hypothesis, or an equation, or a rule" (Harvey, 1969:145). This seems to be a particular way of formulating a process concept. Models have proved to be of great utility in geographical studies (Chorley and Haggett, 1967; Cole and King, 1968:463–520). See also A. N. Strahler, 1980. "Systems Theory in Physical Geography." *Physical Geography* 1:1–27.

[15]Peter Haggett quotes Hartshorne's *Nature of Geography* as follows: ". . . no universals need be evolved, other than the general law of geography that all its areas are unique" (p. 468) (Haggett, 1966:2–3). In the preceding paragraph, which is not quoted, Hartshorne writes: "For the interpretation of its findings it [regional geography] depends upon generic concepts and principles developed in systematic geography. Furthermore, by comparing different units of area that are in part similar, it can test and correct the universals developed in systematic geography" (p. 467). Both Hettner and Hartshorne insisted that geography is both idiographic and nomothetic, as indeed almost all other fields of learning must be (Hartshorne, 1959:146–172).

Such a distinction would never have occurred to anyone experienced in the out-of-door study of the face of the earth. Of course, every observation has to do with things that are unique in time and place. But it is not even possible to identify any one feature as unique until there is some kind of empirical generalization with which to compare it. Geographers have always been concerned with the question of what things to observe. How does one select specific features to observe out of the complex fabric of interwoven strands that makes up the face of the earth? How does one identify things that are significant or relevant? Only in relation to concepts: hypotheses, empirical generalizations, and, hopefully, some kind of generalization that could be treated as if it were a law (Burton, 1963:156). A very fundamental part of the scientific method consists in learning how to distinguish the relevant from the irrelevant, and this cannot be done without a framework of ideas. Geographers have always observed unique things, but they have also sought to formulate those illuminating concepts that make sense out of the apparent disorder of indirectly related parts.

Description and Explanation

There is confusion, too, over the meaning of the words *description* and *explanation*. Davis used the term *explanatory description* to refer to the use of his model of an ideal cycle of erosion as a frame of reference for the description of landforms. To describe means to transfer observations of things and events on the face of the earth into symbols—either word symbols, or cartographic symbols, or mathematical symbols. Robert Brown defines the word *explanation* as an effort to remove impediments of understanding, an effort "to deprive puzzles, mysteries, and blockages of their force, hence of their existence" (Brown, 1963:40–44). Percy Bridgman wrote that "explanation consists of analyzing our complicated systems into simpler systems in such a way that we recognize in the complicated systems the interplay of elements already so familiar to us that we accept them as needing no explanation" (Bridgman, 1964:63). David Harvey puts it more simply: "The purpose of an explanation may be regarded as making an unexpected outcome an expected outcome, of making a curious event seem natural or normal" (Harvey, 1969:13).

There are, of course, many different ways to make a curious event seem normal or to change the confused complexity of things on the face of the earth into some kind of order. There was a time when a satisfactory explanation for observed features could be provided by reference to God's divine plan, but the more that is known about the complexity of man's universe, the more rigid become the requirements for acceptable explanations. Robert Brown lists seven quite different kinds of explanation, for human behavior (Brown, 1963). David Harvey, in a major contribution to the clarification of geographic ideas, analyzes the explanatory procedures available to geographers (Harvey, 1969). It seems, however, that it might be possible to include all these procedures under two categories: temporal and conceptual.

In the temporal approach, explanation is given by adding the time dimension to descriptions of observed patterns and locations. In some cases this is adequately done by showing the origin of what is to be explained—the genetic form of explanation. In other cases it is more satisfactory to trace the changes through time—the evolutionary or developmental form of explanation. In some cases both of these are combined, as when one goes back to origins and traces developments. The difficulty is that actual identification of an origin is seldom possible, for there are always antecedents that must help to explain so-called origins. Basically, the temporal approach shows the changes through time of which the observed features at any moment of time are like snapshots or like the frames in a moving picture. It has been accepted practice to explain the geographical arrangements at any one time by describing the sequences of events of which they are the momentary results. This is the method of historical geography (A. H. Clark in James and Jones, 1954:70–105). This type of explanation provides an important means of testing hypotheses and checking on generalizations based on observation, and as such it is an important and indeed an essential part of the whole program of geographic scholarship.[16]

In the 1950s and 1960s a strong reaction against the temporal method of seeking explanations in favor of the conceptual approach set in. The movement seems to have started at certain universities in the United States (Washington, Northwestern, and the Department of Regional Science at Pennsylvania) and then spread to Cambridge and Bristol in England, with strong support from Lund in Sweden. The temporal method was relegated to a less sophisticated stage of scholarship both by Harvey (Harvey, 1969) and by Haggett (Haggett, 1966). The difference between the temporal and conceptual approaches is illustrated by the story of Newton and the apple. Newton, the story goes, was faced with alternatives when the apple fell on his head:

> Had he asked himself the obvious question: why did that particular apple choose that unrepeatable instant to fall on that unique head, he might have written the history of an apple. Instead of which he asked himself why apples fell and produced the theory of gravitation. The decision was not the apple's but Newton's (quoted in Haggett, 1966:2).[17]

The story is clever and, for the uninitiated, clearly convincing. Of course, one should not bother with unique events but only seek general theory. Yet the fact is that some individual scholars are deeply and productively concerned with the search for valid explanations in geography through studies in historical perspective, most of which are not properly labeled as trivial. Others are as deeply concerned with the search for theory. All scholars are inclined to think that the road they are following toward the horizon is the only straight road. Yet we must acknowledge that for every Newton or Darwin there are thousands of "lesser" scholars searching for sequences of events by which the validity of theory may be tested. From among the many scholars so engaged have come most of the hypotheses whose implementation has brought progress (Haggett, 1966:277). The continued development of geography as a field of learning would not be served by eliminating or denigrating either the temporal or the conceptual methods.

---

[16]Examples of this kind of explanation may be found in Jordan, 1967, 1969; Kniffen, 1965; Meinig, 1962, 1968, 1986, 1993, 1998; and Schroeder, 2002.

[17]The story, which was included in a monograph on geomorphology by R. J. Chorley, was written by M. Postan, "The Revulsion from Thought," *Cambridge Journal* 1(1948): 395–408.

Explanation through the conceptual approach at one time focused on the search for repeated cause and effect relations. A prior cause had to be related to a later effect. In the eighteenth century David Hume argued that it was not possible to demonstrate such a relationship except through experience with repeated examples of the same sequence of events (Ducasse, 1969).[18] Furthermore, the philosophers and metaphysicians became enmeshed in hopeless discussions of the problem of free will as opposed to determinism.[19] The principle of causation is by no means discarded, but it is now generally recognized that simple cause and effect connections do not exist and that, in fact, the interconnections among things and events on the earth are much more complex than was formerly appreciated.

A dangerous but sometimes very useful kind of conceptual explanation is found in the use of analogy (Chorley, 1964). This is what Herbert Spencer did when he developed the concept of social Darwinism and what Ratzel did when he described the state as a pseudoorganism. Explanations based on analogy in these cases had the effect of obscuring rather than clarifying things. On the other hand, Faraday's description of the electromagnetic field in terms of the motions of fluids was a highly successful clarification. In contemporary times geographers who make use of gravity models, employing mathematical formulas derived from Newton's law of gravitation to predict the volume of interconnections between two cities, offer an apparently successful use of the analogy method. Another stimulating use of analogy is suggested by William Bunge in his search for a single rule to predict shifts of highways, rivers, or shopping centers (Bunge, 1966:27–33). The high land values along a main artery of transportation can be likened to the natural levees along a graded river. Imaginative suggestions of this kind, whether or not they prove valid, are urgently needed in geographical studies.

The search for explanation through the conceptual approach has focused on functional relations as opposed to cause and effect relations (Manners and Kaplan, 1968:212–216). The existence of systems of functionally interconnected parts has been known for thousands of years, but until the age of the electronic computer they could not be handled operationally. A system can be defined logically as a set of elements so closely interconnected that a change in any one of them results in changes in all the others. When one element of a system can be identified as spatial—that is, related to location and extension on the face of the earth—it becomes a spatial system. It is now widely accepted that a major concern of geography is with spatial systems. The area occupied by a functionally interconnected system is, by definition, one kind of region, as Robert S. Platt pointed out in 1928. There have been many stimulating studies along these lines and at various resolution levels. Fundamental to the study of systems in geography is the use of

---

[18] David Hume, *Enquiry Concerning Human Understanding* (London, 1748).

[19] There is a long list of books and papers that discuss the problem of determinism in geography, especially environmental determinism. See Chappell, 1970; Glacken, 1967; Herbst, 1961; Hofstadter, 1955; Lewthwaite, 1966; Martin, 1973; Montefiore and Williams, 1955; R. Peet, 1985; Popper 1984; R. S. Platt, 1948; Spate, 1952, 1958; and Sprout and Sprout, 1956.

probability theory. As long as a system is self-perpetuating because at least one element remains in a steady state, the various components of the system can be treated in terms of functions. Stochastic models of such systems have opened new ranges of understanding.

Throughout the 1970s and 1980s the innovation of much specialization and philosophical pluralism characterized and facilitated progress. Specialization quickly became apparent with the specialty groups of the AAG and the proliferation of periodical titles, leading to what has been called "the journal supermarket." Geographers borrowed from the works of other social scientists to shore up their own body of concepts and models, such as borrowing from German economists. Conversely, other social scientists borrowed ideas from the geographers' offerings.

With a new and methodologically more complex geography, there came new geographic epistemologies. In the 1970s there emerged structuralist thought recognizing the invisible role of political and economic structures influencing individual behavior. There followed realist approaches (Sayer, 1993), which recognized dominant conceptual structures but argued that theory could account for different outcomes in different environments. Then, in an ongoing approximate chronology (with much overlap), there came the interpretive approach, which was concerned with the values of the investigator that vary from individual to individual and from group to group (Jackson, 1989). Feminist thinking had emerged strongly by the 1990s, urging that the role and work of women were hardly considered in traditional geographic investigation and that the contemporary posture amounted to bias (Rose, 1993). And dating from the 1980s to the present is the ongoing dialogue concerning postmodernism, with its belief that geographic values and phenomena are social constructions and critical of knowledge production (Harvey, 1989; Soja, 1989). These philosophies change and yet collectively are ongoing.

GEOGRAPHY IN THE UNITED STATES: ITS HISTORY (TRADITION) AND DIVIDES (INNOVATION)

Prior to the 1880s geography in the United States was largely a mix of geology, physiography, exploratory observation, and a wealth of thought published by individuals in the United States and abroad. Truly, it was a subject matter in search of a discipline. An organizing concept was offered in 1859 with the notion of Darwin's evolutionary principle.

It was not until 1883 that W. M. Davis seized on this organizing principle and established the cycle of erosion. This innovation was to become tradition. It provided a functioning model for physical geographers who wished to go into the field and comprehend what they saw. It also provided a model around which a literature could develop and that could be taught in the classroom, thereby giving birth to an embryonic discipline. Students of Davis became his disciples and spread his conception of geography across the United States and elsewhere. Davis himself took his system to Britain, France, and Germany.

Then in 1893 the Committee of Ten was formed (see Chapter 15) in an attempt to establish a place for geography in the national curriculum. The Davisian viewpoint made its way into the majority report of this committee, and Davisian geography was taught in many states. However, a lack of teacher preparedness did not permit a successful conclusion. Geography thereupon began to turn its attention to a form of human geography.

Realizing that a human geography might fit the needs of the situation, Davis offered his notion of ontography, an innovation that permitted a formal version of geographical determinism (Davis, 1906). This work ushered in the innovative causal notion, and for the next 25 years environmentalism dominated academic geography in the United States. Many absurd statements were advanced under this thesis, which brought the point of view into disrepute. Yet a lot of intelligent thought was published in this area. The largest problem confronting this tradition was the difficulty (and then virtually the impossibility) of measuring the impact of component parts of a physical environment on an individual, a group, or a society.

American geography essentially removed itself from environmentalism when it accepted Carl Sauer's innovation, "the morphology of landscape." This was clearly an import from Germany. Now geographers could study humans on the land devoid of the necessity of "influences." Geographers began to study regions small enough to be traversed by foot, bicycle, or sometimes automobile. Sauer's innovation had changed the scale at which geographers worked. Geographers in the United States now studied land use and the small unit area. Field work was necessary for this undertaking.

This point of view was interrupted by World War II, another type of innovation. Geographers worked in government as never before, and many offices were staffed to a considerable degree by them. As was the case in World War I, the map became very important, giving geography a standing it did not seem to enjoy in peacetime. It also encouraged geographers to realize that their regional specialists relating to other parts of the world could become very valuable assets in time of conflict. It also reaffirmed to geographers that no other discipline could produce such specialists. Regional study had long since become one of the traditions.

Following the end of the war, the regional tradition continued but was met by the innovation that was taking place at the University of Washington in Seattle. The story of Garrison and his (slightly) younger enthusiasts at work on a numerate and theoretical geography was related in Chapter 17. This work spread across the United States and to many other parts of the world. Certainly, it changed the way of American geography. By the 1970s the conflicts created by the imposition of a new way had quieted and had settled in as tradition.

During the last 40 years the innovation of eclecticism has imposed itself and has now itself become a tradition. Fragmentation, frequently typified by more than 50 specialty groups in the AAG, has been accompanied by a search for appropriate philosophy and methodology. The quest for comprehension of the nature of geography continues but seems ever more elusive.

The discipline once had a core and a periphery; today we have a strong periphery but one devoid of a traditional core. Perhaps this provides greater mobility of thought. Innovation seems to have become a tradition. Where once the library and knowledge were the armaments of the scholar, now there is the world of the computer. Perhaps this means less acquaintance with the literature, though a larger output of publication. Whereas once rather simplistic philosophy was the order of the day, now well-defined philosophical postures are made explicit in large amounts of literature. Once there was an appreciation of past accomplishment and a willingness to dwell in the present; now there is considerable rejection of the past and anticipation of the future. A large female contingent and concomitant studies in feminism and gender study are now part of the tradition. Today there is much more scope for avant-garde thinking. Indeed, innovation itself has now become a tradition in our undertaking.

Traditions do not die easily and therefore are additive. In a study of the history of geography, these traditions have only intellectual significance. But in geography as a practicing field, geographers, encompassing many different ages and different traditions, perhaps all teaching in the same university department, tend to remind one another that there is no right way to comprehend the human undertaking on earth's surface. Innovations that become traditions are a vital part of the process by which the discipline advances.

This brief summary of American geography post 1850 has seen our field evolve from a subject matter with little structure to a discipline. This discipline made possible a large and active profession for practitioners, who in turn shaped and advanced the field, making it a science. Since the innovations and traditions of the last 40-odd years are so close to us and have so impressed themselves on how we practice geography, their reappraisal is in order.

REAPPRAISAL AND PROSPECT

After World II and especially since 1960, there was a worldwide emphasis on science and mathematics at the expense of history, language, and literature. The traditional teaching of Latin in the schools was either dropped or greatly reduced (Meynier, 1969:118; Warntz, 1959a). In the United States the National Defense Education Act of 1958 provided federal funds to the schools to improve the teaching of mathematics, foreign languages, and sciences, which provide a background for engineering. In 1964 the act was amended to include six additional fields, one of which was geography.

This widespread emphasis on science and mathematics had a major impact on geography. All over the world there was an increasing use of quantitative techniques and a renewed search for useful theoretical models. Gould (1969) described the 1960s as one of the greatest periods of intellectual ferment in the history of geography. Others went further, proclaiming that a new mode for geographical study had replaced the traditional verbal and descriptive studies (Chojnicki, 1970:213).

It should be remembered that the decade of the 1960s was one of social, economic, and political turmoil. The new geography of the late 1960s and the 1970s began to reveal expressions of disillusionment. Something of a retreat from the spatial analysis theme occurred, and the quantitative approach began to emerge. Reappraisal of the geography of the 1960s had led to this. Findings in urban geography had led some of the leading theoreticians to assume professional posts in the realm of planning. But planning for whom and for what? Whose values should be adopted? (Harvey, 1973, 1974.) Geographers were now brought face to face with the substance of power. And it was realized perhaps more clearly than ever before that geography is value laden (Smith, 1977). Dissent was expressed with the founding of *Antipode: A Radical Journal of Geography* and the founding of Socially and Ecologically Responsible Geographers (SERGE). This part of the radical geography movement sought to revise the geography of capitalism using the traditional though ever changing instruments of the discipline (Peet, 1977, 1978). Another sector of the radical geography movement, believing that a different foundation of values was a necessary prerequisite for erecting the alternative geography, founded the Union of Socialist Geographers. These geographers were led to the assumptions of Marxist theory (Folke, 1972).

A more conventional reaction to scientism and the enthusiasm for hard science that resulted was a return to a thoroughgoing humanism within human geography. Historical geography, studies in perception, inquiry into new areas of geosophy, [a study of people's nonscientific geographical beliefs (Lowenthal and Bowden, 1976)], a reappraisal of geography as human ecology, and a newfound appreciation of the regional concept have also characterized geography in the United States since the middle 1960s. Geographers began to study the recent history of their discipline to see if the path of intellectual evolution could reveal order in eclectic and plural complexity. This excursus into the history of geographical thought may do much to bring order to geographers' thoughts and perspectives (Freeman, 1961; Gregory, 1978; Martin, 2004).

Swings of the academic pendulum can be observed in the interplay between the two basic traditions—mathematical and literary. It would be quite erroneous to equate the use of mathematics with a law-seeking objective and the use of language with a descriptive objective. Actually, mathematics in many cases provides a notably more precise descriptive method. Studies in literary form may provide exciting innovative approaches to the formulation of concepts. Verbalized and non-quantitative studies in historical geography have nevertheless led the way in the approaches to environmental perception (Brookfield, 1969; Meinig, 1962:207). In the early 2000s the professional periodicals carry contributions in both the mathematical and the literary tradition—neither seemed about to be abandoned.

By this time geography in the United States and the United Kingdom had converged, and an Anglo-American geography could be written about (Johnston, 1997). (Early in the history of academic geography, the United States imported texts, ideas, concepts, and periodicals from the United Kingdom. Beginning in the 1960s the United States exported some of the same to the United Kingdom.) Travel to meetings across the Atlantic, giving lectures, holding university posts for a year or more, and publishing in each other's journals have become commonplace activities. Geographers from both countries developed a spatial science disciplinary matrix, and many human geographers from both countries rebelled against the machinelike properties of spatial science that did not allow for individuality. Humanism, with its embrace of subjectivity, provided an avenue for one group of human geographers. A second group of human geographers formed and formalized a radical geography. This group sought to solve societal problems. It was driven by a very different set of values than those of the humanities, and it frequently embraced Marxist methodology.

Throughout the period from the 1970s to the present, geography was in a state of turmoil. Given the philosophical dispositions of geographers and the multiplicity of specialty groups, it is hardly surprising that the field "danced but did not march." Individualism reigned supreme, and geographers wrote on whatever they pleased, bending it to a geographic viewpoint. Their works were frequently so overwritten that it was sometimes hard to find substance in the verbiage. Citation indexes further encouraged the concept of "invisible colleges," and networks arose which might exert influence within the profession. With budgets tightening and academic posts ever harder to obtain, the race to tenure, once an academician was hired, became a scramble for publications. The result was that novitiates may have published too much too soon, thus leading to a possible diminution in the worth of some periodicals in the United States.

This sketch does not suggest intellectual anarchy. A lot of good work was also accomplished and is in progress. The absence of a core, the new mobility, and the breakdown of traditional boundary lines between inquiries has released imaginations and enthusiasms. This has been revealed in Gould, *The Geographer at Work* (1985) and *Becoming a Geographer* (1999). In many cases experimentation has led to excellent studies. A new order has arisen in which no one point of view or specialty group of the AAG dominates. Gaile and Willmott edited an encompassing anthology entitled *Geography in America* (1989) and a second volume, *Geography in America at the Dawn of the Twenty-first Century* (2003), which, together with *Geography's Inner Worlds* (1992) and *Rediscovering Geography* (1997), represent extended summaries of content and philosophy explaining both innovation and tradition.

Meanwhile, the field of geography remains too little known to the general public and to workers in other scholarly fields. Although trained geographers are being sought in increasing numbers,[20] the great majority of Americans still have only the vaguest idea about what geographers do. Frequently the newspapers run special articles on the geographical illiteracy of American young people, and there is a flurry of excitement in the schools to see that pupils memorize more place names. This kind of teaching was outmoded in Ritter's time. Although geographers are necessarily concerned with place names (unless their work is purely theoretical), place names by themselves cannot be called geography. (Their study is properly referred to as toponymy.) In the public mind, too, geography is another name for "popular description and travel." Marvin W. Mikesell has reported on how geography is viewed by other social scientists:

> it is probably fair to say that most of the geographic works known to scholars in other fields are not regarded by geographers themselves as indicative of their current interests. Among anthropologists and historians, the voice of American geography is undoubtedly Ellsworth Huntington, whose theories of climatic influence have been obsolete for more than thirty years. Among political scientists, geography is most commonly identified with the various schools of "geopolitics" that flourished in the 1940s and are no longer taken seriously by political geographers. It would be difficult to prove that geographic thought is more clearly perceived by economists, although central-place theory and other formulations of

---

[20]The Association of American Geographers lists many opportunities in its *Newsletter*, from local, state and federal positions to college positions in the United States and abroad.

location theory have some currency among economists concerned with urban and regional planning. Sociologists are perhaps unique among social scientists in having a more accurate perception of modern trends in geographic research, although this awareness is confined very largely to the work of urban geographers (Mikesell, 1969:240–241).

Looking at how geographers themselves view their own field, we find that diversity of approach has always characterized geographic study. There was plenty of diversity in ancient Greece, where most of the traditional currents of geographic thought originated. Diversity was emphasized in the summary of geographic thought provided by Humboldt and Ritter. And for the past century, when the nature of geography as a field of learning and its relation to other fields of learning have been debated at length, differences in the way geography is verbally described have become even more diverse. This state of affairs has bothered some scholars, and from time to time efforts have been made to provide narrow definitions of the field that exclude considerable numbers of active workers, past and present, from membership in the profession. These efforts have not been successful. In 1956 Carl O. Sauer wrote:

> We continue properly to be, as I have said that we have been always, a diverse assemblage of individuals, hardly to be described in terms of dominance of any one kind of aptitude or temperament, mental faculty, or emotional drive, and yet we know that we are drawn together by elective affinity. It is about as difficult to describe a geographer as it is to define geography, and in both cases I am content and hopeful. With all shortcomings as to what we have accomplished, there is satisfaction in knowing that we have not really prescribed limitations of inquiry, method, or thought upon our associates. From time to time there are attempts to the contrary, but we shake them off after a while and go about doing what we most want to do. . . .
>
> It seems appropriate therefore to underscore the unspecialized quality of geography. The individual worker must try to gain whatever he can of special insights and skills in whatever most absorbs his attention. Our overall interests, however, do not prescribe the individual direction. We have a privileged status which we must not abandon. Alone or in groups we try to explore the differentiation and interrelation of the aspects of the earth. We welcome whatever work is competent from whatever source, and claim no proprietary rights. In the history of life the less specialized forms have tended to survive and flourish, whereas the functional self-limiting types have become fossils. Perhaps there is meaning in this analogy for ourselves, that many different kinds of minds and bents do find congenial and rewarding association, and develop individual skills and knowledge. We thrive on cross-fertilization and diversity (Sauer, 1956:292–293).

The decision regarding which approaches to geography will survive and which will disappear can only be made by the coming generations of scholars. As Stephen E. Toulmin points out, progress and change in any scientific field do not take place because the great scholars of an older generation change their minds; rather, they are the result of the younger generation's breaking with the traditions of their teachers (Toulmin, 1967). If a particular approach—whether mathematical or literary, genetic or conceptual—is to flourish, it must be attractive to younger scholars who are in training. If large numbers of new students are attracted to a particular approach to the study of geography, that aspect of the field will be nourished and will progress rapidly; those aspects of geography that do not attract young people must eventually disappear.

What makes some branches of a field like geography attractive? To be attractive it must do one of two things. First, it must make a clear contribution, widely recognized, to the overriding problems with which humankind is faced in the twenty-first century. It must help find solutions to problems of poverty, hunger, injustice, violence, and warfare. If it fails to have any relevance to these problems, it cannot long attract the attention of young scholars. If the scholars who are devoted to the formulation of abstract concepts turn away from the real world and contemplate only self-images, the continued growth of a much-needed conceptual framework will be disastrously affected. Julian Jaynes puts it this way:

> . . . as science folds back on itself and comes to be scientifically studied, it is being caricatured into a conformity which is nonsense, into a neglect of its variety which is psychotic, into a nagging and insistent attention to its cross-discipline similarities which are of trivial importance (1966:94).

Or if geographers are ever satisfied with writing mere descriptions of particular places without reference to broadly unifying concepts, the field will be doomed. A powerful way to make geographic study attractive to the younger generation is to demonstrate clearly the nature of its contribution to the solution of major problems. To fail to come to grips with the need for making practical application of abstract concepts is to face the danger of permitting geographic work to become trivial.

We have a long and dignified heritage of geographic study—of effort to identify order in occupied space on the face of the earth in terms of the symbols we adopt to guide our thinking. We must move forward without repeating the unnecessary errors of the past, yet always with the courage to formulate new hypotheses and to see hypotheses we formulate challenged and perhaps destroyed. There can never be an end to this, for the kind of order we conceive changes with the change of symbols and with the kinds of questions we ask. Always there is the challenge to see whether there is another new world to describe and explain lying just beyond the horizon.

REFERENCES: CHAPTER 20

Ackerman, E. A. 1963. "Where Is a Research Frontier?" *Annals AAG* 53:429–440.

———. 1965. *The Science of Geography*. Washington, D.C.: National Academy of Sciences/National Research Council, Publication No. 1277.

Aitken, S. C., S. L. Cutter, K. E. Foote, and J. S. Sell. 1989. "Environmental Perception and Behavioral Geography." In Gaile and Willmott, eds, 1989. *Geography in America*. Toronto & London: Merrill Publishing Co.

Amedeo, D., and Golledge, R. G. 1986. *An Introduction to Scientific Reasoning in Geography*. New York: John Wiley & Sons.

Barnes, T. J. 2000. "Inventing Economic Geography in Anglo-America, 1889–1960." In E. Sheppard and T. J. Barnes, eds. *A Companion to Economic Geography*. Oxford: Blackwell, 11–26.

Barnes, T. J. 2003. "The Place of Locational Analysis: A Selective and Interpretive History." *Progress in Human Geography* 27:169–195.

Barrows, H. H. 1923, "Geography as Human Ecology." *Annals AAG* 13:1–14.

Bell, D. 1973. *The Coming of Post-Industrial Society*. New York: Basic Books.

Berry, B. J. L. 1961. "An Inductive Approach to the Regionalization of Economic Development." In N. Ginsburg, ed., *Essays on Geography and Economic Development*. Pp. 78–107. Chicago: University of Chicago, Department of Geography. Research Paper No. 62.

———. 1964. "Approaches to Regional Analysis, A Synthesis." *Annals AAG* 54:2–11.

———. 1973. "A Paradigm for Modern Geography." In R. J. Chorley, ed., *Directions in Geography*. Pp. 3–22. London: Methuen.

Berry, B. J. L., and Garrison, W. L. 1958a. "Alternate Explanations of Urban Rank-Size Relationships." *Annals AAG* 48:83–91.

———. 1958b. "A Note on Central Place Theory and the Range of a Good." *Economic Geography* 34:304–311.

Berry, B. J. L., J. B. Parr, B. J. Epstein, A. Ghosh, and R. H. T. Smith. 1988. *Market Centers and Retail Location: Theory and Applications*. Englewood Cliffs, NJ: Prentice Hall.

Berry, B. J. L., and Pred, A. 1961. *Central Place Studies, A Bibliography of Theory and Applications*. Philadelphia: Regional Science Research Institute.

Bertalanffy, L. von. 1951a. "General System Theory: A New Approach to the Unity of Science." *Human Biology* 23:303–361.

———. 1951b. "An Outline of General System Theory." *British Journal for the Philosophy of Science* 1: 134–165.

———. 1956. "General System Theory." *General Systems* 1:1–10.

———. 1962. "General System Theory—A Critical Review." *General Systems* 7:1–20.

———. 1965. "On the Definition of the Symbol." In J. R. Royce, ed., *Psychology and the Symbol: An Interdisciplinary Symposium*. New York: Random House.

———. 1968. *General System Theory, Foundations, Development, Applications*. New York: George Braziller.

Bird, J. H. 1975. "Methodological Implications for Geography from the Philosophy of K. R. Popper." *Scottish Geographical Magazine* 91:153–163.

———. 1977. "Methodology and Philosophy." *Progress in Human Geography* 1:104–10.

———. 1978. "Methodology and Philosophy." *Progress in Human Geography* 2:133–140.

Bird, J. H. 1989. *The Changing Worlds of Geography: A Critical Guide to Concepts and Methods*. Oxford: Clarendon Press.

Blaut, J. M. 1979. "The Dissenting Tradition." *Annals AAG* 69:157–164.

Bongaarts, J. and R. Bulatao eds. 2000. *Beyond Six Billion: Forecasting the World's Population*. Washington DC: National Academy Press.

Boulding, K. E. 1956. "General Systems Theory—The Skeleton of a Science." *Management Science* 2:197–208.

Braithwaite, R. B. 1968. *Scientific Explanation*. New York: Harper Bros. (Harper Torchbooks); Cambridge: Cambridge University Press, 1968.

Bridgman, P. 1964. *The Nature of Physical Theory*. Princeton, N.J.: Princeton University Press.

Bronowski, J. 1966. "The Logic of the Mind." *American Scientist* 54:1–14.

Brookfield, H. C. 1969. "On the Environment as Perceived." *Progress in Geography* 1:51–80.

Brown, L. A. 1968. *Diffusion Process and Location, A Conceptual Framework and Bibliography*. Philadelphia: Regional Science Research Institute.

Brown, L. A. 1981. *Innovation Diffusion: A New Perspective*. London: Methuen.

———. 1999. "Change, Continuity, and the Pursuit of Geographic Understanding." *Annals of the Association of American Geographers*. 89, 1–25.

Brown, L. A., and Moore, E. G. 1969. "Diffusion Research in Geography: A Perspective." *Progress in Geography* 1:121–157.

Brown, R. 1963. *Explanations in Social Science*. Chicago: Aldine.

Brush, J. E. 1953. "The Hierarchy of Central Places in Southwestern Wisconsin." *Geographical Review* 43:380–402.

Buckley, E., ed. 1968. *Modern Systems Research for the Behavioral Scientist, A Sourcebook*. Chicago: Aldine.

Bunge, W. 1966. *Theoretical Geography*. Lund: University of Lund. (Lund Studies in Geography.)

Burns, B. 1992. *Percepts, Concepts, and Categories: The Representation and processing of Information*. North Holland: Amsterdam.

Burton, I. 1963. "The Quantitative Revolution and Theoretical Geography." *Canadian Geographer* 7:151–162.

Chamberlin, T. C. 1892. "The New Geology." *University of Chicago Weekly* 1:7–9.

Chappell, J. E., Jr. 1970. "Climatic Change Reconsidered: Another Look at 'The Pulse of Asia.'" *Geographical Review* 60:347–373.

Chojnicki, Z. 1970. "Prediction in Economic Geography." *Economic Geography* 46:213–222.

Chorley, R. J. 1962. *Geomorphology and General Systems Theory*. Washington, D.C.: U.S. Geological Survey, Professional Paper 500–B: 1–10.

———, ed. 1973. *Directions in Geography*. London: Methuen.

Chorley, R. J., and Haggett, P., eds. 1965. *Frontiers in Geographical Teaching*. London: Methuen.

———. 1967. *Models in Geography*. London: Methuen.

Christaller, W. 1933. *Die zentralen Orte in Süddeutschland*. Jena: Gustav Fischer. Trans. C. W. Baskin, 1966. *Central Places in Southern Germany*. Englewood Cliffs, N.J.: Prentice Hall.

Cliff, A. D. and J. K. Ord. 1973. *Spatial Autocorrelation*. London: Pion.

Cohen, S. B., and Rosenthal, L. D. 1971. "A Geographical Model for Political Systems Analysis." *Geographical Review* 61:5–31.

Cole, J. P., and King, C. A. M. 1968. *Quantitative Geography, Techniques and Theories in Geography*. New York: John Wiley & Sons.

Curry, L. 1964. "The Random Spatial Economy: An Exploration in Settlement Theory." *Annals AAG* 54:138–146.

Curry, L. 1998. *The Random Spatial Economy and Its Evolution*. Aldershot: Ashgate.

Dacey, M. 1974. *One-dimensional Central Place Theory*. Northwestern University Studies in Geography 21. Evanston, Ill.: Department of Geography, Northwestern University.

Davis, W. M. 1888. "Geographic Methods in Geologic Investigation." *National Geographic Magazine* 1:11–26.

———. 1906. "An Inductive Study of the Content of Geography." *Bulletin of the American Geographical Society* 38:67–84.

DeLaubenfels, D. J. 1970. *A Geography of Plants and Animals*. Dubuque, Iowa: W. C. Brown.

Dickenson, J. P., and Clarke, C. G. 1972. "Relevance and the 'Newest Geography.'" *Area* 4:25–27.

Dow, M. W. 1974. "The Oral History of Geography." *The Professional Geographer* 26:430–435.

Folke, S. 1972. "Why a Radical Geographer Must Be Marxist." *Antipode* 4:13–18.

Freeman, T. W. 1961. *A Hundred Years of Geography*. London: Gerald Duckworth.

Gaile, G. L., and Willmott, C. J., eds. 1989. *Geography in America*. Toronto and London: Merrill Publishing Co.

Gaile, G. and C. Willmott eds. 2003. *Geography in America at the Dawn of the Twenty-First Century*. Oxford and New York: Oxford University Press.

Garrison, W. L. 1959–1960. "The Spatial Structure of the Economy." *Annals AAG* 49:232–239, 471–482; 50:357–373.

Gober, P. and J. A. Tyner. 2003. "Population Geography." Pp. 185–199. In Gaile and Willmott, 2003. *Geography in America at the Dawn of the 21st Century*. Oxford: Oxford University Press.

Goodchild, M. F. and D. G. Janelle. 1988. "Specialization in the Structure and Organization of Geography." *Annals of the Association of American Geographers* 78:1–28.

Goodchild, M. F. 1992. "Analysis." In *Geography's Inner Worlds: Pervasive Themes in American Geography*. Pp. 138–162. Eds. R. F. Abler, M. G. Marcus, J. M. Olson. New Brunswick, N.J.: Rutgers University Press.

Gould, P. R. 1969. "Methodological Developments since the Fifties." *Progress in Geography* 1:1–50.

———. 1985. *The Geographer at Work*. London: Routledge & Kegan Paul.

Gregory, D. 1978. *Ideology, Science and Human Geography*. New York: St. Martin's Press.

Grigg, D. B. 1965. "The Logic of Regional Systems." *Annals AAG* 55:465–491.

Guyot, A. 1872. *Physical Geography*. New York: Scribner, Armstrong & Co.

Hägerstrand, T. 1967. *Innovation Diffusion as a Spatial Process*. Trans. A. Pred. Chicago: University of Chicago Press (first published in Sweden in 1953).

Haggett, P. 1966. *Location Analysis in Human Geography*. New York: St. Martin's Press.

Haggett, P. 2003. "Peter Robin Gould, 1932–2000." *Annals of the Association of American Geographers* 93:925–934.

Hartshorne, R. 1939. *The Nature of Geography, a Critical Survey of Current Thought in the Light of the Past*. Lancaster, Pa.: Association of American Geographers.

———. 1959. *Perspective on the Nature of Geography*. Chicago: Rand McNally.

Harvey, D. 1969. *Explanation in Geography*. London: Edward Arnold.

———. 1973. *Social Justice and the City*. London: Edward Arnold.

———. 1974. "What Kind of Geography for What Kind of Public Policy?" *Transactions, Institute of British Geographers* 63:18–24.

———. 1989. *The Condition of Postmodernity*. Oxford: Blackwell.

Herbst, J. 1961. "Social Darwinism and the History of American Geography." *Proceedings of the American Philosophical Society* 105:538–544.

Hofstadter, R. 1955. *Social Darwinism in American Thought*. Boston: Beacon Press.

Isard, W. 1956. "Regional Science, the Concept of Region, and Regional Structure." *Papers and Proceedings, Regional Science Association* 2:13–39.

———. 1960. "The Scope and Nature of Regional Science." *Papers of the Regional Science Association* 6:9–34.

James, P. E. 1967. "On the Origin and Persistence of Error in Geography." *Annals AAG* 57:1–24.

———. 1966. *A Geography of Man*. 3rd ed. Waltham, Mass.: Ginn & Co.

James, P. E., and Jones, C. F., eds. 1954. *American Geography, Inventory and Prospect*. Syracuse, N.Y.: Syracuse University Press.

James, P. E., and Martin, G. J. 1979. *The Association of American Geographers: The First Seventy-Five Years, 1904–1979*. Washington, D.C.: Association of American Geographers.

Jaynes, J. 1966. "The Routes of Science." *American Scientist* 54:94–102.

Johnson, L. J. 1971. "The Spatial Uniformity of a Central Place Distribution in New England." *Economic Geography* 47:156–170.

Johnston, R. J. 1970. "Grouping and Regionalizing: Some Methodological and Technical Observations." *Economic Geography* 46:293–305.

————. 1997. *Geography and Geographers: Anglo-American Human Geography Since 1945*. 5th ed. London: Arnold. New York: Oxford University Press.

————. 2003. "Geography as a Discipline in Distance." 303–345. In *A Century of British Geography*, R. J. Johnston and M. Williams, 2003.

Jordan, T. G. 1967. "The Imprint of the Upper and Lower South on Mid-nineteenth Century Texas." *Annals AAG* 57:677–690.

————. 1969. "The Origin of Anglo-American Cattle Ranching in Texas: A Documentation of Diffusion from the Lower South." *Economic Geography* 45:63–87.

Kalesnik, S. V. 1964. "General Geographic Regularities of the Earth." *Annals AAG* 54:160–164.

Kansky, K. J. 1963. *The Structure of Transportation Networks: Relations between Network Geometry and Regional Characteristics*. Chicago: University of Chicago, Department of Geography, Research Paper No. 84.

Kasperson, R. E., and Minghi, J. V. 1969. *The Structure of Political Geography*. Chicago: Aldine.

King, L. J. 1969. *Statistical Analysis in Geography*. Englewood Cliffs, N.J.: Prentice-Hall.

Kniffen, F. B. 1965. "Folk Housing: Key to Diffusion." *Annals AAG* 55:549–577.

Kurath, H. 1949. *A Word Geography of the Eastern United States*. Ann Arbor: University of Michigan Press.

Lankford, P. M. 1969. "Regionalization: Theory and Alternative Algorithms." *Geographical Analysis* 1:196–212.

Lee, H. N. 1972. *Percepts, Concepts and Theoretical Knowledge*. Memphis, TN: Memphis State U. Press.

Lewthwaite, G. R. 1966. "Environmentalism and Determinism: A Search for a Clarification." *Annals AAG* 56:1–23.

Linsky, A. S. 1965. "Some Generalizations Concerning Primate Cities." *Annals AAG* 55:506–513.

Lösch, A. 1940. *Die räumliche Ordnung der Wirtschaft*. Jena: Gustav Fischer. 2nd ed., 1944. Trans. W. H. Woglom, *The Economics of Location*, 1954. New Haven, Conn.: Yale University Press.

Lowenthal, D., ed. 1967. *Environmental Perception and Behavior*. Chicago: University of Chicago, Department of Geography.

Lowenthal, D., and Bowden, M. J., eds. 1976. *Geographies of the Mind: Essays in Historical Geosophy in Honor of John Kirtland Wright*. New York: Oxford University Press.

Lukermann, F. 1958. "Towards a More Geographic Economic Geography." *The Professional Geographer* 10:2–10.

————. 1961. "The Role of Theory in Geographical Inquiry." *The Professional Geographer* 13:1–6.

Machamer, Peter. 2002. *The Blackwell Guide to the Philosophy of Science*. Malden, Mass.: Blackwell.

Manners, R. A., and Kaplan, D., eds. 1968. *Theory in Anthropology, a Sourcebook*. Chicago: Aldine.

McColl, R. W. 1969. "The Insurgent State: Territorial Bases of Revolution." *Annals AAG* 59:613–631.

McKinney, W. M. 1968. "Carey, Spencer, and Modern Geography." *The Professional Geographer* 20:103–106.

MacNeish, R. S. 1964. "Ancient Mesoamerican Civilization." *Science* 143:531–537.

Marbut, C. F. 1925. *Annals of the Association of American Geographers* 15:1–29.

Martin, G. J. 2003. "The History of Geography." In Gaile and Willmott, 2004. *Geography in America at the Dawn of the 21st Century*. Oxford: Oxford University Press. 550–561.

Mathewson, K. and M. S. Kenzer. Eds. 2003. *Culture, Land, and Legacy: Perspectives on Carl O. Sauer and Berkeley School Geography*. Baton Rouge: Geoscience Publications, Louisiana State University.

Meinig, D. W. 1968. *The Great Columbia Plain, A Historical Geography, 1805–1910*. Seattle: University of Washington Press.

———. 1969. *Imperial Texas, an Interpretive Essay in Cultural Geography*. Austin and London: University of Texas Press.

———. 1988. *The Shaping of America: A Geographical Perspective on 500 Years of History*. Vol. 1. *Atlantic America, 1492–1800*. New Haven: Yale University Press.

———. 1993. Vol. 2. *Continental America, 1800–1867*. New Haven: Yale University Press.

———. 2000. Vol. 3. *Transcontinental America, 1850–1915*. New Haven, Conn.: Yale University Press.

Mikesell, M. W. 1969. "The Borderland of Geography as a Social Science." In M. Sherif and C. W. Sherif, eds., *Interdisciplinary Relationships in the Social Sciences*. Pp. 227–248. Chicago: Aldine.

Montefiore, A. C., and Williams, W. M. 1955. "Determinism and Possibilism." *Geographical Studies* 2:1–11.

Morrill, R. L. 1970. "The Shape of Diffusion in Space and Time." *Economic Geography* 46:259–268.

Nystuen, J., and Dacey, M. F. 1961. "A Graph Theory Interpretation of Nodal Regions." *Papers of the Regional Science Association* 7:29–42.

Olsson, G. 1965. *Distance and Human Interaction, A Review and Bibliography*. Philadelphia: Regional Science Research Institute.

Parkes, D. 1980. *Times, Spaces, and Places: A Chronogeographic Approach*. New York: Halsted Press.

Parr, J. B., and Denike, K. G. 1970. "Theoretical Problems in Central Place Analysis." *Economic Geography* 46:568–586.

Pattison, W. D. 1964. "The Four Traditions of Geography." *Journal of Geography* 63:211–216.

Peet, J. R. 1977. "The Development of Radical Geography in the United States." *Progress in Human Geography* 1:240–263.

———. 1978. *Radical Geography*. London: Methuen.

Philbrick, A. K. 1957. "Principles of Areal Functional Organization in Regional Human Geography." *Economic Geography* 33:299–336.

Platt, R. S. 1948. "Determinism in Geography." *Annals AAG* 38:126–132.

———. 1957. "A Review of Regional Geography." *Annals AAG* 47:187–190.

Preston, R. E. 1971. "The Structure of Central Place Systems." *Economic Geography* 47:136–155.

Raven, P., ed. 2000. *Nature and Human Society: The Quest for a Sustainable World*. Washington, D.C.: National Academy Press.

Sauer, C. O. 1925. "The Morphology of Landscape." *University of California Publications in Geography* 2:19–54.

———. 1952. *Agricultural Origins and Dispersals*. New York: American Geographical Society.

———. 1956. "The Education of a Geographer." *Annals AAG* 46:287–299.

———. 1966. "On the Background of Geography in the United States." In *Heidelberger Studien zur Kulturgeographie, Festgabe für Gottfried Pfeifer, Heidelberger Geographische Arbeiten*.

Schlier, O. 1937. "Die zentralen Orte des Deutschen Reichs." *Zeitschrift der Gesellschaft für Erdkunde zu Berlin*: 161–170.

Schroeder, W. A. 2002. *Opening the Ozarks: A Historical Geography of Missouri's Ste. Genevieve District 1760–1830*. Columbia and London: University of Missouri Press.

Siddall, W. R. 1961. "Two Kinds of Geography." *Economic Geography* 37:189.

Siebert, H. 1967. *Zur Theorie des regionalen Wirtschaftswachstums*. Tübingen: J. C. B. Mohr.

Smith, D. M. 1977. *Human Geography: A Welfare Approach*. London: Edward Arnold.

Soja, E. W. 1968. *The Geography of Modernization in Kenya*. Syracuse, N.Y.: Syracuse University Press. Geographical Series No. 2.

———. 1971. *The Political Organization of Space*. Washington, D.C.: Commission on College Geography, Resource Paper No. 8.

———. 1989. *Postmodern Geographies*. London: Verso.

Sprout, H., and Sprout, M. 1956. *Man-Milieu Relationship Hypothesis in the Context of International Politics*. Princeton, N.J.: Center for International Studies.

Stoddart, D. R. 1967. "Growth and Structure of Geography." *Transactions, Institute of British Geographers* 41:1–19.

Taaffe, E. J., ed. 1970. *Geography*. Englewood Cliffs, N.J.: Prentice-Hall.

Taylor, P. J. 1976. "An Interpretation of the Quantification Debate in British Geography." *Transactions, Institute of British Geographers*, NSI. 129–142.

Thompson, J. H., ed. 1966. *Geography of New York State*. Syracuse, N.Y.: Syracuse University Press.

Tocalis, T. R. 1978. "Changing Theoretical Foundations of the Gravity Concept of Human Interaction." In B. J. L. Berry, ed. *The Nature of Change in Geographical Ideas*. De Kalb: Northern Illinois University Press, 65–124.

Toulmin, S. E. 1967. "The Evolutionary Development of Natural Science." *American Scientist* 55:456–471.

Trewartha, Glenn T. 1953. "A Case for Population Geography." *Annals AAG* 43:71–97.

———. 1969. *A Geography of Population: World Patterns*. New York: John Wiley & Sons.

Ullman, E. L. 1941. "A Theory of Location for Cities." *American Journal of Sociology* 46:853–864.

———. 1953. "Human Geography and Area Research." *Annals AAG* 43:54–66.

von Thünen, J. H. 1826. *Der isolierte Staat in Beziehung auf Landwirtschaft und Nationalökonomie*. Hamburg. Trans. Carla M. Wartenburg; ed., Peter Hall, 1966. *Von Thünen's Isolated State*. Oxford: Pergamon Press.

Wallace, W. L., ed. 1980. *Sociological Theory, an Introduction*. Chicago: Aldine.

Ward, D. 1964. "A Comparative Historical Geography of Streetcar Suburbs in Boston, Massachusetts, and Leeds, England, 1850–1920." *Annals AAG* 54: 447–489.

Warntz, W. 1959a. "Geography at Mid-Twentieth Century." *World Politics* 11:442–454.

———. 1959b. *Toward a Geography of Price: A Study in Geo-Econometrics*. Philadelphia: University of Pennsylvania Press.

———. 1964. "A New Map of the Surface of Population Potentials for the United States, 1960." *Geographical Review* 54:170–184.

Weber, A. 1909. *Ueber den Standort der Industrien*. Tübingen. Trans. C. J. Friedrich, 1966. *Alfred Weber's Theory of the Location of Industry*. Chicago: University of Chicago Press.

White, S. E., L. A. Brown, W. A. V. Clark, P. Gober, R. Jones, K. McHugh, and R. L. Morrill. "Population Geography." Pp. 258–289. In Gaile and Willmott, 1989. *Geography in America*. Toronto and London: Merrill Publishing Co.

Wilbanks, T. J., and Symanski, R. 1968. "What Is Systems Analysis?" *The Professional Geographer* 20:81–85.

Wilson, A. G. 1974. *Urban and Regional Models in Geography and Planning*. Chichester: John Wiley.

Woldenberg, M. J. 1968. "Energy Flow and Spatial Order: Mixed Hexagonal Hierarchies of Central Places." *Geographical Review* 58:552–574.

Woods, R. 1982. *Theoretical Population Geography*. London, New York: Longman.

Yuill, R. S. 1966. *A Simulation Study of Barrier Effects in Spatial Diffusion Problems*. Ann Arbor: Michigan Inter-University Community of Mathematical Geographers.

Zelinsky, W. 1970. "Beyond the Exponentials: The Role of Geography in the Great Transition *Economic Geography* 46:498–535.

Zelinsky, W., Kosiński, L. A., and Prothero, R. M., eds. 1970. *Geography and a Crowding World*. New York: Oxford University Press.

Zipf, G. K. 1965. *Human Behavior and the Principle of Least Effort: An Introduction to Human Ecology*. New York: Hafner (first published 1949).

Zobler, L. 1958. "Decision Making in Regional Construction." *Annals AAG* 48:140–148.

INDEX OF NAMES

Note: This index includes not only the names of individuals cited in the text, but also names of other persons who may be of interest to the geographer.

Al-Masudi (–956), Arab scholar who described the monsoons and the relation of evaporation to rainfall, 48

Al-Rashid, Harun (700s), Caliph, eighth-century Muslim who was patron to the works of translating Greek works into Arabic, 48

Albertus Magnus (1193–1280), German scholar who joined the Dominican Order in 1223; lectured at Paris and Cologne; introduced Aristotelian ideas to Medieval Europe; his major geographical work was *De Natura Locorum* (*The Nature of Places*), 41

Alexander, Lewis M. (1921–), political geographer. Emeritus at the University of Rhode Island.

Alexander the Great (356–323 B.C.), pupil of Aristotle; King of Macedonia; led Greek army eastward to the Indus River, 6, 24, 27–28

Alexey, Andreyevich Tillo (1839–1900), Russian geographer, geophysicist, geodesist, and cartographer; active in the Russian Geographical Society.

Allenby, Lord (1860–1936), British soldier, statesman, and geographer.

Almagia, Roberto (1884–1962), Italian geographer; Professor at Rome (1915–1959).

Amundsen, Roald (1872–1928), Norwegian explorer; first to reach the South Pole (1911), 157, 279, 303

Anaximander (610–547 B.C.), disciple of Thales in Miletus; introduced the gnomon to the Greeks; drew map of the known world to scale, xv, 15–19

Anderson, James R. (1919–1980), Ph.D., University of Maryland. Applied geographer; worked for USDA, University of Florida and USGS.

Andersson, Gunnar (1865–1928), Swedish geographer; for 25 years secretary of the Swedish Society of Anthropology and Geography, 302

Andree, Richard F. (1835–1912), geographer, ethnographer, and cartographer. Noted for the *Hand Atlas*.

Andrews, Howard F. (1944–1988), Doctorate, University of Sussex; spe-

cialist in evolution of human geography in France, 209

Annenkov, Vladimir V. (1936–), Ph.D., Moscow; Institute of Geography, Moscow University. Specialist in world problems and Africa, 437

Antevs, Ernst V. (1888–1974), Swedish authority on glaciation in U.S.A.; associated with the University of Stockholm and later the University of Arizona, 304

Anuchin, Dmitry N. (1843–1923), Russian geographer; founded Department of Geography at Moscow in 1887, 252, 253, 258, 260, 265

Anuchin, Vsevolod A. (1913–), Soviet geographer who argued for unity of physical and economic geography, 265

Apian, Peter (1495–1592), German mathematician and cartographer who was the teacher of Gerhard Kremer, 74, 75

Applebaum, William (1906–1978), American geographer; specialist in marketing geography, 405, 467

Arago, François (1783–1853), French physicist who developed principle of magnetism by rotation; close friend of Alexander von Humboldt, 117

Arbos, Philippe (1882–1956), Doctorate, University of Grenoble, 1922; French specialist on rural life; at the University of Clermont-Ferrand (1919–1952).

Arctowski, Henryk (1871–1958), Polish scientist resident in US from c. 1916. Authority on sun spots and climate. Worked for The Inquiry.

Arden-Close, Charles Frederick (1856–1952), British geographer: director general of the Ordnance-Survey (1927–1930); elected president of the I.G.U. (1934).

Arduino, Giovanni (1713–1795), an Italian scholar who classified rocks in 1760, 94

Aristarchos of Samos (ca. 300–ca. 230 B.C.), believed that the earth rotates on its own axis and revolves around the sun, 29, 86

Aristotle (384–322 B.C.), Greek philosopher who believed the world was developing toward final perfection; taught the theory of natural places for earth, air, fire, and water; founder of the Lyceum at Athens, 4, 6, 13, 23–27, 29, 30, 35, 36, 41, 51, 58, 86, 87, 133, 134, 364, 465, 499

Armstrong, Patrick (1941–), Ph.D., Durham, UK. At the University of Western Australia; authority on Darwin and geography; authored *The Changing Landscape* (1975) and *Charles Darwin in Western Australia* (1985), 158, 407

Arnold, Thomas (1823–1900), taught at Oxford, 1841; pupils included A. P. Stanley and E. A. Freeman, 231

Arrowsmith, John (1790–1873), British cartographer who made maps of Australia, Africa, and other continents.

Arsenyev, K. I. (1789–1865), Russian geographer who pioneered regional studies and classified cities by functions, 251, 252

Ashmead, Percy H., organized a field study for the AGS of the Guatemala-Honduras boundary (1919), 454

Atwood, Wallace W. (1872–1949), Ph.D., Chicago, in geology; Professor of Physiography at Harvard (1913–1920); President, Clark University (1920–1946), 302, 308, 371, 383, 449

Aurousseau, Marcel (1891–1981), Australian polymath; secretary of the Committee on Geographical names in London (1936–1955), 392, 515

Austin, Oscar Phelps (1847–1933), American economist; Chief, Bureau of Statistics, Department of Commerce; charter member A.A.G., 353

Avicenna or **Ibn-Sina** (980–1037), described formation of mountains by stream erosion, 49, 53, 94

Babcock, William H. (1849–1922), author of *Legendary Islands of the Atlantic*, 77

Bacon, Francis (1561–1626), English philosopher and author; helped define induction in science.

Bacon, Roger (1220–1292), "Nothing of importance can be known about things unless the places where they are, are also known"; English philosopher, 45

Baedeker, Karl (1801–1859), German traveler and publisher credited with compiling many of the Baedeker Guide books (the first was published in 1827)

Bagrow, Leo (1881–1957), historian of cartography who worked in Russia until 1918, in Berlin (1918–1945), and in Stockholm (1945–1957); his monumental *Geschichte der Kartographie* was published in 1951 (Bagrow and Skelton, 1964), 316, 329

Bain, Alexander (1818–1903), Scottish philosopher at Aberdeen; Rector of the university (1880–1887), 388

Baker, Alan R. H. (1938–), chairman and historical geographer at Cambridge University, 238

Baker, J. N. L. (1894–1971), British specialist on the history of geography; at Oxford (1923–1962), 90, 239

Baker, Oliver E. (1883–1949), Ph.D., Wisconsin; specialist in agricultural geography, U.S. Department of Agriculture; Professor at Maryland (1942–1949), 285, 392, 450, 515

Banks, Joseph (1743–1820), sailed with James Cook on his first voyage, 134

Banse, Ewald (1883–1953), German geographer and freelance writer who demonstrated the regional unity of North Africa and Southwest Asia on basis of landscape similarity, 176, 308

Baranskiy, Nikolai N. (1881–1963), Soviet geographer, friend of Lenin, who helped to establish role of geography after the October Revolution of 1917, 260, 261, 262, 263, 265, 267, 270

Barnes, Carleton P. (1903–1962), Ph.D., Clark; soil specialist in U.S. Department of Agriculture, 405

Barnes, Trevor (1956–), Ph.D., Minnesota; professor, University of British Columbia. Specialist in history and philosophy of geography, 442

Barrett, Robert L. (1871–1969), American geographer and explorer; charter member A.A.G., 352

Barrows, Harlan H. (1877–1960), one of the leaders in the development of American geography between 1919 and 1942, when he was chairman of the Department of Geography at Chicago, 371, 372, 373, 374, 382, 384, 385, 390, 450, 460, 501

Bartels, Dietrich (1931–1983), Ph.D., University of Hamburg; argued for change to a new geography, 186

Bartholomew, John (1831–1893), Scottish cartographer; head of map publishing firm (1856–1893), 140

Bartholomew, John (1890–1962), Scottish cartographer; head of map publishing firm (1929–1962), 140

Bartholomew, John C. (1923–), Scottish cartographer; became head of map publishing firm in 1962, 140

Bartholomew, John George (1860–1920), Scottish cartographer; head of map publishing firm (1893–1920), 140

Bartlett, Robert Abram (1875–1946), American Arctic explorer.

Barton, Thomas F. (1905–1985), professor of geography at Indiana and editor, *Journal of Geography* (1950–1965), 407

Baruth, Christopher (1948–), Ph.D., University of Wisconsin at Milwaukee. Curator AGS special collections and authority on history of cartography, cartography and history of geography, xvi

Bassin, Mark (1953–), American geographer, Ph.D. Berkeley. At University College, London. Interests include geography of national identity, ideologies of nature and the role of space in the political process, 189, 275

Bates, Henry W. (1825–1892), British naturalist; explorer of the Amazon, 134

Batschelet, Clarence E. (1889–1974), worked for the U.S. Bureau of the Census (1914–1957), 515

Bauer, Louis A. (1865–1932), American specialist on terrestrial magnetism; Chief, Division of Terrestrial Magnetism, U.S. Coast and Geodetic Survey (1899–1906); charter member A.A.G., 353

Baugh, Ruth E. (1889–1973), member of the U.C.L.A. geography department (1913–1956).

Baulig, Henri (1877–1962), studied with Vidal in Paris and W. M. Davis at Harvard; brought American geomorphology to Paris; taught at Strasbourg (1919–1947), 206, 261, 281, 352, 512

Beaufort, Francis (1774–1858), British Naval officer who developed techniques of maritime sounding; best known for the Beaufort scale of wind force.

Beaujeu-Garnier, Jacqueline (1917–1995), French geographer; specialist on the geography of population; Professor at the Sorbonne since 1960,

Beauregard, Ludger (1923–), French Canadian geographer who has written much on Montreal, 288, 295

Beaver, Stanley H. (1907–1984), author of numerous textbooks with L. D. Stamp. Authority on the midlands of England, 220, 237

Beazley, Sir Charles Raymond (1868–1955), British historian of geography and exploration; author of *The Dawn of Modern Geography*, 59

Beck, Hanno (1923–), German specialist in the history of geographical thought at Bonn, 139, 158, 189

Beckinsale, Robert P. (1908–1998), British authority on physical geography at Oxford (retired); coauthor with R. J. Chorley of *The History of the Study of Landforms* (3 vols.), 234, 239, 241

Behaim, Martin (1436 or 1459–1507), advisor to King Joao II of Portugal on navigation after the death of Prince Henry; made first world globe at Nuremberg, 74

Bell, Alexander Graham (1847–1922), Scottish born inventor of the telephone, 433

Bellingshausen, Fabian (1778–1852), noted leader of Russian expedition to the Antarctic (1819–1821). One of the founders of the Russian Geographical Society, 250

Bengtson, Nels A. (1879–1963), M.S., Nebraska, Ph.D., Clark; Professor at Nebraska (1929–1949), 449, 451

Bennett, Hugh Hammond (1881–1960), with the U.S. Department of Agriculture

Fahrenheit, Gabriel D. (1686–1736), Dutch physicist who laid down the Thermometric scale now in use.

Fairchild, Wilma B. (1915–1983), Editor of *Geographical Review*, 1949–1972.

Fairgrieve, James (1870–1953), author of *Geography and World Power* (1915), 302

Farabee, William C. (1865–1925), anthropologist; author of *A Pioneer in Amazonia* (1917), 452

Faraday, Michael (1791–1867), English chemist and physicist, 149, 523

Fassig, Oliver L. (1859–1936), Ph.D., Johns Hopkins University. Served with the US Weather Bureau, 1883–1936.

Fawcett, Charles B. (1883–1952), British geographer; identified regions of Great Britain based on trade areas of cities, 228, 236

Febvre, Lucien P. V. (1878–1956), Doctorate, Paris; French author of *A Geographical Introduction to History*, 198, 207

Fellmann, Jerome D. (1926–), Ph.D., Chicago. Prof. Emeritus, University of Illinois, 240

Fenneman, Nevin M. (1865–1945), Ph.D., Chicago (geology); taught at Colorado, Wisconsin, and Cincinnati (1907–1937 at Cincinnati); charter member A.A.G., 307, 364, 388

Ferguson, Adam (1723–1816), Scottish philosopher and historian, 135

Ferrel, William (1817–1891), American meteorologist who was on the U.S. Coast Survey (1867–1882); in the Signal Office (1882–1886), 150

Fewkes, J. Walter (1850–1930), with the Bureau of American Ethnology, 1895–1928. Authority on Pueblo and Hopi Indians

Field, Henry (1902–1986), anthropologist and an administrator in the M Project, 467

Finch, Vernor C. (1883–1959), Ph.D., Wisconsin; taught at Wisconsin (1911–1954), 399, 414, 450, 459, 487

Finley, John H. (1863–1940), President and Honorary President of the American Geographical Society.

Fischer, Eric (1898–1985), Ph.D., University of Vienna. Austrian. At University of Virginia and George Washington University, 184

Fischer, Theobald (1846–1910), trained in history at Bonn (1876); Professor at Marburg (1883–1910), 171

Fitzgerald, Walter (1898–1949), British geographer; author of *Africa: A Social, Economic, and Political Geography of its Major Regions*, 220

Fitzroy, Robert (1805–1865), captain of *HMS Beagle* during Darwin's voyage; hydrographer; writer on geography, meteorology, and anthropology.

Flamsteed, John (1646–1719), first director of the Greenwich Observatory (1675), 92

Fleming, Sanford (1827–1915), railroad engineer, surveyor, 282

Fleure, Herbert J. (1877–1969), D.Sc., University of Wales; British geographer and anthropologist; taught at Aberystwyth and University of Manchester, 223, 224, 229, 329

Forry, Samuel (1811–1844), American climatologist, 150

Forster, Georg (1754–1794), son of Johann Reinhold Forster; was on Cook's second voyage around the world; influenced Alexander von Humboldt to study geography, 80, 100, 101, 109, 134, 141

Forster, Johann Reinhold (1729–1798), accompanied Cook on his second voyage around the world, 80, 100, 101, 134

Franck, Harry A. (1881–1962), writer of numerous travel books.

Franklin, Benjamin (1706–1790), American printer, author, inventor, diplomat, and scientist, xiii, 11, 18, 143, 338

Frazier, John W. Jr (1947–), Ph.D., Kent State University Specialist in applied geography, 469

Freeman, Edward A. (1823–1892), English historical geographer.

Freeman, T. W. (1908–1988), Professor Emeritus, University of Manchester;

author of several books on the history of geography, 239

Frémont, John Charles (1813–1890), American explorer ("The Pathfinder") of Rocky Mountains and the West, 151

Freshfield, Douglas William (1845–1934), mountaineer and geographer; president of the Royal Geographical Society (1898–1911).

Friis, Herman R. (1905–1989), American geographer; retired Director, Center for Polar Archives, National Archives, Washington, D.C.

Fröbel, Friedrich (1782–1852), German educational reformer. Founder of the Kindergarten system, 121

Fröbel, Julius (1805–1893), taught geography and mineralogy at Zurich (1832–1842); exiled from Germany for revolutionary activity (1848); field studies in western North America (1848–1858); recognized chorology in writings of Kant and Humboldt, 124, 174

Frye, Alexis E. (1859–1936), author of school geographies and educational administrator, 302

Fuchs, Roland J. (1933–), Ph.D., Clark; specialist in urbanization in Asia; Director, East-West Center, University of Hawaii and Vice-Rector UN University; President of the IGU; Director International "Start" Secretariat, 434, 437

Füchsel, Georg Christian (1722–1773), identified rock strata of the Thüringerwald (1762), 94

Fukuzawa, Yukichi (1834–1901), Japanese scholar whose book, *Countries of the World*, was first source of information about the outside world (1869), 324

Gaile, Gary L. (1945–), Ph.D. UCLA, 1976. Currently at University of Colorado. Co-editor with C. Willmott of decennial volumes of US geography, 424, 426, 430, 488, 501, 528

Galileo (1564–1642), first to formulate the concept of a universal mathematical order and to demonstrate correctness of the ideas of Copernicus (1623, 1632), 86, 91, 92

Gallais, Jean (1926–1998), Doctorate, Rouen. French geographer noted for studies in colonial Africa.

Gallois, Lucien (1857–1941), Doctorate, Paris; close associate of Vidal; editor, *Géographie Universelle*; editor, *Annales de Géographie* (1898–1919), 198, 206, 223, 302, 512

Galton, Francis (1822–1911), British geneticist who was also interested in geographical problems; he made the first weather map with isobars, 11, 214, 215

Galvani, Luigi (1737–1798), Italian physician and physicist, 110

Gannett, Henry (1846–1914), American geographer; topographer with Hayden Survey; Geographer of the Tenth, Eleventh, and Twelfth Censuses; chief geographer, U.S. Geological Survey (1882); charter member A.A.G., 154, 353

Ganong, William F. (1864–1941), botanist, geographer, and authority on eastern Canada, 288

Garcia-Ramon, M. (1943–), Ph.D., Barcelona. Specialist in history of geography and gender studies; at University of Barcelona, 436

Garrison, William L. (1924–), Ph.D., Northwestern University (1950); pioneer in the use of statistical models for the study of transportation problems, 417, 418, 419, 485, 488, 490, 508, 525

Gatterer, Johann Christoph (1727–1799), teacher of geography in Germany who popularized Buache's ideas regarding the division of the earth's surface into naturally defined basins (1773–1775), 102, 139

Gauss, Carl Friedrich (1777–1855), German mathematician and astronomer.

Gauthier, Howard L., Jr. (1935–), Ph.D., Northwestern (1966); Emeritus Professor, Ohio State University.

Gautier, Emile-Felix (1864–1940), French geographer. Authority on the Sahara.

Gay, Edwin F. (1876–1946), a student of transportation in economic history, 450

Jones, Wellington D. (1886–1957), American geographer; Ph.D., Chicago; taught at Chicago (1913–1945), 373, 374, 397, 457, 520

Jordan, Terry G. (1938–2003), Walter Prescott Webb Professor and historical geographer at the University of Texas; author of *Texas Graveyards* (1982), *American Log Buildings* (1985), and others.

Kalesnik Stanislav V. (1901–1977), Soviet geographer; professor of geography at Leningrad, 265

Kane, Elisha Kent (1820–1857), American polar explorer, 148

Kant, Edgar (1902–1978), Estonian geographer; studied central place hierarchy in Estonia using mathematical procedures; taught at Lund (1950–1970), 312, 314, 509

Kant, Immanuel (1724–1804), German philosopher who rejected the doctrine of final causes; lectured on physical geography at Konigsberg (1755–1796); provided an early statement of geography as chorology, 103–104, 108, 120, 123, 125, 132, 136, 174, 295

Kapp, Ernst (1808–1896), born in Germany; came to the US in 1849, returned to Germany, 1865. Wrote *General Comparative Geography*, published in Germany, 1869, 338

Karaska, Gerald J. (1933–), American geographer; Ph.D., Pennsylvania State University (1962); editor, *Economic Geography*; professor at Clark.

Karlsefni, traveled with a crew to the coast of North America in 1003, 52

Kates, Robert W. (1929–), Clark University (1962–1986) and Brown University (independent scholar); specialist in hazards and hunger, 423

Keasbey, Lindley M. (1876–1946), pioneer in economic geography in North America.

Keckermann, Bartholomaus (1572–1609), a German geographer who developed the terms "general" and "special" geography before Varenius, 90, 91

Keller, Albert G. (1874–1956), at Yale worked with sociologist W. G. Sumner, 285

Keltie, John Scott (1840–1927), secretary, Royal Geographical Society (1892–1915), 215, 238, 448

Kemp, John (1763–1812), professor of geography at Columbia College (1795–1812), 143

Kendall, Henry M. (1901–1966), urban and population geographer at Miami University, Ohio.

Kendall, Maurice G. (1907–1983), British statistician; author of *Advanced Theory of Statistics*, 1946–1948 (2 vols.), 484

Kendrew, Wilfred George (1890–1962), British geographer; taught at Oxford; author of *The Climates of The Continents* (1922), 220

Kenzer, Martin S. (1950–), historian of geography and historical geographer; editor of *Carl O. Sauer: A Tribute* and *Applied Geography: Issues, Questions and Concerns*, 386, 414, 429, 470

Kepler, Johannes (1571–1630), German astronomer who rejected Ptolemy's concept of circular orbits in favor of elliptical orbits, 86, 91

Kesseli, John E. (1895–1980), born in France; studied in Germany (Halle), University of California, 1932–1962 (Ph.D. 1938)

Kiepert, Heinrich (1818–1899), classical historian; taught at Berlin (1859) as docent replacing Ritter; professor (1874), 164

Kimble, George H. T. (1908–), Ph.D., Montreal; British geographer; Director. American Geographical Society, 1950–1953, 49, 51, 57, 234, 286

King, Charles F. (1843–), member of Committee of Ten (1892), 350

King, Clarence (1842–1901), American geologist and explorer; headed King Survey along the 40th parallel; first director, U.S. Geological Survey (1879), 151, 154

King, Leslie J. (1934–), New Zealand geographer; Ph.D. Iowa; Emeritus,

Pitts, Forrest R. (1924–), American geographer; Ph.D., Michigan (1955); professor emeritus at the University of Hawaii; authority on Korea, diffusion studies, and their incorporation into computer programs, 429, 488, 489, 490

Pius II, Pope (Aeneas Silvius Piccolomini) (1404–1464), wrote geography of Europe and Asia; doubts Ptolemy on enclosed Indian Ocean and on uninhabitability of equatorial regions, 46, 66

Plato (428–348 B.C.), Greek philosopher; believed the world had been created in perfection and was in process of deteriorating; stressed deductive method; accepted idea of a round earth; founder of the Academy at Athens, 4, 5, 6, 23–27, 40, 53, 98, 144

Platt, Elizabeth T. (1900–1943), Librarian of the American Geographical Society.

Platt, Raye R. (1891–1973), American geographer; in charge of the Millionth Map of Hispanic America project at the American Geographical Society (1923–1938)

Platt, Robert S. (1891–1964), American geographer; Ph.D., Chicago; taught at Chicago (1919–1957), enthusiastic field worker and expert, 237, 265, 392, 393, 414, 415, 502, 512, 523

Playfair, John (1747–1819), Scottish geologist who clarified the ideas of James Hutton, 94

Pleva, Edward G. (1912–), Ph.D. Minnesota, Emeritus U. Western Ontario, 286

Plewe, Ernst (1907–1986), Doctorate, University of Greifswald; German authority on the history of geography

Pliny the Elder (A.D. 23–79), Roman geographer; compiled a treatise on geographical writings; died during eruption of Vesuvius, 35, 36, 40, 41

Pokshishevskiy, Vadim V (1905–), Soviet geographer; professor at the Institute of Ethnography at Moscow

Polo, Marco (1254–1323), Venetian traveler who visited China (1271–1295); his book describes the many places he visited in eastern Asia, 43–45, 46, 58, 63, 66, 67, 72, 87, 99, 131

Polybius (205–125 B.C.), Greek historian, 33

Pomponius, Mela (1st century A.D.), first Latin writer of geography, 36, 40

Pond, Peter (1739–1807), fur trader, explorer, map maker; first to outline features of Mackenzie river system, 279

Popper, Karl (1902–1994), Austrian philosopher widely read by geographers, 384

Posidonius (135–50 B.C.), Greek geographer; estimated circumference of the earth to be smaller than Eratosthenes had thought; believed highest temperatures are along the tropics, not along the equator; one of the earliest philosophers to face the apparent dichotomy between factual descriptions of unique phenomena and the formulation of theoretical explanations, 33–34, 35, 37, 45, 66, 96

Postnikov, Alexei V. (1939–), Doctorate, Moscow; much published Russian specialist in the histories of cartography and geography, xvi, 272, 273

Powell, John Wesley (1834–1902), American explorer and pioneer geographer; first to explore the Grand Canyon; leader of U.S. Geographical and Geological Survey of the West; second director U.S. Geological Survey (1880–1894), 151, 155, 156, 158, 342, 343, 344, 362, 373, 459, 460

Powell, Joseph M. (1938–), born in England, moved to Australia in 1964; writer on the human impact on the Australian environment; author, *Environmental Management in Australia* (1976), *Water in Garden State* (1989).

Pred, Allan R. (1936–), American geographer; Ph.D., Chicago (1962); teaching at Berkeley; specialist in social theory and local and regional transformation, 313, 315, 511

Price, Archibald Grenfell (1892–1977), prominent Australian geographer; with the University of Adelaide (1925–1957); author of *White Settlers in the Tropics* (1939).

Prince, Hugh (1927–), specialist in historical geography. Emeritus, London, 238

Pruitt, Evelyn L. (1918–2000), Office of Naval Research, 448

Ptolemy (A.D. 90–168), summarized Greek knowledge of astronomy; gazeteer of places located by latitude and longitude; world map showed Indian Ocean enclosed on south by terra incognita; accepted smaller earth circumference from Marinus, 6, 20, 36–37, 40, 45, 51, 58, 64, 66, 67, 71, 73, 77, 79, 86, 87, 90, 102, 133

Pumpelly, Raphael (1837–1923), American geologist and explorer; surveyed resources along route of the Northern Pacific Railroad (1882); exploration in Central Asia (1903–1904), 342

Putnam, Donald F. (1903–1977), Ph.D., (botany) Toronto; taught at the University of Toronto, 1938–1966. With L. J. Chapman, author of *The Physiography of Southern Ontario*, 286, 287, 289, 293, 416

Puzachenko, Yuriy G. Doctorate, Moscow, Specialist in Pacific and eco-science, 273, 274

Pythagoras (6th century B.C.), Greek mathematician; believed in concept of a round earth, xiv, 5, 25, 26

Pytheas (d. 285 B.C.), explorer of Great Britain and coast of Western Europe, 24, 28–29, 31, 87

Quam, Louis O. (1906–2001), Ph.D., Clark University; retired in 1970's as chief scientist at the office of Polar Programs, NSF, 406

Quetelet, Lambert (1796–1874), a Belgian astronomer; founder of statistical science and early formulator of the laws of probability, 96

Raisz, Erwin J. (1893–1968), Ph.D., Columbia; cartographer at Harvard University (1931–1950); author of *General Cartography* (1938) and *Principles of Cartography* (1962).

Ramón, Father Manuel, eighteenth-century Jesuit missionary who observed bifurcation of the Orinoco, 113

Ramusio, Gian Battista (1485–1557), collected and published accounts of voyages of exploration, including a new edition of Marco Polo, 87

Ransome, Jack C. (1917–), American geographer; Ph.D., Harvard (1955); Emeritus professor at University of Windsor in Ontario

Rapoport, Anatol (1911–), Ph.D., (math) Chicago; interested in "systems," peace, and model building, 502

Rasmussen, Knud (1879–1933), Danish authority on northern lands

Ratcliffe, Michael R (1962–), U.S. Bureau of the Census, 470, 471

Ratzel, Friedrich (1844–1904), educated at Heidelberg, Jena and Berlin; Ph.D., Heidelberg (zoology), 1868; traveled as roving reporter, 1871–1875, including visit to America; taught geography at the Technische Hochschule in Munich, 1875–1886, and at Leipzig, 1886–1904, 167–170, 184, 198, 253, 302, 305, 309, 324, 333, 364, 365, 367, 390, 455, 510, 523

Ravenstein, Ernst Georg (1834–1913), British geographer and cartographer; studied population movements, 506, 514

Ray, John (1627–1705), early classifications of plants and animals and detailed description of the hydrologic cycle, 94, 95

Reclus, Elisée (1830–1905), student of Carl Ritter; left France in 1851 and was banished in 1872; professor at Université Nouvelle in Brussels, 1894–1905, 122, 141–142, 252, 281, 338

Redway, Jacques Wardlaw (1849–1942), author of numerous excellent school geographies, 350

Reeves, Edward A. (1862–1945), served on the staff of the Royal Geographical Society for 55 years.

Reid, Harry Fielding (1859–1944), Ph.D., Johns Hopkins; Geologist and geomorphologist. Specialist on glaciers and earthquakes; taught at Johns Hopkins; charter member AAG, 353

British Tudor and Stuart geography, 76, 232

Taylor, T. Griffith (1880–1963) D.Sc. Sydney; British geographer; introduced modern geography in Australia (1920) and Canada (1935); left Australia for United States in 1928, went to Canada in 1935, and returned to Australia in 1952, 285, 287, 288, 295, 429, 512

Teggart, Frederick J. (1870–1946), wrote on theory and processes of history and geographical determinism.

Teleki, Paul (1879–1941), Doctorate, Budapest; Hungarian geographer; specialist in political boundaries; three times Prime Minister of Hungary, 329

Tenzing, Norgay (1916–1986), mountaineer; climbed Mount Everest (1953), 157, 222

Tezuka, Akira (1923–), human geographer; translated Humboldt, Ritter, Hettner and others, 333

Thales of Miletus (ca. 624–ca. 548), Greek merchant and scholar; originator of several basic theorems of geometry, 6, 15–19, 26, 31, 484

Theophrastus of Eresos (ca. 372–ca. 288), succeeded Aristotle as head of Lyceum in Athens; wrote over 200 books on many subjects, including a treatise on cultivated plants; identified as the "father of botany," 29

Thoman, Richard S. (1919–), Ph.D., Chicago; Professor Emeritus at California State University, Hayward, 293

Thomas, Bertram (1892–1950), explorer and authority on Arabia.

Thomas, William L., Jr. (1920–2002), American geographer; Ph.D., Yale (1955); professor of geography and anthropology, California State University, Hayward, 431

Thompson, David (1770–1857), mapped much of Western Canada and northwestern US, 279

Thornthwaite, C. Warren (1899–1963), Ph.D., Berkeley; American authority on climate; developed a classification of climates, 468

Thralls, Zoe A. (1888–1965), founder, Geography Department, University of Pittsburgh.

Thrift, Nigel (1949–), Ph.D., University of Bristol; interests are various but include meanings of space and time, 424

Thrower, Norman (1919–), Ph.D., Wisconsin; Professor Emeritus, UCLA; authority on history of cartography; director of Columbus Quincentennial Program at UCLA.

Thünen, Johann Heinrich von (1783–1850), German economist and author of *Der Isolierte Staat*; developed the concept of agricultural zones around a market center, 237, 310, 314, 328, 417, 421, 437, 509

Ting, Wen-kiang (1887–1936), Chinese geologist; founded Department of Geology at Peking and National Geological Survey of China (1916).

Tinkler, Keith J. (1942–), Ph.D., Liverpool, UK; geomorphologist and author of *A Short History of Geomorphology*. Professor, Brock University, Canada.

Tobler, Waldo R. (1930–), Ph.D., Washington (1955); professor at University of California, Santa Barbara; pioneer in creative map-making, 488, 490

Tomkins, George S. (1920–1985), Ph.D., University of Washington; active in geographic education, 295

Törnquist, Gunnar Evald (1933–), Swedish geographer; student of Sven Godlund; second professor of geography at Lund, 314

Torrieri, Nancy K. (1948–), Ph.D., Maryland, U.S. Bureau of the Census, 470, 471

Toscanelli, Paolo da Pozzi (1397–1482), astronomer whose work may have influenced Columbus, 58, 67, 74

Toulmin, Stephen E. (1922–), philosopher and educator; Ph.D., Cambridge (1948); professor at Leeds (1955–1959); professor, Brandeis University (1965–69); at Michigan State University (1969–1974), and later at the University of Chicago, 529

INDEX OF SUBJECTS